Remote Sensing Applications in Environmental and Earth System Sciences

Remote Sensing Applications in Environmental and Earth System Sciences

Nicolas R. Dalezios

CRC Press

Taylor & Francis Group

Boca Raton London New York

CRC Press is an imprint of the
Taylor & Francis Group, an **informa** business

First edition published 2021 by
CRC Press
6000 Broken Sound Parkway NW, Suite 300, Boca Raton, FL 33487-2742

and by
CRC Press
2 Park Square, Milton Park, Abingdon, Oxon, OX14 4RN

CRC Press is an imprint of Taylor & Francis Group, LLC

Library of Congress Cataloging-in-Publication Data
Names: Dalezios, Nicolas R., author.
Title: Remote sensing applications in environmental and earth system sciences / authored by Nicolas R. Dalezios.
Description: First edition. | Boca Raton, FL : CRC Press, 2021. | Includes bibliographical references. | Summary: "This book is a contemporary, multi-disciplinary, multi-scaling, updated and upgraded approach of applied remote sensing in the environment. It aims to cover the whole spectrum of current environmental issues related to atmosphere, land, and water. This book starts with an overview of remote sensing technology, and then throughout the book, the author explains with many case studies, the types of data that can be used and the image processing and analysis methods that can be applied to each type of application. This book is an excellent resource for anyone who is interested in remote sensing technologies and their use in Earth systems, natural resources, and environmental science"-- Provided by publisher.
Identifiers: LCCN 2020055144 (print) | LCCN 2020055145 (ebook) | ISBN 9781138054561 (hardback) | ISBN 9781315166667 (ebook)
Subjects: LCSH: Environmental sciences--Remote sensing. | Earth sciences--Remote sensing.
Classification: LCC GE45.R44 D35 2021 (print) | LCC GE45.R44 (ebook) | DDC 363.7/063--dc23
LC record available at https://lccn.loc.gov/2020055144
LC ebook record available at https://lccn.loc.gov/2020055145

ISBN: 978-1-138-05456-1 (hbk)
ISBN: 978-0-367-76800-3 (pbk)
ISBN: 978-1-315-16666-7 (ebk)

Typeset in Times
by SPi Global, India

Dedication

"Εν πάσι γαρ τοις φυσικοίς ενεστί τι θαυμαστόν"
"For everything in nature, there is something worth of wonder"

Aristotle, 384–322 BC

Contents

PART 1   Prolegomena

PART 2   Applications in Environmental Sciences

PART 3   *Applications in Earth System Sciences*

Preface

Remote sensing is the acquisition of physical data of an object without coming into contact with it. As humans have come to understand the physical universe better, efforts have been made to design tools and instrumentation and to provide access to ever more data. The limitations of the visible region of the total electromagnetic spectrum were recognized early on. However, it was not until significant technological breakthroughs were made, around the time of World War II, that it was possible to obtain the useful range of data available from the extravisible regions of the electromagnetic spectrum. At the present time, Earth Observation (EO) provides one of the most promising avenues for providing information at global, regional and even basin scales related to hydrometeorological hazards. The general circumstances that make EO technology attractive for this purpose in comparison with traditional techniques include its ability to provide inexpensive, repetitive and synoptic views of large areas in a spatially contiguous fashion without disturbing the area to be surveyed and without site accessibility issues.

The understanding of remote sensing lacks restrictions. Indeed, the environmental fields and the electromagnetic spectrum are vehicles for the detection and identification of data concerning physical environmental features. The only limitation to the objectives of this book, which the author follows, is to restrict the treatment to environmental and earth system sciences applications of remote sensing. The writing of this book is motivated by the scientific challenges emerging from the requirements to develop a capability to support research, practical applications and decision-making from local to larger scales in environmental and earth system sciences. In all cases, the need to improve relevant observations and modeling capabilities of parameters related to the above-mentioned fields is mandatory in order to overcome the current drawbacks and the limitations faced by the scientific and operational communities.

Such a book is necessary due to the importance of environmental and earth system sciences today globally, impacting not only human lives but also the global economy, with environmental issues having increased over the past decade or so. This book integrates decades of research conducted by the author in the field and has been designed with a range of readers in mind. This book constitutes the outcome of diachronic upgrading of university lectures in several similar courses. Indeed, valuable help and comments received over the years from colleagues and associates in the teaching and learning of environmental remote sensing have been incorporated into the book. Potential readers may come from a wide spectrum of scientific backgrounds, such as environmental sciences, environmental economics and management, physical and natural sciences, applied sciences, engineering, geology, ecology, earth sciences, hydrology, meteorology, soil science, agricultural sciences, and geography. Due to the unique structure of the book, consisting of a series of independent parts, its use can be adapted to meet the specific needs of different readers and potentially used for teaching and research purposes alike. The different chapters can be viewed as even smaller units which can be combined with other materials if required.

This book is divided into the following three parts: Part 1: Prolegomena consists of one chapter, namely Chapter 1: Remote Sensing Concepts and Systems. Part 2: Applications in Environmental Sciences consists of five chapters: Chapter 2: Precipitation; Chapter 3: Meteorology and Climate; Chapter 4: Hydrology and Water Resources; Chapter 5: Marine and Coastal Ecosystems and Chapter 6: Environmental Hazards. Finally, Part 3: Applications in Earth System Sciences consists of five chapters: Chapter 7: Agriculture; Chapter 8: Forestry; Chapter 9: Geology; Chapter 10: Renewable Energy Sources and Chapter 11: Land Use and Land Cover (LULC).

This book attempts primarily to serve as a reference book and to present current applications of quantitative remote sensing methodologies in environmental and earth system sciences. It is also intended to be used as a textbook and also to serve as a "cookbook" in the broad fields of environmental and earth system sciences for senior undergraduate students, graduate students, researchers and professionals of the previously cited fields. As mentioned, the emphasis is placed on remote sensing applications of methodological approaches based on geoinformatics technologies and data, as well as simulation.

Nevertheless, the need for such a book comes from the author's own experience in teaching such courses and conducting research on the subject for several decades. The author hopes this preface successfully provides an insight into the breadth of the topics covered in this book. Readers are encouraged to adapt the book to fit their needs and to help them better understand the capabilities and potentials of EO technology in the field.

Nicolas R. Dalezios
Volos, Greece

Acknowledgements

This book has been completed thanks to generous assistance from many sources. I wish to acknowledge the comprehensive work carried out by Dr. Athanassios Ganas for his major contribution in Chapter 9: Geology. Moreover, I wish to acknowledge the valuable support and help in several chapters by research associates, such as Dr. Ioannis Faraslis, Dr. Marios Spiliotopoulos, Dr. Stavros Sakellariou, Dr. Christos Vasilakos and Ms. Anna Blanta. In addition, several individuals have provided useful unpublished information. The successful completion of the book is also due to the invaluable editorial advice by Taylor and Francis. Irma Britton has been particularly important in offering continuous encouragement and practical advice. I would also like to thank my secretary Ms. Konstantina Giannousa for her dedicated effort to successfully complete this book. Finally, I would like to thank my family for providing a relaxed home environment appropriate for writing. Every effort to identify and acknowledge the original sources has been made, however, if there are any omissions or errors, the author and the publisher apologize to those concerned.

Nicolas R. Dalezios
Volos, Greece

About the author

Nicolas R. Dalezios is Professor of Agrometeorology and Remote Sensing at the University of Thessaly (UTH), Volos, Greece. Member of the Greek Agricultural Academy. Professor and founding Director of the Laboratory of Agrometeorology, University of Thessaly, Volos Hellas (1991–2011). Postgraduate degrees in Meteorology (Athens, 1972) and in Hydrological Engineering (Delft, 1974) and a PhD in Civil Engineering, University of Waterloo, Canada, in 1982. His research interests include agrometeorology, agrohydrology, modeling, remote sensing, environmental hazards, climate change and risk assessment. He is author and/or co-author of more than 450 refereed publications and technical reports. He is a reviewer and editorial board member for international scientific journals. He is an editor or co-editor of 15 edited publications, co-author of 35 book chapters, and author of two recent books: Agrometeorology: Analysis and Simulation (2015) (Publisher: Association of Greek Universities) and Environmental Hazards Methodologies for Risk Assessment and Management (2017) (Publisher: International Water Association (IWA), London, UK).

Part 1

Prolegomena

1 Remote Sensing Concepts and Systems

1.1 INTRODUCTION: REMOTE SENSING DEFINITIONS AND SYSTEMS

Remote sensing can be considered as the science of obtaining information about objects on the surface of the Earth through data analysis, which is collected using special instruments without physical contact with these objects. Thus, remote sensing can also be attributed as the identification of a distant object (Avery and Berlin, 1992). Similarly, remote sensing can be defined as the art and science to obtain information about a phenomenon, object or area from the data analysis obtained from a medium, which is not in contact with the phenomenon, object or area under consideration (Lillesand and Kiefer, 2000). Moreover, Mather (1999) provides a definition in a narrower sense by stating that environmental remote sensing includes the measurement and imprinting of electromagnetic energy, which is emitted or reflected from the atmosphere and the Earth's surface. Remote sensing has a dual dimension, including the following two parameters, which are directly linked: (1) the data acquisition technique by means of a remote medium from the object, and (2) data analysis for the interpretation of the object.

Remote sensing sensors are divided into two major categories, the recorders, e.g. the spectrometer, and the imaging sensors (Avery and Berlin, 1992). The recorders measure the intensity of radiation under continuous wavelength fluctuations, and simultaneously record different spectral bands. On the other hand, the imaging sensors record in two dimensions (length and width), in order to compose the image of the scanning objects. Most remote sensing applications require information from different spectral regions (multispectral-multiband), which are obtained with different sensor (multi-sensor), or a single sensor operating simultaneously in different spectral regions (multispectral or multiband sensor). Furthermore, earth imaging systems, i.e. sensors, can be subdivided into three basic types, namely cameras, scanners and radar. Cameras and scanners are optical systems and are passive sensors, whereas radars are active sensors.

At the present time, there are several remote sensing systems, which are increasing year by year. The classification of these systems can be based on various criteria. A key criterion is the wavelength of the electromagnetic radiation, which classifies the systems into sensitive areas of the spectrum, the infrared and the microwave area (Avery and Berlin, 1992). Another criterion is the way in which the signal is transmitted and reflected back and the systems are classified into active systems and passive systems. Nevertheless, the most important attributes of satellite imagery are their spatial and temporal resolution. The choice of data depends on the type of application. Specifically, when it is necessary to use successive images in a short period of time, data with high temporal resolution can be selected, albeit lagging in spatial resolution. The opposite can happen when it is necessary to study a region on a large scale at a given time.

1.2 ELECTROMAGNETIC ENERGY

1.2.1 ELECTROMAGNETIC RADIATION

Electromagnetic radiation (EMR) is the basic energy quantity, which can produce work measured in joules. Electromagnetic energy is expressed in the form of mechanical, kinetic, chemical, electrical or thermal energy. Energy transfer is implemented through conduction, advection or radiation. Energy transfer through radiation is exploited by remote sensing, where energy is transferred from a body to a receiver or sensor, which receives and records the signal. Two models describe the most significant features of electromagnetic energy, namely the wave model and quantum theory (Barrett and Curtis, 1992). In the wave model, every molecule with a temperature above absolute zero vibrates and then becomes a radiation source for the magnetic and electrical fields around this molecule. These fields are perpendicular to each other. The number of waves that pass through a point in one second is called frequency, thus, long wavelength radiation is characterized by small frequency and vice versa. The following equation is valid:

$$\lambda f = c \tag{1.1}$$

where λ is the radiation wavelength, f is the frequency and c is the constant speed of light $c = 3 \times 10^9$ m/s through vacuum. Quantum theory emphasizes the radiation behavior and states that it consists of many small particles, the quanta of light, which Einstein called photons. The above justifies that light also has the properties of a body. The energy E of a photon is given by the equation:

$$E = hxf \tag{1.2}$$

where f is the wave frequency and h is Planck's constant, $h = 6.625 \times 10^{-27}$ erg/s or 6.626×10^{-34} J/s. The ratio of the energy of a photon over its frequency is constant (Plank's constant h). Thus, in photon's scale, high frequency light (short wavelength) is of high energy, whereas low frequency (long wavelength) is of low energy.

1.2.2 ELECTROMAGNETIC SPECTRUM

The entire range of electrical radiation is the electromagnetic spectrum (EMS-Electromagnetic Spectrum). The spectrum is divided into spectral bands, which in turn are composed of small groups of continuous spectra lines. These spectral bands are ultraviolet (UV), Visible, infrared (IR) and microwave (MW), of which the Visible is more clearly defined by human vision (Figure 1.1). The remote sensing systems are designed to specifically detect EMR, between one or more subdivisions, but this is not possible with only one sensor. Thus, the sensor groups can collect information through the spectral fluctuation, which is, of course, millions of times larger than the visible channel.

The Ultraviolet (UV) channel is located between X-rays and visible in the electromagnetic spectrum, with a wavelength of 0.01–0.40 μm and distinguished by far (UV UV, 0.01–0.20 μm), middle (UV UV, 0.20–0.30 μm) and near (near UV, 0.30–0.40 μm).

The visible band, with a wavelength of 0.40–0.70 μm, is limited by human vision. The white color of light comes from a mixture of six colors, which are, following the order of wavelength, violet (0.4–0.446 μm), blue (0.446–0.500 μm), green (0.500–0.578 μm), yellow (0.578–0.592 μm), orange (0.592–0.620 μm) and red (0.620–0.7 μm) (CCRS, 1998; NASA, 1998). But the basic colors of the visible are red, green and blue, also called RGB, and from which all the others arise through combination.

The IR part of the EMS ranges from red of the visible band to microwaves with a wavelength of 0.70–1000 μm. It is classified in Near-infrared (0.70–1.50 μm), Mid-infrared (1.50–5.60 μm) and Far-infrared (far IR, 5.60–1000 μm). The infrared is also divided into reflected IR (from 0.70 to 3.00 μm) and thermal (thermal IR, from about 3.00–1000 μm or 0.1 cm).

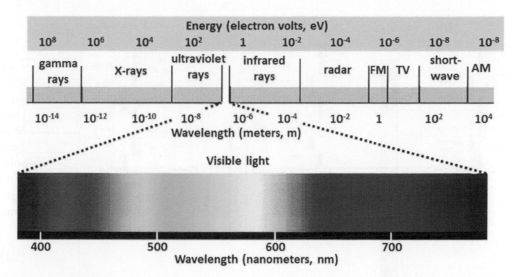

FIGURE 1.1 Range of electromagnetic spectrum and its channels, with the wavelength, the frequency and energy of the photons.

The microwave band is located between the infrared and the radio waves, with a wavelength of 0.1 cm to 1 m, where both passive and active sensors operate. Microwave (MW) radiation can penetrate clouds and various surface objects depending on the wavelength. Passive sensors (receivers) record the energy coming from an object (target), whereas the active sensors emit a signal that is sent to the target and record the returned signal by appropriate sensors.

1.3 SOLAR RADIATION AND EARTH ENVIRONMENT

1.3.1 Radiation Interaction with the Atmosphere and Targets

The areas of the spectrum that include wavelengths that can unobtrusively penetrate the atmosphere are called atmospheric windows or transmission bands. Thus, wavelengths can be defined that can be effectively used for remote sensing (Figure 1.2). The visible portion of the spectrum coincides with the peak energy level of the sun, as well as an atmospheric window. Moreover, heat energy emitted by the Earth coincides with a window around 10 μm in the thermal IR portion of the spectrum, whereas the large window, which is at wavelengths beyond 1 mm, is related to the microwave region.

Besides scattering, absorption is the other mechanism, in which electromagnetic radiation interacts with the atmosphere and results in the absorption of energy by molecules in the atmosphere at various wavelengths. Absorption dominates the infrared part (1–20 μm) of the spectrum, where there are various absorbent zones. Indeed, carbon dioxide, water vapor and ozone are the three main atmospheric constituents which absorb radiation. Only microwave radiation, with a wavelength greater than 0.90 mm, can completely penetrate the clouds. In the visible or near-visible infrared spectrum, the diffusion of light from the atmosphere is the main cause of the reduction in electromagnetic energy caused by gas molecules, such as O_2 and N_2, and clouds, whereas for wavelengths larger than 18 μm, in the microwave area, there is no great attenuation of the electromagnetic energy due to the atmosphere.

After its initial passage through the atmosphere, radiation reaches the surface. There are three potential forms of interaction of energy with the surface, namely absorption, transmission and reflection. The two extreme types of reflection of radiation from surface objects are mirror reflection and diffusion. Absorption is the phenomenon where radiation is bound by the object and then re-transmitted to longer wavelengths of the thermal infrared channel.

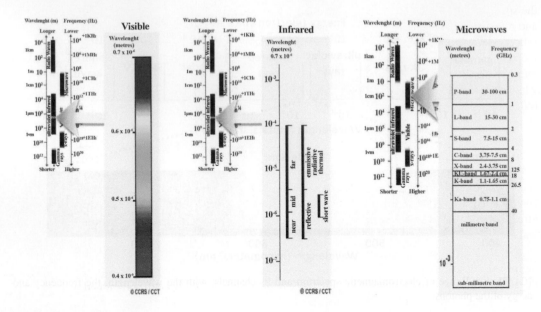

FIGURE 1.2 Main spectral areas used by Remote Sensing with the range of atmospheric absorption and Remote Sensing Systems (CCRS, 1998).

1.3.2 SPECTRAL SIGNATURES

The amount and spectral distribution of the emitted and reflected radiation from an object is used as a means of identifying that object. Indeed, this property is referred to as spectral response or spectral signature of the object and is recorded by the satellite sensors (Mather, 1999) (Figure 1.3).

In Figure 1.3, the spectral signatures of soil, vegetation and water are shown with high reflectance in the visible and infrared. Specifically, vegetation shows low values in the visible portion of

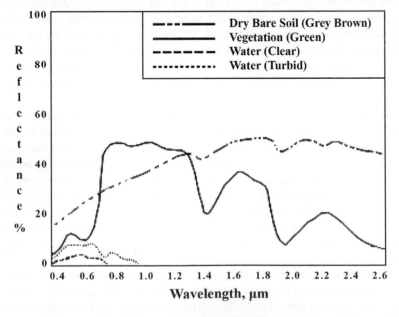

FIGURE 1.3 Typical spectral reflectance curves of vegetation, water and soil in different wavelengths of the electromagnetic spectrum (Lillesand and Kiefer, 2000).

the spectrum, since chlorophyll absorbs energy to a great extent (70%–90%) in the blue (0.45 μm) and red (0.67 μm) regions of the spectrum, which is necessary for the photosynthesis process, and reflects energy in the green region. In the near-infrared region (about 0.75 μm), the reflection of the healthy vegetation shows significant increase and vegetation appears very bright in the infrared region. In the region 0.7–1.3 μm, a plant leaf reflects about 40%–50% of the receiving radiation, whereas absorption is about 5%, and the remainder is transmitted. Beyond the region of 1.3 μm, the reflection of leaves depends on the wetness of the leaf (high absorption) and the thickness of the leaf (less absorption). Plant stress due to drought or the existence of minerals in the soil may have an impact on the spectral signatures of leaves, i.e. plant aging.

Water is absorbed in the near-infrared region (Figure 1.3). The increase of chlorophyll concentration in the water increases the energy reflection in the green region, resulting in the green color of water and reduces it in the blue region. These changes can be used to monitor and assess algae concentrations with remote sensing data. In Figure 1.3 the soil curve is rather smooth, since the factors affecting soil reflection have a similar impact in the different spectral regions. The main factors are soil moisture, texture, roughness, organic matter in the soil and the incident angle of solar radiation.

A useful quantitative expression of the reflection of different objects is the reflection coefficient or spectral reflection or albedo, which expresses the percentage of incident radiation reflected by the objects, e.g. with 50% cloud cover the spectral reflection is 35%. Therefore, albedo affects how objects appear when observing Earth from space. The brightest spots are the clouds and the darkest the water bodies.

1.4 ACTIVE REMOTE SENSING SYSTEMS

Satellite systems are classified into active and passive systems. Specifically, active remote sensing systems include energy-efficient remote sensing imaging systems that emit energy and record its reflection, whereas passive systems record the natural reflected or emitted radiation. Active remote sensing systems are SARs (Synthetic Aperture Radar) and weather radars and are considered all-weather systems (Dalezios, 2015; Dalezios et al., 2018). SARs perform in the microwave portion of the electromagnetic spectrum. In image processing and analysis of environmental issues, two types of passive remote sensing systems are considered, namely meteorological and environmental or resource satellites, respectively. There are differences between the two types of satellites, which are based on their spatial and temporal resolutions (Dalezios, 2017). Specifically, meteorological satellites have high temporal re-occurrence, but a rather coarse spatial resolution, which results in their suitability mainly for operational and monitoring meteorological applications. On the other hand, environmental satellites have usually low temporal re-occurrence, but fine spatial resolution, thus, being essentially suitable for land-use land-cover (LULC) classification.

It is important to mention the new European Copernicus satellite system, namely the Sentinel satellites, which was launched in 2014 essentially to replace ENVISAT (ESA, 2014), and expected to continue for several years. Table 1.1 presents a brief description of the functions of the different Sentinel satellites, which include active and passive satellite systems.

1.4.1 SATELLITE RADARS

Restrictions on passive systems can be overcome by active remote sensing systems. Since the wavelengths are large (between 1 cm and 1 m), microwaves have significant properties for remote sensing. Radiation corresponding to long wavelengths can penetrate clouds, fog, dust and heavy rainfall day and night, since long wavelengths are not subject to atmospheric diffusion (Gillespie et al., 2007).

Radar images are created by different types of sensors (Dalezios, 2015). Specifically, the SAR sensors are mainly found on satellites, e.g. SEASAT, whereas the side-looking airborne radar (SLAR) is an airborne side-by-side radar system. The sensor emits and receives signals as it moves and the signals received are combined to create the image. Since 1978, very successful radar satellite

TABLE 1.1

European Copernicus System with Sentinel Satellites

Satellite	Type of sensor	Applications
Sentinel-1A Sentinel-1B	C-band radar medium spatial resolution	Land, ocean, ice, oil spill, ship detection, flood, land surface, forest, water, soil, agriculture
Sentinel-2A Sentinel-2B	Optical multispectral medium to moderate spatial resolution instruments	Land-cover, detection maps, biophysical parameter maps, risk mapping, disaster relief
Sentinel-3A Sentinel-3B	Optical ocean and land color, sea and land temperature radiometer, altimeter and microwave radiometer	Sea, land, SST, LST, sea-surface topography, land ice topography, coastal zones
Sentinel-4A Sentinel-4B	Optical and microwave low-resolution sensors	Geostationary atmospheric chemistry missions
Sentinel-5A Sentinel-5B	UV-VIS-NIR-SWIR push broom grating spectrometer - TROPOMI moderate resolution sensor	Low earth orbit atmospheric chemistry missions
Sentinel-6A Sentinel-6B	Altimetry sensors	Altimetry mission

- https://sentinel.esa.int/web/sentinel/home
- https://www.esa.int/Applications/Observing_the_Earth/Copernicus/Overview4sentinel.esa.int/web/sentinel/home
- https://en.wikipedia.org/wiki/Copernicus_Programme
- https://sentinel.esa.int/web/sentinel/missions/sentinel-1
- https://directory.eoportal.org/web/eoportal/satellite-missions/c-missions/copernicus-sentinel-1
- https://www.satimagingcorp.com/satellite-sensors/other-satellite-sensors/sentinel-2a/
- https://earth.esa.int/web/sentinel/missions/sentinel-2
- https://space.skyrocket.de/doc_sdat/sentinel-2.htm
- https://en.wikipedia.org/wiki/Sentinel-2
- http://www.esa.int/Our_Activities/Observing_the_Earth/Copernicus/Sentinel-2_overview
- https://sentinels.copernicus.eu/web/sentinel/missions/sentinel-5p
- https://earth.esa.int/web/guest/missions/esa-eo-missions/sentinel-5p
- https://directory.eoportal.org/web/eoportal/satellite-missions/c-missions/copernicus-sentinel-5p
- https://sentinel.esa.int/web/sentinel/missions/sentinel-3/overview/mission-summary
- https://www.eumetsat.int/website/home/Satellites/CurrentSatellites/Sentinel3/index.html

missions have been launched. An overview of the main Radar satellite missions is presented in Figure 1.4. Radar satellite missions are used in environmental remote sensing. A brief description of these missions follows.

SEASAT Satellite. SEASAT was the first civilian satellite carrying onboard an imaging radar instrument (SAR). It was launched by NASA's Jet Propulsion Laboratory (JPL) in 1978. During its 105 days of operation in Earth orbit, SEASAT collected data, such as sea-surface temperatures (SST) and winds, wave heights, length and direction, atmospheric water content, sea ice morphology and dynamics.

ERS satellites. Earth Remote Sensing Satellites (ERS-1 and ERS-2) were ESA's (European Space Agency) first satellites in polar track and rotate around the Earth every 100 minutes. The main task of these satellites was to monitor oceans (waves), SST, coastal zones, land ecology, geology, forestry, natural disasters, mainly floods and earthquakes, global weather, such as wind direction and velocity, as well as polar ice. The ERS-1 and ERS-2 satellites consist of a series of instruments: an AMI (Active Microwave Instrument) (SAR/a wave and wind-scatterometer) to obtain information from the sea surface; a radar altimeter for the height of significant waves above the oceans, as well as the wind speed above them; a precise equipment for range and range rate (PRARE); a laser retroreflector array; an along-track scanning radiometer (ATSR); a microwave radiometer (MWR in ERS-2 only); and a global ozone measurement experiment (GOME, ERS-2 only).

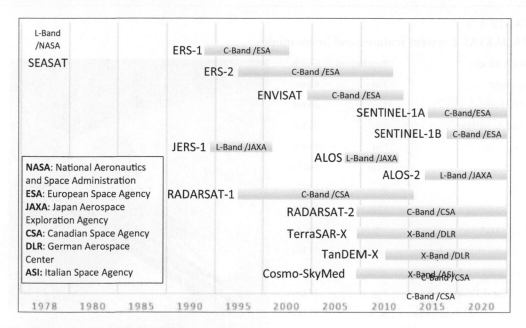

FIGURE 1.4 Radar satellites timeline.

ENVISAT. ESA's successor to ERS was ENVISAT (Environmental Satellite) satellite. ENVISAT was launched in 2002 and its mission ended in 2012. It was a multi-mission satellite carrying onboard ten scientific instruments and a high-performance radar antenna. These instruments enabled more advanced: (1) Atmospheric monitoring (clouds, aerosols, water vapor, temperature); (2) Oceanography; (3) Sea ice monitoring; and (4) Land classification (discrimination between vegetation and bare soil, or forested and clear-cut regions).

RADARSAT-1 and 2. RADARSAT-1 and 2 are environmental satellites built in Canada as a result of public-private partnerships under the auspices of the Canadian Space Agency (CSA) (Dicati, 2017). RADARSAT-1 and 2 were put into orbit in November 1995 and December 2007, respectively. The RADARSAT-1 mission ended in May 2013. Their products are useful for applications in land-use land-cover and planning, agriculture, cartography, geology, mining, forestry, hydrology, oceanography, ice studies, oil-gas, coastal monitoring and DEM generation (InSAR). The main features of RADARSAT-2 are described in Table 1.2. The next generation is expected to be RADARSAT Constellation Mission (RCM), consisting of three satellites and improving the spatial (<1.3 m) and the temporal resolutions (four days).

SENTINEL-1. Sentinel-1 is a constellation of two satellites, SENTINEL-1A and SENTINEL-1B, which carry a C-SAR sensor (ESA, 2014). The main features and applications of Sentinel-1 products are presented in Table 1.1.

JERS-1 and its successors (ALOS – ALOS-2). JERS-1(Japanese Earth Resources Satellite) satellite was launched in February 1992 by Japan Aerospace Exploration Agency (JAXA) carrying optical and radar sensors. Its data are used for: (1) surveying land-use and geological phenomena; (2) monitoring environment and disaster; and (3) change detection analysis using SAR interferometry. The operation of JERS-1 was terminated in October 1998.

ALOS (Advanced Land Observing Satellite) was the continuity of JERS-1. It carried three instruments: (1) the panchromatic sensor (PRISM) for digital elevation mapping; (2) the Near-Infrared Radiometer (AVNIR-2) for land-cover mapping; and (3) the L-band SAR (PALSAR) land observation in all weathers (Ravi, 2017). ALOS satellite was launched in February 2006 and retired in April 2011. The second Advanced Land Observation Satellite (ALOS-2) was launched in May 2014.

TABLE 1.2

RADARSAT-2 system features and beam modes

Beam Modes	Resolutions	Swath
Spotlight	3 m	8 km
Ultra Fine	3 m	20 km
Multi-Look Fine	8 m	50 km
Fine	8 m	50 km
Standard	30 m	100 km
Wide	30 m	150 km
ScanSAR Narrow	50 m	300 km
ScanSAR Wide	100 m	500 km
Extended High Incidence	18–27 m	75 km
Extended Low Incidence	30 m	170 km
Fine Quad-Polarization	8 m	25 km
Standard Quad-Polarization	30 m	25 km

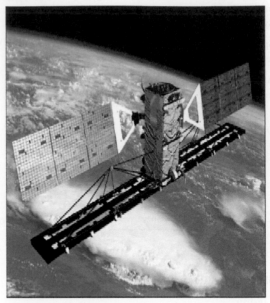

RADARSAT-2 (image credit MDA)
Satellite Orbital Features
Orbit: Sun-synchronous
Launch Date: December 2007
Altitude: 798 km
Revisit Time: 24 days

- http://www.asc-csa.gc.ca/eng/satellites/radarsat2/Default.asp
- https://directory.eoportal.org/web/eoportal/satellite-missions/r/radarsat-2

Soil Moisture Ocean Salinity (SMOS) mission. ESA deployed the SMOS Earth Explorer mission in order to provide global observations of soil moisture over land surfaces and ocean salinity over the oceans (Raizer, 2017). These two components are useful to understand the exchange processes between the Earth's surface and the atmosphere in the water cycle framework. To attain these measurements a satellite has been in orbit since 2009 (Table 1.3).

Soil Moisture Active Passive (SMAP) mission. SMAP mission is designed to measure soil moisture and freeze/thaw state for non-liquid water on Earth surfaces. Applications include climate change, hydrology, carbon-energy cycle, weather forecasts or crop-yield predictions. Moreover, the SMAP mission data can help climate models to estimate future trends in water resource availability (Table 1.4).

Global Precipitation Measurement (GPM) Core Observatory. NASA and JAXA (Japanese Aerospace Exploration Agency) built a science satellite called GMP core observatory (Table 1.5). GPM can improve climate, weather and hydrological forecasting and can help to understand the mechanism of tropical cyclones, extreme weather, floods and landslides.

The Tropical Rainfall Measuring Mission (TRMM). The ancestor of GMP mission was the TRMM, which was also a collaboration between NASA and JAXA (Andronache, 2018). The TRMM satellite was launched in 1997 and conducted precipitation measurements, on Earth, in tropical and subtropical areas helping to better understand of lightning-storm relationships, tropical cyclone structure, human impacts on rainfall and similar aspects. Its mission ended in April 2015. The 18-year data set included applications such as flood and drought monitoring and weather forecasting. The TRMM satellite carried a Precipitation Radar (PR) and four additional observation instruments: (1)

TABLE 1.3
SMOS main technical satellite features

Instrument: Microwave Imaging Radiometer using Aperture Synthesis (MIRAS)

Specifications: L band, 21 cm. Frequency 1.4 GHz, Spatial resolution 35–5

Applications: Soil moisture for hydrology studies and salinity for understanding of ocean circulation.

Satellite Orbital Features: Launch Date: November 2, 2009. Orbit: Sun-synchronous, polar. Repeat cycle: 18 days, 3-day subcycle. SMOS (image credit ESA)

- https://earth.esa.int/web/guest/missions/ esa-operational-eo-missions/smos
- https://directory.eoportal.org/web/eoportal/ satellite-missions/s/smos

SMOS (image credit ESA)

TABLE 1.4
SMAP main technical satellite features

Instruments for: Soil moisture	Instruments	Applications
Radar L-band (VV, HH, and HV polarizations Frequency: 1.26 GHz. Resolution: 1	Radar L-band Polarization: HH, Friquency: 1.26 GHz Resolution (SAR): 1–3 km	Soil moisture its freeze/ thaw state. Estimates much water is in the top layer of soil.
Radiometer L-band Frequency: 1.41GHz. Polarization: V, H, U, Resolution: 4,		

SMAP (image credit NASA)
Satellite Orbital Features
Launch Date: January 31, 2015
Orbit: Sun-synchronous
Orbit Altitude: 685 km
Repeat cycle: 8 days

- https://smap.jpl.nasa.gov/observatory/specifications/
- https://directory.eoportal.org/web/eoportal/satellite-missions/content/-/article/smap

TRMM Microwave Imager (TMI) (Yang and Wu, 2010); (2) Visible Infrared Radiometer (VIRS); (3) Cloud and Earth Radiant Energy Sensor (CERES); and (4) Lightning Imaging Sensor (LIS). All these instruments provided observations about the precipitation column, the water vapor, the cloud water, and the rainfall intensity in the atmosphere.

1.4.2 WEATHER RADAR SYSTEMS

Radar is the acronym for Radio detection and ranging. Radar is an electronic system for sensing, detecting and tracking objects or targets (Figure 1.5). Weather radars or meteorological radars operate in frequencies sensitive to raindrops, cloud droplets, ice particles, snowflakes, atmospheric nuclei, insects, birds, and regions of large index-of-refraction gradients. Radar was developed and used during the World War II to monitor the atmosphere (Battan, 1973; Berne and Krajewski, 2013).

TABLE 1.5

GPM technical satellite features and instruments

GPM Microwave Imager (GMI)	Dual-frequency Precipitation Radar (DPR)	Applications
13 microwave channels Frequency: 10–183 GHz	KuPR: Ku-band (13.6 GHz) Spatial Resolution (Nadir) KaPR: Ku-band (35.5 GHz) Spatial Resolution (Nadir)	Precipitation, heavy to light rain, falling snow. Enhanced Prediction Skills for Weather and Climate

GPM (image credit NASA)
Satellite Orbital Features
Launch Date: February 27, 2014
Orbit: Sun-synchronous
Altitude: 407 km
Repeat cycle: 3 hours

- NASA and JAXA Launch the GMP Satellite - SpaceRef
- https://www.nasa.gov/mission_pages/GPM/spacecraft/index.html
- https://pmm.nasa.gov/GPM/flight-project/GMI

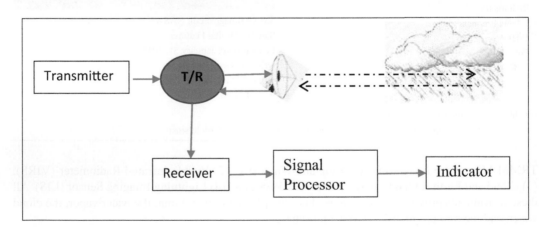

FIGURE 1.5 Basic components of a weather radar system.

There are also the so-called atmospheric radars with antenna targeting vertically upwards for studies of the upper layers of the atmosphere.

Conventional or non-coherent radars have been developed since the early 1950s, and they do not consider the phase of the returned radar wave in relation to the phase of the transmitted wave, which means that what is detected is the location and the reflectivity of a target (Battan, 1973). During the

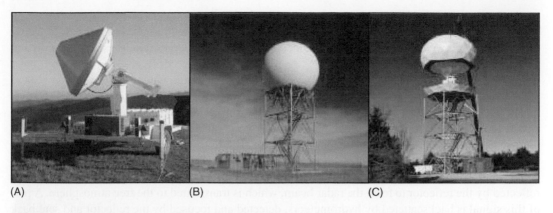

FIGURE 1.6 Different types of radar. A) Voluminal Radar, France; B) Doppler Radar, USA; C) Polarimetric Radar, USA (from Richards et al., 2010).

last 30 years, voluminal and Doppler radars have been developed, as well as polarimetric coherent radars (Bringi and Chandrasekar, 2004; Richards et al., 2010) (Figure 1.6). Doppler radars also measure the speed of the target along the radar beam axis detecting the rate of change in the phase difference between transmitted and returned signals.

There are advantages and disadvantages of conventional radars, which constitute most current operational systems, as well as polarimetric radars, which are the future operational systems (Berne and Krajewski, 2013). The typical radar measurement in rainfall estimation is the horizontal polarization reflectivity (Z_H in dBZ), which is derived from low frequency (S-band or C-band), and high-power weather radar systems (Chandrasekar et al., 2004; Michaelides et al., 2009). However, the so-called Z-R relationship is required, which is a power law function with two coefficients a and b, to convert reflectivity to rainfall rate.

Latest developments involve the upgrade of the single polarization systems to include dual-polarization capability, which moderates the effect of Z-R spatiotemporal variability and radar calibration (Bringi and Chandrasekar, 2004; Dalezios, 2017). The use of cost-effective X-band radar units is particularly stressed in cases of regions prone to localized severe weather phenomena, such as tornadoes and flash floods, and over mountainous basins not well covered by operational weather radar networks due to terrain blockage. Polarimetric radar systems constitute recent advances in weather radar technology, which are more suitable for hydrometeorological and hydrological applications. Initially, rainfall estimation was developed using the anisotropy information, which originates from the oblateness of raindrops (Seliga and Bringi, 1976). This approach led to the development of new parameters, such as the differential reflectivity (ZDR) for automatic distinction between rain and hail and the differential phase shift (ΦDP) for the estimation of hydrometeor size distributions and for the quantification of rain-path attenuation in radar observations by short wavelength weather radars (Michaelides et al., 2009). Indeed, longer wavelengths, such as S-band weather radars, are more attractive in the quantification of rainfall, as compared to shorter wavelengths, such as X- and C-band weather radars, due to the rain-path attenuation (Dalezios, 2017). For illustrative purposes, Figure 1.6 presents different types of radar, namely: A) Voluminal Radar, France; B) Doppler Radar, USA; C) Polarimetric Radar, USA (from Richards et al., 2010).

1.4.2.1 Conventional Radar: Non-coherent Radar

Radar basics. Conventional or non-polarimetric Doppler radars normally transmit horizontally polarized electromagnetic (EM) waves at the speed of light, 2.998×10^8 m s^{-1}, and is considered to travel along straight lines until a reflection is obtained and determine the direction to the reflected object. These radars only provide a measure of the horizontal dimension of the cloud particles, such as snow, ice pellets, hail and rain (Berne and Krajewski, 2013). Moreover, the distance to the object can be easily calculated by measuring the time interval between the transmission of the radio energy and the reception of the reflected signal. This one-dimensional picture makes it difficult to assess the difference between precipitation types, since ice particles and larger raindrops are generally not spherical.

Radar Components. The basic radar components are: transmitter, antenna, receiver and indicator (Figure 1.5). The transmitter produces power at the radar frequency and generates an EM wave, which travels through the waveguide connecting the transmitter with the radar antenna. The antenna, which radiates the power and intercepts the reflected signals, is composed of three components: the pedestal, the dish (or reflector) and the feedhorn. The reflector is mechanically rotated by the pedestal in the azimuth direction and vertically in elevation. The EM wave, after the waveguide, is conveyed to the feedhorn, which converts the alternative current wave into a radio wave. This is then reflected by the reflector to form the radar beam, which is transmitted to the free atmosphere. A part of this signal is backscattered by hydrometeors, detected and focused by the reflector and sent back to the feedhorn. This energy is called the backscatter. The receiver detects, amplifies and transforms the received signals into video form. Specifically, the radio wave is converted in volts and this energy travels down to the receiver, where it is amplified and processed. There is also an indicator on which the returned signals can be displayed. The reflected power depends on the hydrometeor and radar characteristics, such as frequency, antenna and distance to the radar. As a result, radar is considered a system of active remote sensing.

The Radar Beam. The relationship between radio frequency (f), wavelength (λ) and velocity at the speed of light (c) is the following:

$$c = f\lambda \tag{1.3}$$

where c is taken to be 3×10^8 m s^{-1}, f is in cycles per second or Hertz (Hz), and λ is in m. Figure 1.7 illustrates how the beam volume (or distance between the upper and lower part of the beam) increases with range and is referred to as beam broadening or beam spreading (Wang and Yang, 2014). This characteristic is one of the major limiting factors of radar measurements taken at far range. Table 1.6 presents the correspondence between radar frequencies, wavelength and band. Notice that the smaller the size of the hydrometeors, the smaller the required wavelength to detect them, i.e. weather radar with wavelength of 3 cm (or X-band) can detect smaller-size hydrometeors.

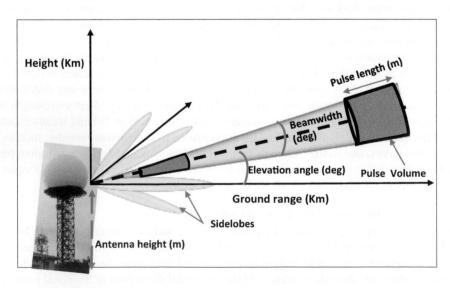

FIGURE 1.7 Radar beam propagation through the atmosphere.

TABLE 1.6
Frequencies of weather radar

Frequencies (GHz)	Wavelength λ (cm)	Band
18–27	1	K
8–12	3	X
4–8	5	C
2–4	10	S
1–2	20	L

The Radar Pulse. The range r from the radar to the target is determined by the following Equation (1.4), considering the two-way travel of the pulse to the target and back to the receiver:

$$r = cT/2 \tag{1.4}$$

where c is the speed of light (2.998×10^8 m s^{-1}) and T in sec is the elapsed time between a transmitted pulse and the reception of the backscattered energy from the same pulse. The pulse repetition frequency (PRF) in sec^{-1} is the number of pulses the radar transmits per second. The reciprocal of the PRF is the pulse repetition time (PRT), which is the elapsed time from the beginning of one pulse to the next. The pulse duration (τ) in sec is how long it takes to transmit a single pulse of energy. For PRF = 1000 Hz the distance is about 30 and, thus, the maximum range of radar is about 15. For typical antenna sizes, such as from 2 to 8 m, depending on the frequency, the radar beam width is typically about 1°, leading to a radar sampling volume that linearly increases with the distance from the radar (Figure 1.7) (Berne and Krajewski, 2013). As a result, final Cartesian radar products are commonly provided at a resolution in the order of 1×2 and 5 min.

Radar equation. The radar equation, known as the Marshall-Palmer equation (Marshall and Palmer, 1948), relates the average received power Pr (W) to the transmitted power Pt (W) via the following equation:

$$P_r = \left[\left(\pi^3 c \right) / \left(1024 \ln 2 \right) \right] \left[\left(P_t \tau G^2 \Theta^2 \right) / \lambda^2 \right] \left[|K|^2 \left(Z / r^2 \right) \right] \tag{1.5}$$

where G denotes the antenna gain (dimensionless), Θ denotes the half power beam width (rad), h denotes the pulse width (m), K denotes the complex dielectric factor of the targets (dimensionless), Z denotes the reflectivity factor (mm^6 m^{-3}), λ denotes the wavelength of the radar (cm), and r denotes the distance to the target (km). In arriving at Equation (1.5), it is assumed that the beam is uniformly filled with targets, multiple scattering may be ignored, the total average power equals the sum of the scattered power from individual particles, the beam is Gaussian-shaped and the targets are Rayleigh scatterers (Battan, 1973). The radar reflectivity factor Z depends only on the hydrometeor features, namely dropsize distribution (DSD) and shape. If these assumptions are met and the targets are known (water, ice, etc.) to specify the complex dielectric factor accurately, then Z can be estimated from Equation (1.5).

Radar meteorology. Weather radar is ground radar, which sweeps the atmosphere at preselected relatively low elevation angles, that is from the horizon or 0° to about 25° with steps of 0.5° or 1°. The scanning is conducted in polar coordinates up to 360° for each elevation angle per 0.1° and every 300 or 600 m distance from the radar site according to the radar operating frequency. Quantitative analysis can be made for radar signals up to a range or distance of about 15. The reflectivity of Z is associated with the distribution of the water droplets size in the radar volume sample with the relation:

$$Z = \int_0^\infty D^6 \cdot N_v \cdot (D) dD \tag{1.6}$$

where $N_v(D)$ dD refers to the average number of water droplets with equivalent spherical diameters between D and D + dD (mm) per unit volume of air. Note that the units of $N_v(D)$ are mm^{-1} m^{-3}. Since there is a very wide range of Z values, the radar logarithmic reflectivity is used, defined as 10 logZ, which is expressed in dBZ units (Battan, 1973). If the turbulence due to wind and its impact on the water droplets are not considered, then the rainfall intensity R (mm/hr) is associated with the water droplets distribution $N_v(D)$ through the equation:

$$R = 6 \cdot \pi \cdot 10^{-4} \int_0^\infty D^3 \cdot v(D) \cdot N_v(D) dD \qquad (1.7)$$

where v(D) is the functional relationship between the final velocity of the water droplets and the equivalent spherical water droplets diameter D (mm), which is given by:

$$v(D) = c \cdot D^\gamma \qquad (1.8)$$

where c = 3.778 and $\gamma = 0.67$ for a raindrop range of $0.5 \le D \le 5$ mm. Based on measurements on the ground of water droplets magnitude distributions and the Equation (1.8) for v(D), then empirical Z-R relations can be derived. In general, these relationships follow the form (Battan, 1973):

$$Z = a \cdot R^b = \Sigma (D/2)^6 \qquad (1.9)$$

where a and b are coefficients that change in space and time, but remain independent of the rainfall intensity R. These coefficients express the climatic characteristics of a region and vary with the type of precipitation.

Radar indicators. There are two basic modes of operating a rotating pedestal: plan position indicator (PPI) mode and range height indicator (RHI) mode, respectively. PPIs are obtained by spinning the antenna in the azimuthal direction, while keeping it fixed at a constant elevation angle. A full 360° rotation in PPI mode constitutes a surveillance scan and yields a tilt of data. On the other hand, an RHI keeps the azimuth fixed and varies the elevation angle. RHI mode is useful for interrogating specific storms, whereas a surveillance scan in PPI mode is more practical for operational precipitation estimation. For illustrative purposes, Figure 1.8 shows an RHI mode (A), a PPI mode (B) and a CAPPI mode (C), respectively.

Weather radar sources of error. There are non-meteorological sources of error (Battan, 1973; Dalezios, 1988), such as: (1) electronic radar calibration; (2) beam propagation as a function of distance from the radar site affecting the height and width of the beam; (3) anomalous propagation of the signal due to the curvature of the Earth; (4) ground clutter caused by terrain echoes, such as mountains, as a function of range and height; (5) shielding behind the ground clutter location; (6) wavelength issues; (7) rotational speed; (8) beam interception of the freezing level; (9) frequency of radar data collection; (10) reflections from various non-meteorological targets, such as insects, birds, aircrafts, chaffs, solar radiation or ships.

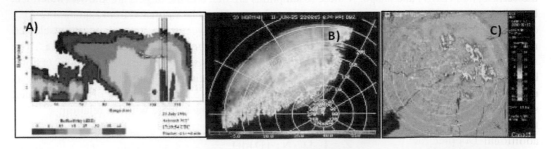

FIGURE 1.8 Radar images A) RHI; B) PPI; C) CAPPI (from Dalezios, 2015).

There are also several meteorological sources of error (Wilson and Brandes, 1979; Zawadzki, 1984), such as: (1) signal attenuation with the range from the radar site as a function of frequency and wavelength; (2) non-uniform radar beam filling as a function of the range from the radar site; (3) variations in the snow crystal type with an impact on the reflectivity of the target signal; (4) wind effect between the measuring height and the ground as a function of the range from the radar site (Dalezios, 1988); (5) updraft causing evaporation or downdraft causing growth of precipitation between the measuring height and the ground; (6) presence of hail or other hydrometeors; (7) variations in the drop-size distribution affecting the Z-R relationships; (8) variability in the vertical profile of reflectivity decreasing with height; (9) orographic enhancement of precipitation; (10) dielectric factor as a function of precipitation type; (11) shape of rainfall droplets (spherical); (12) averaging errors of received power in regions of strong precipitation gradients; (13) additional attenuation by water on the radome; (14) identification of the precipitation phase by bright band detection; (15) presence of hail or other hydrometeors.

1.4.2.2 Conventional Radar: Coherent or Doppler Radar

Two new coherent weather radar types developed mainly during the last 30 years, namely voluminal radar and Doppler radar, are now briefly described (Figure 1.6).

Voluminal radar. This is a conventional radar, which can delineate storm intensity and the total precipitation amount within the 4-D precipitation structure. To achieve this, it uses consecutive radar PPIs. The advantages of voluminal radar compared to the classical conventional radar are: identification of precipitation location and beam as related to its intensity based on the 4-D structure; anchor or elbow echo for severe mesoscale cyclonic systems; correction of the vertical reflectivity profile; discrimination between precipitation types and improvement of the quantitative precipitation estimation (Figure 1.6A).

Doppler radar. It is useful to know the velocity of the backscattered signals with respect to a fixed point on the ground. A non-coherent radar cannot provide such information except in a few special circumstances. The term "Doppler radar" has been given to the class of radar sets which measures the shift in microwave frequency caused by moving targets (Figure 1.6B) (Zohuri, 2020). Doppler radars, in addition to obtaining the data collected by a non-coherent radar, also measure the velocity of the targets along the radar beam axis by noting the rate of change of the difference in phase between the outgoing and received signals. As the target moves, the phase changes at a rate proportional to the velocity of the target toward or away from the radar. From the measurement of the radial velocity of the target it is easy to extract information on the 1-D wind speed, since only one component of the wind vector can be measured, which is the radial component.

Doppler capability allows radar systems to provide information on the range of the target and the radial velocity of the backscatters, also called the "Doppler velocity" (Berne and Krajewski, 2013). Doppler velocities are more commonly used for severe weather detection, such as rotation with supercell thunderstorms, and less so for quantitative precipitation estimation. There are two measured variables, the mean Doppler radial velocity v (m s^{-1}), which is the mean of the distribution of the velocity of the hydrometeors within the sampling volume along the radar beam direction and the Doppler spectrum width σ_v (m s^{-1}), which is the standard deviation of the distribution of the velocity of the hydrometeors being sensitive to process, such as atmospheric turbulence that can significantly modify the velocity distribution.

The basic Doppler radar characteristics and advantages are as follows: (1) identification of the beam, location and intensity of precipitation, i.e. the 4-D structure of precipitation; (2) recording of tornadoes and mesoscale convective systems (MCSs); (3) calculation of the vertical wind shear, i.e. the vertical wind change and profile; (4) information on the 1-D structure of the wind field and patterns; (5) delineation of the 2D and 3D wind fields through simulation or use of additional Doppler radars. Examples of typical Z-R relationships are the classical Marshall-Palmer Z-R relationship (Equation 1.10) for non-coherent radars (Marshall and Palmer, 1948) and a commonly used relationship for coherent Doppler and polarimetric radars (Cifelli and Chandrasekar, 2010) (Equation 1.11):

$$R_{MP}(Z_h) = 0.0365 Z^{0.625} \left(mm\, h^{-1} \right) \tag{1.10}$$

$$R_{WSR}(Z_h) = 0.017 Z^{0.714} \left(mm\, h^{-1} \right) \tag{1.11}$$

where R is rain rate and Z_h is the radar reflectivity factor in $mm^6\, m^{-3}$ at horizontal polarization.

1.4.2.3 Polarimetric Radar

Polarimetric radars are dual-polarization systems, which allow for dual-channel reception and transmission of horizontally and vertically polarized electromagnetic waves (Figure 1.6C and Figure 1.9) (Cifelli and Chandrasekar, 2010; Bringi and Chandrasekar, 2004). Conventional radar measures only backscattered power, whereas polarimetric radar systems provide additional information about the scatterer characteristics (Berne and Krajewski, 2013). For precipitation, these characteristics are related to the size, shape, phase state (liquid or ice) and orientation of hydrometeors in the radar resolution volume, as well as the polarization of the transmitted wave (Cifelli and Chandrasekar, 2010; Hong and Gourley, 2015). Thus, dual-polarization radar can significantly improve the accuracy of precipitation estimation, can be used to more accurately retrieve characteristics of the drop-size distribution (DSD), can distinguish between very heavy rain and hail, attenuation effects, partial beam blocking and can reduce the effects of non-weather scatterers on radar displays (Bringi et al., 1984). Although dual-polarization techniques have been applied to S-band and C-band radar systems for several decades, polarization diversity at higher frequencies, including X-band, are now widely available to the radar community due to lower cost and the improvements in attenuation correction with polarimetry (Berne and Krajewski, 2013).

Polarimetric radar principles. Raindrops with an equivolume diameter larger than about 1 mm are not spherical, but oblate with their horizontal dimension larger than their vertical one. Their shape is therefore described as an oblate spheroid with semi-major and semi-minor axes a and b, respectively, where oblateness is usually expressed by the ratio (b/a) (Cifelli and Chandrasekar,

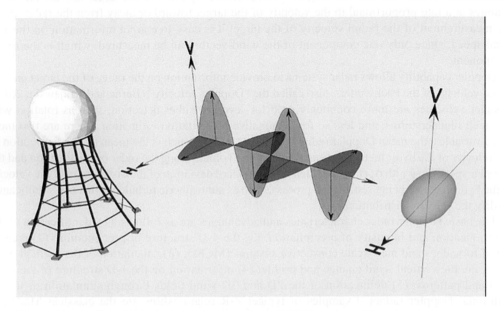

FIGURE 1.9 Polarization of Polarimetric radar (http://arrc.ou.edu).

2010). These differential observations constitute the basis of weather radar polarimetry (Berne and Krajewski, 2013). The simplest form is a linear relation:

$$(b/a) = 1.03 - \beta D, 1 \leq D \leq 9 \, mm \tag{1.12}$$

A polarimetric radar is a dual-polarization Doppler radar system, since it also has the Doppler capability (Berne and Krajewski, 2013). In addition, hail can be better detected with dual-polarization than with conventional radar and helps eliminate the need for vertical integrated liquid (VIL) or similar products (Ryzhkov and Zrnic, 2019).

Polarimetric radar characteristics. The main characteristics of polarimetric radar are the following: (1) identification of the type of hydrometeor, which means discrimination between ice, hail (including size), snow and rain, even in winter storms; (2) improvement in the estimation of rainfall intensity, even with signal attenuation; (3) improvement of the estimation of snow intensity, including drop diameter; (4) recognition of the type of target, i.e. discrimination between meteorological (e.g. storms) and non-meteorological targets (e.g. terrain echoes); (5) identification and correction of radar sources of error, such as meteorological targets, anomalous propagation, bright band region and signal attenuation.

1.5 PASSIVE REMOTE SENSING SYSTEMS

Passive remote sensing systems depend on the Earth's "light" from sunlight to record the reflected and emitted radiation. Their performance is limited by the presence of cloud, fog, smoke and darkness. Meteorological and environmental satellites belong to passive remote sensing systems.

1.5.1 METEOROLOGICAL SATELLITES

There are two classes of meteorological satellites. The first class, which is the majority, operates at heights between 800 and 1,50. These satellites are called low orbit and consist of the polar or near-polar orbit satellites with a sun-synchronous orbit, since they cross the equator at the same time, e.g. the series of NOAA-N (NOAA, 2006; Dalezios et al., 2018). The second class operates at approximately 36,00 and are considered high orbit satellites. This second class are called geostationary satellites, since they appear to be "stationary" at specific locations over the equator and move in the same direction and at the same rate that the Earth spins (Emery and Camps, 2017). There are five such satellites, e.g. METEOSAT and GOES, which cover the Earth and they can be used mainly in operational meteorology, weather forecasting and monitoring.

The USA's National Oceanic and Atmospheric Administration (NOAA) uses meteorological satellite systems for weather advisory services (NOAA, 2006). At the present time, NOAA's operational weather satellite system is composed of polar-orbiting environmental satellites (POES) and geostationary operational environmental satellites (GOES) (Yang and Wu, 2010). Both types of satellites provide data continually and are used for short-range warning and for long-term forecasting. The NOAA's satellite data are also used for applications, such as climate research, e.g. water surface temperatures, snow or monitoring of sea ice extents, environmental hazards monitoring, volcanic eruption monitoring, forest fire detection and rescue operations, among others (Ilčev, 2019). The capability of meteorological satellites, along with the increasing remote sensing reliability and the continuous technological advances, enables tracking of the atmosphere, ensuring real-time coverage of short-term dynamic events, such as local storms and MCSs, and detecting meteorological parameters quantitatively, such as precipitation, humidity, wind and temperature, among others (Ilčev, 2019).

Geostationary satellites: METEOSAT. The first generation of satellites (METEOSAT-1 to -7) began with the launch of the METEOSAT satellite on November 27, 1977 (Capderou, 2014). The second-generation series started with the launch of the MSG-1 (METEOSAT Second Generation) in 2002 (METEOSAT-8 to -11). The first generation carried a MVIRI (METEOSAT Visible and InfraRed Imager) sensor and the MSG had onboard two radiometers as presented in Table 1.7 (Ilčev, 2019).

NOAA polar orbit satellites: Polar orbit satellite systems, such as NOAA, provide continuous data over the globe and, thus, they are potentially efficient and relatively inexpensive tools for regional applications, such as monitoring vegetation conditions, agriculture and crop-yield assessment with daily coverage and data acquisition compared with conventional weather data (Dalezios, 2015). A long series of geosynchronous, polar-orbiting meteorological satellites NOAA/AVHRR data sets already exist. Since 1978, NOAA satellites have been carrying onboard AVHRR (Advanced Very High-Resolution Radiometer), a scanner that senses in the visible, near-infrared, mid-infrared and thermal infrared portions of the electromagnetic spectrum. The last series of POES are NOAA-15 to NOAA-19, which carry onboard the latest generation of AVHRR/3 and TOVS with six channels and the following sensors (Table 1.8): (1) AMSU-A1, -A2 and -B (Advanced Microwave Sounding Units), which can calculate vertical temperature/moisture profiles from the Earth's surface; (2) HIRS/4 (High-Resolution Infrared Radiation Sounder), which measures high-resolution temperature profiles under cloudless conditions (Guo et al., 2019); (3) SBUV/2 (The Solar Backscatter Ultraviolet Spectral Radiometer), which measures stratospheric ozone; (4) SEM-2 (Space Environment Monitor), which measures the flow of charged particles at the satellite's altitude, thus contributing to our understanding of the "sun-earth" system; and (5) SAR (Search and Rescue Repeater and Processor Support), which receives emergency signals sent from ships or airplanes.

TERRA polar orbit satellites: The TERRA satellite (formerly called EOS AM-1) was launched by NASA in December 1999, as the Earth-Observing System (EOS) platform. Its mission is a global monitoring of the Earth's state of atmosphere, ocean, ice, land and energy flows (Ilčev, 2019). The data are used for applications, such as: (1) disaster-water-coastal-carbon-ecological management; (2) agricultural efficiency; and (3) security-public health. The technical features of MODIS-ASTER are provided in Table 1.9. In addition, the TERRA spacecraft carries

TABLE 1.7
Technical features of METEOSAT Second Generation (MSG)

Band	Bandwidth (µm)	Resolution (km)
Visible (VIS)	0.56–0.71	3
Visible (VIS)	0.74–0.88	3
High-Res. VIS	0.5–0.9	1
Infrared (IR)	1.5–1.78	3
IR	3.48–4.36	3
IR	8.3–9.1	3
IR	9.8–11.8	
IR	11–3	
Water vapor	5.35–7.15	
WV	6.85–7.85	
IR	9.38–9.94	
IR	12.4–14.4	

METEOSAT-8 (image credit EUMETSAT)
Satellite Orbital Features
Orbit: Geosynchronous (geostationary)
Altitude: 36000 km
Revisit Time: METEOSAT-1 to -7: 30 min. MSG: 15 min

- https://space.skyrocket.de/doc_sdat/meteosat-1.htm
- https://directory.eoportal.org/web/eoportal/satellite-missions/m/meteosat-first-generation
- https://www.eumetsat.int/website/home/Satellites/CurrentSatellites/Meteosat/index.html
- https://earth.esa.int/web/eoportal/satellite-missions/m/meteosat-second-generation

TABLE 1.8
Technical features of the AVHRR/3 system and the NOAA satellites

AVHRR/3 (NOAA-15 through 19)

Band	Bandwidth (μm)	Resolution (km)
1	0.58–0.68 (Red)	1.1
2	0.73–0.98 (Near-IR)	1.1
3a	1.58—1.6 (Mid-IR)	1.1
3b	3.54–3.87 (High-Temp TIR)	1.1
4	10.3–11.3 (TIR)	1.1
5	11.5–12.4 (TIR)	1.1

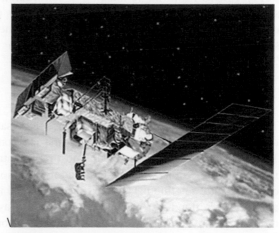

POES NOAA-19 (image credit NASA)
Satellite Orbital Features
Orbit: Sun-synchronous. Altitude: 85. Two NOAA
operational satellites twice-daily global coverage.

- http://www.ospo.noaa.gov/Operations/POES/status.html
- https://directory.eoportal.org/web/eoportal/satellite-missions/n/noaa-poes-series-5th-generation

TABLE 1.9
Technical features of MODIS-ASTER spectroradiometer in TERRA satellite

Band	Instrument	Resolution
MODIS		
1–2	VNIR	250 m
3–7	VIS - SWIR	500 m
8–36	VNIR-SWIR- TIR	1000 m
ASTER		
1–3	VNIR	15 m
4–9	SWIR	30 m
10–14	TIR	90 m

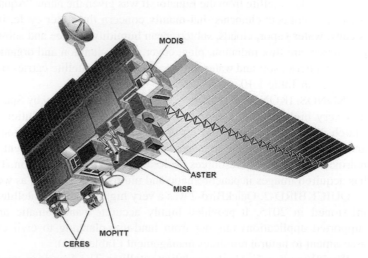

TERRA satellite (image credit NASA)

- https://eospso.gsfc.nasa.gov/missions/terra
- https://modis.gsfc.nasa.gov/about/
- https://www.nasa.gov/mission_pages/terra/spacecraft/index.html
- https://www.satimagingcorp.com/satellite-sensors/other-satellite-sensors/aster/

the following instruments: CERES. Clouds and Earth's Radiant Energy System (CERES): two CERES instruments measure Earth's fluxes from the surface to the top of the atmosphere (Ilčev, 2019). MISR. Multi-angle Imaging Spectroradiometer: this instrument measures the amount of sunlight that is scattered under natural conditions (Neeck et al., 1998). MOPITT. Measurements of Pollution in the Troposphere (Dicati, 2017): measures the distribution, transport, sources and sinks of carbon monoxide in the troposphere using three spectral bands with spatial resolution of 2 at nadir.

1.5.2 ENVIRONMENTAL SATELLITES

There is a long list of environmental or resource satellites, which, as already mentioned, are polar or near-polar low orbit satellites. The first environmental satellite, LANDSAT, was launched in 1971, and is considered one of the most representative and successful satellites of this type, still in orbit and providing valuable data and information (Dalezios, 2015). Existing successful uses of environmental satellite data in the monitoring and management of environmental hazards include, among others, flood area mapping, snow-cover measurements for runoff estimation and flash floods, drought quantification and detection of several drought features including areal extent, as well as drought assessment and monitoring (Dalezios, 2015).

A-train Satellite Constellation. A-train (from Afternoon Train) is a satellite constellation that flies close together and follows each another at the same speed. Their instruments monitor the Earth's changing climate, measuring clouds, aerosols, atmospheric chemistry, etc. The name A-train stands for satellites that have equator crossings in the early afternoon and in the middle of the night (about 1:30 p.m. and 1:30 a.m., respectively). At the present time, there are six active Earth Observation satellites belonging to A-train: Aqua, Aura, CALIPSO, CloudSat, GCOM-W1 and OCO-2 (Guo et al., 2019. A-train satellite constellation complements the morning satellite constellation, like TERRA, Landsat-7 and 8, which cross the equator in the morning (around 10:30 a.m.) and evening (about 10:30 p.m.) (Dicati, 2017).

Aqua Earth-observing satellite mission. The low-polar-satellite Aqua is part of NASA's EOS (Earth Observation System). Originally the mission was named EOS PM to signify the afternoon passage of the satellite from the Equator. It was given the name "Aqua", which means water in Latin, because it collects elements that mainly concern the water cycle, including evaporation from the oceans, water vapor, clouds, subterranean humidity, sea ice and snow cover. Additional measurable parameters are flux radiation, plant cover, phytoplankton and organic matter in the oceans, as well as atmospheric, soil and water temperatures. Aqua satellite carries onboard six instruments that are described in Table 1.10.

IKONOS. IKONOS was launched on September 24, 1999 by Space Imaging and was the world's first very high-resolution civilian satellite to offer 1 m resolution. Its 15-year mission ended in March 2015. Its data are used for urban cartography (visualization of buildings, roads), land-use and planning, agricultural management, environmental assessment, natural disaster management, marine studies, natural resources exploration, or DEM generation. The very high-resolution sensor has acquired images in panchromatic and multispectral modes, as well as stereo images products.

QUICKBIRD-2. QuickBird-2 was a very high-resolution satellite, launched in 2001 and decommissioned in 2015. It provided highly accurate panchromatic and multispectral imagery and supported applications ranging from land-use planning to civil engineering and environmental assessment to natural resources management (Table 1.11).

WorldView/GeoEye high-resolution satellites. The American commercial company DigitalGlobe provides a constellation of high-resolution commercial earth observation imaging satellites. These are: WorldView-1 (launched in September 2007), GeoEye-1 (September 2008), WorldView-2 (October 2009), WorldView-3 (August 2014) and WorldView-4 (November 2016). This constellation of very high and high-resolution earth observation satellites provides new and enhanced applications, such as: land-cover mapping (soil/vegetation analysis), change detection analysis, feature

TABLE 1.10
Aqua satellite technical features

Instruments	Specifications	Applications
AIRS: Atmospheric Infrared Sounder	2378 IR, 4 VNIR channels	Temperature profiles, Cloud etc.
AMSR-E: Advanced Microwave Scanning Radiometer for EOS	12-channel microwave radiometer	Monitor, cloud water, water vapor, sea ice sea-surface winds/ temperature, etc.
AMSU-A: Advanced Microwave Sounding Unit-A	15-channel microwave sounder	Monitor temperature profiles in the upper atmosphere
CERES: Clouds and the Earth's Radiant Energy System	3-channel, radiometer	Measure major elements of Earth's radiation budget.
HSB: Humidity Sounder for Brazil	4-channel microwave sounder	Measures humidity profiles and detects heavy precipitations.
MODIS: Moderate Resolution Imaging Spectroradiometer	36 VIS-IR channels	Measures, aerosol properties, sea ice, land vegetation, etc.

Aqua satellite (image credit NASA)
Satellite Orbital Features
Orbit: Near-polar, sun-synchronous
Altitude: 705 km
Launch Date: May 4, 2002
Revisit Time: 16 days

- https://eospso.gsfc.nasa.gov/sites/default/files/mission_handbooks/Aqua.pdf
- https://aqua.nasa.gov/

TABLE 1.11
Technical features of Quickbird-2

Band	Bandwidth	Resolution (GSD at nadir)
Panchromatic imagery		
1	0.405–1.053 μm	0.61 m
Multispectral imagery		
1 Blue	0.43–0.545 μm	2.44 m
2 Green	0.466–0.620 μm	2.44 m
3 Red	0.590–0.710 μm	2.44 m
3 NIR	0.715–0.918 μm	2.44 m

Quickbird-2 (image credit Digital Globe)

Satellite Orbital Features: Orbit: Sun-synchronous polar. Altitude: 450 km. Pixel quantization: 11 bit. Revisit Time: 2.5 days

- https://directory.eoportal.org/web/eoportal/satellite-missions/q/quickbird-2
- https://www.satimagingcorp.com/gallery/quickbird/

extraction (detection of man-made materials), environmental monitoring, coastal monitoring. The constellation satellite features are provided in Table 1.12.

SENTINEL–2. The Earth observation satellite "Sentinel-2" consists of two identical satellites, Sentinel-2A and Sentinel-2B, constructed by ESA (European Space Agency) as part of the Copernicus Program. The description is presented in Table 1.1.

TABLE 1.12

WorldView and GeoEye satellite features

WV-1. Bands (nm): Panchromatic: (400–900). Resolution: 0.5 m.

GeoEye-1. Bands (nm): Panchromatic: (450–800). Resolution: 0.41 m. Bands (nm): 4 Multispectral: (450–510), (510–580), (655–690), (780–920). Resolution: 1.65 m.

WV-2. Bands (nm): Panchromatic: (450–800). Resolution: 0.5 m. Bands (nm): 8. Multispectral: (400–450), (450–510), (510–580), (585–625), (630–690), (705–745), (770–895), (860–1040). Resolution: 1.85 m.

WV-3. Bands (nm): Panchromatic: (450–800). Resolution: 0.31 m.
Bands (nm): 8 Multispectral: (397–454), (445–517), (507–586), (580–629), (626–696), (698–749), (765–899), (857–1039). Resolution: 1.24 m.
Bands (nm): 8 SWIR: (1184–1235), (1546–1598), (1636–1686), (1702–1759), (2137–2191), (2174–2232), (2228–2292), (2285–2373). Resolution: 3.7 m. Bands (nm): 12CAVIS: (405–420), (459–509), (525–585), (635–685), (845–885), (897–927), (930–965), (1220–1252), (1365–1405), (1620–1680), (2105–2245), (2105–2245). Resolution: 30 m.
WV-4. Bands (nm): Panchromatic: (450–800). Resolution: 0.31 m. Bands (nm): 0.31 m
4 Multispectral: (450–510), (510–580), (655–690), (780–920). Resolution: 1.24 m.

WorldView-4 Orbital Features: Orbit: Sun-synchronous
Altitude: 617 km. Dynamic Range: 11-bits per pixel.
Repeat Cycle: 1 day
- https://www.euspaceimaging.com/about/satellites/
- http://www.landinfo.com/WorldView2.htm
- http://www.landinfo.com/WorldView1.htm

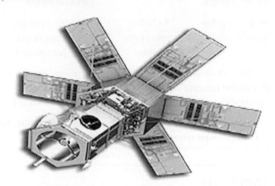

WV-4 (image credit Digital Globe)

SPOT. The first SPOT (Satellite Pour l'Observation de la Terre) French Satellite (SPOT-1) was launched by the French Space Agency CNES (Centre national d'études spatiales) in 1986 and are sun-synchronous, nearly-polar orbit, at a height of 83. The last generation of SPOT satellites are SPOT 6 and 7, which are supported by AIRBUS Defense and Space. The characteristics of these satellites are presented in Table 1.13, and the ten-year life span assures image data continuity up to 2024. Finally, their co-orbit with very high-resolution Pléiades 1A and 1B constellation offers detailed and broader regional coverage.

PLÉIADES. The Pléiades program was launched by French organization CNES (Centre national d'études spatiales) in 2003, as a continuation of the SPOT program satellites. The two satellites Pléiades-1A and Pléiades-1B with daily global revisits offer a large range of monitoring applications: disaster areas, natural hazards, evacuation and rescue operations, coastline changes and environmental incidents. The Pléiades-1A and Pléiades-1B satellite characteristics are provided in Table 1.14.

Indian Earth Observation Satellites (IEOS). Indian Space Research Organization (ISRO) has launched many Earth Observation (EO) satellites for over 25 years. The Indian Remote Sensing (IRS) satellite program includes four generations of satellites ensuring the continuity of medium and high-resolution data supply for the observation and management of natural resources: (1) first generation IRS-1A/1B; (2) second-generation IRS-1C/1D; (3) third-generation Resourcesat-1/2 and Cartosat-1; and (4) fourth-generation Cartosat-2/2A/2B/2C/2D/2E (Gupta et al., 2016). The various missions in the IRS satellite program employ various sensors: Linear Imaging Self Scanning Sensor (LISS-I/II/III/IV), Advanced Wide Field Sensor (AwiFS), among others, in the visible and

TABLE 1.13
SPOT 6–7 features

SPOT 6 & 7:
Band (μm): Panchromatic (0.45–0.71). Resolution: 1.5 m.
Band (μm): Blue (0.45–0.52). Resolution: 6 m.
Band (μm): Green (0.53–0.59). Resolution: 6 m.
Band (μm): Red (0.61–0.68). Resolution: 6 m.
Band (μm): NIR (0.76–0.89). Resolution: 6 m.
SPOT 6–7 Orbital Features:
Orbit: Sun-synchronous, polar. Altitude: 694 km. Dynamic
Range: 12 bits per pixel. Daily revisit capacity with SPOT
6 & 7

- http://gisgeography.com/
  spot-satellite-pour-observation-terre/
- https://earth.esa.int/web/eoportal/satellite-missions/s/
  spot-6-7

SPOT-7 (image credit AIRBUS Defense & Space)

TABLE 1.14
Pléiades -1A and Pléiades-1B features

Pléiades-1A&1B:
Bands (μm): Panchromatic (0.48–0.83). Resolution: 0.5 m.
Bands (μm): Blue (0.43–0.53). Resolution: 2 m.
Bands (μm): Green (0.49–0.61). Resolution: 2 m.
Bands (μm): Red (0.6–0.72). Resolution: 2 m.
Bands (μm): Near-Infrared (0.75–0.95). Resolution: 2 m.
Pléiades -1A,1B Orbital Features:
Orbit: Sun-synchronous, polar. Launch Date: 2011 &
2012 (1A,1B). Altitude: 694 km. Dynamic Range: 12 bits
per pixel. Revisit Interval: Daily

- https://earth.esa.int/web/guest/missions/3rd-
  party-missions/current-missions/
  pleiades-hr
- http://gisgeography.com/pleiades-satellite/
- https://www.satimagingcorp.com/satellite-sensors/
  pleiades-1/
- https://www.satimagingcorp.com/satellite-sensors/
  pleiades-1b/

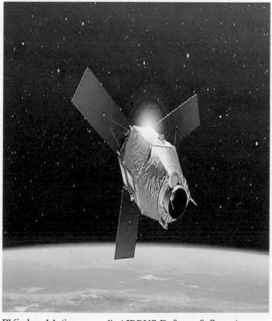

Pléiades -1A (image credit AIRBUS Defense & Space)

near-infrared range of the spectrum (Pandya et al., 2013). The applications cover a wide range, which includes land cover, agriculture, hydrology (water resources), geology, urban planning, rural development, environmental pollution, forestry, drought-flood management, cartography, coastal zones, ocean resources, national security.

LANDSAT. The Landsat Program consists of four generations of satellite missions starting with the launch of Landsat-1 in 1972. It is the longest running program for acquiring satellite imageries of the Earth. Over these years of the Program, eight Landsat satellites have been launched, monitoring the Earth surface, jointly managed by NASA and the US Geological Survey (Guo et al., 2019). The Landsat program provides valuable data, available for free, for land-use research and a high range of applications in agriculture (crop classification, forecasting), forestry (timber types), geology, hydrology, environmental monitoring, land-cover change detection analysis, coastal management, regional planning, cartography, surveillance, education. The first generation was Landsat 1 to 3 carrying optical instruments, a multispectral sensor (MultiSpectral Scanner or MSS) and a series of video cameras (Ravi, 2017). The second-generation satellites, Landsat 4 and 5, were equipped with a multispectral scanner (MSS) and a Thematic Mapper (TM). The third-generation satellite, Landsat 7, is equipped with Enhanced Thematic Mapper Plus (ETM+). Finally, the fourth-generation satellite, Landsat 8, is equipped with the Operational Land Imager (OLI) sensor, and the thermal sensor. Moreover, Landsat 9 is planned to be launched in 2023. Landsat 8 is presented in Table 1.15.

AMD-Aeolus. ESA, in the framework of Living Planet Program, has developed the weather satellite Aeolus or AMD-Aeolus (Atmospheric Dynamics Mission Aeolus). The objective is to perform wind observations to improve weather forecasts and understand atmospheric dynamics and climate. Table 1.16 presents the characteristics of Aeolus.

Sentinel-5 Precursor (S-5P). S-5P satellite is a collaboration between ESA and the Netherlands Space Office, as part of the Copernicus Program. The objective of S-5P is to monitor anthropogenic and natural emissions as: formaldehyde, aerosol, carbon monoxide, carbon dioxide and sulphur dioxide and greenhouse gases, such as ozone and methane (Kohler et al., 2018; Liang, 2018). The description is presented in Table 1.1.

Sentinel-3. This is a constellation of three satellites by ESA, in the framework of the Copernicus Program. The main objectives are: (1) to measure sea-surface topography, sea-water quality, SST and LST; (2) to monitor sea and land ice topography; (3) to monitor land-use changes; (4) to detect fire; and (5) to forecast weather. The description is presented in Table 1.1.

JPSS SATELLITES. Joint Polar Satellite System (JPSS) is a constellation of new generation polar-orbiting operational environmental satellites. The JPSS program is a collaboration between National Aeronautics and Space Administration (NASA) and the National Oceanic and Atmospheric Administration (NOAA). The overarching goal of the satellites is to gather measurements of atmospheric (temperature, water vapor, ozone), terrestrial (vegetation, snow, ice cover, fire locations) and oceanic (sea temperature) conditions. The JPSS satellites are: Suomi NPP, JPSS-1, JPSS-2, JPSS-3, JPSS-4 and TCTE, where two satellites, namely (1) Suomi National Polar-orbiting Partnership (Suomi NPP), and (2) NOAA-20 (formerly JPSS-1), are in orbit. Table 1.17 presents the characteristics of Suomi NPP and NOAA-20.

Global Change Observation Mission - Climate "SHIKISAI" (GCOM-C). The GCOM-C belongs to Japanese (Japan Aerospace Exploration Agency-JAXA) program consisting of two satellites, GCOM-W and GCOM-C, respectively. The GCOM-C was launched in December 2017, its primary goal to measure the global warming mechanisms related to fluctuations in radiation budget, carbon cycles and similar aspects (Liang, 2018). The satellite carries the SGLI (Second-Generation Global Imager) sensor, which has 19 spectral bands from near-ultraviolet to thermal infrared region. The SGLI sensor can observe 15 Essential Climate Variables (ECV), such as cloud, aerosols, vegetation, biomass, chlorophyll-a photosynthesis and similar aspects.

(GOSAT-2) Greenhouse gases Observing SATellite-2, "IBUKI-2". Japan Aerospace Exploration Agency (JAXA): GOSAT-2 aims to measure, with higher levels of accuracy, greenhouse gases in order to provide data to environmental administrations to mitigate global warming and monitor

TABLE 1.15

Technical features of Landsat 7 & 8

Landsat satellite	Spectral Bands (µm)	Spatial Resolution
Landsat 7		
Enhanced Thematic Mapper Plus (ETM+)	Blue (0.441–0.514)	30 m
	Green (0.519–0.601)	30 m
	Red (0.631–0.692)	30 m
	NIR (0.772–0.898)	30 m
	SWIR1 (1.547–1.749)	30 m
	Thermal (10.31–12.36)	60* m
	SWIR2 (2.064–2.345)	30 m
	Panchromatic (0.515–0.896)	15 m
Landsat 8	Ultra Blue (0.435–0.451)	30 m
Operational Land Imager (OLI)andThermalInfraredSensor(TIRS)	Blue (0.452–0.512)	30 m
	Green (0.533–0.590)	30 m
	Red (0.636–0.673)	30 m
	NIR (0.851–0.879)	30 m
	SWIR1 (1.566–1.651)	30 m
	SWIR2 (2.107–2.294)	30 m
	Panchromatic (0.503–0.676)	15 m
	Cirrus (1.363–1.384)	30 m
	TIRS-1 (10.60–11.19)	100** m
	TIRS-2 (11.50–12.51)	100** m

Landsat 8 (image credit NASA)

Landsat 8 Orbital Features

Orbit: sun-synchronous, near-polar

Altitude: 70

Dynamic Range: 12 bit

Repeat Cycle: 16 days

*Resampling at 30 meters

**Resampling at 30 meters

- https://www.usgs.gov/land-resources/nli/landsat/landsat-8?qt-science_support_page_related_con=0#qt-science_support_page_related_con
- https://www.usgs.gov/land-resources/nli/landsat/landsat-7?qt-science_support_page_related_con=0#qt-science_support_page_related_con
- https://directory.eoportal.org/web/eoportal/satellite-missions/l/landsat-8-ldcm

TABLE 1.16
Technical features of Aeolus (Atmospheric Dynamics Mission)

Instrument: ALADIN
Atmospheric LAser Doppler INstrument

ALADIN: ultraviolet laser lidar
The instrument provides 60 mJ pulses of ultraviolet light at 355 nm.
ALADIN instrument contains Mie and Rayleigh backscattering receiver assembly.
The Mie receiver consists of Fizeau interferometer with a resolution of 100 MHz (equivalent to 18 m/s).
The Rayleigh receiver consists of Fabry–Pérot interferometer with a 2 GHz resolution and 5 GHz spacing.

ADM-Aeolus (image credit ESA/ESTEC)
ADM-Aeolus Orbital Features
Orbit: Sun-synchronous
Launch Date: August 22, 2018
Altitude: 320 km.
Repeat Interval: 7 days

- https://earth.esa.int/web/eoportal/satellite-missions/a/adm-aeolus
- http://www.esa.int/Enabling_Support/Operations/Aeolus_operations

TABLE 1.17
Technical features of Suomi NPP and NOAA-20 (JPSS)

(VIIRS) Visible Infrared Imaging Radiometer Suite. Specifications: Resolution: 750 m. 22 spectral bands from 412 nm to 12 µm. Applications: Snow-ice, clouds, aerosol, vegetation health, smoke, fire, fog, phytoplankton, chlorophyll, dust.

(OMPS) Ozone Mapping and Profiler Suite. Specifications: Nadir Spatial Resolution: 50 km Mapper & 250 km Profiler. Mapper: 0.3–0.38 µm. Profiler: 0.25–0.31 µm. Applications: Ozone layer and the concentration of ozone.

(CrIS) Cross-track Infrared Sounder. Specifications: Nadir Spatial Resolution: FOV 1 diam. vertical layer 1305 spectral channels from 3.92 µm to 15.38 µm. Applications: Temperature, moisture.

(CERES) Clouds and the Earth's Radiant Energy System Specifications: Nadir Spatial Resolution: 2 3 channels: 0.3 to 5 µm - 8 to 12 µm - 0.3 to > 50 µm. Applications: Solar energy reflected by Earth, the heat the planet emits.

(ATMS) Advanced Technology Microwave Sounder. Specifications: Nadir Spatial Resolution: 15.8–74..22 bands from 23 GHz to 183 GHz. Applications: Atmospheric temperature and moisture profiles. Collects microwave radiation from the Earth's atmosphere and surface.

- https://directory.eoportal.org/web/eoportal/satellite-missions/n/noaa-20
- http://rammb.cira.colostate.edu/projects/npp/
- https://www.nasa.gov/mission_pages/NPP/mission_overview/index.html
- https://www.nesdis.noaa.gov/content/jpss-series-satellites-noaa-20-spacecraft

Suomi NPP (image credit BATC)
Suomi NPP Orbital Features Orbit: Sun-synchronous.
Launch Date: October 28, 2011. Altitude: 824 km.
Revisit time: 16 days

NOAA-20 (image credit Ball Aerospace)
NOAA20 Orbital Features Orbit: Sun-synchronous.
Launch Date: November 18, 2017. Altitude: 824 km
Revisit time: 20 days

the impacts of climate change. It carries two main sensors: (1) (TANSO-FTS), Thermal And Near-infrared Sensor for carbon Observation – Fourier Transform Spectrometer for measuring, CO2, CH4, O3, H2O, CO, NO2, and (2) (TANSO-CAI), Thermal And Near-infrared Sensor for carbon Observation – Cloud and Aerosol Imager (Liang, 2018).

GEO-KOMPSAT 2A and 2B constitute a South Korean geostationary meteorological satellite developed by KARI (Korea Aerospace Research Institute). GEO-KOMPSAT-2 (GK-2A) is a meteorological mission, its overarching goal to improve the accuracy of weather observation and weather forecasting. GEO-KOMPSAT-2B (GK2B) is a meteorological mission to monitor the ocean and the environment. The GK-2A carries two sensors for space weather observations: (1) AMI, Advanced Meteorological Imager, and (2) KSEM, Korean Space Environment Monitor. The GK-2A and GK-2B were launched in October 2018 and in February 2020, respectively, with an expected life span of ten years.

GOES-R. Geostationary Operational Environmental Satellites. The GOES-R series is NOAA's latest generation of geostationary weather satellites, which include a four-satellite program, namely GOES-R, GOES-S, GOES-T and GOES-U. The GOES-R series provides measurements of Earth's weather, oceans and environment, real-time mapping of total lightning activity and improved monitoring of solar activity. They will improve accurate forecasts of phenomena, such as severe storms, fog, fire and aerosols (Goodman et al. 2020). Two satellites are active: (1) GOES-R (GOES-16) and (2) GOES-S (GOES-17). Table 1.18 presents the technical features of GOES-16 and GOES-17.

MetOp-C (Meteorological operational satellite-C). This satellite is the last of three polar-orbiting meteorological satellites (Metop-A, Metop-B) developed by the European Space Agency (ESA) in collaboration with Eumetsat. Metop-C carries eight instruments, which provide observations of the

TABLE 1.18
Technical features of GOES-16 & GOES-17

ABI (Advanced Baseline Imager). Specifications & Applications: 16 bands. Spatial resolution: 0.5–2 km. Cloud formation, atmospheric motion. Tropical cyclone forecasts.

EXIS (Extreme Ultraviolet and X-ray Irradiance Sensors) Specifications & Applications: Extreme Ultraviolet Sensor (EUVS), X-Ray Sensor (XRS), Monitoring solar irradiance in the upper atmosphere.

GLM (Geostationary Lightning Mapper). Specifications & Applications: Single band 777.4 nm, Spatial resolution nadir, Measures total lightning activity. Forecasts severe storms.

Magnetometer. Specifications/Applications: Measurements of the space environment magnetic field.

SUVI (Space environment In-Situ Suite). Specifications & Applications: Four sensors, Monitor proton, electron, and heavy ion fluxes in the magnetosphere.

SUVI (Solar Ultraviolet Imager). Specifications & Applications:
Observations of solar flares and solar eruptions. Early warning of impacts to Earth's environment.

GOES16&17 (image credit NOAA)
GOES16&17, Orbital Features
Orbit: Geostationary
Launch Date: GOES16: November 19, 2016, GOES17: March 1, 2018 Altitude: 35,800 km

- https://www.goes-r.gov/mission/mission.html
- https://en.wikipedia.org/wiki/List_of_GOES_satellites

global atmosphere, oceans and continents, the overarching goal to increase the accuracy of weather forecasting (Liang, 2018). Table 1.19 presents the technical characteristics of MetOp-C.

FormoSat-5 is an Earth Observation satellite supported by National Space Organization (NSPO) of Taiwan (Ravi, 2017). FORMOSAT-5 carries an optical payload and a science payload to execute remote sensing missions and perform science research. The Remote Sensing Imager sensor provides 2 m GSD (Ground Sample Distance) for Pan imagery and 4 m GSD for multispectral imagery on a swath of 2. The Advanced Ionospheric Probe (AIP) instrument measures ionospheric plasma concentrations, velocities, temperatures and ambient magnetic fields (Sarris, 2019). The satellite was launched on July 19, 2017.

The SAOCOM-1 (SAtélite Argentino de Observación COn Microondas) mission was managed by Argentina's Space Agency. The SAOCOM-1 is composed of two satellites (SAOCOM-1A and -1B). The first (SAOCOM-1A) was launched on October 8, 2018 and carries the SAOCOM-SAR instrument. It provides all-weather, day/night polarimetric L-band SAR observations, satisfying applications in the fields of agriculture, desertification, forestry, weather, droughts, landslides and similar aspects.

TABLE 1.19
Technical characteristics of MetOp-C

Metop-C (image credit EUMETSAT)
Metop-C, Orbital Features: Orbit: Polar, sun-synchronous
Launch Date: November 7, 2018
Altitude: 817 km
- https://www.tryo.es/metop-c-launch/
- https://www.eumetstat.int/website/home/ MetopCLaunch/Status/index.html

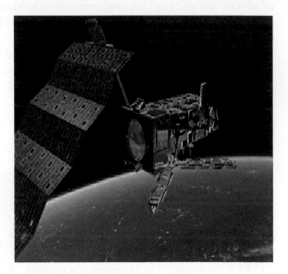

A-DCS (Advanced Data Collection System). Specifications & Applications: Transponder of messages. Receiver frequency: 401.65 MHz. In-situ Data collection & Platforms.

AMSU-A (Advanced Microwave Sounding Unit-A). Specifications & Applications: 15 bands. Temperature, cloud ice & liquid water.

ASCAT (Advanced Scatterometer). Specifications & Applications: C-band: 5.255 GHz. Resolution: 50–25 km. Surface wind speed-vector.

AVHRR/3 (Advanced Very High-Resolution Radiometer/3. Specifications & Applications: 6-band radiometer ranging from VIS to TIR. Resolution: 1.1 km. Cloud cover, type, CTT.

GOME-2. Specifications & Applications: 4096 channels from UV-NIR. Resolution: 40 km. Ozone vertical profile.

GRAS (GNSS Atmospheric Sounding). Specifications & Applications: Resolution: 30 horizontal, 0. vertical. Temperature, humidity.

IASI (Infrared Atmospheric Sounding Interferometer). Specifications & Applications: 8461 channels. Resolution: 48 x 4². Temperature, humidity, water vapor.

MHS (Microwave Humidity Sounding). Specifications & Applications: 5-channels. Resolution: 16 km, IFOV, liquid water. humidity, cloud-ice.

TABLE 1.20
Technical features of FY-4

AGRI (Advanced Geosynchronous Radiation Imager). Applications & Specifications: VNIR – 3 bands, SWIR – 3 bands. MWIR – 2 bands, Water Vapor- 2 bands, LWIR – 4 bands. Clouds, aerosol, water vapor, SST.

GIIRS (Geostationary Interferometric Infrared Sounder). Applications & Specifications: LWIR-S/MWIR–VIS, precipitation, cloud clusters.

GLI (Geostationary Lightning Imager). Applications & Specifications: Charge-Coupled Device camera: 777.4 nm. Observe regional lightning activity in China.

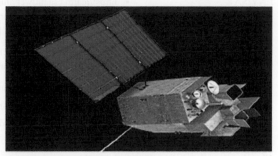

FY-4A (image credit: CMA/NSMC)
FY-4A Orbital Features
Orbit: Geostationary. Launch Date: December 10, 2016
Altitude: 35786 km

- https://www.wmo-sat.info/oscar/satellites
- https://directory.eoportal.org/web/eoportal/satellite-missions/f/fy-4

The VENµS (Vegetation and Environment monitoring on a New MicroSatellite) satellite is a collaboration between Israel (ISA) and France (CNES) administrations (Meygret et al., 2007). The VENµS is a minisatellite carrying two sensors, the VENµS Super-spectral Camera (VSSC) and the Israeli Hall Effect Thruster (IHET). The VSSC provides 12 spectral bands digital imagery for monitoring, analysis and modeling of land surface.

FY-4 China's Meteorological Administration second-generation geostationary meteorological satellite series. The FY (FenYun, in Chinese) means ""winds and clouds or storm". The new generation FY-4 satellites follow on from the FY-2 series (Ilčev, 2019). It is used for monitoring small- and medium-scale weather systems. Table 1.20 presents the technical characteristics of FY-4.

GRACE-FO (Gravity Recovery and Climate Experiment Follow-On) mission is a partnership between NASA and the German Research Centre for Geosciences (GFZ). It consists of twin satellites, which follow each other in orbit around the Earth. Their mission is to monitor changes in underground water storage, the amount of water in large lakes and rivers, soil moisture, ice sheets and sea level changes. The satellites were successfully launched on May 22, 2018 (Ilčev, 2019).

PRISMA (Hyperspectral Precursor of the Application Mission) satellite belongs to the Italian Space Agency ASI (Agenzia Spaziale Italiana). PRISMA is an Earth observation satellite, which delivers medium-resolution hyperspectral products. Its mission is to monitor land cover, crop status, pollution in inland waters, status of coastal zones and atmospheric characteristics. The satellite was launched on March 22, 2019.

Table 1.21 contains additional links of satellite systems.

TABLE 1.21
Additional links of satellite systems

SEASAT	• https://www.jpl.nasa.gov/missions/seasat/
	• https://directory.eoportal.org/web/eoportal/satellite-missions/s/seasat
ERS-1	• https://earth.esa.int/web/sppa/mission-performance/esa-missions/ers
	• https://directory.eoportal.org/web/eoportal/satellite-missions/e/ers-1
ERS-2	• https://m.esa.int/Enabling_Support/Operations/ERS-2
	• https://directory.eoportal.org/web/eoportal/satellite-missions/e/ers-2

(continued)

TABLE 1.21　　(continued)

ENVISAT	• https://earth.esa.int/web/guest/missions/esa-operational-eo-missions/envisat • https://directory.eoportal.org/web/eoportal/satellite-missions/e/envisat • https://en.wikipedia.org/wiki/Envisat
RADARSAT-1	• http://www.asc-csa.gc.ca/eng/satellites/radarsat1/Default.asp • https://directory.eoportal.org/web/eoportal/satellite-missions/r/radarsat-1
JERS-1	• https://earth.esa.int/web/guest/missions/3rd-party-missions/historical-missions/jers-1
TRMM	• https://trmm.gsfc.nasa.gov/ • https://pmm.nasa.gov/TRMM/mission-overview • http://global.jaxa.jp/projects/sat/trmm/
Classification of environmental satellites based on spatial resolution	• https://www.geoimage.com.au/satellite/rapideye • https://www.geoimage.com.au/satellites/very-high-resolution-satellites
SPOT	• https://en.wikipedia.org/wiki/SPOT_(satellite) • https://gisgeography.com/spot-satellite-pour-observation-terre/ • https://earth.esa.int/web/guest/missions/3rd-party-missions/current-missions/spot
SPOT 1–5	• https://en.wikipedia.org/wiki/SPOT_(satellite) • http://gisgeography.com/spot-satellite-pour-observation-terre/ • https://crisp.nus.edu.sg/~research/tutorial/spot5.htm
IRS	• http://uregina.ca/piwowarj/Satellites/IRS.html • https://directory.eoportal.org/web/eoportal/satellite-missions/i/irs • http://space.skyrocket.de/doc_sdat/cartosat-2.htm • https://directory.eoportal.org/web/eoportal/satellite-missions/c-missions/cartosat-2e
GCOM-C	• https://en.wikipedia.org/wiki/Global_Change_Observation_Mission • https://global.jaxa.jp/projects/sat/gcom_c/topics.html#topics14198 https://eoportal.org/web/eoportal/satellite-missions/content/-/article/gcom
GOSAT-2	• https://global.jaxa.jp/projects/sat/gosat2/ • https://directory.eoportal.org/web/eoportal/satellite-missions/g/gosat-2#sensors
GEO-KOMSAT-2A	• https://space.skyrocket.de/doc_sdat/geo-kompsat-2a.htm • https://directory.eoportal.org/web/eoportal/satellite-missions/g/geo-kompsat-2
Formosat-5	• https://www.nspo.narl.org.tw/inprogress.php?c=20030301&ln=en • https://directory.eoportal.org/web/eoportal/satellite-missions/f/formosat-5
SAOCOM-1A	• https://space.skyrocket.de/doc_sdat/saocom-1.htm • https://directory.eoportal.org/web/eoportal/satellite-missions/s/saocom
VenμS	• https://directory.eoportal.org/web/eoportal/satellite-missions/v-w-x-y-z/venus • https://space.skyrocket.de/doc_sdat/venus.htm
GRACE-FO	• https://gracefo.jpl.nasa.gov/microwave-instrument/ • https://directory.eoportal.org/web/eoportal/satellite-missions/g/grace-fo • https://gracefo.jpl.nasa.gov/mission/overview/
PRISMA	• https://directory.eoportal.org/web/eoportal/satellite-missions/p/prisma-hyperspectral • http://prisma-i.it/index.php/en/
Landsat 1–5	• https://landsat.gsfc.nasa.gov/a-landsat-timeline/ • https://www.usgs.gov/media/images/landsat-missions-timeline • https://www.usgs.gov/faqs/what-are-band-designations-landsat-satellites?qt

1.6 CURRENT TRENDS IN SATELLITE SYSTEMS

The main characteristic of the series of new meteorological and environmental or resource satellites is the gradual improvement of their spatial resolution and the increase of the spectral bands throughout the years (Dalezios, 2015; Dalezios et al., 2018). A long list of such satellites has already been presented. Indeed, the trend is to further improve the spatial resolution, reaching the level of micro-remote sensing in the order of one meter or less with new satellites, such as Quick bird or Rapid-eye, among others, and very recently WV-2 or WV-3 or WV-4. Similarly, there is a very recent tendency to increase the number of available bands in these satellites, resulting in new and valuable information covering a range of environmental and meteorological applications (Ilčev, 2019; Liang, 2018); Dalezios, 2017). Table 1.22 presents an indicative list of recent and upcoming Meteorological/Environmental satellites.

TABLE 1.22
Indicative list of recent and upcoming Meteorological/Environmental satellites

Sensor/Platform: FY 4B. Administration: CHINA Meteorological Administration, Launch date: until 2025. Applications: Monitoring small- and medium-scale weather systems. https://directory.eoportal.org/web/eoportal/satellite-missions/f/fy-4

Sensor/Platform: JPSS-2. Administration: NOAA-NASA, USA. Launch date: 2021
Applications: Atmospheric, terrestrial and oceanic conditions: SST, LST, vegetation, clouds. http://www.jpss.noaa.gov/mission_and_instruments.html

Sensor/Platform: JPSS-3. Administration: NOAA-NASA, USA. Launch date: 2026.
Applications: Atmospheric, terrestrial and oceanic conditions: SST, LST, vegetation, clouds. http://www.jpss.noaa.gov/mission_and_instruments.html

Sensor/Platform: JPSS-4. Administration: NOAA-NASA, USA. Launch date: 2031.
Applications: Atmospheric, terrestrial and oceanic conditions: SST, LST, vegetation, clouds. http://www.jpss.noaa.gov/mission_and_instruments.html

Sensor/Platform: EarthCARE. Administration: ESA, Europe. Launch date: 2021.
Applications: Global profiles of clouds and aerosols. Solar/thermal radiation reflected/emitted. https://earth.esa.int/web/guest/missions/esa-future-missions/earthcare

Sensor/Platform: Metop-SG- A& B Second-Generation. Administration: EUMETSAT ESA. Launch date: 2022–2043.
Applications: Climate monitoring and atmospheric chemistry.

Sensor/Platform: TARANIS. Administration: CNES, France. Launch date: November 2020. Applications: Transient luminous events (TLEs) and terrestrial gamma-ray flashes.
https://taranis.cnes.fr/en/TARANIS/index.htm

Sensor/Platform: GOES-T. Administration: NASA, USA. Launch date: 2021. Applications: Accurate forecasts of environmental phenomena like: severe storms, fog, fire and aerosols. https://www.goes-r.gov/mission/mission.html

Sensor/Platform: GOES-U. Administration: NASA, USA. Launch date: 2024. Applications: Accurate forecasts of environmental phenomena like: severe storms, fog, fire and aerosols. https://www.goes-r.gov/mission/mission.html

Sensor/Platform: SAOCOM-1B. Administration: CONAE- Argentina. Launch date: August 2020. Applications: Hydrology and land observations.

Sensor/Platform: SAOCOM-2A, 2B. Administration: CONAE- Argentina. Launch date: Planned ≥2020. Applications: Hydrology and land observations.

Sensor/Platform: EnMAP: Environmental Mapping and Analysis Program. Administration: DLR-Germany. Launch date: Planned ≥2020. Applications: Environmental, agricultural, land use, water management and geological monitoring. http://www.enmap.org/

Sensor/Platform: ALOS-3. Administration: JAXA – Japan. Launch date: Planned ≥2020. Applications: Disaster monitoring. Updating topographic maps, land use and vegetation. Survey of crops. https://directory.eoportal.org/web/eoportal/satellite-missions/a/alos-3

(continued)

TABLE 1.22   (continued)

Sensor/Platform: Urthedaily constellation. Administration: Urthecast-Canada, USA, Spain. Launch date: Planned ≥2020. Applications: Global change detection and analysis on unprecedented scale. https://space.skyrocket.de/doc_sdat/urthedaily.htm

Sensor/Platform: KOMPSAT-6. Administration: KARI – Korea. Launch date: 2021. Applications: Ocean & Land management, disaster monitoring and environment monitoring. https://directory.eoportal.org/web/eoportal/satellite-missions/k/kompsat-6

Sensor/Platform: Landsat 9. Administration: NASA, USA. Launch date: 2021. Applications: Monitoring and managing land resources. https://landsat.gsfc.nasa.gov/landsat-9/

Sensor/Platform: BIOMASS. Administration: ESA, Europe. Launch date: 2021. Applications: Measurements of the amount of carbon stored in the world's forests. https://earth.esa.int/web/guest/missions/esa-future-missions/biomass

Sensor/Platform: NISAR. Administration: NASA, USA. Launch date: 2021. Applications: Studying hazards and global environmental changes. https://nisar.jpl.nasa.gov/nisarmission/mission/

Sensor/Platform: SWOT-Surface Water and Ocean Topography. Administration: NASA, USA. Launch date: 2021. Applications: Global survey of Earth's surface water. https://swot.jpl.nasa.gov/applications/

Sensor/Platform: Tandem-L. Administration: DLR- Germany. Launch date: 2022. Applications: Global monitoring of forest biomass/deformations of the Earth's surface. https://directory.eoportal.org/web/eoportal/satellite-missions/t/tandem-l

Sensor/Platform: FLEX. Administration: ESA, Europe. Launch date: 2022. Applications: Vegetation mapping to quantify photosynthetic activity. https://earth.esa.int/web/guest/missions/esa-future-missions/flex

Sensor/Platform: PACE. Administration: NASA, USA. Launch date: 2022. Applications: Studying phytoplankton diversity, research and measure atmospheric particles. https://pace.gsfc.nasa.gov/

Sensor/Platform: UrtheDaily. Administration: UrtheCast-Canada. Launch date: 2022. Applications: Global change detection and analysis in the fields of agriculture, ecology, climate change, security, insurance. https://www.urthecast.com/missions/urthedaily/

Sensor/Platform: SAR-XL. Administration: UrtheCast-Canada. Launch date: 2023. Applications: Maritime Surveillance, Oil Spill Tracking, forestry biomass estimation https://www.urthecast.com/missions/sar-xl/

1.7 SUMMARY

Remote sensing systems have been presented in this chapter and classified into active and passive systems, respectively. Specifically, active remote sensing systems represent energy-efficient imaging systems, which emit energy and record its reflection, whereas passive systems record the natural reflected or emitted radiation. Indeed, active remote sensing systems are SARs and weather radars, which operate in the microwave portion of the electromagnetic spectrum and are considered all-weather systems, since there is no signal attenuation when they penetrate clouds. On the other hand, passive satellite systems just record the naturally reflected or transmitted radiation. Specifically, passive remote sensing systems include, as already mentioned, meteorological and environmental satellites. The current trend is the continuous improvement of the spatial resolution with new satellites reaching the level of micro-remote sensing in the order of one meter or less. Similarly, a very recent trend is also to provide additional bands in these satellites resulting in new and valuable information.

REFERENCES

Andronache, C. (ed.), 2018. *Remote Sensing of Clouds and Precipitation*. Springer, 282.
Avery, T.E., and Berlin, G.L., 1992. *Fundamentals of Remote Sensing and Airphoto Interpretation*, 5th ed. Macmillan Publishing Company, New York

Barrett, C.E., and Curtis, F.L., 1992. *Introduction to Environmental Remote Sensing*, 3rd ed. Chapman & Hall, London.

Battan, L.J., 1973. *Radar Observations of the Atmosphere*. University of Chicago Press, Chicago, USA.

Berne, A., and Krajewski, W.F., 2013. Radar for hydrology: Unfulfilled promise or unrecognized potential? *Advances in Water Resources*, 51, 357–366.

Bringi, V.N., and Chandrasekar, V., 2004. *Polarimetric Doppler Weather Radar*, 2nd ed. Cambridge University Press, UK.

Bringi, V.N. Seliga, T.A., and Aydin, K., 1984. Hail detection with a differential reflectivity radar. *Science*, 225, 1145–1147.

Capderou, M., 2014. *Handbook of Satellite Orbits: From Kepler to GPS*. Springer, 922.

CCRS (Canada Centre for Remote Sensing), 1998. "Fundamentals of Remote Sensing," tutorial, http://www.ccrs.nrcan.gc

Chandrasekar, V., Lim, S., Bharadwaj, N., Li, W., McLaughlin, D., Bringi, V.N., and Gorgucci, E., 2004. *Principles of networked weather radar operation at attenuating frequencies. 3rd European Conference on Radar Meteorology and Hydrology*, Visby, Sweden, pp. 67–73.

Cifelli, R., and Chandrasekar, V., 2010. Dual-Polarization Radar Rainfall Estimation, pp. 105–125, in: *Rainfall: State of the Science*. Editors: F.Y. Testik and M. Gebremichael, Geophys. Monograph 191, AGU, Washington, DC, 287.

Dalezios, N.R., 1988. Objective rainfall evaluation in radar hydrology. *Journal of Water Resources Planning and Management (JWRMDS), ASCE*, 114, 5, 531–546.

Dalezios, N.R., 2015. *Agrometeorology: Analysis and Simulation (in Greek)*. KALLIPOS: Libraries of Hellenic Universities (also e-book), 481.

Dalezios, N.R. (ed.), 2017. *Environmental Hazards Methodologies for Risk Assessment and Management*. IWA, London UK, 534

Dalezios, N.R., Dercas, N., and Eslamian, S., 2018. Water Scarcity Management: Part 2: Satellite-based Composite Drought Analysis. *International Journal of Global Environmental Issues*, 17, 2/3, 267–295.

Dicati, R., 2017. *Stamping the Earth from Space*. EBOOK Paradise, 443.

Emery, W., and Camps, A., 2017. *Introduction to Satellite Remote Sensing: Atmosphere, Ocean, Cryosphere and Land Applications*. Elsevier, 856.

ESA, "Sentinel, Earth online – ESA", <https://earth.esa.int/web/guest/missions/esa-future-missions/sentinel-1>, 06/04/2014.

Gillespie, T.W., Chu, J., Frankenberg, E., and Thomas, D., 2007. Assessment and prediction of natural hazards from satellite imagery. *Progress in Physical Geography*, 31, 5, 459–470.

Goodman, S.J., Schmit, T.J., Daniels, J., and Redmon, R.J. (eds.), 2020. *The GOES-R Series: A New Generation of Geostationary Environmental Satellites*. Elsevier, 279.

Guo, H., Fu, W., and Liu, G., 2019. *Scientific Satellite and Moon-Based Earth Observation for Global Change*. Springer, 618

Gupta, S., Karnatak, H., and Raju, P.L.N., 2016. *Geoinformatics in India: Major Milestones and Present Scenario*. ISPRS International Archives of the Photogrammetry, Remote Sensing and Spatial Information Sciences

Hong, Y., and Gourley, J.J., 2015. *Radar Hydrology: Principles, Models and Applications*. Taylor and Francis Group, Boca Raton, FL, USA, 174.

Ilčev, S.D., 2019. *Global Satellite Meteorological Observation (GSMO) Applications. Volume 2*, Springer, 401.

Kohler, P., Frankenberg, C., Troy S., Guanter, L., Joiner, J., and Landgraf, J., 2018. Global retrievals of solar-induced chlorophyll fluorescence with TROPOMI: First results and intersensor comparison to OCO-2. *Geophysical Research Letters*, 10456–10463.

Liang, S., (Editor-in-Chief), 2018. *Comprehensive Remote Sensing*. Elsevier, 3134.

Lillesand, T.M., and Kiefer, R.W., 2000. *Remote Sensing and Image Interpretation*, 4th ed. J. Wiley & Sons, USA, 750

Marshall, J.S., and Palmer, W.M.K., 1948. The distribution of raindrops with size. *Journal of Meteorological Research*, 5, 165–166.

Mather, M.P., 1999. *Computer Processing of Remotely-Sensed Images*, 2nd ed. J. Wiley & Sons, UK, 292.

Meygret, A., Hagolle, O., Hillairet, E., Dedieu, G., Crebassol, P., Ferrier, P., and Xiong, J., 2007. VENµS (vegetation and environment monitoring on a new micro satellite) image quality. *Earth Observing Systems XII*, n1, 66771D.1–66771D.8.

Michaelides, S., Levizzani, V., Anagnostou, E., Bauer, P., Kasparis, T., and Lane, J.E. 2009. Precipitation: Measurement, remote sensing, climatology and modeling. *Atmospheric Research*, 94, 4, 512–533.

NASA, 1998. *"The Remote Sensing Tutorial": An Online Handbook*, Goddard Space Flight Center.

Neeck, S.P., Scolese, C.J., and Bordi, F., 1998. *EOS AM-1. Proc. SPIE 3498, Sensors, Systems, and Next-Generation Satellites II*. Editor: H. Fujisada, doi: 10.1117/12.333619

NOAA, 2006. NOAA Polar-orbiting Operational Environmental Satellites (POES) Radiometer Data. 1-828-271-4800, ncei.orders@noaa.gov

Pandya, M.R., Murali, K.R., and Kirankumar, A.S., 2013. Quantification and comparison of spectral characteristics of sensors on board Resourcesat-1 and Resourcesat-2 satellites. *Remote Sensing Letters*, 4, 306–314.

Raizer, V.(ed.), 2017. *Advances in Passive Microwave Remote Sensing of Oceans*. Taylor & Francis, 251.

Ravi, S.D., 2017. *Remote Sensing of Soils*. Springer, 500.

Richards, M.A., Scheer, J.A., and Holm, W.A., (eds.), 2010. *Principles of Modern Radar*. SciTech Publishing.

Ryzhkov, A.V., and Zrnic, D.S., 2019. *Radar Polarimetry for Weather Observations*. Springer, 504.

Sarris, T.E., 2019. Understanding the ionosphere thermosphere response to solar and magnetospheric drivers: status, challenges and open issues. *Philosophical Transactions of the Royal Society A: Mathematical, Physical and Engineering Sciences*, 377, 2148, 20180101.

Seliga, T.A., and Bringi, V.N., 1976. Potential use of radar differential reflectivity measurements at orthogonal polarizations for measuring precipitation. *Journal of Applied Meteorology*, 15, 69–76.

Wang, L.K., and Yang, C.T., (eds.), 2014. *Modern Water Resources Engineering*. Springer, 866.

Wilson, J.W., and Brandes, E.A., 1979. Radar measurement of rainfall – a summary. *Bulletin of American Mathematical Society*, 60, 9, 1048–1058.

Yang, S.-E., and Wu, B.-F., 2010. *Calculation of monthly precipitation anomaly percentage using web-serviced remote sensing data. Proceedings, IEEE, 2nd International Conference on Advanced Computer Control*, China. 10.1109/ICACC.2010.5486796

Zawadzki, I.I., 1984. *Factors affecting the precision of radar measurements of rain. 22nd Conference on Radar Meteorology*, Zurich, Switzerland, 251–256.

Zohuri, B., 2020. *Radar Energy Warfare and the Challenges of Stealth Technology*. Springer, 310

Part 2

Applications in Environmental Sciences

Part 2

Applications in Environmental Sciences

2 Precipitation

2.1 INTRODUCTION: REMOTE SENSING PRECIPITATION CONCEPTS

Precipitation is considered a fundamental component of the global water cycle and is the most significant variable with regards to atmospheric circulation in meteorological and climate studies (Testik and Gebremichael, 2010; Kidd and Levizzani, 2011; Fabry, 2015; Sun et al., 2017). Accurate estimation of precipitation and reliable time series of precipitation records are considered essential for the study of climate variability and change, as well as for water resources management and hydrometeorological monitoring (Larson and Peck, 1974; Yilmaz et al., 2005; Jiang et al., 2012; Liu et al., 2017). Nevertheless, the spatiotemporal variability is considered one of the major features of the precipitation phenomenon (Michaelides et al., 2009). Measurement of global precipitation uses networks of conventional instrumentation, such as rain (or snow) gauges, as well as remote sensing systems, namely weather radar systems, if available, and/or satellite systems (Michaelides et al., 2009; Kidd and Levizzani, 2011). Specifically, recent developments in technologically sophisticated devices, placed either aboard space platforms (e.g. radar, microwave sensors), or on the Earth's surface (e.g. radars, disdrometers) have resulted in more detailed spatiotemporal precipitation analysis, which in turn have led to a better view of the structure and the associated physical processes of precipitation (Michaelides et al., 2009). Thus, remote sensing of precipitation has significantly contributed to a better understanding of several precipitation features.

Accurate quantitative precipitation measurements are essential for improving hydrometeorological now-casting and forecasting, or monitoring water requirements in agriculture, especially at small temporal (minutes) and spatial (hundreds of meters to few kilometers) scales. This has led to an increased interest in hydrometeorology and hydrology for the weather radar's capability to monitor precipitation at high spatiotemporal scales (Michaelides et al., 2009). On the other hand, satellite sensors provide large-scale observations, suitable for regional to global scale studies of the hydrological cycle (Michaelides, 2008). Nevertheless, the accuracy of satellite precipitation retrievals is still the subject of current research, especially when ground validation based on weather radar and other in situ measurements are considered to assess the retrieval uncertainty (Chandrasekar et al., 2008).

2.2 RADAR-BASED QUANTITATIVE PRECIPITATION ESTIMATION (QPE)

Rainfall is a critical variable in the analysis of atmospheric and terrestrial patterns over a wide range of spatiotemporal scales. This justifies the great significance of accurate quantitative precipitation estimation to a wide range of applications in Environmental and Earth system sciences, as well as engineering disciplines (Michaelides, 2008). Indeed, rain gauges are considered poor instruments for instantaneous rain rate measurement, however, their measurement error diminishes rapidly with integration time (Creutin and Obled, 1982). Indeed, point observations based on in situ sensors, such as rain gauges, may provide only very limited information about the spatiotemporal distribution of rainfall, depending on the required spatiotemporal scale of the analysis, being inadequate to show the extremely large spatiotemporal precipitation variability (Seo et al., 2010). Numerous review papers and textbooks on hydrometeorology and rainfall analysis and modeling techniques have emerged over the years (e.g. Creutin and Obled, 1982; Michaelides, 2008; Rakhecha and Singh, 2009; Gebremichael and Hossain, 2010; Testik and Gebremichael, 2010; Sene, 2016).

The invention and development of weather radar has changed the aforementioned picture dramatically by filling the spatiotemporal observation gap of rain gauges and providing spatially continuous precipitation estimates at small temporal sampling intervals (Marshall and Palmer, 1948; Battan,

1973; Seo et al., 2010). However, weather radar records the backscattered power in terms of volumetric reflectivity of hydrometeors aloft and does not provide direct surface precipitation measurements. As such, radar rainfall estimation is inherently subject to various sources of error. Indeed, there are several meteorological and non-meteorological sources of error, listed earlier, mainly due to technological limitations of the radar systems. To improve the quality of radar-based rainfall estimates, it is therefore necessary to understand, assess, reduce, quantify and account for these errors. It is recognized that weather radars can delineate the areal extent and coverage of precipitation very accurately, as well as the peak time of a flood hydrograph. However, the accuracy of radar-based QPE is reduced compared to point precipitation measurements by rain gauges. Moreover, there is a sampling problem, since radar signals detect the 3D structure of a storm and radar measurements are averaged over areas ranging from several hundred square meters to several square kilometers depending on the distance from the radar site, whereas rain gauges represent an area that is roughly at least ten orders of magnitude smaller. Especially in convective precipitation, very steep horizontal gradients of precipitation are observed, and the precipitation information provided by rain gauges can be misleading.

Radar rainfall measurement techniques can be broadly classified as: (1) physically based and (2) statistical/engineering based. Physically based rainfall algorithms (as defined here) rely on physical models of the rain medium without any feedback from rain gauge observations, whereas statistical/engineering solutions are derived using such feedback either directly or indirectly. It should be emphasized that radar and rain gauges are not competitive, but complementary sensors. The best results are achieved by combining the information from both measurement systems. This improves not only the quality of the data, but continues to provide information on precipitation when one of the two systems malfunctions. The optimum solution in radar-based QPE is the combination of the two general approaches. The main advantage of using radar for precipitation estimation is that radars can obtain measurements over large areas (about 40,000 km^2 – radar radius of 150-km) with high temporal and spatial resolution over a 150-km radius. These measurements are sent to a central location at the speed of light by "natural" networks. In addition, radars can provide rapid updates of the three-dimensional structure of precipitation (5–6 minutes for the radar volume scan). As a result, radar measurements of precipitation continue to enjoy widespread operational usage.

In this section, certain sources of error are considered in single-polarization radars. Then, rain gauge-adjusted radar-based QPE methods are presented, namely an empirical adjustment procedure and the bivariate statistical analysis, as well as precipitation retrieval methods for coherent Doppler radar and polarimetric radar, respectively, over regions of interest or watersheds. This section builds on previous review papers (Wilson and Brandes, 1979; Delrieu et al., 2009; Berne and Krajewski, 2013), and textbooks (Battan, 1973; Bringi and Chandrasekar, 2004; Chandrasekar et al., 2004; Doviak and Zrnic, 2006; Hong and Gourley, 2015; Richards, Scheer and Holm, 2010; Beck and Chau, 2012; Fukao and Hamazu, 2014; Fabry, 2015).

2.2.1 Conventional Radar-based QPE

The quantitative use of weather radar variables is the most demanding and thus requires careful processing and error considerations to convert the radar signal to a useful measurement of precipitation rates. The greatest benefit of weather radar usage is its potential to estimate rainfall rates at high spatiotemporal resolution (i.e. 1 km/5 min), in real time, within a radar radius of approximately 250 km. The conventional single-polarized Doppler radar uses the measurement of radar reflectivity, radial velocity and storm structure to infer some aspects of hydrometeor types and amounts. This section explains the basic procedures needed to get from reflectivity measured by conventional radar to precipitation rates with uncertainty estimates.

2.2.1.1 Radar Calibration

The calibration of radar has a major influence on the accuracy of rainfall rates. A miscalibration of only 1 dB results in bias in rainfall rates of 15%. Transmit and receive components can be jointly calibrated by using a reflective target with known scattering properties within the field of view of the radar. A disdrometer primarily measures the diameter of individual droplets for the precipitation drop size distribution (DSD) and the radar reflectivity Z. These in situ measurements can be compared to the radar-measured equivalent Z values in rain to identify biases due to miscalibration of the radar (Michaelides et al., 2009). However, the sample volumes between a typical radar pixel aloft and a disdrometer orifice at ground level differ by about eight orders of magnitude (Droegemeier et al., 2000). Furthermore, the radar samples precipitation at some height above the disdrometer, which depends on the range of the instrument from the radar, elevation differences and beam propagation paths. This leads to a space-time lag between the measured raindrops and its entering the disdrometer. The disdrometer approach to radar calibration is useful for individual radars, especially in a field experiment setting, but they are often not feasible for calibrating large radar networks. Moreover, even if the radars are calibrated within 0.5 dB of each other, lines of data discontinuities or **radar artifacts** often arise, where a precipitation estimation algorithm switches using data from one radar to the neighboring one. These artifacts are most noticeable for long-term accumulations of rainfall, such as daily accumulations. This problem can be dealt with in the precipitation algorithm by spatially interpolating or smoothing data between neighboring radars. Another approach is to compare the Z values in rain from neighboring radars at these equidistant lines to identify relative calibration differences. This approach of comparing remote sensing observations to one another, as opposed to in situ data, has been quite useful for calibrating radar networks (Gourley et al., 2003).

2.2.1.2 Quality Control

Every single bin of radar data must be scrutinized to remove deleterious effects from non-meteorological scatterers on the ground, biota in the atmosphere, planes, chaff, etc. The basic approaches of quality control are presented.

1. **Signal processing.** The first level of screening takes place at the spectral level, which is prior to the stage at which reflectivity is estimated. Radar measures several independent samples within a given range bin. If the samples are associated with no Doppler shift, i.e. a radial velocity of 0 m sec^{-1}, then this indicates a stationary target, probably from the ground or a building. Conversely, there are numerous non-meteorological targets that are non-stationary. All biota in the atmosphere such as birds, bats and insects are generally in flight and are affected by the prevailing wind. Even certain ground targets like trees and windmills have nonzero velocities. Since each and every bin around a radar must be scrutinized, automated algorithms must be developed, tuned and implemented to screen out non-meteorological echoes. The first indication of ground clutter is the echoing with zero Doppler velocity in regions away from the zero isodop. These echoes are first evident at the spectral level of processing and are subsequently removed.

2. **Fuzzy logic.** There is a need to discriminate non-meteorological from meteorological echoes by employing algorithms. These algorithms range in complexity from simple thresholds placed on variables, thresholds applied to multiple variables as in decision tree logic, fuzzy logic, neural networks and combinations therein. Fuzzy logic algorithms are well suited for radar observations because they incorporate information from multiple variables with different weights and are less susceptible to misclassifications due to noise in the measured variables. It is outside the scope of this section to cover a fuzzy logic algorithm.

2.2.1.3 Precipitation Rate Estimation

Calibrated reflectivity values describing the size, shape, state and concentration of the hydrometeors within the radar sampling volume are used next to compute precipitation rate. If there are no beam blockages in the vicinity of the radar, then reflectivity measured at the lowest elevation angle,

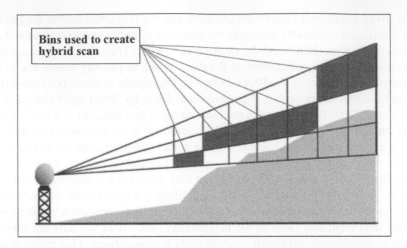

FIGURE 2.1 Construction of the hybrid scan using multiple elevation angles over complex terrain. Criterion 1: Beam center > 50 m above underlying terrain. Criterion 2: Beam blockage < 60% (from Hong and Gourley, 2015).

usually 0.5 deg, should be used for QPE. In regions with complex terrain, bins are selected at multiple elevation angles to construct the **hybrid scan** for QPE, where the beam blockage between the bin and the radar at the same elevation angle must be less than 60% (Figure 2.1) and 1 dB is added to each measured Z value per 10% blockage (2 dB added for 20% blockage and so on). Note that the blockages are computed using a digital elevation model (DEM). Reflectivity from the hybrid scan is used to compute two-dimensional fields of precipitation rate in spherical coordinates (range, azimuth). The general form of reflectivity-to-rainfall relationships, or $Z–R$ equations (Equation 1.9 of Chapter 1).

Vertical profile of reflectivity (VPR). Assuming all non-meteorological echoes have been successfully removed, Z is well calibrated within 1.0 dB and the parameters of the $Z–R$ relation yield accurate precipitation rates, there is the unavoidable circumstance that the radar bin volume and beam heights increase with range from the radar. This results in **range-dependent errors.** A model can be developed to describe the **vertical profile of reflectivity (VPR)** of a storm to quantify the variability of reflectivity with height. This VPR model can be used to correct for sampling height- and range-dependent errors, namely biases. This permits Z measured by radar at a given height and range to be adjusted so that it better represents what is occurring at the surface. Figure 2.2 shows general VPR models for (1) convection and (2) stratiform precipitation, the most elementary level of storm segregation. The convective VPR in Figure 2.2 illustrates that the profile is more upright, which means that the reflectivity does not vary greatly with height. This implies that a reflectivity sample taken aloft (e.g. 5 km) is approximately representative of rainfall rates experienced at the surface. VPR for stratiform rainfall, on the other hand, shows how reflectivity has little variability with height in the rain region. Then, it increases significantly in the bright band. Rainfall estimates have been shown to be overestimated up to a factor of ten when using uncorrected measurements taken within the bright band (Wilson and Brandes, 1979; Zawadzki, 1984). Above this layer, reflectivity decreases again in the pristine ice region. Once the vertical structure is modeled, the next step is to account for beam spreading. The impact of the bin volumes increasing with range causes the radar to observe a smoothed version of the actual VPR. Two of these apparent VPRs are shown as gray curves in Figure 2.2. The next step is to correct the radar-observed reflectivity measured at some height so that it represents the reference reflectivity value at or very near the surface. The method uses the apparent VPR, which depends on the bin volume at that range and then estimates the beam center height from equation. These factors and the VPR model determine the amount of reduction needed for measurements taken near the melting layer and the increases needed for

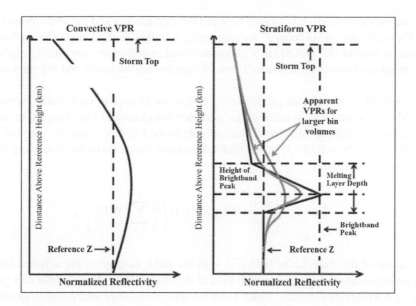

FIGURE 2.2 Model vertical profiles of reflectivity for convective and stratiform echoes. These models can be designated by the parameters listed in the text. The gray curves correspond to radar sampled VPRs at longer ranges from radar, illustrating the smoothing effect caused by beam broadening (from Hong and Gourley, 2015).

measurements taken above it. There is generally a vertical limit to which corrections can be applied. Stratiform systems tend to be much shallower than convective storms, thus there is a finite range to which corrections can be applied.

Rain gauge adjustment. Rain gauges provide useful measurements of rainfall to a point. For this reason, they are essential for evaluating and improving radar-based QPE algorithms, despite their significantly large sampling volume differences from radar bins. The two general approaches for correcting QPEs using rain gauges are: (1) mean field bias correction and (2) spatially variable bias correction. In this section, two rain gauge adjustment methods are presented, namely Brandes field-adjustment procedure, which is an empirical adjustment method and the bivariate statistical analysis.

Brandes field-adjustment procedure. A field-adjustment model has been developed by using existing dense rain gauge networks and radar data sets (Brandes, 1975; Dalezios, 1988). At first, based on the rain gauge to radar rainfall ratios for each grid element, a weighted adjustment factor is computed by the equation:

$$F_1 = \left[\sum_{i=1}^{N} \left[(WT)_i \times (RT)_i \right] \right] / \left[\sum_{i=1}^{N} (WT)_i \right] \quad (2.1)$$

where F_1 is the adjustment factor for a grid element, N is the number of rain gauges, $(WT)_i$ is the weight of rain gauge i for the same grid element, and $(RT)_i$ is the ratio $(G/R)_i$ of the rain gauge rainfall amount G over the spatially averaged radar estimate R for rain gauge i. Indeed, for each rain gauge site, the raw radar data are averaged within 6 km x 6 km centered around the site. The weight $(WT)_i$ is expressed by the equation:

$$(WT)_i = \exp\left[-d_i^2 (EP)_1 \right] \quad (2.2)$$

where d_i is the previously defined distance between the rain gauge i and the grid element in kilometers and $(EP)_1$ is the area of "influence" for rain gauge i. In this method, a maximum distance of 45 km and a square area of $(EP)_1 = 20$ km × 20 km are used to eliminate very small weights from the process. Rain gauge records of less than 1 mm of rain are not included and the ratio ranges from 0.1 to 10.

In the second stage, at each rain gauge site the difference D_i is computed, based on the form $D_i = (RT)_i - F_1$, where the factor F_1 is taken at the grid element closest to rain gauge i. In addition, the previously defined weighting function (Equation 2.2) is modified by reducing the size of the area $(EP)_1$ by half, i.e. $(EP)_2 = [(EP)_1/2]$. The final adjustment factor takes the following form and is assigned to each grid element:

$$F_2 = F_1 + \left[\sum_{i=1}^{N} \left[(WT)_i \times (D_i) \right] \right] / \left[\sum_{i=1}^{N} (WT)_i \right] \tag{2.3}$$

Multiplication of this final factor field (F_2) with the radar field gives the adjusted radar rainfall field. The corresponding rain gauge rainfall field is developed from the available rain gauge dense network using a weighting function, which is the reciprocal-distance interpolation model, given by the following equation:

$$w_i(x,y) = (1/d_i^n) / \sum (1/d_i^n) \tag{2.4}$$

where $w_i(x,y)$ is the weighting factor of rain gauge i of the grid element with midpoint coordinates (x,y), d_i is the distance in kilometers to rain gauge i from the midpoint of the element and the exponent n is an empirical coefficient usually taking the value of 2. At grid points coinciding with a rain gauge site, the rain gauge-derived field receives a weight of 100%, which linearly decreases to 0%, and the radar-field weight increases to 100% as the distance from the rain gauge increases to 10km.

Bivariate statistical analysis. This is an objective method, which combines radar and rain gauge data fields to produce hourly surface rainfall estimates (Crawford, 1979; Dalezios, 1988; Dalezios, 1990; Dalezios and Kouwen, 1990). For the bivariate predictor set, the bivariate statistical objective analysis may be accomplished by:

$$Y = X_r b_r + X_g b_g \tag{2.5}$$

where Y is an estimate of the unknown "true" surface rainfall Y, X_r is the radar rainfall estimate at a grid element, X_g is the estimated rainfall amount at the same grid element from the rain gauge field, and b_r, b_g are the estimates of the regression coefficients for the radar and rain gauge data sets, respectively. The reciprocal-distance interpolation model (Equation 2.17) is used to generate the rain gauge field over the area.

The regression coefficients b_r and b_g must be estimated at each time step. In regression analysis, it is often the case that the variances of the observations are not all equal, since there might be interdependence of the correlation of spatial observation data sets. Moreover, residual variance can be non-stationary and/or the residuals can be autocorrelated in time and space, or even the data may be statistically dependent. As a result, the estimators are not of minimum variance, leading to the use of weighted least-squares as follows:

$$b_w = \left(X^T W X \right)^{-1} \left(X^T W Y \right) \tag{2.6}$$

where b_w is a 2x1 vector, W is a 2x2 weighted matrix, $C = X^T W X$ is a 2x2 covariance matrix and $c = X^T W Y$ is a 2x1 cross-covariance vector of the unknown "true" surface rainfall. A fundamental restriction is that the weighted matrix W should be symmetric, positive definite to assure unique non-singularity, which is accomplished by the following expression:

$$W = R^{-1}, \text{where } R = E\left[e^T e \right] \tag{2.7}$$

where R is the error covariance matrix and E(.) is the expectation operator. The correlation field should depend on the structure of the covariance matrix, which can be achieved by modeling the two-dimensional random field e(x.y). In this study, the error covariance matrix is modeled using a two-dimensional, Gaussian, isotropic, unit-variance and homogeneous spatial function of the form:

$$r(d) = \exp\left(-a\,d^2\right) \tag{2.8}$$

where $d^2 = u^2 + v^2 = (x_1 - x_2)^2 + (y_1 - y_2)^2$ is the square distance of any pair of points in space (x_1, y_1) and (x_2, y_2), and a is the correlation parameter, i.e. the decorrelation distance at which the autocorrelation coefficient declines to the value of 1/e. This correlation parameter is a time-dependent function of meteorological parameters, which varies from storm to storm, as well as within storms and in previous studies has taken an empirical value of 0.15 (Greene et al., 1980). Finally, there is a need to model the cross-covariance vector $c = X^T W Y$, since the "true" surface rainfall Y is unknown. In this study, a triangular correlation function is used, based on the following form:

$$\rho(|x - x_i|) = \begin{cases} 1 - \gamma|x - x_i|, & \text{for } |x - x_i| \le 1/\gamma \\ 0 \end{cases} \tag{2.9}$$

where γ takes the value of 0.1.

2.2.1.4 Space-time Aggregation

Radars can measure rainfall rates on their native spherical grid with bin resolution depending on the operating characteristics of the radar. This resolution is nominally 1 deg in azimuth by 1 km in range, and estimates are computed on a 5 min basis. These rainfall rates must be totaled typically to hourly accumulations in order to be adjusted by rain gauges. Additional aggregation takes place to compute 3-, 6-, 12-, and 24-hour, 72-hour, 10-day, monthly, etc. accumulations. Rainfall estimates from neighboring radars comprising a network can be combined or mosaicked to create rainfall maps covering a larger spatial domain. Because the native radar coordinates are centered around each radar, it is more practical to resample the rainfall estimates onto a common Cartesian grid. The simplest mosaicking approach merely chooses the rainfall estimate from the radar closest to each grid point. This method creates linear discontinuities in the rainfall fields at points equidistant (at the same range) from neighboring radars. These discontinuities can result from calibration differences between the radars, different Z-R relationships, or different treatment of the various radar sources of error. Tabary (2007) proposed a mosaicking scheme that weights precipitation estimates from adjacent radars based on their quality as judged by their influence by ground clutter, degree of beam blockage and altitude of the beam. Once the weights are empirically estimated for a given grid point, the mosaicking scheme selects the rainfall estimate from the radar with the highest weight.

2.2.1.5 Uncertainty Estimation

Using deterministic radar QPEs without considering their underlying distribution assumes that they are error-free or that their uncertainties are negligible in the application for which they are being used. In general, it is good practice to estimate and utilize uncertainties associated with radar QPE. Several radar rainfall error models have been proposed. An error model is the first step toward computing the rainfall uncertainty associated with a deterministic QPE, probabilistic QPE (PQPE), and generating ensembles of QPE. The first error modeling approach attempts to describe all the individual errors specific to a given radar's hardware, operating characteristics controlling the resolution

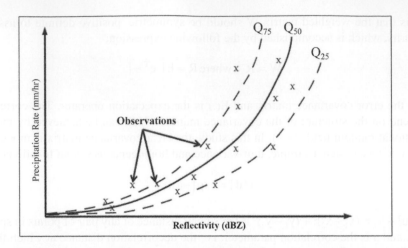

FIGURE 2.3 Illustration of method for computing probabilistic QPE at 1 km/5 min scale based on gauge-corrected radar observations and radar reflectivity factor (from Hong and Gourley, 2015).

and sensitivity of the estimated rainfall fields and software used to estimate rainfall. A more general approach in error modeling involves correcting the radar QPEs to the best extent possible using the methods outlined in this chapter and then modeling the remaining error, or the residual, using an independent reference from rain gauges.

A recent method for PQPE was proposed (Kirstetter et al., 2013) to quantify the radar rainfall uncertainty at the measurement scale of 1 km/5 min. Data points (indicated by x in Figure 2.3) of the filtered reference rainfall rate, R (mm hr−1) are plotted as a function of Z (dBZ) for each precipitation type. The data points suggest a power-law relationship between Z and R, which is typically assumed. The next step involves fitting an error model to the observed values of Z and R. The generalized additive model for location, scale and shape (GAMLSS; Rigby and Stasinopoulos, 2005) technique is used to create the smooth curves shown in Figure 2.3. These can be considered as empirical fits to the data points that now describe the distribution of R for a given radar observation of Z and precipitation type. Figure 2.3 shows the median of the distribution, or the 50% quantile denoted as Q50, as well as the 25% and 75% quantiles (Q25, Q75). The distance between the latter two quantiles can be used to estimate the uncertainty for a rainfall rate associated with a given measurement of Z and precipitation type. The error model describes the entire data distribution and can be used to accommodate the intended application of radar rainfall estimates.

2.2.2 Improved Rain Rate Retrieval Based on Polarimetric Radar

Several rain rate estimators based on polarimetric radar variables contribute to the improvement of radar rain-rate estimation for different applications (Cifelli and Chandrasekar, 2010; Gourley et al., 2010; Hong and Gourley, 2015). Z_{DR} and K_{dp} are commonly used to complement Z_h in multiple power laws used in rain rate estimation. Moreover, polarimetric estimators, combined with more accurate hydrometeor identification, can improve rain rate estimation and help to distinguish hail and rain in convective storms (Giangrande and Ryzhkov, 2005; Ryzhkov and Zrnic, 2019). Based on the above, it can be stated that acceptable echo classification, artifact-free rainfall products, better identification of microphysical processes and the potential to estimate space-time variability of the DSD are among the promising achievements of dual polarization (Bringi and Chandrasekar, 2004). Nevertheless, there remain several sources of uncertainty that could significantly affect the quality of rain rate estimates. In summary, dual-polarization radar is a critical tool for weather research applications, including rainfall estimation, and is at the verge of being a key instrument in operational meteorology and hydrology. Figure 2.4 presents a flow diagram indicating the stages of data

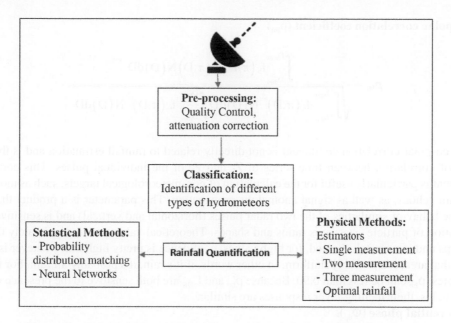

FIGURE 2.4 Flow diagram indicating the stages of data processing where dual-polarization radar can make important contributions and classification of techniques utilized in radar rainfall estimation.

processing, where dual-polarization radar can make important contributions, including a classification of techniques utilized in radar rainfall estimation into statistical and physical.

2.2.2.1 Polarimetric Radar Variables

Several polarimetric radar variables are used in QPE.

Radar reflectivity factors at horizontal and vertical polarizations ($Z_{h,\upsilon}$ or $Z_{H,V}$) are given by the following equations, and the **DSD** can be illustrated in $R(Z_h)$

$$Z_{h,\upsilon} = \frac{4\lambda^4}{\pi^4 K^2} \int_{D_{min}}^{D_{max}} \left| f_{h,\upsilon}(\pi,D) \right|^2 N(D) dD \left(mm^6\, m^{-3} \right) \tag{2.10}$$

$$Z_{H,V} = 10\log_{10}\left(Z_{h,\upsilon}\right)\left(dBZ\right) \tag{2.11}$$

where K is the dielectric factor of water, $\sigma_{h,\upsilon}$ is the backscattering cross-sections at horizontal (h) and vertical (υ) polarization, λ is the wavelength, N(D) is the DSD, and D is the particle equivalent diameter. If the Rayleigh approximation is valid, the reflectivity factor $Z_{h,\upsilon}$ can be expressed as for non-coherent radars. The assumption of gamma distribution for DSD can be represented with three parameters as:

$$N(D) = N_0\, D^\mu \exp(-\Lambda D) \tag{2.12}$$

where N_0 (mm $^{[-1-\mu]}$ m $^{-3}$) is a scaling parameter, μ is a distribution shape parameter, Λ (mm $^{-1}$) is a slope term and D (mm) is the spherical equivalent volume diameter.

Differential reflectivity (Z_{DR}). The differential reflectivity (Z_{dr} or Z_{DR}) is defined as (Seliga and Bringi, 1976):

$$Z_{dr} = Z_h\, /\, Z_\upsilon \tag{2.13}$$

$$Z_{DR} = 10\log_{10}\left(Z_h\,/\,Z_\upsilon\right) = Z_H - Z_V\left(dB\right) \tag{2.14}$$

Co-polar correlation coefficient ($\rho_{h\upsilon}$)

$$\rho_{h\upsilon} = \frac{\int_{D_{min}}^{D_{max}} f_h(\pi,D) f_\upsilon(\pi,D) N(D) dD}{\sqrt{\int_{D_{min}}^{D_{max}} |f_h(\pi,D)|^2 N(D) dD \int_{D_{min}}^{D_{max}} |f_\upsilon(\pi,D)|^2 N(D) dD}} \qquad (2.15)$$

The co-polar correlation coefficient is not directly related to rainfall estimation and is the coefficient of correlation between time series of Z_h and Z_υ of the individual pulses. This correlation coefficient is particularly useful for the calculation of non-meteorological targets, such as mountains or terrain echoes, as well as signal anomalous propagation. This parameter is a product that measures the behavior similarties of the two radar pulses (horizontal and vertical) and is sensitive to the distribution of particle sizes, axis ratios and shapes. Theoretical values are approximately 0.99 for raindrops and ice crystals. Indeed, for high values, confidence is pretty high that whatever is occurring in that area, it's probably uniform. In other words, if it's rain, it's probably all rain. For melting aggregates, $\rho_{h\upsilon}$ can be less than 0.90. Because $\rho_{h\upsilon}$ and L_{DR} are both sensitive to the presence of large wet particles, their melting-layer responses are similar.

Differential phase (Φ_{dp}).

$$\Phi_{d,p}(r_g) = 2\int_0^{r_g} K_{dp}(r) dr \quad (\text{deg}) \qquad (2.16)$$

The differential phase shift (Φ_{dp}) [deg] is the difference in phase between the horizontally (h) and vertically (v) polarized signals. In rain, Φ_{dp} increases with distance and drop oblateness depending on both the shape and concentration of the hydrometeors. For raindrops with an oblate spheroid shape, h polarized waves propagate more slowly than v polarized waves. Φ_{dp} is an important component for rainfall estimation.

Specific differential phase (K_{dp}). The spatial derivative of Φ_{dp} is called the specific differential phase K_{dp}, [deg km^{-1}], given by:

$$K_{dp} = \frac{180\lambda}{\pi} \int_{D_{min}}^{D_{max}} R_e f_h(0,D) - f_\upsilon(0,D) N(D) dD \quad (\text{deg km}^{-1}) \qquad (2.17)$$

K_{dp} is closely related to the DSD through the rainwater content and is relatively insensitive to ice in the radar volume. For Rayleigh conditions:

$$K_{dp} = (180/\lambda)10^{-3} C W(1-r_m) \quad (\text{deg km}^{-1}) \qquad (2.18)$$

where C is a constant with magnitude ~3.75, λ is the radar wavelength in meters, W is the water content in g m^{-3}, and r_m is the mass weighted mean axis ratio.

Specific attenuation at horizontal or vertical polarization (A_H or A_V).

$$A_{H,V} = 8.686\lambda \int_{D_{min}}^{D_{max}} \text{Im} f_{h,\upsilon}(0,D) N(D) dD \quad (\text{dB km}^{-1}) \qquad (2.19)$$

Specific differential attenuation (A_{DP}).

$$A_{DP} = A_H - A_V \quad (\text{dB km}^{-1}) \qquad (2.20)$$

Linear depolarization ratio (L_{DR}). Linear depolarization ratio (L_{DR}) is not directly related to rainfall estimation. L_{DR} (dB) is defined as the logarithm of the ratio of cross-polar and co-polar signal returns, that is:

$$L_{DR} = 10\log_{10}\left(Z_{ex} / Z_h\right)(dB) \tag{2.21}$$

where Z_{ex} is the signal received at the vertical polarization (cross-polar return) for a transmitted horizontally polarized wave.

In Equations (2.10) to (2.21), λ is the radar wavelength; $K = (\varepsilon_r - 1)/(\varepsilon_r + 2)$, where ε_r is the complex dielectric constant of water; D denotes the effective diameter of particle (i.e. hydrometeor); Dmax (or Dmin) indicates the maximum (or minimum) D within a radar resolution volume; and N(D) is the particle size distribution (PSD) of all these particles; $f_{h,v}$ represents the complex scattering amplitude at horizontal or vertical polarization and the parameters 0 and π for $f_{h,v}$ denote the forward-scattering and backward-scattering components, respectively; the notation |.| signifies the complex norm and Re (or Im) indicates the real (or imaginary) part of a complex number; and r denotes the range from radar and r_g is the range for a given range gate.

Z_h represents the energy backscattered by precipitating hydrometeors and depends on their concentration, size and phase, which are closely linked to precipitation rate and water content. Z_{dr} is directly related to the median size of observed hydrometeors, a parameter used to describe the DSD, and thus provides valuable supplementary information for QPE. K_{dp} is dependent on the raindrop number concentration, but it is less sensitive to the size distribution than Z_h. It is independent of radar calibration and partial beam blockage and relatively immune to hail contamination in rain estimation. Positive K_{dp} values result from a phase lag in the horizontally polarized wave compared with the vertical one. Oblate raindrops, (i.e. larger horizontal dimensions than vertical) basically cause a slight phase delay, which is more pronounced at horizontal polarization.

These three polarimetric measurements, namely Z_h, Z_{dr} and K_{dp}, can be directly applied to estimate rainfall. The correlation coefficient (ρ_{hv}) indicates how well the backscatter amplitudes at vertical and horizontal polarization are correlated. It is a good indicator of hydrometeor phase (homogenous vs. mixed phase) and data quality. This variable is used for classifying the hydrometeor species of the radar echo, which benefits QPE. Precipitation can cause strong attenuation (power loss) in radar measurements, depending on the frequency of the radar wave. Specific attenuation (A_V, A_H) and specific differential attenuation (A_{DP}) are two important variables to address how much power has been lost in Z_h, Z_v, or Z_{dr}, though they are not directly measured. If the attenuation effect is not negligible such as with C-, X-, and Ka/Ku-band radars, attenuated Z_h and Z_{dr} need to be corrected to avoid underestimation in QPE. Values of A_H and A_{DP} also have a strong correlation with precipitation rate.

2.2.2.2 Polarimetric Radar Data Quality Control

Careful data quality control is required with radar measurements, especially for QPE. For example, a 3dB error in reflectivity may cause 100% overestimation of precipitation.

System noise mitigation. System noise is one of most common error sources. For example, the ρ_{hv} of rainfall or dry snow or ice should be close to 1. System noise can reduce the value of ρ_{hv} to below 0.9, which might lead to an incorrect interpretation that raindrops are either mixed-phase precipitation or even ground clutter. A common approach to mitigate the noise is the smoothing of moment data. For example, the estimation of K_{dp} is usually done by a gradient calculation of averaging Φ_{DP} over multiple range gates (Ryzhkov et al., 2005). Other measurements, such as Z_h and Z_{dr}, are usually smoothed (e.g. over 1 km range), as well before they are used for the QPE. The noise effect can also be mitigated by advanced radar signal processing, such as the conventional autocorrelation/cross-correlation function (ACF/CCF) method, which gives the moment estimation mainly based on lag-0 of ACF/CCF. To sufficiently apply the information of ACF and CCF, the multi-lag correlation estimator has been proposed for radar moment data estimation (Lei et al., 2012).

Ground clutter consideration. Ground clutter generally comes from stationary targets and has a small radial velocity in its radar measurements. To remove it, conventional radar systems usually apply various "notch" filters such as finite/infinite impulse response (FIR/IIR) filters to detect echoes with 0 Doppler velocity (Torres and Zrnić, 1999). Moreover, advanced clutter filtering techniques are mostly based on spectrum analysis. The most popular technique is the Gaussian model adaptive processing (GMAP) algorithm (Siggia and Passarelli, 2004). These advanced algorithms are superior to clutter filters because they can preserve the weather components while removing ground clutter, especially when clutter and weather components have similar radial velocities. Clutter identification is highly desirable for efficient filtering. A typical algorithm is the clutter mitigation decision (CMD) algorithm developed by the National Center for Atmospheric Research (NCAR), which is mainly based on the phase of clutter signal (Hubbert et al., 2009). The spatial continuity of weather signals in the range-spectrum space has also been applied to identify clutter (Moisseev and Chandrasekar, 2009). The detection and removal of ground clutter are of great significance to the application of polarimetric radar data.

Correction of attenuation. Precipitation attenuation is an unavoidable problem in radar QPE. It is a major problem for shorter than S-band wavelength radars (e.g. C, X, Ku, and Ka band), and thus requires correction. Previous algorithms to correct attenuation losses with single-polarization radars are mainly based on algorithms, which rely on the empirical power-law relation between attenuation and radar reflectivity (Delrieu et al., 2000). The general form of the equations for the first-order corrections to Z_H and Z_{DR} due to precipitation attenuation is related to the path-integrated differential phase measurements (Φ_{dp}) as:

$$\Delta Z_H = a\Phi_{dp} \quad (dB) \tag{2.22}$$

$$\Delta Z_{DR} = b\Phi_{dp} \quad (dB) \tag{2.23}$$

where the coefficients a and b (in dB deg^{-1}) depend mainly on the radar frequency. The constants a and b increase with shorter radar wavelength. At X band, values of 0.22 dB deg-1 and 0.032 dB deg^{-1} for a and b, respectively, were found (Matrosov et al., 2002).

Polarimetric radar calibration. Polarimetric radar QPE is more sensitive to data quality than the single-polarization radar QPE. A change of only several tenths of a dB in Z_{DR} may cause significant changes in QPE. Therefore, calibrating the systematic bias is extremely important for Z_{dr}. The basic method is the engineering calibration, which will accurately measure and compare the gain/damping within the receiving paths for two polarimetric channels. The antenna can be pointed vertically in light precipitation. Then, the Z_{DR} should approach zero at vertical incidence. Another method is to track the sun because Z_{DR} should be zero as well for sun signals. The National Severe Storm Laboratory (NSSL) has applied this method for the calibration of the polarimetric NEXRAD network (Zrnić et al., 2006).

Self-consistency check. Z_H in rain could be approximated from Z_{DR} and K_{dp} measurements using the following relation:

$$Z_H = a + b\log(K_{dp}) + c Z_{DR} \tag{2.24}$$

where the coefficients a, b and c depend on radar wavelength and prevalent raindrop shape. These coefficients are also supposed to be relatively insensitive to the raindrop size distribution. Similarly, K_{dp} could be expressed as a function of Z_H and Z_{DR}. The self-consistency check uses the dependence between Z_H, Z_{DR} and K_{dp} to assess the calibration of the radar. For example, K_{dp} can be quite noisy in light rain, and the calibration may be suitable for moderate or heavy rain only. A drop shape model was used (Brandes et al., 2002), assuming a drop temperature of 0°C, using a normalized gamma

TABLE 2.1

Coefficients for a Third-Degree Polynomial Fit to the Polarimetric Consistency Relations at Three Weather Radar Frequencies (from Hong and Gourley, 2015)

Frequency	a0	a1	a2	a3
X-band	11.74	−4.020	−0.140	0.130
C-band	6.746	−2.970	0.711	−0.079
S-band	3.696	−1.963	0.504	−0.051

model for the DSD, and computed third-order polynomial regressions for the consistency relations at X-, C- and S-band frequencies (Gourley et al., 2009). Their regression takes the following form:

$$\left(K_{dp} / Z_h\right) = 10^{-5}\left(a_0 + a_1 Z_{DR} + a_2 Z_{DR}^2 + a_3 Z_{DR}^3\right) \tag{2.25}$$

where K_{dp} is one way in units of deg km^{-1}, Z_h is in linear units (mm6 m^{-3}) and Z_{DR} is in dB. Table 2.1 provides the values for the coefficients as a function of the radar frequency.

2.2.2.3 Classification of Hydrometeors

Appropriate scattering models should be applied to the different hydrometeor species to yield accurate QPE.

Characteristics of polarimetric radar echoes. For those scatterers that are approximately spherical in shape (e.g. small raindrops) or behave like isotropic scatterers (e.g. dry, tumbling hail), Z_{DR} and *Kdp* values are close to zero. Z_{DR} and *Kdp* values increase as the particle sizes increase. Taking S-band polarimetric radar for example, Z_{DR} (*Kdp*) values normally increase from 0 to 5 dB (3 deg km−1) for drizzle, tropical rain, weak convective rain, stratiform rain and intensive convective rain. The increase of Z_{DR} and *Kdp* values follows increases in the median size and concentration of the raindrops. Table 2.5 gives some typical ranges of polarimetric variables (S-band) for different radar echoes.

Z_{DR} is independent of the hydrometeor concentration and strongly depends on the shape of the reflectors, i.e. it is close to 0 for nearly spherical shapes, positive for oblate shapes ($Z_h > Z_v$) with horizontal return and negative for prolate shapes ($Z_h < Z_v$) with vertical return. Z_{DR} ranges from -8 to 8, though values rarely get that extreme on either side. In rain, as Z_{DR} increases, drop size increases. For raindrops bigger than ~0.5 mm, Z_{DR} ranges from ~0.3 to 3dB. Z_{DR} over 3 is indicative of big drops. Although hail is usually not spherical in shape, Z_{DR} values drop to near zero. However, if hail is small and/or water coated, Z_{DR} gets usually very high, 5 or 6, but this hail is almost never severe in size by the time it reaches the surface. Conversely, negative Z_{DR} can be indicative of very large hail. For snow/ice, Z_{DR} can vary significantly. In general, dry snow is closer to 0 Z_{DR}, with wet snow increasing in amount. Non-weather returns can have a Z_{DR} of literally almost anything. Hail is typically characterized as having high reflectivity (> 55dBZ), low Z_{DR} (< 1dB) and a local reduction in co-polar correlation coefficient CC (0.95 < CC < 0.98). However, when hail diameters become significantly large (> 5cm), Mie scattering begins to affect the dual-pol base products. Specifically, CC becomes significantly low (< 0.9, to as low as 0.7). Table 2.2 presents typical ranges of polarimetric variables (S-Band) for different radar echoes.

Methods of classification. Popular algorithms include the radar echo classifier (REC) developed by NCAR (Vivekanandan et al., 1999), the polarimetric hydrometeor classification algorithm (HCA) developed by NSSL (Park et al., 2009) and the hydrometeor classification system (HCS) developed by Colorado State University (Lim et al., 2005). These algorithms generally classify more than ten distinct species of radar echoes, such as rain, snow, hail, clutter, and so on. The HCA uses six radar variables for classification: (1) Z_H, (2) Z_{DR}, (3) ρ_{hv}, (4) K_{dp}, (5) texture of Z_H, and (6) texture of Φ_{dp}.

TABLE 2.2

Typical Ranges of Polarimetric Variables (S-Band) for Different Radar Echoes (from Hong and Gourley, 2015)

Category	Z_H (dBZ)	Z_{DR} (dB)	K_{dp} (degree/km)	ρ_{hv}
Rain (light, moderate, heavy)	5–55	0–5	0–3	0.98–1.0
Graupel	25–50	0–0.5	0–0.2	0.97–0.995
Dry hail	45–75	−1–1	−0.5–0.5	0.85–0.97
Melting hail	45–75	1–7	−0.5–1	0.75–0.95
Ice crystal	<30	<4	−0.5–0.5	0.98–1.0
Dry snow aggregate	<35	0–0.3	0–0.05	0.97–1.0
Wet snow aggregate	<55	0.5–2.5	0–0.5	0.9–0.97
Ground clutter	20–70	−4–2	very noisy	0.5–0.95
Biological scatterer	5–20	0–12	low & very noisy	0.5–0.8

The HCA discriminates between ten classes of radar echo: (1) ground clutter including anomalous propagation (GC/AP); (2) biological scatterers (BS); (3) dry aggregated snow (DS); (4) wet snow (WS); (5) crystals of various orientations (CR); (6) graupel (GR); (7) big drops (BD); (8) light and moderate rain (RA); (9) heavy rain (HR); and (10) a mixture of rain and hail (RH). HCA considers several error factors, including radar miscalibration, attenuation, nonuniform beam filling (NBF), partial beam blockage (PBB), ρhv, and signal-to-noise ratio (SNR), which are either sources or indicators of measurement errors (Bringi and Chandrasekar, 2004; Giangrande and Ryzhkov, 2005).

The classification of hydrometeors is greatly enhanced by considering the temperature profile. This information helps guide the algorithm in terms of precipitation phases. Solid and liquid hydrometeors occasionally have similar polarimetric signatures, but they exist at much different temperatures. A good example can be seen with the radar measurements of dry snow aggregates and light rain in Table 2.2. Beam broadening and beam center height increasing in altitude with range complicates the HCA functioning. Within a specific range rb, the radar only measures the rain region (below the melting layer) while beyond a specific range rt (rt > rb), the radar only measures solid hydrometeors above the melting layer. Within the range between rb and rt, the hydrometeors that the radar measures may come from the rain region, melting layer, and/or ice region. This may increase the ambiguity of the hydrometeor classification within this range. To reduce the classification error, HCA has implemented several rules to confine the radar echo classes as a function of range. With the physical constraint guided by the height of the melting layer, the ambiguity of classification can be largely reduced.

2.2.2.4 Polarimetric Radar-based QPE

The liquid water content W and the precipitation rate (R) can be theoretically computed using the following equations, respectively:

$$W = \frac{\pi}{6} \times 10^3 \rho \int_{D_{min}}^{D_{max}} D^3 N(D) dD, \quad (gm^{-3}) \tag{2.26}$$

$$R = 6\pi \times 10^{-4} \int_{D_{min}}^{D_{max}} D^3 \upsilon(D) N(D) dD, \quad (mm\,h^{-1}) \tag{2.27}$$

According to Equations (2.26) and (2.27), liquid water content W and precipitation rate R have different moments of the DSD. Therefore, power-law relations can usually be found between radar variables (Z_H, K_{dp}, or A_H) and bulk variables (W or R), providing an empirical approach to precipitation estimation. Many Z_h–R relations have been reported for different rain types, seasons and

TABLE 2.3

Parameters for Several Common Polarimetric Radar Rainfall Estimators (from Hong and Gourley, 2015)

	a	b	c	Notes
R(Zh,Zdr)	6.7×10^{-3}	0.927	-3.43	S-band (10 cm)
	5.8×10^{-3}	0.91	-2.09	C-band (5.5 cm)
	3.9×10^{-3}	1.07	-5.97	X-band (3 cm)
R(Kdp)	50.7	0.85		S-band (10 cm)
	24.68	0.81		C-band (5.5 cm)
	17.0	0.73		X-band (3 cm)
R(Kdp,Zdr)	90.8	0.93	-1.69	S-band (10 cm)
	37.9	0.89	-0.72	C-band (5.5 cm)
	28.6	0.95	-1.37	X-band (3 cm)

locations (Rosenfeld and Ulbrich, 2003). Polarimetric variables can be used to observe DSD variability and subsequently improve the accuracy of QPE. In general, rainfall estimators based on polarimetric radar variables have the following forms:

$$R\left(Z_h, Z_{dr}\right) = a Z_h^{\ b} Z_{dr}^{\ c} \tag{2.28}$$

$$R\left(K_{dp}\right) = a K_{dp}^{\ b} \tag{2.29}$$

$$R\left(K_{dp}, Z_{dr}\right) = a K_{dp}^{\ b} Z_{dr}^{\ c} \tag{2.30}$$

where R is in mm h^{-1}. Z_h and K_{dp} have linear units in mm^6 m^{-3} and deg km^{-1}, respectively. Z_{dr} is a dimensionless linear ratio. Table 2.3 gives the parameters a, b and c for several common polarimetric radar rainfall estimators at S-, C- and X-bands.

Each polarimetric estimator has its own advantages and disadvantages. The use of Z_{dr} gives a better estimation of raindrops representing the median of the DSD, i.e. those that contribute the majority to the total rainfall amount. Z_{dr} is a relative measurement and must be combined with either Z_h and/or K_{dp} for rainfall estimation. In general, the measurement error of Z_{DR} is on the order of a few tenths of a dB. K_{dp} is a phase measurement and is immune to any error in the absolute calibration of the radar. It is unaffected by precipitation attenuation along the propagation path and less affected by mixed-phase precipitation such as rain mixed with hail. However, since K_{dp} is derived from Φ_{dp} measurements over a given path length, K_{dp} estimation error increases rapidly as the path length decreases below 2 km (Bringi and Chandrasekar, 2004; Hong and Gourley, 2015). This results in a trade-off between the accuracy and range resolution of K_{dp}. In general, K_{dp} can be estimated to an accuracy of around 0.3–0.4 deg km^{-1} and has a smaller estimation error for heavy rain than for light rain. Therefore, when rainfall is intense and/or mixed with hail, $R(K_{dp})$ is more suitable than other estimators, while in light rain it is not appropriate to apply $R(K_{DP})$ relations. Figure 2.5 presents a flow chart describing the CSU-HIDRO optimization algorithm logic.

The measurement error of Z_h, Z_{dr} and K_{dp} may propagate into the final rainfall estimate. The total estimation error is attributed to two terms: εm is the error propagating from the measurement and εp is the parametric error of the estimator (Bringi and Chandrasekar, 2004; Hong and Gourley, 2015).

$$\hat{R} = R + \varepsilon_m + \varepsilon_p \tag{2.31}$$

$$\sigma_R^{\ 2} = \sigma_m^{\ 2} + \sigma_p^{\ 2} \tag{2.32}$$

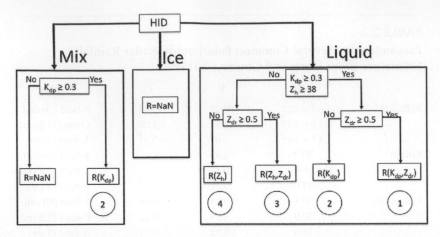

FIGURE 2.5 Flow chart describing the CSU-HIDRO optimization algorithm logic. The rainfall estimators corresponding to the circled numbers are described in the text (from Hong and Gourley, 2015).

where R denotes the true rainfall rate; notation $\wedge$ indicates the estimation; σ_2 is the error variance; and subscripts R, m and p indicate the variances associated with the estimation error, measurement error and parametric error, respectively. From the error analysis of different estimators, it can be concluded that the use of polarimetric measurements may enlarge the measurement error effect in the rainfall rate estimators but can effectively reduce the parametric error effect.

2.3 SNOWFALL AND HAIL ESTIMATION

The estimation of snowfall and hail presents difficulties compared to rainfall estimation. The main reason is that snow aggregates and ice crystals have quite different particle shapes from the oblate spheroids associated with raindrops. Remote sensing tools, such as weather radar and satellite surveys, are used to observe and track hailstorms and, in general, severe storms. Hailstorms result from energy imbalances in the atmosphere, which produce storms generating huge power. Radar observations open a new dimension in the detection, measurement and forecasting of hailstorms and provide detailed information on the structure and formation of such storms with considerable precision.

Conventional radar-based snowfall estimation and hail detection. The snowfall rate estimator can be expressed as a power-law relation as:

$$Z_e = aR_s^b \tag{2.33}$$

where Z_e is the equivalent radar reflectivity factor (in $mm^6\,m^{-3}$) of water drops and R_s is the snowfall rate expressed as the snow water equivalent (SWE) per unit time (in $mm\,h^{-1}$). With radar, the SWE is the variable that is estimated rather than the snow depth. The latter variable depends on the snow density, which varies with temperature and moisture. Temperature may also affect the Z–S parameters for SWE estimates. A widely used snowfall estimator is the $Z_e = 1780\,Rs^{2.21}$ (Sekhon and Srivastava, 1970).

Hail Detection. The height of the 45 dBZ contour (a radar echo-intensity level) was a criterion tested in a Swiss hail research program (Waldvogel et al., 1979). This research found that all hailstorms had 45 dBZ contours above the altitude of the $-5°C$ temperature level. There was a False Alarm Rate (FAR) of 50%, largely because some strong rainstorms also met the criterion. However, it is much preferable to make an error and assume that a heavy rainstorm is going to produce hail than to mistakenly believe that a hailstorm is only going to produce heavy rain. The criteria used by the National Hail Research Experiment in the USA (1972–1974) for a declared hail day was defined by radar maximum reflectivity greater than 45 dBZ above the -5°C level. A radar reflectivity (≥45 dBZ) implies that significant supercooled liquid water exists at temperatures cold enough for large hail growth. Waldvogel et al. (1979) conducted tests using the following simple criterion:

$$H_{45} > H_o + 1.4 \, Km \qquad (2.34)$$

where H_{45} is the height of the 45dBZ radar echo contour and H_o is the zero-degree level height. Vertical Integrated Liquid (VIL) is a radar derived estimate of the total mass of precipitation within a vertical column in the clouds. VIL is a function of reflectivity and converts reflectivity data into an equivalent liquid water content value based on studies of drop size distribution and empirical studies of reflectivity factor and liquid water content (Amburn and Wolf, 1997). VIL has units of mass divided by area (kg/m²). A higher VIL means there is more precipitation mass in the column of air. VIL values are often seasonally and regionally dependent. VIL can be used as a "guide" for assessing hail size and intensity of precipitation, but it has limitations. The VIL value correlation to hail size can depend on season, synoptic environment, elevation, storm speed, storm structure or hail reflectivity characteristics.

Polarimetric radar measurements for snow and hail. Polarimetric radar measurements are generally useful in the identification of snowfall, but not used for the quantitative estimation of snowfall rate. The major reason is the complexity of natural snowflakes, which creates difficulties to accurately model the scattering properties of snowflakes. The result is that Z_{dr} and other polarimetric variables cannot be easily applied. Polarimetric radar measurements can be used to distinguish hail from rain (Depue et al., 2007; Hong and Gourley, 2015). The hail differential reflectivity (H_{DR}) is defined as:

$$H_{DR} = Z_H - f(Z_{DR}), \qquad (2.35)$$

where

$$f(Z_{DR}) = \begin{cases} 27, & \text{for } Z_{DR} \le 0 \text{ dB} \\ 19Z_{DR} + 27, & \text{for } 0 \le Z_{DR} \le 1.74 \text{ dB} \\ 60, & \text{for } Z_{DR} > 1.74 \text{ dB} \end{cases} \qquad (2.36)$$

There is evidence that the H_{DR} thresholds of 21 dB and 30 dB were reasonably successful in respectively identifying the regions, where large and structurally damaging hailstones were reported. The vertically integrated liquid water content (VIL) is also a good indicator of hail existence and hail size (Amburn and Wolf, 1997). A substantial increase of severe hail (size >19 mm) is usually associated with VIL densities greater than 3.5 g m⁻³. It is noted that VIL can be derived from the empirical power-law relation between radar Z_h and liquid water content.

There is not a widely applicable Z_h–R relation for hail estimation. A relation between equivalent rainfall rate of hail R_H (mm h⁻¹) and the PSD parameter Λ (mm⁻¹) was derived (Torlaschi et al., 1984), which is given by:

$$\Lambda = \ln(88 / R_H) / 3.45 \qquad (2.37)$$

According to the Rayleigh approximation and Equation (2.37), the empirical relation between radar reflectivity Z (mm⁶ m⁻³) and R_H (mm h⁻¹) is given by:

$$Z = 5.38 \times 10^6 \left[\ln(88 / R_H) \right]^{-3.37} \qquad (2.38)$$

Hail Forecasting. Hail forecasting can be based on general severe thunderstorms forecasting along with the assessment of maximum hailstone size (Dalezios and Papamanolis, 1991) or could be based on radar data for storm tracking (Rigo and Camen Llasat, 2016), the latter being presented in the quantitative precipitation forecasting (QPF) section. Indeed, deep moist convection is necessary for hail forecasting, which could be supported by an adequate updraft to sustain the growing hailstone aloft for some time. Moreover, enough supercooled water near the hailstone is required to foster growth during an updraft, along with an initial piece (embryo) of ice or dust for it to grow upon. It should be mentioned that most severe thunderstorms may produce hail aloft, though it may melt before reaching the ground. Specifically, multi-cell thunderstorms produce numerous hailstones, however, not necessarily the largest hailstones, since the mature stage is not long enough to allow for hailstone growth. On the other hand, supercell thunderstorms can sustain updrafts, which support the formation of large hail, since continuous hailstone lifts into the very cold air at the top of the thunderstorm cloud. Indeed, the stronger the updraft the larger the hailstone can grow. In all cases, the hail falls when the thunderstorm's updraft can no longer support the weight of the ice. Supercells usually have hail larger than 5 cm in diameter, although non-supercell storms can produce hail the size of a golf ball.

Several important factors must be considered when forecasting the probability and size of hail, such as the fact that hailstone size is maximized by high convective available potential energy (CAPE), large wind shear, low freezing levels, low precipitable water (PW) and dry mid-level air.

2.4 RADAR-BASED QUANTITATIVE PRECIPITATION FORECASTING (QPF)

A largely unsolved problem in mesoscale convective system (MCS) forecasting involves anticipating the timing and location of convective initiation. Experimental forecasters can produce successful daily forecasts of deep convection. However, since storms often form within a range of 10–20 km of mesoscale boundaries, which may themselves be meso-γ-scale or narrower in width, variability in the boundary layer is probably critical for the storm initiation process along with formation, structure and movement. A brief review of radar-based storm tracking and forecasting methods follows (Krauss and Dalezios, 2017). These methods include extrapolation methods, knowledge-based now-casting methods, numerical models, neural network models and further approaches, such as probability forecasts and modified turning band (MTB) models.

2.4.1 EXTRAPOLATION METHODS

Extrapolation methods include steady-state assumption methods, which involve cross-correlation and feature tracking methods, as well as echo size and intensity trending methods.

2.4.1.1 Steady-state Assumption Methods

Forecasting radar echo motion requires determination of their velocity and direction of displacement. Several storm tracking algorithms based on weather radar data are used, either by tracking specific features of the radar echo (Browning et al., 1982), or by determining the echo motion through cross-correlation (Austin and Bellon, 1974). There are short-term and small-scale variations of the radar echo pattern, such as orographic impacts and radar technical limitations, which may cause tracking algorithms to malfunction. The technological trend leads to wind measurements from the steering level based on Doppler radar data, which can be used for radar echo advection (Andersson, 1991).

Cross-correlation method. In the cross-correlation method, similar patterns of radar echoes are detected by comparing tracking areas in consecutive scans, i.e. considering the overall motion of radar echoes. Then, the correlation coefficient justifies the best fit between the tracking areas. Moreover, the displacement vector is determined from the distance between the tracking area and the time lag of the scans. Figure 2.6 shows how the cross-correlation method works through an idealized sequence of radar-detected precipitation patterns at times $(t_1 - \Delta t)$, t_1 and $(t_1 + \Delta t)$. The vector AB represents the forecasted motion of the center of gravity between $(t_1 - \Delta t)$ and t_1. Specifically, the direction of the vector AB is found by selecting the highest cross-correlation among all the computed cross-correlations between the center of gravity A and all the adjacent pixels and the distance

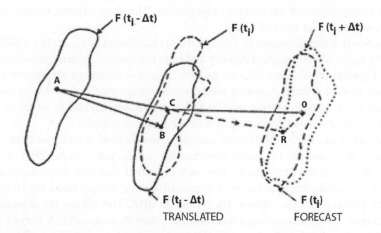

FIGURE 2.6 Cross-correlation storm tracking method (from Austin and Bellon, 1974).

AB is found based on the pattern's speed. The vector BC corresponds to the correction obtained using the cross-correlation technique leading to AC as the measured displacement vector. Note that the forecasted shape of the pattern remains the same, which is being adjusted with each measurement. CR is the forecast translation vector, which in this case is equal to AC. Similarly, RO is the difference between the forecast vector CR and the actual motion CO.

There have been some additions to the initial methodology. Specifically, the 2D continuity equation was applied to identify sources of error in the derived displacement vector field, such as clutter, shielding and rapid changes in the radar pattern (Lí et al., 1995). A geometric algorithm was also added, aiming at the detection of merging and splitting cells (Dixon and Wiener, 1993). Moreover, an approach was implemented combining cross-correlation and feature identification (du Vachat et al., 1995).

There are several software packages based on the cross-correlation method. **TREC** (Tracking Radar Echo by Correlation) (Rinehart and Garvey, 1978) emphasizes the best fit between arrays of radar reflectivity in consecutive scans by optimizing a test quantity, e.g. the correlation coefficient of a minimum absolute difference. Similarly, **COTREC** (Continuity of TREC vectors), being an extension of the TREC method (Li et al., 1995), uses a variational technique to address inconsistencies of the displacement vectors, which are usually related to clutter, shielding and rapid changes in the radar pattern of divergent components of the motion field.

TITAN (Thunderstorm Identification, Tracking, Analysis and Now-casting): **TITAN** is based on both cross-correlation and mass centroid approaches (Dixon and Wiener, 1993; Mecklenburg et al., 2000). The mass centroid approach has been used to identify storms as "three-dimensional entities". Moreover, geometric algorithms can detect merging and splitting cells. A weighted linear fit is used from the storms' history to forecast both position and size.

SHARP (Short-term Precipitation Forecasting Procedure) (Austin and Bellon, 1974): SHARP is an automated operational now-casting system, which is based on radar CAPPI data in Cartesian coordinate system as input. Specifically, the displacement vector of the rainfall patterns is linearly extrapolated using the cross-correlation approach between consecutive radar images. In the initial method routines have been added to address problems caused by new echoes over the edge of the radar image and missing, due to ground clutter elimination (Austin and Bellon, 1974). The forecasting horizon used is 3 hours.

HYRAD (HYdrological RADar system) (Moore et al., 1994): HYRAD system consists of procedures starting from pre-processing calibration of radar and leading to radar rainfall forecasting and catchment averaging. For the movement of the rainfall field, an advection model is used, which is based on linear extrapolation of the current radar pattern. The displacement vectors are derived by considering consecutive radar rainfall fields.

VSRF (Very-Short-Range Forecast of Precipitation) (Kunitsugu et al., 2001): VSRF system was developed by the Japan Meteorological Agency and is an operational system providing up to 6-hour forecasts with a spatial resolution of 5 km. At first, 3-hour forecasts of displacement vectors are derived based on pattern matching, which also consider orographic enhancement and dissipation of rain. Then, an extrapolation method is used for up to 6 hours lead time, which is merged with forecasts derived from a mesoscale numerical weather prediction model.

2PiR (du Vachat et al., 1995): 2PiR is an extrapolation method, which has been developed by the Meteo-France and used in the Aspic now-casting operational system. 2PiR is incorporated into the French Aramis (radar) composite. For each identified rain cell, displacement vectors are derived based on cross-correlation to boxes of consecutive radar images (time lag 10 minutes) using three reflectivity threshold levels, namely 10, 20 and 28 dBZ. This allows the depiction of intense rain cells movement. If the boxes exceed a certain size, then they are split. A Barnes-type analysis scheme is considered to develop displacement fields of scattered motion vectors. Moreover, ground clutter is considered for each radar pixel at low elevation angles.

Feature tracking method. The mass centroid method is a widely used feature tracking approach, which computes the displacement vector between consecutive radar scans based on the distance of the mass centers of two corresponding radar echoes. The mass centroid method provides detailed analysis of cell tracks and features. This "structured" approach determines the radar echo motion, which consists of echo identification, characterization, matching and forecasting (Einfalt et al., 1991). Several features are identified and recognized, such as mass, size and centroid, as well as orientation, elongation, intensity distribution, previous size, moment vectors and the number of previous recognitions of the radar echo. Alternatively, Fourier transformation (Blackmer et al., 1973), or the maximal echo diameter and the moments of inertia on the X and Y axes (Wolf et al., 1977) can be used.

Moments of inertia method (Wolf et al., 1977). Moments of inertia are the second moments of an area with respect to axes X and Y. It is possible to determine the moments of inertia with respect to inclined axes, where, in such case, the moments of inertia can be obtained by formal integration (Figure 2.7). From the derived equations, it is possible to describe accurately the shape and orientation of the cloud. Then, the following information can be obtained: the location of centroid with respect to radar location; the angle a, which represents the rotation of the perpendicular axes U and V to the original axes X and Y and is positive in a counter-clockwise direction; the major and minor principal moments of inertia; the ratio Imax/Imin. From these parameters, the first cloud is calculated. The following steps include: calculation of Imax, Imin and angle a by using the derived equations for the time step (t-dt, t); estimation of the above at t+dt by using either spline surfaces or least square method; regeneration of the rainfall distribution; estimation of the forecast vector V_{t+1}.

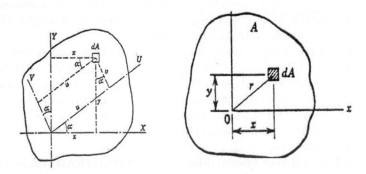

FIGURE 2.7 Schematic diagram of the Moments of inertia method.

Echo Clustering. The clustering of radar echo might be defined either by considering pixels as one echo, if they touch only directly or diagonally, or by adding the closest neighboring pixels, if they differ by no more than one level from each other (Blackmer et al., 1973). In the matching process, feature tracking is used to recognize echoes in consecutive radar scans. Indeed, the matching criteria are based on either a minimal shape difference, or a minimal squared distance, or by optimizing a quality criterion, which includes several features to obtain the displacement vectors. Radar echo tracking algorithms, which are used in operational now-casting, rely mainly on three-dimensional data sets and tend to combine both approaches, namely cross-correlation and feature tracking. There are several software packages, such as TITAN, SCIT, SCOUT, TRACE3D, PARAPLUIE and Automatic tracking of convective cells and cell complexes from lightning and radar data, which are briefly presented.

SCIT (Storm Cell Identification and Tracking) (Johnson et al., 1998). This is an operational pattern recognition approach, which has been developed to support the WSR-88D (Weather Surveillance Radar-1988 Doppler) storm series algorithm. SCIT considers a mass centroid approach with seven reflectivity thresholds (30, 35, 40, 45, 50, 55, 60 dBZ), instead of using one reflectivity threshold (30 dBZ) of the WSR-88D algorithm. The result is a significant improvement in storm identification. SCIT detects isolated cells, as well as clustered and line storms, improving, thus, the WSR-88D algorithm. Each motion vector is calculated by applying a linear least square fit to the ten previous, equally weighted, storm locations. Moreover, attributes of the storm cells of the previous ten volume scans are stored, such as the maximum reflectivity value and height, storm base and top, VIL based on cells, height of the mass center and similar parameters. For trailing stratiform or mesoscale convective systems, an area-based tracking algorithm is usually considered.

SCOUT (Einfalt et al., 1991). This is a feature tracking approach to determine echo motion. The method consists of echo definition, characterization, matching and the forecast step. Radar echoes in consecutive scans are identified through previously presented feature tracking parameters. A certain threshold is used to identify an echo. A degree of change in each of the features is the criterion for the recognition of consecutive radar scans in the matching process. As a result, echo tracking is considered, and splitting and merging of echoes can also be detected. Moreover, isolated displacement vectors are used to extrapolate each echo separately. SCOUT was initially developed to control the sewer network in suburban Paris.

TRACE3D (Handwerker, 2001). This is a tracking method of convective cells, which is based on polar volume scan data. The development includes two steps, namely cell identification and tracking. Algorithms for radar error sources are developed and parameters are computed. Certain criteria are required for cell identification, such as reflectivity thresholds to distinguish from the background and the distance between two cells should take a certain value. For tracking, the velocity of cells is calculated by applying a weighted average of all previous velocities, and merging and splitting is also computed. An individual velocity for each cell is used to determine the cell position in the next scan.

PARAPLUIE (Bremaud and Pointin, 1993). This feature tracking approach includes four steps, namely tracked entity definition, characterization, matching and forecast through extrapolation. Tracked entities, called "CEL echoes", are determined based on varying reflectivity thresholds according to the rain type. As a result, tracking of heavy rain cells can be achieved throughout their life cycle. Characterization of tracked entities is based on area, reflectivity distribution characteristics, coordinates of the centroid, elongation of the inertia matrix, inclination of the principle matrix of inertia and the previous detected motion. Moreover, the matching process of the tracked entities is controlled by optimizing confidence criteria. Finally, the extrapolation of CEL echoes is based on an arithmetic mean of the all the speeds of the CEL echoes.

Automatic tracking of convective cells and cell complexes from lightning and radar data (Steinacker et al., 1999). A lightning detection network consisting of eight sensors and four C-band Doppler radars are used, where the spatial resolution used is 4x4 km^2 and the temporal resolution is 20 minutes, respectively. The lightning system records intense convection based on cloud electrification and radar detects precipitation types. This combined system contributes to the distinction between convective and stratiform precipitation, as well as to the recognition of the stage of the life cycle of the convective cell. The algorithm works in three steps as follows: at first, the input data are smoothed using a Gaussian filter; then, isolated maxima are identified as cells with reflectivities above both, a pre-defined threshold and the values of the neighboring cells; finally, the choice of the most likely displacement vector is based on several criteria, e.g. wind velocity at 600 hPa, increasing/decreasing rainfall rate, or deviation from mean vector. The system was initially implemented within the HERA (Heavy Precipitation in the Alpine Area) project in 1996.

2.4.1.2 Echo Size and Intensity Trending Method

A joint approach for echo size and intensity trending consists of weighted or unweighted fits of past developments of the radar echo. Specifically, a now-casting methodology includes the evolution of individual cells to forecast the motion of precipitation areas by applying a nonlinear extrapolation to the intensity and echo size, e.g. increase echo size for 15 minutes then decrease (Tsonis and Austin, 1981). However, initiation and dissipation of precipitation pattern, as well as local orographic impacts, constitute the main limiting factors to intensity forecasts. The most appropriate time interval seems to be 30 minutes for fitting the extrapolation scheme on the basis that a trend should be established for a developing cell. It is recognized that each event has its own course of growth and decay, even if events are similar in large-scale conditions. The TITAN and SCIT algorithms, which were previously presented, fit into this methodology.

2.4.2 Knowledge-based Now-casting Methods

Knowledge-based now-casting systems rely on observations. Specifically, several data sources and conceptual models are combined to obtain information about the initiation, development and dissipation of precipitation. Conceptual models used in knowledge-based now-casting essentially utilize the relation between convergence lines and enhanced convection. The causes for initiation of thunderstorms along a dryline have been studied, including vegetation, which often causes thunderstorm initiation when its thermal circulation intersects a dryline (Hane et al., 1997). Similarly, storm initiation along convergence lines has also been investigated showing significant impact of local differences in the surface moisture and the moisture profile in the planetary boundary layer (PBL) (Mueller et al., 1993). The initiation of a squall line was observed in Kansas by propagating radar-detected convergence.

Furthermore, radar-based surface rainfall and VIL (Vertically integrated liquid water content) has been used as input to a conceptual short-term rainfall prediction model for urban catchments (Thielen et al., 2000). The model uses the water mass balance within air columns and their spatial advection, which is produced from consecutive time steps. The study is based on simulated data from a mesoscale meteorological model, where the focus has been on the usefulness of VIL information for rainfall forecasting. Moreover, a rainfall forecast model has been used, which combines mass balance of mean VIL derived from three-dimensional radar data, surface meteorological measurement and radiosonde data with the prediction of residual VIL. This model combines physically and statistically based components (Seo and Smith, 1992). The predicted VIL is converted into rainfall by using the relations between VIL, water content at cloud bottom and echo-top height. In a similar study, radar and satellite data have been used to define the model state and boundary conditions, as well as to conduct an uncertainty analysis using a Kalman filter approach (French and Krajewski, 1994). Moreover, an advection equation was developed to determine the updraft strength and included spatial and temporal information about the strength of the convection. There are software packages, such as Auto-Now-cast (AN), GANDOLF, NIMROD, CARDS, which are briefly presented.

Auto-Now-cast System (AN) (Mecklenburg et al., 2000). This system was developed at the National Center for Atmospheric Research (NCAR). This is a complicated system based on several algorithms, as well as WSR-88D radar and GOES satellite data, surface mesonet, sounding, lightning and surface elevation data (Saxen et al., 1999). The intensity forecast of the AN is based on the following algorithms: (1) **oldGrowth** predicts the growth of cumulus. The cloud type is determined using the VIS and IR channels of the GOES satellite. Specifically, in regions of cumulus occurrence, the rate of cooling in the IR temperature provides information about vertical development ("growth") (Roberts, 1997); (2) COLIDE (Convergence line detection and extrapolation) identifies, tracks and extrapolates convergence boundaries based on radar data; (3) a combination of the WSR-D88 algorithm and TITAN, both using radar data, predicts the storm volume and growth. Furthermore, the **BrightBand** algorithm minimizes the contamination of radar data, due to melting snow and graupel below the freezing level, TREC is applied to calculate boundary layer winds and convergence and TITAN is used to identify and extrapolate storms. The final forecast is developed by a user-defined weighted combination of all algorithms and is carried out for lead times up to 60 minutes.

GANDOLF (Pierce et al., 2000). This is a now-casting system for non-frontal convective rain (showers), which was jointly developed by the Met Office and the Environment Agency, UK (Fox et al., 2001). The system is based on a conceptual life cycle model of a convective cell, which simulates the growth and decay of showers in a physically realistic manner. At first, a convective cell analysis produces a convective rain analysis field. Then, vertical reflectivity profiles, which are produced from volumetric or multi-beam radar data, are used to identify the developmental state of all existing convective cells. This conceptual model uses object-oriented programming techniques, known as the Object-Oriented Model (OOM) (Hand and Conway, 1995). Then, a mesoscale NWP model is used for the cell advection, thus, generating a now-cast, where each identified cell follows the life cycle, which is conceptualized in the OMM. The main outputs of GANDOLF consist of rainfall rate and accumulation, however, large hail and peak convective gust may also be produced.

NIMROD (Now-casting and Initialization for Modeling using Regional Observation Data) (Fox et al., 2001). This system produces analyses and short-range forecasts of several meteorological parameters including rainfall and has been operational at the UK Met Office since 1995 (Golding, 1998). NIMROD accepts inputs, such as radar data, infrared and visible satellite imagery, surface synoptic weather reports and forecast fields from a mesoscale NWP model. The rainfall now-casting scheme is based on object-oriented advection methods and NWP mesoscale rainfall prediction. Cross-correlation vectors or wind vectors from a NWP model forecasts are used for the advection of precipitation "objects" in order to generate a short-range rain forecast. The outputs include precipitation type, instantaneous rain rate fields and rain accumulation. At longer lead times greater weight is given to the NWP component of the rainfall forecast.

CARDS (CAnadian Radar Decision System) (Donaldson et al., 2001). This is a modular system, operational at Environment Canada. The system monitors meteorological extremes, such as hail, VIL based downburst/gust, severe weather and mesocyclones, using different types of radar data. Specifically, the cross-correlation approach is used for radar echo tracking. Moreover, the empirical hail algorithm incorporates information about the freezing level, VIL and the maximum height of the 50 dBZ echo to predict rain (Treloar, 1998). Furthermore, the downburst algorithm uses a low-level divergent radial shear detection algorithm and the VIL downdraft algorithm, which is an empirical relationship between VIL and the downdraft outflow (Lapczak et al., 1999).

In addition, the mesocyclone algorithm is based on the pattern vector approach and can be used by C-Band radar data with reduced Nyquist velocities or with dual PRF data (May and Joe, 2001). The multi-height analysis is used to detect the location and severity of the mesocyclone (Burgess et al., 2001).

2.4.3 NUMERICAL MODELS

Numerical models for thunderstorms are usually based on large-scale simulation to derive low-level convergence. This means that mainly large-scale convective events could be the subject of successful and reliable forecast. Indeed, observational information that could be used to define initial conditions is not included in numerical models (Wilson et al., 1998). To overcome this deficiency in numerical models, observed storm data should be used. However, only radar reflectivity and velocities of a single Doppler radar are usually available operationally, whereas explicit initialization for thunderstorms requires a full dynamic, thermodynamic and microphysical field. There is, thus, a need for the so-called adjoint methods, i.e. from observed fields, such as reflectivity or Doppler velocity, to make inferences about unobserved fields, such as temperature or microphysical properties. This can be achieved by combining numerical models without and with initialization in order to benefit from both of their advantages, leading to improved thunderstorm forecasting (Wilson et al., 1998). Specifically, numerical models should be run over domains, large enough to include the large-scale forcing and should have a high spatial resolution, enabling them to be initialized with explicit data. Existing software packages are NIMROD and VSRF, which have been previously presented.

2.4.4 NEURAL NETWORK MODELS

Precipitation forecasting is a difficult subject to be expressed mathematically. For example, thunderstorms can be forecasted based on surface meteorological data, or data from a numerical weather prediction model and upper air soundings to predict precipitation. On the other hand, artificial neural networks (NN) are a suitable tool for solving complex problems that might be difficult to express in mathematical terms, such as precipitation forecasting. Indeed, a mathematical rainfall simulation model has been used to provide input to the NN (Moriyama and Muneo, 2000). In general, the simplest form of neural network is a three-layer network, which consists of the input, the hidden and the output layer. Weights are used to connect neurons from different layers. The network is trained on learning sets, which consist of several variables corresponding to input and output fields. Specifically, the connection weights could be adjusted during the training, which means that the magnitude of the error between real and forecasted output is optimized until a desired accuracy of time limit is reached, known as back propagation. Then, independent inputs are used to test NN for verification. Following successful training and testing, the network can eventually predict output fields based on any newly provided input.

2.4.5 STATISTICAL AND STOCHASTIC APPROACHES

Statistical and stochastic approaches for rainfall forecasting based on radar data as input involve probabilities of accumulated spot precipitation, namely probability forecasts or statistical approaches, such as the Modified Turning Bands (MTB) Model (Mellor et al., 2000) or the spectral model, namely **SPROG** (SpectraI PROGnosis) (Seed and Keenan, 2001).

MTB (Modified Turning Bands) model (Mellor et al., 2000). This is a stochastic model of the rainfall process, which is used to generate future rainfall scenarios or ensembles on four different scales, namely rain cells, cluster potential regions, rainbands and storms in a synoptic context. The forecast lead time ranges from minutes to several hours, since, for each scale, different advection velocities and evolution rates are allowed. Indeed, radar data provide the current storm observations, which constitute the basis for the future rainfall scenarios. The MTB model is especially capable for short-term rainfall forecasts, because it combines deterministic modeling at large scales and stochastic modeling at smaller scales, i.e. it considers the variability of the rainfall process and the uncertainties of future evolutions.

SPROG (SpectraI PROGnosis) (Seed and Keenan, 2001). This now-casting system, operational at the Australian Bureau of Meteorology, is mainly based on advection of precipitation and uses radar data. The basic idea is that the scale of the precipitation features drives the evolution of the precipitation features. In this spectral model, the precipitation field is investigated on different scales and the evolution of each scale is calculated using a simple autoregressive model.

2.5 SATELLITE-BASED PRECIPITATION ESTIMATION METHODS

Satellite-based precipitation estimation methods consist of information from low Earth orbiting (LEO) satellites and/or geostationary Earth orbiting (GEO) satellites. GEO satellites are in such an orbit that they rotate around the Earth at the same speed as the Earth rotates, thus appearing stationary relative to a location on the Earth. GEO satellites can provide images every 5–30 minutes in multiple spectral bands, but their spectral coverage is limited to visible and infrared wavelengths (Nguyen et al., 2018). GEO satellites share several common attributes: they typically carry visible (VIS) and infrared (IR) sensors with nadir resolutions of about 1×1 km and 4×4 km, respectively, acquiring images nominally every 30 minutes. GEO satellites provide an unrivaled platform for continual observation, however, they are limited in the Polar Regions by the unfavorable viewing angle at high latitudes (Kidd and Levizzani, 2011). On the other hand, LEO satellites generally cross the Equator at the same local time on each orbit, providing about two overpasses per day. LEO satellites carry a range of instruments capable of precipitation retrievals, including multichannel VIS/IR sensors, and passive microwave (PMW) sounders and imagers. LEO satellites can provide passive microwave (PMW) information about the hydrometeors directly relevant to surface precipitation rates (Kidd and Levizzani, 2011). Current operational polar-orbiting satellites include EUMETSAT's MetOp series and the National Oceanic and Atmospheric Administration (NOAA) series of satellites with NOAA-19. Observations made by these satellites are typified by wide swaths (2800 km) with resolutions of about 1 km.

Sensors onboard satellites are currently the only instruments that can provide homogeneous and global precipitation measurements. The sensors can be classified into three categories: visible/infrared (VIS/IR) sensors on GEO and LEO satellites; passive MW (PMW) sensors on LEO satellites; and active MW sensors on LEO satellites (Michaelides et al., 2009; Nguyen ct al., 2018). Corresponding rainfall intensity retrieval methodologies from space-borne sensors fall primarily into three main categories based upon the type of observation, namely VIS/IR techniques, PMW techniques (also active MW only with TRMM PR) and multi-sensor techniques on GEO and LEO satellites (Kidd and Levizzani, 2011). The first precipitation radar has been operating in space since the launch of the Tropical Rainfall Measuring Mission (TRMM) in 1997 (Kummerow et al., 1998). The availability of active precipitation detection from space has laid the groundwork for space-borne estimations from radar and radiometers (Michaelides et al., 2009).

The measurement resolution is another important limiting factor, which is around 50×50 km^2 over ocean and no better than 10×10 km^2 over land. Several methods assume that the rainfall field is homogeneous across the instrument's field of view (FOV), which disregards the three-dimensional structure of the rain system (Michaelides et al., 2009). This assumption, along with the nonlinear response of brightness temperatures to rain rates, leads to the well-known problem of beam filling (e.g. Kummerow et al., 1998). Moreover, microwave frequencies are more directly responsive to cloud internal processes and thus to precipitation formation mechanisms, because in this portion of the electromagnetic spectrum the precipitating hydrometeors are the main source of the attenuation (Michaelides et al., 2009). Indeed, low-frequency microwave measurements (below 40 GHz) over ocean, thus have the capability of detecting liquid hydrometeors and they can be combined with the sensitivity of higher frequencies (above 60 GHz) over land to precipitation size ice. However, there are more independent variables in the rain, although the channels are not independent of each other.

A general overview of algorithms, sensors and methods used in rainfall retrievals from space has been presented (Levizzani and Cattani, 2019). Early efforts for the development of methods to estimate precipitation from satellites have been discussed (Barrett and Martin, 1981; Hsu et al., 1997). These efforts include the analysis of individual pixel information (Arkin and Meisner, 1987), as well as the analysis of cloud image types and their variations in time (Gado et al., 2017; Nguyen et al., 2018). The TRMM has signified the beginning of a new era for operational satellite-based precipitation products. It carried the first orbital rainfall radar, which was used to calibrate passive microwave sensors on other satellites, resulting in significant improvements in rainfall retrievals over the tropical regions of the globe (Simpson et al., 1987; Kummerow et al., 1998).

The TRMM has been followed by the Global Precipitation Measurement (GPM) mission, which deployed an enhanced dual-frequency radar sensor. The objective of the GPM program is to combine observations from multiple passive microwave sensors mounted on both pre-existing and newly deployed satellites. The GPM satellite constellation has global coverage in the range (68_ S–68_ N) with a return interval of 3 h. Today, several agencies and institutes provide satellite-based data sets, each produced by different algorithms. These products include the NOAA Climate Prediction Center (CPC) morphing technique CMORPH (Joyce et al., 2004), NASA TRMM Multi-Satellite Precipitation Analysis (TMPA) (Huffman et al., 2007), NASA Integrated Multi satellite Retrievals for GPM (IMERG) (Huffman et al., 2014), NRL-Blend satellite rainfall estimates from the Naval Research Laboratory (NRL) (Turk et al., 2010) and the Precipitation Estimation from Remotely Sensed Information using Artificial Neural Networks (PERSIANN) family of products (Hsu et al., 1997).

The main objective of the satellite rainfall assessment is to provide information on the appearance, size and distribution of rainfall on the planet for the full range of environmental applications and uses. However, it is only recently that the new orbiting satellites and the programmable platforms and sensors have allowed continuous global coverage through reliable business applications. Of course, business applications require quantitative rainfall estimation as a synthesis of various meteorological systems such as rain gauge, weather radar and satellites, with different microphysical and dynamic characteristics that constitute non-unique solutions. Satellite estimates of precipitation can be derived from a range of observations from many different sensors. At the present time, global coverage of meteorological satellites exists with long data records and low data acquisition costs, as well as with terrestrial rainfall networks on the planet. Consequently, it is possible to combine these two systems for a reliable quantitative estimate of regional rainfall at a planetary level, and the weather radar, where available, can be used both operationally and as an alternative to the certification of rainfall estimates.

A brief presentation and description of the existing satellite-based precipitation retrieval methods and techniques follows. The methods are classified into visible (VIS) and infrared (IR) methods, which include cloud indexing methods, cloud climatology methods, life history methods, bispectral methods and cloud model-based methods. Then, passive microwave (PMW) methods, active microwave (AMW) methods and combined satellite methods are briefly presented.

2.5.1 VISIBLE (VIS) AND INFRARED (IR) PRECIPITATION RETRIEVAL METHODS

Satellite visible (VIS) and infrared (IR) methods are classified into two classes, namely those which are based on one system only, e.g. VIS, and combined methods based on more than one system. There is extensive literature on the methods assessing precipitation from satellites (Barrett and Martin, 1981; Kidder and Vonder Haar, 1995; Levizzani et al., 2002; Dalezios, 2015). A brief description of the methods and techniques of the two categories follows.

2.5.1.1 Cloud Indexing Methods

Rainfall intensity is categorized for each type of cloud, which is identified in the satellite image. There are several cloud indexing methods, the most typical of which are described below.

GOES Precipitation Index (GPI). GPI was developed during the GATE (GARP Atlantic Tropical Experiment). Note that GARP stands for Global Atmospheric Research Program (Arkin and Meisner, 1987). The GPI relies on the high correlation between radar rainfall and the percentage of the infrared area, which is colder than 235° K. This determines areas with constant rainfall of 3 mm/h, suitable for tropical rainfall in areas of 2.5° × 2.5°. The GPI has been standardized for climatic analysis ranging from five days to a month.

Bristol Methods. Several cloud indexing methods have been developed at the University of Bristol, UK, initially in polar orbit satellites (NOAA) and later in geostationary satellites (METEOSAT). In

TABLE 2.4
Probabilities and precipitation intensity in relation to satellite cloud classification

Categories of Clouds	Rainfall Probability (0–1.0)	Rainfall Intensity (0—.0)
Cumulonimbus (Cb)	0.90	0.80
Stratus (St)	0.50	0.50
Cumulus (Cu)	0.10	0.20
Stratocumulus (Sc)	0.10	0.01
Cirrus (Ci)	0.10	0.01
cloudless sky	---	---

these methods, rainy days are initially determined in a region based on brightness temperature (T_B) values in the IR part of the spectrum, which are lower than a threshold value. Then, rainy days are combined with spatial rainfall variability per day in order to produce final rainfall estimates for periods of more than ten days (Barrett and Martin, 1981). Indicatively, cumulative rainfall for a time period based on the following form:

$$R = f\left(c, i\left(A\right)\right) \tag{2.39}$$

where R is the cumulative rainfall for a time period in a pixel or cell, c is the cloud area, i is the cloud type and A is the altitude.

Table 2.4 is indicative of satellite-based cloud classification and corresponding rainfall probability of occurrence and intensity.

National Environmental Satellite Service (NESS) of NOAA-USA Methods. NESS methods are based on afternoon polar orbit satellite imagery (NOAA). The mean 24-hour rainfall R is calculated from the relationship:

$$R = \left(K_1 A_1 + K_2 A_2 + K_3 A_3\right) / A_0 \tag{2.40}$$

where A_0 is the area under consideration, A_1, A_2 and A_3 are regions of A_0, covered by the three most important types of clouds, causing rain (Cumulonimbus, Nimbostratus, cumulus), and K_1, K_2 and K_3 are empirical coefficients. The method has been successfully applied to tropical and subtropical areas and is improved after continuous calibration with rain gauges (Follansbee, 1973).

2.5.1.2 Cloud Climatology Methods

The basis of these methods is a correlation between the cold cloud tops frequency from satellite data and mean rainfall measurements by rain gauges. A typical cloud climatology method is the TAMSAT method or cold cloud duration (CCD) method (Domenikiotis et al., 2005).

TAMSAT Method. Cold Cloud Duration (CCD) is defined as the duration in hours at which the cloud temperature in an area remains lower than a threshold value. The process requires adaptation to the conditions of each region and an estimate of cumulative duration. Correlations have been calculated between ten-day rainfall amounts and cold cloud duration. The selection of the threshold temperature is considered critical, since it depends on both the satellite data and the area of interest. The relationship is:

$$R = a D + b \tag{2.41}$$

where R is the rainfall, e.g. ten days in mm, D is the duration in hours and a, b are the regression coefficients. Equation (2.41) is applied to each pixel in the region, with R = 0, if D = 0. These methods, although quite simple, are highly accurate over long duration periods or large areas.

2.5.1.3 Life History Methods

The family of life history methods uses geostationary satellite imageries. These methods are based on the identification of storm clouds, the time series monitoring of the evolution and the detailed analysis of the lifecycle of mostly vertical growth clouds, which cause convective precipitation. The following life history methods are presented.

Method of Stout-Martin-Sikdar. In this method (Stout et al., 1979), the amount of rainfall produced by Cumulonimbus (Cb) clouds or cloud systems is calculated as the sum of the regional extent and the change in regional extent from the relationship:

$$R_v = a_0 A_c + a_1 (dA_c / dt) \qquad (2.42)$$

where R_v is the volumetric rainfall rate for a specific cloud in m^3/s, A_c is the regional extent of the cloud in m^2 at time t, dA_c/d_t is the change in the regional extent of the cloud and a_0 (in m/s) and a_1 (in m) are empirical coefficients. This method has been successfully used in the mapping and illustrating of convective rainfall in the GATE experiment.

Griffith-Woodley Method. This empirical method (Griffith et al., 1978; Negri et al., 1984) focuses on the estimation of surface rainfall by deep tropical convective systems. It is based on the observation that in tropical areas of deep convection rainfall occurs with brighter clouds in the visible portion of the spectrum and colder in the corresponding infrared than in non-active regions. It is also observed that the cumulonimbus intensity is a function of the cloud development stage. This method provides rainfall estimates beyond the range of calibrated radars, whose reflectivity is associated with regional coverage of cumulonimbus and varies with the evolution of the cloud. Finally, the reflectivity region is linearly related to the volumetric rainfall rate, which can be calculated from a sequence of measurements of the cloud area. The relationship is:

$$R_v = R \times f \left[A_e, A_m \right] \qquad (2.43)$$

where R_v is the volumetric rainfall rate in m^3/s, A_e is the reflectivity region with a threshold of about 1 mm/h, A_m is the maximum cloud cover area and R is a variable coefficient that connects the reflectivity region to the volumetric rainfall rate and depends on the increase or decrease of this region. The f is a higher-order function, delineated by tables or figures. Rainfall estimates can be conducted in visible or infrared. The method has similarities to the previous one, except for the addition of the reflectivity region and the possibility of changes and variations in this reflectivity region.

Scofield-Oliver method. The method focuses on the estimation of convective rainfall intensity at specific locations or stations and is based on the following basic assumptions: (1) rainfall is favored by clouds of high brightness in the visible portion of the spectrum; (2) heavy precipitation is favored by low cloud top temperatures in the infrared, from growth and convergence, whereas light rainfall is favored by decaying clouds with warmer tops; (3) rainfall is centered on the windward side of a convective system or the anvil of cumulonimbus clouds. This method is essentially a hierarchical decision support system based on the aforementioned three assumptions and simulates the vertical storm structure through sounding by adding conditions and factors aiming to improve the final quantitative rainfall estimate. It is considered one of the most reliable methods for estimating convective rainfall from the infrared spectrum of geostationary satellites (Scofield and Oliver, 1977). The method has the flexibility for direct interaction with meteorologists. Finally, the method is operationally used in the USA for short-term forecasting of severe storms and flash floods.

2.5.1.4 Bispectral Methods

In bispectral methods, combined infrared and visible spectrum satellite images are used for mapping the extent and distribution of precipitation. It is well known that infrared sensors provide information on temperature and as a result the cloud tops. Correspondingly, visible spectrum sensors provide information on cloud thickness, geometry and composition. In general, the bispectral methods rely on the very simple relationship between cold and bright clouds and the high probability of precipitation, which is characteristic of the cumulonimbus (Cb) clouds. An operational difficulty arises from the fact that images in the visible portion of the spectrum are not available during the night.

RAINSAT Method. This method is a two-dimensional (2D) pattern matching approach that has been developed to delineate areas of rain-producing clouds from satellite images in both the visible (VIS) and infrared (IR) portions of the spectrum and uses radar data for ground validation (Lovejoy and Austin, 1979). The aim is to minimize an objective function of two categories (R = rain, N = no rain) in a two-dimensional field, namely infrared and visible with the following four variables: N_N = estimation of no rain and confirmed no rain, N_R = estimation of no rain but confirmed rain, R_N = estimation of rain but confirmed no rain, R_R = estimation of rain and confirmed. The general loss function, which must be minimized, is based on the Bayes decision theory and is given by the relationship:

$$f = \left(l_R \, R_N + l_N \, N_R\right)/\left(N+R\right) \tag{2.44}$$

where $N = N_N + N_R$, $R = R_N + R_R$ and l is the weighted coefficient for false estimate. If l_R and l_N are equal to one (unit), then the above relationship becomes:

$$f = \left(R_N + N_R\right)/\left(N+R\right) \tag{2.45}$$

which corresponds to the percentage of errors. It can be assumed that L_R = (N+R)/R and N_N = (N+R)/N, thus, the weighted coefficients for each class are inversely proportional to the class sample, leading to a loss function of the form:

$$f_f = \left(R_N \, / \, R\right) + \left(N_R \, / \, N\right) \tag{2.46}$$

For each pair of images there is an infrared and visible reflectivity function g, corresponding to the minimum of f or f_f. The f_f function defines an optimal dividing line between the satellite-based rain or non-rain classes. When f_f is calculated, then the g function can be used to generate satellite-based rainfall maps. If rain is considered to be the unit and no rain is considered to be zero, then the correlation coefficient ρ for each map is given by the relation:

$$\rho = \left(R_R \, N_N - R_N \, N_R\right)/\, N \, R \tag{2.47}$$

2.5.1.5 Cloud Modeling Methods

These methods aim to exploit cloud physic to improve the natural description of the rain formation process. Specifically, in the Convective Stratiform Technique (CST) (Adler and Negri, 1988; Bendix, 1997) a one-dimensional (1-D) cloud model initially connects the cloud top temperature to the rainfall rate and areal extent. Then, a slope parameter S is calculated for each minimum temperature T_{min} of the pixel (i, j) in the infrared (Figure 2.8), and is given by the relation:

$$S = T_{1-6} - T_{min} \tag{2.48}$$

where T_{1-6} is the average temperature of the six adjacent trace elements, also given by the relation:

$$T_{1-6} = \left(T_{i-2,j} + T_{i-1,j} + T_{i+1,j} + T_{i+2,j} + T_{i,j+1} + T_{i,j-1}\right)/6 \tag{2.49}$$

Removing the T_{min} from the Equation (2.48) excludes the cirrus clouds, which do not produce rain, as shown in Figure 2.8, and the equation of the dividing linear slope is given by the relation:

$$S = 0.568\left(T_{min} - 217\right) \tag{2.50}$$

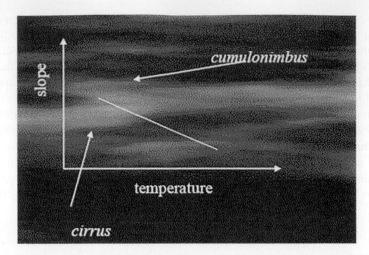

FIGURE 2.8 Slope parameter and minimum temperature (from Dalezios, 2015).

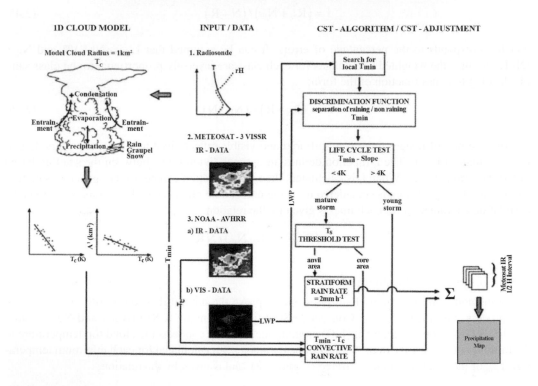

FIGURE 2.9 CST Cloud model (from Bendix, 1997).

There are modifications to the CST method with the inclusion of meso-scale numerical models along with the one-dimensional cloud model, in order to be applied to sub-tropics and mid latitudes beyond the tropics. The cloud models are based on the development and adjustment process, such as the CST model. The performance of cloud models is considered equivalent to that of the empirical models, however, the use of cloud models is expected to increase in the future, as well as their performance. Figure 2.9 presents the structure of the CST cloud model.

2.5.2 Passive Microwave (PMW) Methods

In passive microwave frequencies, particles or water droplets are the main source of attenuation of the returned radiation. As a result, microwave methods are naturally and directly related to rainfall estimation than visible/infrared (VIS/IR) based methods. Specifically, radiation emission from atmospheric particles or droplets results into an increase in the received signal from the satellite sensor, whereas, at the same time, hydrothermal refraction limits the radiation flux. The type and size of the measured hydrometeors depends on the frequency of returned radiation (Table 2.5). In different frequencies, radiometers observe different parts of the rain column. In the other portions of the spectrum, microwave radiation (MW) is absorbed, however, it is diffused from particles or cloud droplets, water vapor and oxygen, which makes it potentially difficult to estimate precipitation based on absorption.

The particles or precipitation droplets interact strongly with microwave radiation (MW) and are calculated by radiometers without the intense infrared voltage. However, the biggest disadvantage is the coarse spatial and temporal resolution. Specifically, the spatial resolution is coarse, since, due to refraction, the ground distinction for a given microwave satellite antenna is restricted, and the temporal frequency is affected by the fact that the microwave sensors follow polar tracks. In addition, the emission of radiation from the sea surface is relatively constant and low ($\varepsilon = 0.4$), resulting in an increase in the radiation emitted by water ($\varepsilon = 0.8$) as recorded by the sensor. In contrast, land surfaces exhibit variable and high radiation emission ($\varepsilon = 0.7$-0.9), such as precipitation, as well as low polarimetry, since the emission depends on surface characteristics, such as vegetation and moisture content. As a result, rainfall over land surfaces increases the flux of returned radiation, although, at the same time, it absorbs radiation, creating significant difficulties to determine the regional extent of rainfall. Therefore, the solution to microwave rainfall estimation methods over land surfaces lies on refraction, and the 85.5 GHz channel of SSM/I is very sensitive to refraction by small particles or droplets.

Significant instruments and sensors used in microwave rainfall estimation are SSM/I, a scanning instrument or sensor that measures microwave radiation in a swath of 1,400 km in four bands (Table 2.5). The radiometer operates in dual polarization (horizontal and vertical) at any frequency, except in the water vapor frequency (22.235 GHz), where it operates only in vertical polarization. Moreover, the TMI instrument in the TRMM system constitutes an evolution of SSM/I at a new frequency of 10.7 GHz of horizontal and vertical polarization and displacement of the water vapor frequency at 21.3 GHz. Finally, the SSMI/S constitutes an upgrade of SSM/I with 24 frequencies between 19 and 183 GHz in simultaneous coverage, a conical scanning system, sounding of up to 70 km and a wider swath of 1,700 km.

TABLE 2.5
Microwave characteristics of the passive SSM/I sensor

Bands SSM/I	Frequency	Spatial Resolution	Possible Uses
1,2	19.35 GHz	25 km	Sea ice and Rain Soil moisture Vegetation cover
3	22.235 GHz	25 km	Vapor in the oceans
4,5	37.0 GHz	25 km	Sea ice and Rain wind on the sea surface extreme rain on the mainland Snow cover
6,7	85.5 GHz	12.5 km	Rainfall Non-rainy clouds in the oceans Sea ice and snow cover

Today's evolution has led to the Advanced Microwave Sounding Unit (AMSU-B). The AMSU-B has 12 frequencies in the oxygen zone between 5-60 GHz, four frequencies in the 183 GHz water vapor zone and four frequency windows at 23.8, 31.4, 50.3 and 89 GHz, respectively. The series of AMSU refers to temperature sounding, but the frequencies can be used for measurements of water, water vapor, liquified cloud, snow cover, sea ice concentration and surface humidification. Based on data from AMSU systems, various rainfall estimation methods have been developed. Reference is given to the following algorithm (Grody et al., 1999), based on SI (Scattering Index) and altered according to instrument characteristics (AMSU-A). The two indicators for land and sea rainfall estimations are respectively:

$$SI_{land} = T_B(23) - T_B(89) \tag{2.51}$$

$$SI_{ocean} = -113.2 + \left[2.41 - 0.0049 \, T_B(23) \right] T_B(23) + 0.454 \, T_B(31) - T_B(89) \tag{2.52}$$

where T_B is the brightness temperature for the corresponding frequency, $SI_{land} \geq 3$ and $SI_{ocean} \geq 9$. After calibration with rainfall data from weather radar, then the precipitation is calculated. In particular, the equation for rain intensity R over land is:

$$R = 0.005 \left[SI_{land} + 18 / 1.3 \right] \tag{2.53}$$

In addition, for the AMSU-B, the frequencies of 89 and 150 GHz are used, respectively, in order to extract higher resolution intensity rain, i.e. 16 km at nadir compared to 48 km of AMSU-A. After data processing and analysis the corresponding refractive indices (SI) become:

$$SI_{land} = \left[42.72 + 0.85 T_B(91) \right] - T_B(150) \tag{2.54}$$

$$SI_{oceaTRMMn} = 0.013 \left\{ T_B(91) + 33.58 \ln \left[300 - T_B(150) \right] - 341.17 \right\} \tag{2.55}$$

Based on the above Equations (2.54) and (2.55) and following the previous procedure based on weather radar, the equation for rain intensity, which corresponds to Equation (2.53), can be calculated.

2.5.3 Active Microwave (AMW) Methods

In active microwave methods, the rainfall estimation is based on satellite radar systems, that is, installation of weather radar on satellites. The most typical application is the USA-Japanese mission, known as the TRMM (Tropical Rainfall Monitoring Mission). The characteristics of the TRMM system are listed in Table 2.6. The TRMM system of SSM/I sensors with a temporal resolution twice

TABLE 2.6

Microwave characteristics of the active SSM/I system

Band	Frequency (GHz)	Polarity	Spatial Resolution (km)
1	10.7	V, H	38.3
2	19.4	V, H	18.4
3	21.3	H	16.5
4	37.0	V, H	9.7
5	85.5	V, H	4.4

a day has shown positive results in rainfall estimation over marine areas, as well as in the identification of mesoscale convective systems and tropical cyclones, but not equally satisfactory over land at least in the early stages. The evolution and upgrade of SSM/I, SSMI/S and SSM/T2, respectively, as well as their combined use, provides a great opportunity to identify and estimate rainfall over continental areas. There are already positive results, however, applied research continues to improve algorithms and to add new sensors.

2.5.4 COMBINED SATELLITE METHODS FOR RAINFALL ESTIMATION

Low-Orbit microwave data can be combined with geosynchronous infrared radiation using various approaches, with significant time-scale variations, from instantaneous to monthly. Today, the general conclusion is that passive microwave (MW) methods are considered superior for instant applications in oceans, whereas infrared (IR) or IR/MW combinations methods show improved monthly rainfall estimates, mainly due to the high temporal frequency of geosynchronous observations. Specifically, with the use of microwave and infrared data, the GPI index has been upgraded to a new UAGPI (Universally Adjusted GOES Precipitation Index) method that provides consistent monthly rainfall estimates across different spatial scales (Xu et al., 1999). Moreover, new instantaneous rainfall estimation strategies have been developed with the combined use of microwave and infrared data (Vicente et al., 1998). In addition, the combined use of MW, VIS and Infrared (IR) data that is available on the same satellite, such as the TMI and VIRS sensor systems on the TRMM system, provide further possibilities for improving the instantaneous estimate (Bauer et al., 1998). Reference should also be made to the new SEVIRI multispectral sensors in MSG or GOES, which have led to new research programs, such as EURAISAT (Levizzani et al., 2002). In addition, the pattern of rainfall algorithms of hybrid MW-PR is the TRMM "day-1" system, which also uses the rain profile approach, giving equal weight to TMI and PR (PR: Precipitation Radar) measurements (Haddad et al., 1997). Lastly, reference is made to the Global Precipitation Mission (GPM), an international program by NASA and NASDA, aiming to measure global rainfall with satisfactory quality, as well as improving weather forecasting and specific components of the global water balance.

2.6 SUMMARY

This chapter began with a presentation of remote sensing precipitation concepts. Then, radar-based quantitative precipitation estimation (QPE) was considered, which includes conventional radar-based QPE and improved rain rate retrieval based on polarimetric radar.

This was followed by snowfall and hail estimation. The next section involved methods of radar-based Quantitative Precipitation Forecasting (QPF). Finally, satellite-based estimation of precipitation was presented, which includes Visible (VIS) and Infrared (IR) methods, Passive microwave (PMW) methods, Active microwave (AMW) methods, as well as Combined satellite methods.

REFERENCES

Adler, R.F., and Negri, A.J., 1988. A satellite infrared technique to estimate tropical convective and stratiform rainfall. *Journal of Applied Meteorology*, 27, 30–51.

Amburn, S.A., and Wolf, P.L., 1997. VIL density as a hail indicator. *Weather and Forecasting*, 12, 473–478.

Andersson, T., 1991. An Advective Model for Probability Nowcasts of Accumulated Precipitation using Radar, pp. 325–330, in *Hydrological Applications of Weather Radar*. Editors: I.D. Cluckie and C.G. Collier. Ellis Horwood, England.

Arkin, P.A., and Meisner, B.N., 1987. The relationship between large-scale convective rainfall and cold cloud over the western hemisphere during 1982–84. *Monthly Weather Review*, 115, 1, 51–74.

Austin, G.L., and Bellon, A., 1974. The use of digital weather radar records for short-term precipitation forecasting. *Quarterly Journal of the Royal Meteorological Society*, 100, 658–664.

Barrett, E.C., and Martin, D.W., 1981. *The Use of Satellite Data in Rainfall Monitoring*. Academic Press, 340.

Battan, L.J., 1973. *Radar Observations of the Atmosphere*. University of Chicago Press, 324.

Bauer, P., Schanz, L., Bennartz, R., and Schlüssel, P., 1998. Outlook for combined TMI-VIRS algorithms for TRMM: Lessons learned from the PIP and AIP projects. *Journal of Atmospheric Sciences*, 55, 1714–1729.

Beck, J., and Chau, J.L., (eds.), 2012. *Doppler Radar Observations-Weather Radar, Wind Profiler, Ionospheric Radar, and other advanced applications*. InTech Publications, Rijeka, Croatia, 470.

Bendix, J., 1997. Adjustment of the Convective-Stratiform Technique (CST) to estimate 1991/93 El Nino rainfall distribution in Equador and Peru by means of Meteo-3 IR data. *International Journal of Remote Sensing*, 18, 6, 1387–1394.

Berne, A., and Krajewski, W.F., 2013. Radar for hydrology: Unfulfilled promise or unrecognized potential? *Advances in Water Resources*, 51, 357–366.

Blackmer, R.H., Duda, R.O., and Reboh, E., 1973. *Application of Pattern Recognition Techniques to Digitized Weather Radar Data* Stanford Research Institute, Menlo Park, CA.

Brandes, E.A., 1975. Optimizing rainfall estimates with the aid of radar. *Journal of Applied Meteorology*, 14, 1339–1345.

Brandes, E.A., Zhang, G., and Vivekanandan, J., 2002. Experiments in rainfall estimation with a polarimetric radar in a subtropical environment. *Journal of Applied Meteorology*, 41, 674–685.

Bremaud, P.J., and Pointin, Y.B., 1993. Forecasting heavy rain from cell motion using radar. *Journal of Hydrology*, 142, 373–389.

Bringi, V.N., and Chandrasekar, V. 2004. *Polarimetric Doppler Weather Radar: Principles and Applications*, 2nd edn. Cambridge Univ. Press, 664.

Bringi, V.N., Tang, T., and Chandrasekar, V., 2004. Evaluation of a new polarimetrically based Z–R relation. *Journal of Atmospheric and Oceanic Technology*, 21, 612–623.

Browning, K.A., Collier, C.G., Larke, P.R., Menmuir, P., Monk, G.A., and Owens, R.G., 1982. On the forecasting of frontal rain using a weather radar network. *Monthly Weather Review*, 110, 534–552.

Burgess, D.W., Wood, V., and Brown, R., 2001. *Mesocyclone evolution statistics. 12th Conference on Severe Local Storms*, San Antonio, Texas, 422.

Chandrasekar, V., Lim, S., Bharadwaj, N., Li, W., McLaughlin, D., Bringi, V.N., and Gorgucci, E., 2004. *Principles of networked weather radar operation at attenuating frequencies, 3rd European Conference on Radar Meteorology and Hydrology*, Visby, Sweden, 67–73.

Chandrasekar, V., Hou, A., Smith, E., Bringi, V.N., Rutledge, S.A., Gorgucci, E., Petersen, W.A., and Jackson, G., 2008. Potential role of dual-polarization radar in the validation of satellite precipitation measurements: Rationale and opportunities. *Bulletin of the American Meteorological Society*, 89, 1127–1145.

Cifelli, R., and Chandrasekar, V. 2010. Dual-Polarization Radar Rainfall Estimation, pp.105–125, in *Rainfall: State of the Science*. Editors: F.Y. Testik and M. Gebremichael. AGU, Washington.

Crawford, K.C., 1979. Consideration for the design of a hydrologic data network using multivariate sensors. *Water Resources Research*, 15, 6, 1752–1762.

Creutin, J.D., and Obled, C., 1982. Objective analysis and mapping techniques for rainfall fields: An objective comparison. *Water Resources Research*, 18, 2, 413–431.

Dalezios, N.R., 1988. Objective rainfall evaluation in radar hydrology. *Journal of Water Resources Planning and Management (JWRMDS), ASCE*, 114, 5, 531–546.

Dalezios, N.R., 1990. Digital processing of weather radar signals for rainfall estimation. *International Journal of Remote Sensing*, 11, 9, 1561–1569.

Dalezios, N.R., 2015. *Agrometeorology: Analysis and Simulation*. KALLIPOS: Libraries of Hellenic Universities, 481

Dalezios, N.R., and Kouwen, N., 1990. Radar signal interpretation in warm season rainstorms. *Nordic Hydrology*, 21, 1, 47–64.

Dalezios, N.R. and Papamanolis, N.K., 1991. Objective assessment of instability indices for operational hail forecasting in Greece. *Meteorology and Atmospheric Physics*, 45, 87–100.

Delrieu, G., Andrieu, H., and Creutin, J.D., 2000. Quantification of path-integrated attenuation for X- and C-band weather radar systems operating in Mediterranean heavy rainfall. *Journal of Applied Meteorology*, 39, 840–850.

Delrieu, G., Braud, I., Berne, A., Borga, M., Boudevillain, B., and Fabry, F., 2009. Weather radar and hydrology preface. *Advances in Water Resources*, 32, 7, 969–974.

Depue, T.K., Kennedy, P.C., and Rutledge, S.A., 2007. Performance of the hail differential reflectivity (HDR) polarimetric radar hail indicator. *Journal of Applied Meteorology and Climatology*, 46, 1290–1301.

Dixon, M., and Wiener, G., 1993. TITAN: Thunderstorm identification, Tracking, Analysis. and Now-casting-A Radar-based Methodology. *Journal of Atmospheric and Oceanic Technology*, 10, 785–797.

Domenikiotis, C., Spiliotopoulos, M., Galakou, E., and Dalezios, N.R. 2005. *Assessment of Cold Cloud Duration (CCD) Methodology for Rainfall Estimation in Central Greece. Proc., Intern. Symp. On "GIS and Remote Sensing: Environmental Applications"*. Volos Greece, 185–194.

Donaldson, N., Pierce, C., Sleigh, M., Seed, A., and Saxen, T., 2001. *Comparison of forecasts of windspread precipitation during the Sydney 2000 Forecast Project. 30th International Conference on Radar Meteorology*, Munich, Germany, 503.

Doviak, R., and Zrnic, D., 2006. *Doppler Radar and Weather Observations*, 2nd ed. Dover Publications,

Droegemeier, K., Smith, J., Businger, S., et al., 2000. Hydrological aspects of weather prediction and flood warnings: Report of the Ninth Prospectus Development Team of the US Weather Research Program. *Bulletin of the American Meteorological Society*, 81, 11, 2665–2680.

du Vachat, R., Thomas, P., Bocrie, E., Monceau, G., Cosentino, P., Senesi, S., Tzanos, D., and Boichard, J.-L., 1995. *The precipitation nowcast scheme in the Aspic project*. In *Proc. Second European Conf. on Applications of Meteorology*, Toulouse, France, 29–32.

Einfalt, T., Denoeux, T., and Jaquet, G., 1991. The development of the SCOUT 11.0 rainfall forecasting method, pp. 359–367, in *Hydrological Applications of Weather Radar*. Editors: I.D. Cluckie and C.G. Collier, Ellis Horwood, England

Fabry, F., 2015. *Radar Meteorology: Principles and Practices*. Cambridge University Press, 256.

Follansbee, W.A., 1973. Estimation of Average Daily Rainfall from Satellite Cloud Photographs. NOAA Tech. Memo NESS 44, Washington D.C., 39pp.

Fox, N.I., Sleigh, M.W., and Pierce, C.E., 2001. Forecast demonstration project Sydney 2000 - Part I: An overview of the project and the participating systems. *Weather*, 56, 397–404.

French, M.N., and Krajewski, W.F., 1994. A model for real-time quantitative rainfall forecasting using remote sensing. Part I. Formulation. *Water Resources Research*, 30, 1075–1083.

Fukao, S., and Hamazu, K., 2014. *Radar for Meteorological and Atmospheric Observations*. Springer, Japan, 537.

Gado, T.A., Hsu, K., and Sorooshian, S., 2017. Rainfall frequency analysis for ungauged sites using satellite precipitation products. *Journal of Hydrology*, 554, 646–655.

Gebremichael, M., and Hossain, F. 2010. *Satellite Rainfall Applications for Surface Hydrology*. Springer, New York, 327.

Giangrande, S.E., and Ryzhkov, A.V., 2005. Calibration of dual-polarization radar in the presence of partial beam blockage. *Journal of Atmospheric and Oceanic Technology*, 22, 1156–1166.

Golding, B.W., 1998. Nimrod: A system for generating automated very short-range forecasts. *Meteorological Applications*, 5, 1–16.

Gourley, J.J., Kaney, B., and Maddox, R.A., 2003. *Evaluating the calibrations of radars: A software approach. 31st International Conference on Radar Meteorology*, Seattle, WA, American Meteorological Society, 459–462.

Gourley, J. J., Illingworth, A.J., and Tabary, P., 2009. Absolute calibration of radar reflectivity using redundancy of polarization observations and implied constraints on drop shapes. *Journal of Atmospheric and Ocean Technology*, 26, 689–703.

Gourley, J.J., Maddox, R.A., Howard, K.W., and Burges, D.W., 2010. Impacts of polarimetric radar observations on hydrologic simulation. *Journal of Hydrometeorology*, 11, 3, 781–796.

Greene, D.R., Hudlow, M.D., and Johnson, E.R., 1980. *A test of some objective analysis procedures for merging radar and raingage data. 19th Conference on Radar Meteorology*, Miami, FL, 470–479.

Griffith, C.G., Woodley, W.L., Grube, P.G., Martin, D.W., Stout, J., and Sikdar, D.N., 1978. Rain estimation from geosynchronous satellite imagery – Visible and infrared studies. *Monthly Weather Review*, 106, 1153–1171.

Grody, N.C., Weng, F., and Ferraro, R.R., 1999. Application of AMSU for obtaining hydrological parameters, pp. 339–351, in *Microwave Radiometry and Remote Sensing of the Environment*. Editors: P. Pampaloni and S. Paloscia. VSP International Science Publishers, Utrecht (The Netherlands).

Haddad, Z., Smith, E.A., Kummerow, C.D., Iguchi, T., Farrar, M., Darden, S., Alves, M., and Olson, W., 1997. The TRMM 'Day-1' radar/radiometer combined rain-profile algorithm. *Journal of the Meteorological Society, Japan*, 75, 799–808.

Hand, W.H., and Conway, B., 1995. An object-oriented approach to now-casting showers. *Weather and Forecasting*, 10, 327–341.

Handwerker, J., 2001. Cell Tracking with TRACE3D – a new algorithm. Atmospheric Research, accepted.

Hane, C.E., Bluestein, H.B., Crawford, T.M., Baldwin, M.E., and Rabin, R.M., 1997. Severe thunderstorm development in relation to along-dryline variability: A case study. *Monthly Weather Review*, 125, 231–251.

Hong, Y., and Gourley, J.J., 2015. *Radar Hydrology: Principles, Models and Applications*. Taylor and Francis Group, Boca Raton, FL USA, 174.

Hsu, K.L., Gao, X.G., Sorooshian, S., and Gupta, H.V., 1997. Precipitation estimation from remotely sensed information using artificial neural networks. *Journal of Applied Meteorology and Climatology*, 36, 1176–1190.

Hubbert, J.C., Dixon, M., and Ellis, S.M., 2009. Weather radar ground clutter. Part II: Real-time identification and filtering. *Journal of Atmospheric and Oceanic Technology*, 26, 1181–1197.

Huffman, G.J., Adler, R.F., Bolvin, D.T., Gu, G.J., Nelkin, E.J., Bowman, K.P., et al., 2007. The TRMM multi-satellite precipitation analysis (TMPA): quasi-global multiyear, combined-sensor precipitation estimates at fine scales. *Journal of Hydrometeoroly*, 8, 3, 8–55.

Huffman, G.J., Bolvin, D.T., Braithwaite, D., Hsu, K., Joyce, R., Kidd, C., Nelkin, E.J., and Xie, P., 2014. *NASA Global Precipitation Measurement (GPM) Integrated Multi-satellitE Retrievals for GPM (IMERG)*, Algorithm Theoretical Basis Document (ATBD), NASA/GSFC, Greenbelt, MD, USA, 2014.

Jiang, S., Ren, L., Hong, Y., Yong, B., Yang, X., Yuan, F., and Ma, M., 2012. Comprehensive evaluation of multi-satellite precipitation products with a dense rain gauge network and optimally merging their simulated hydrological flows using the Bayesian model averaging method. *Journal of Hydrology*, 452–453, 213–225.

Johnson, J.T., MacKeen, P.L., Witt, A., Mitchell, E.D., Stumpf, G.J., Eilts, M.D., and Thomas, K.W., 1998. The storm cell identification and tracking algorithm: An enhanced WSR-88D algorithm. *Weather and Forecasting*, 13, 263–276.

Joyce, R.J., Janowiak, J.E., Arkin, P.A., and Xie, P., CMORPH: A method that produces global precipitation estimates from passive microwave and infrared data at high spatial and temporal resolution. *Journal of Hydrometeoroly*, 5, 487–503.

Kidd, C., and Levizzani, V., 2011. Status of satellite precipitation retrievals. *Hydrology and Earth System Sciences*, 15, 4, 1109–1116.

Kidder, S.Q., and Vonder Haar, T.H., 1995. *Satellite Meteorology: An Introduction*. Academic Press, 466.

Kirstetter, P.E., Viltard, N., and Gosset, M., 2013. An error model for instantaneous satellite rainfall estimates: Evaluation of BRAIN-TMI over West Africa. *Quarterly Journal of the Royal Meteorological Society*, 139, 673, 894–911.

Krauss, T., and Dalezios, N.R., 2017. Storms, pp. 95–136, in *Environmental Hazards Methodologies for Risk Assessment and Management*. Editor: N.R. Dalezios. IWA, London, UK

Kummerow, C., Barnes, W., Kozu, T., Shiue, J., and Simpson, J., 1998. The Tropical Rainfall Measuring Mission (TRMM) sensor package. *Journal of Atmospheric and Oceanic Technology*, 15, 809–817.

Kunitsugu, M., Makihara, Y., and Shinpo, A., 2001. *Now-casting system in JMA, 5th International Symposium on Hydrological Application of Weather Radar "Radar Hydrology"*. Proceedings, Kyoto, Japan, 267

Lapczak, S., AIdcroft, E., Stanley-Jones, M., Scott, J., Joε, P., Van Rijn P., Falla M., Gagne A., Ford P., Reynolds, K. and Hudak, D., 1999. *The Canadian National Radar Project, 29th International Conference on Radar Meteorology*, Montreal, Canada, 267327.

Larson, L.W., and Peck, E.L., 1974. Accuracy of precipitation measurements for hydrologic modeling. *Water Resources Research*, 10, 4, 857–863.

Lei, L., Zhang, G., Doviak, R.J., et al. 2012. Multilag correlation estimators for polarimetric radar measurements in the presence of noise. *Journal of Atmospheric and Oceanic Technology*, 29, 772–795.

Levizzani, V., and Cattani, E., 2019. Satellite remote sensing of precipitation and the terrestrial water cycle in a changing climate. *Remote Sensing*, 11, 2301, doi: 10.3390/rs1192301

Levizzani, V., Amorati, R., and Meneguzzo, F., 2002. A Review of Satellite-based Rainfall Estimation Methods. EC Research Project MUSIC. Tech. Report, 66p.

Lim, S., Chandrasekar, V., and Bringi, V., 2005. Hydrometeor classification system using dual-polarization radar measurements: Model improvements and in situ verification. *IEEE Transactions on Geoscience and Remote Sensing*, 43, 792–801.

Liu, X., Yang, T., Hsu, K., Liu, C., and Sorooshian, S., 2017. Evaluating the streamflow simulation capability of PERSIANN-CDR daily rainfall products in two river basins on the Tibetan Plateau. *Hydrology and Earth System Sciences*, 21, 1, 169–181.

Lovejoy, S., and Austin, G.L., 1979. The delineation of rain areas from visible and IR satellite data from GATE and mid-latitudes. *Atmosphere-Ocean*, 17, 77–92.

Li, L., Schmid, W., and Joss, J., 1995. Now-casting of motion and growth of precipitation with radar over a complex orography. *Journal of applied Meteorology and Climatology*, 34, 1286–1300.

Marshall, J.S., and Palmer W.M.K., 1948. The distribution of raindrops with size. *Journal of Meteorology*, 5, 165–166.

Matrosov, S.Y., Clark, K.A., Martner, B.E. et al., 2002. X-band polarimetric radar measurements of rainfall. *Journal of Applied Meteorology*, 41, 941–952.

May, P., and Joe, P., 2001. *The production of high quality Doppler velocity fields for dual PRT weather radar. 30th International Conference on Radar Meteorology*, Munich, Germany, 286.

Mecklenburg, S., Joss, J., and Schmid, W., 2000. Improving the now-casting of precipitation in an Alpine region with an enhanced radar echo tracking algorithm. *Journal of Hydrology*, 239, 46–68.

Mellor D., Sheffield J., O'Connell, P.E., and Metcalfe, A.V., 2000. A stochastic space time rainfall forecasting system for real time flow forecasting I: Development of MTB conditional rainfall scenario generator. *Hydrology and Earth System Sciences*, 4, 603–615.

Michaelides, S., 2008. *Precipitation: Advances in Measurement, Estimation and Prediction*. Springer, 540

Michaelides, S., Levizzani, V., Anagnostou, E., Bauer, P., Kasparis, T., and Lane, J.E., 2009. Precipitation: Measurement, remote sensing, climatology and modeling. *Atmospheric Research*, 94, 4, 512–533.

Moisseev, D.N., and Chandrasekar, V., 2009. Polarimetric spectral filter for adaptive clutter and noise suppression. *Journal of Atmospheric. and Oceanic Technology*, 26, 215–228.

Moore, R.J., Hotchkiss, D.A., Jones, D.A., and Black, K.B., 1994. Local rainfall forecasting using weather radar: The London case study, in *Advances in Radar Hydrology*. Editors: M.E. Almeida-Teixeira, R. Fantechi, R.J. Moore and V.M. Silva. European Commission, 235–241.

Moriyama, T., and Muneo, H., 2000. Quantitative precipitation forecasting using neural networks, http://www.unesco.org.uy/phi/libros/radar/art24.html.

Mueller, C.K., Wilson, J.W., and Crook, N.A., 1993. The utility of sounding and mesonet data thunderstorm initiation. *Weather and Forecasting*, 8, 132–146.

Negri, A.J., Adler, R.F., and Wetzel, P.J., 1984. Rain estimation from satellite: An examination of the Griffith-Woodley technique. *Journal of Applied Meteorology and Climatology*, 23, 102–116.

Nguyen, P., Ombadi, M., Sorooshian, S., Hsu, K., AghaKouchak, A., Braithwaite, D., Ashouri H., and Thorstensen, A.R., 2018. The PERSIANN family of global satellite precipitation data: a review and evaluation of products. *Hydrology and Earth System Sciences*, 22, 5801–5816.

Park, H.S., Ryzhkov, A.V., Zrnić, D.S., and Kim, K-E., 2009. The hydrometeor classification algorithm for the polarimetric WSR-88D: Description and application to an MCS. *Weather and Forecasting*, 24, 730–748.

Pierce, C.E., Collier, C.G.. Hardaker, P.J., and Haggett, C.M., 2000. GANDOLF: a system for generating automated nowcasts of convective precipitation. *Meteorological Applications*, 7, 341–360

Rakhecha, P.R., and Singh, V.P., 2009. *Applied Hydrometeorology*. Springer, Dordrecht, The Netherlands, 384.

Richards, M.A., Scheer, J.A., Holm, W.A. (eds.), 2010. *Principles of Modern Radar*. SciTech Publishing, 962.

Rigby, R.A., and Stasinopoulos, D.M., 2005. Generalized additive models for location, *scale and shape. Journal of the Royal Statistical Society: Series C (Applied Statistics)*, 54, 3, 507–554.

Rigo, T., and Carmen Llasat, M., 2016. Forecasting hailfall using parameters for convective cells identified by radar. *Atmospheric Research*, 169, 366–376.

Rinehart, R.E., and Garvey, E.T., 1978. Three-dimensional storm motion detection by conventional weather radar. *Nature*, 273, 287–289.

Roberts, R.D., 1997. *Detecting and forecasting cumulus cloud growth using radar and multi-spectral satellite data. Preprints, 28th Conference on Radar Meteorology*, Austin, Texas, 408–409.

Rosenfeld, D., and Ulbrich, C.W., 2003. Cloud microphysical properties, processes, and rainfall estimation opportunities. *American Meteorological Society Meteorological Monographs*, 30, 237.

Ryzhkov, A.V., and Zrnic, D.S., 2019. *Radar Polarimetry for Weather Observations*. Springer, 504.

Ryzhkov, A.V., Giangrande, S.E., and Schuur, T.J., 2005. Rainfall estimation with a polarimetric prototype of WSR-88D. *Journal of Applied Meteorology and Climatology*, 44, 502–515.

Saxen, T.R., Mueller C.K., Jumeson, T. C., and Hatfield, E., 1999. *Determining key parameters for forecasting thunderstorms at white sand missile range. 29th Conference on Radar Meteorology*, Montreal, Canada, American Meteorological Society, 9–12.

Scofield, R.A., and Oliver, V.J., 1977. A scheme for estimating convective rainfall from satellite imagery. NOAA Tech. Memo. NESS, 86, Dept. of Commerce, Washington, DC, 47pp.

Seed, A.W. and Keenan, T., 2001. *A dynamic and spatial scaling approach to advection forecasting. 30th International Conference on Radar Meteorology*, Munich, Germany, 492.

Sekhon, R.S., and Srivastava, R.C., 1970. Snow-size spectra and radar reflectivity. *Journal of Atmospheric Science*, 27, 299–307.

Seliga, T.A., and Bringi, V.N., 1976. Potential use of radar differential reflectivity measurements at orthogonal polarizations for measuring precipitation. *Journal of Applied Meteorology and Climatology*, 15, 69–76.

Sene, K., 2016. *Hydrometeorology*, 2nd ed. Springer, New York, 427.

Seo, D-J., and Smith, J.A., 1992. Radar-based short-term rainfall prediction. *Journal of Hydrology*, 131, 341–367.

Seo, D-J., Seed, A., and Delriew, G., 2010. Radar and Multisensor Rainfall Estimation for Hydrologic Applications, pp. 79–104, in *Rainfall: State of the Science*, . Editors: F.Y. Testik and M. Gebremichael. AGU, Washington, DC

Siggia, A., and Passarelli, R., 2004. *Gaussian model adaptive processing (GMAP) for improved ground clutter cancellation and moment calculation. Proceedings of 3rd European Conference on Radar in Meteorology and Hydrology*, Visby.

Simpson, J., Adler, R.F., and North, G.R., 1987. A proposed Tropical Rainfall Measurement Mission (TRMM) satellite. *Bulletin of the American Meteorological Society*, 69, 278–295

Steinacker, R., Dorninger, M., Wolfelmaier, F., and Krennert, T., 1999. Automatic tracking of convective cells and cell complexes from lighting and radar data. *Meteorology and Atmospheric Physics*, 73, 101–110.

Stout, J.E., Martin, D.W., and Sikdar, D.N., 1979. Estimating GATE rainfall with geosynchronous satellite images. *Monthly Weather Review*, 107, 585–598.

Sun, O., Miao, C., Duan, Q., Ashouri, H., Sorooshian, S., and Hsu, K.-L., 2017. *A Review of Global Precipitation Data Sets: Data Sources, Estimation, and Intercomparisons*. AGU Publications, 1–29.

Tabary, P., 2007. The new French operational radar rainfall product. Part I: Methodology. *Weather and Forecasting*, 22, 3, 393–408.

Testik, F.Y., and Gebremichael, M., 2010. *Rainfall: State of Science*. AGU, Washington, DC, USA, 287.

Thielen, J., Boudevillain B., and Andrieu, H., 2000. A radar data based short-term rainfall prediction model for urban areas – a simulation using meso-scale meteorological modelling. *Journal of Hydrology*, 239, 97–114.

Torlaschi, E., Humphries, R.G., and Barge, B.L., 1984. Circular polarization for precipitation measurement. *Radio Science*, 19, 193–200.

Torres, S.M., and Zrnić, D.S., 1999. Ground clutter canceling with a regression filter. *Journal of Atmospheric and Oceanic Technology*, 16, 1364–1372.

Treloar, A., 1998. *Vertically integrated radar reflectivity as an indicator of hail size in the greater Sydney region of Australia. 9th Conference on Severe Local Storms*, Minneapolis, MN, 489.

Tsonis A., and Austin, G.L., 1981. Evaluation of extrapolation techniques for the short-term prediction of rain amounts. *Atmosphere-Ocean, Toronto, Canada*, 19, 54–65.

Turk, J.T., Mostovoy, G.V., and Anantharaj, V., 2010. The NRL Blend High Resolution Precipitation Product and its Application to Land Surface Hydrology, pp. 85–104, in *Satellite Rainfall Applications for Surface Hydrology*. Editors: M. Gebremichael and F. Hossain. Springer, Dordrecht, the Netherlands

Vicente, G.A., Scofield, R.A., and Menzel, W.P., 1998. The operational GOES infrared rainfall estimation technique. *Bulletin of the American Meteorological Society*, 79, 1883–1898.

Vivekanandan, J., Ellis, S.M., Oye, R., et al. 1999. Cloud microphysics retrieval using S-band dual-polarization radar measurements. *Bulletin of the American Meteorological Society*, 80, 381–388.

Waldvogel, A., Federer, B., and Grimm, P., 1979. Criteria for the detection of hail cells. *Journal of Applied Meteorology* 25, 1521–1525.

Wilson, J.W., and Brandes, E.A., 1979. Radar measurement of rainfall – a summary. *Bulletin of the American Meteorological Society*, 60, 9, 1048–1058.

Wilson, J.W., Crook, N.A., Mueller, C.K., Sun, J., and Dixon, M., 1998. Now-casting Thunderstorm: a status report. *Bulleting of the American Meteorological Society*, 79, 2079–2099.

Wolf, D.E., Hall, D.J., and Endlich, R.M., 1977. Experiment in automatic cloud tracking using SMS-GOES data. *Journal of Applied Meteorology*, 16, 1219–1230.

Xu, L., Gao, X., Sorooshian, S., Arkin, P.A., and Imam, B., 1999. A microwave infrared threshold technique to improve the GOES precipitation index. *Journal of Applied Meteorology and Climatology*, 38, 569–579.

Yilmaz, K.K., Hogue, T.S., Hsu, K.L., Sorooshian, S., Gupta, H.V., and Wagener, T., 2005. Intercomparison of rain gauge, radar, and satellite-based precipitation estimates with emphasis on hydrologic forecasting. *Journal of Hydrometeorology*, 6, 4, 497–517.

Zawadzki, I.I., 1984. *Factors affecting the precision of radar measurements of rain. 22nd Conference on Radar Meteorology*, Zurich, Switzerland, AMS, 251–256.

Zrnić, D.S., Melnikov, V.M., and Carter, J. K., 2006. Calibrating differential reflectivity on the WSR-88D. *Journal of Atmospheric and Oceanic Technology*, 23, 944–951.

3 Meteorology and Climate

3.1 INTRODUCTION: REMOTE SENSING CONCEPTS OF METEOROLOGY AND CLIMATE

The operational use of meteorological data requires both a method of measurement or sensing and a method of collecting the data at a single point for computer processing. Meteorological satellites offer an efficient method of rapidly collecting data, and by 1970 satellite data were used routinely to provide information on a global scale. Nevertheless, accurate storm position, size of storm areas, storm intensities or effective integration of this information into numerical analyses are still subjects of current research.

Several atmospheric remote-probing techniques have been considered due to their potential to advance observational capacity and to significantly increase knowledge of atmospheric structures and processes. These techniques include: (1) **Lidar** (Light detection and ranging), where a pulse is transmitted and the backscattered signal is measured as a function of time or range. Intensity backscatter measurements through Lidar can be used to derive air density profiles above 30 km and to map aerosol distribution. (2) **Radar** probing measurements of reflectivity and Doppler shift also include attenuation, measure of total water content in storm, and polarization effects, indicating particle shape and thereby the phase of water. Weather Radar is covered in Chapter 2 (3) **Infrared** optical techniques are useful to measure the Earth's temperature. Cloud tops, the earth surface and other objects radiate at the wavelengths of 10 to 20 μm (in IR) and thus, indicate their temperature. The remote sensing of pollutants has been carried out in IR to a lesser extent. (4) **Microwave** radiometer probing is a passive method of probing, where at 0.5 cm wavelength may reveal clear-air turbulence, and at wavelengths longer than 1.5 cm can provide temperature profiles. (5) **Acoustic** techniques are useful for measuring close to the atmospheric boundary layer. (6) **Sferics** radio disturbances generated by lightning discharges and modified during their propagation toward a measuring station are composed of emissions over a period of about one second. The direction to the source of the emissions and the intensity of the signals have been used to observe regions of thunderstorm activity. (7) **Cross-beam correlation** technique has been developed to investigate turbulent flow characteristics. (8) **Microwave line-of-sight** propagation techniques can result in inferences of atmospheric properties, such as the total oxygen and the total water vapor, the amplitude of the refractive-index fluctuations and the spectral distribution of reflectivity. (9) **Optical and IR line-of-sight** propagation at micrometer wavelengths can infer atmospheric parameters, such as average path temperature, temperature gradient, turbulence, temperature and wind structure, liquid water content and integrated density of a gaseous constituent. The aforementioned remote sensing techniques have great potential, but they do not cover all possibilities, since the total electromagnetic spectrum may be used by remote sensors of meteorology. There are already many satellite-borne sensors operating in selected spectral bands.

In Chapter 4, remote sensing methods are presented for measuring and estimating basic meteorological parameters and concepts, such as temperature, wind, atmospheric physics and clouds, atmospheric chemistry, weather systems, including fronts, storms, hurricanes, cyclones, tornadoes and lightning, as well as climatic analysis and climate change.

3.2 TEMPERATURE

Atmospheric temperature at different altitudes, along with land surface temperature (LST) and sea surface temperature (SST) contribute to satellite temperature measurements through radiometric measurements by satellites. These temperature measurements can significantly contribute to operational meteorology and be used in several applications, such as to monitor the El Niño-Southern Oscillation, locate weather fronts, assess the urban heat islands, determine the strength of tropical cyclones and monitor the global climate, among others. (Moreover, thermal imaging derived from meteorological satellites can be used to identify and detect industrial hot spots, volcanoes or wildfires. Indeed, meteorological satellites measure radiances in various wavelength bands and not temperature directly. Specifically, since 1978 National Oceanic and Atmospheric Administration (NOAA) polar orbiting satellites have measured the intensity of upwelling microwave radiation from atmospheric oxygen through microwave sounding units (MSUs), which is related to the temperature of broad vertical atmospheric layers.

In this section, remote sensing temperature applications are considered, such as Land Surface Temperature (LST), Sea Surface Temperature (SST), Atmospheric Temperature, temperature mapping and monitoring, El Niño and La Niña.

3.2.1 LAND SURFACE TEMPERATURE (LST)

In this study, surface temperature retrievals are used for the calculation of LST from Landsat 8 Operational Land Imager (OLI) Thermal Infrared Sensor (TIRS) of 2015 data (Babita et al., 2018). At first, OLI and TIRS band data are converted to radiance using the radiance rescaling factors provided in the metadata file by using Equation (3.1):

$$L_\lambda = M_L\, Q_{cal} + A_L \tag{3.1}$$

where L_λ temperature of atmosphere spectral radiance, M_L band-specific multiplicative rescaling factor from the metadata (RADIENCE MULTI BAND X, where X is the band number), Q_{cal} is the quantized and calibrated standard product pixel values (DN) and A_L is the band-specific additive rescaling factor from the metadata (RADIENCE_ADD_BAND_X, where X is the band number). In the second step, band data are converted into reflectance using reflectance rescaling coefficients provided in the product metadata file (MTL file). Equation 3.2 is used to convert DN values to TOA reflectance for OLI data:

$$\rho\lambda' = M\rho\, Q_{cal} + A\rho \tag{3.2}$$

where $\rho\lambda'$ is the reflectance, $M\rho$ is the band-specific multiplicative rescaling factor from the metadata (REFLECTANCE_MULT_BAND_x, where x is the band number), Q_{cal} is the quantized and calibrated standard product pixel values (DN) and $A\rho$ is the band-specific additive rescaling factor from the metadata (REFLECTANCE_ADD_BAND_x, where x is the band number). Thirdly, TIRS band data are converted from spectral radiance to brightness temperature using the thermal constants provided in the metadata file using Equation (3.3):

$$BT = K2\,/\ln\left[\left(K1\,/\,L_\lambda\right) + 1\right] \tag{3.3}$$

where BT is Top of Atmosphere Brightness Temperature (deg K), L_λ is the residence, K1 (Wcm^{-2} sr^{-1} μm^{-1}) and K2 (deg K) are pre-launch band-specific thermal conversion (calibration) constants from the metadata. Table 3.1 presents the Landsat 8 TIRS thermal band calibration constants.

TABLE 3.1
Landsat 8 TIRS Thermal Band Calibration
Constants (Source: Zhang et al., 2016)

	Band 10	Band 11
K1 (Wcm 2sr 1 μm^{-1})	774.89	480.89
K2 (deg K)	1321.08	1201.14

The equation for conversion from brightness temperature to land surface temperature follows (Weng et al., 2004; Zareie et al., 2016).

$$LST = BT / \left[1 + \left(\lambda \times BT / \rho \right) \log \varepsilon \right] \tag{3.4}$$

where LST is the Land surface temperature (deg K), λ is the wavelength of emitted radiance (11.5 μm), ε is the Land surface emissivity (typically 0.95), ρ h $*$ c/σ = 1.438*10^{-2} mK (σ = Boltzmann constant = 1.38*10^{-23} J/K, h = Planck's constant = 6.626*10^{-34}Js, c = velocity of light = 2.998*108 m/s). Finally, the land surface temperature in degrees Kelvin was converted to Celsius by subtracting from 273.15 (Equation 3.5):

$$T°C = T(K) - 273.15 \tag{3.5}$$

The preliminary interpretation of the Landsat imageries involved the classification of the study area's land cover into five classes, namely water bodies, mixed forests, wetlands, built-up areas and bare land. The next step consisted of a training class selection through a step-by-step process based on the spectral signatures of each class and ancillary data, which was conducted with ENVI 5.0 software. Then, the parallelepiped method was used for supervised classification of the imageries. After classification, the feature classes were transferred to ArcGIS 10.4 for editing, elimination of spurious clusters and refinement of the output. Figure 3.1 shows the land cover and land surface temperature maps for the same time period in 2015, respectively.

Empirical method for LST. The calculation of LST is complicated for three main reasons. First, surface emission concerns rocky, bare soil, sandy deserts, as well as areas with vegetation cover. The surface emissivity exhibits variability. Consequently, for each pixel, there are two unknowns, the temperature and the transmittance capability, while the sea surface has only the temperature. Second, the surface temperature is not homogeneous in a pixel relative to the corresponding sea level. Even vegetation-free soil surfaces, except for some deserts, are often very heterogeneous. On a surface with vegetation it is difficult to determine the surface temperature. Third, the difference between the surface soil temperature and the air temperature near the ground is greater on land than at sea. The air temperatures just above the sea are quite different from the surface temperatures of the land and it is difficult to use the split-window method, which is widely applied at sea level.

The basis of split-window techniques for calculating heat is the difference in atmospheric absorption at two different wavelengths. Various separation methods have been developed, where the surface temperature, Ts, is expressed in relation to a linear combination of radiometric temperatures in channels 4 (11 μm) and 5 (12 μm) of AVHRR/2 (Table 3.2). Becker and Li (1990) have introduced in the split-window method a correction for the moisture content in the atmosphere, W (gr / m²) along the path between the target and the sensor and the angle of view θ. The coefficients A_0, A_1, A_2 have been calculated with Lowtran-7 software. The result is the relationship:

$$T_s = A_0 + P \frac{T_4 + T_5}{2} + M \frac{T_4 + T_5}{2} \tag{3.6}$$

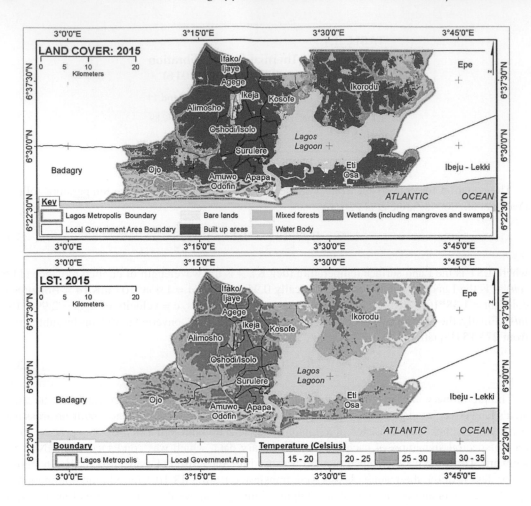

FIGURE 3.1 Lagos metropolis 2015 – Land Cover map (top); Land Surface Temperature map (bottom) (from Obiefuna et al., 2018).

where P = 1.0290 + (0.2106 - 0.0307 cosθW) (1 - ε$_4$) - (0.3696 – 0.0737W)(ε$_4$ - ε$_s$), M = 4.25 + 0.56W + (3.41 + 1.59W)(1- ε$_4$) - (23.85 - 3.89W)(ε$_4$ - ε$_s$),

$$W = \frac{0.259}{\cos\Theta} - 11.352\ln\left(\frac{\sigma_{45}}{\sigma_{44}}\right) - 11.649\cos\Theta\left[\ln\left(\frac{\sigma_{45}}{\sigma_{44}}\right)\right]^2, A_0 = -7.49 - 0.407W$$

The water vapor content in the atmosphere, W (g / cm²), along the path between the target and the sensor, with angle θ, can be calculated from satellite measurements of the brightness temperature fluctuations (σ45) and (σ44), from the channels 4 and 5 of the AVHRR, respectively.

3.2.2 Sea Surface Temperature (SST)

Sea surface temperature (SST) is considered as the water temperature close to the surface of the ocean. There are different explanations of the term surface, which are related to the measurement method used and, in general, the term surface varies between 1 millimeter and 20 meters below the sea surface. There is a diurnal SST variation, namely the air above it, however, there is more

SST variation on calm days than on breezy days. Moreover, long time-scale ocean currents, such as the Atlantic Multidecadal Oscillation (AMO), can have a multi-decadal effect on SSTs. This, in turn, may result in major impacts on the global thermohaline circulation, affecting the average SST significantly throughout most of the world's oceans (US National Research Council (NRC), 2000). SSTs strongly affect air masses in the Earth's atmosphere even within a short distance of the shore. Moreover, turbulent mixing of the upper 30 meters of the ocean, being the result of warm SSTs, may lead to tropical cyclogenesis over the oceans of the Earth, which, in turn, may cause cool wake (Fiolleau and Roca, 2013). Furthermore, upwelling can be produced by offshore winds from coastal SSTs, which may result in significant cooling or warming of nearby landmasses, although shallower waters over a continental shelf are often warmer (Dalezios, 2011). In addition, in areas with fairly stable upwelling, such as the northern coast of South America, onshore winds may cause significant warming (Krien et al., 2018). This, in turn, may result in the formation of sea fog and sea breeze, since SST affects the atmosphere above.

Meteorological satellites have been used to determine SST since 1970s, when the first global composites were produced (Rosenzweig and Hillel, 2008). Since then, SST measurements have been conducted by satellites, along with the assessment of several features, such as temporal and spatial variability. In addition, calibrations have been conducted leading to acceptable agreement between satellite measurements of SST and in situ temperature measurements (Weng, 2003). Indeed, the ocean radiation is sensed in two or more wavelengths within the infrared part of the electromagnetic spectrum or other parts of the spectrum and is related to SST, which constitutes the satellite measurement.

The satellite-based SST measurements provide both a synoptic view of the ocean and a high reoccurrence interval, allowing the examination of basin-wide upper ocean dynamics not possible with ships or buoys. NOAA's GOES (Geostationary Orbiting Earth Satellites) satellites are geo-stationary satellites, which can provide SST data on an hourly basis with only a few hours of lag time. Moreover, since 2000, Moderate Resolution Imaging Spectroradiometer (MODIS) of NASA (National Aeronautical and Space Administration) satellites provide global SST data, which are available with a one-day lag. For illustrative purposes, Figure 3.2a presents a daily global SST data set produced on December 20, 2013 at 1-km resolution (also known as ultra-high resolution) by the JPL (Jet Propulsion Laboratory) ROMS (Regional Ocean Modeling System) group (from JPL GISST, 2013). Figure 3.2b presents global 8-year (2003–2011) SST based on MODIS Aqua data.

Calculation of Sea Surface Temperature (SST). The SST calculation is based on algorithms using thermal channels (e.g. 3, 4 and 5 for NOAA-AVHRR, 6 for LANDSAT TM) to normalize the pixel values (PV: Pixel Values) of the image in degrees Celsius. Usually, the calculation produced by the algorithms of the surface temperature of the water is more precise than that of the ground. Using the calibration information accompanying the satellite data, the natural values of the sun's

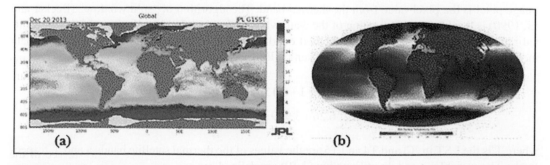

FIGURE 3.2 (a) A daily global Sea Surface Temperature (SST) data set produced on December 20, 2013 at 1-km resolution (also known as ultra-high resolution) by the JPL ROMS (Regional Ocean Modeling System) group (from JPL GISST, 2013). (b) 2003–2011 SST based on MODIS Aqua data.

reflection from the earth's surface (Albedo) and temperature, have been calculated. The calibration coefficients are calculated by a linear combination, converting the PVs to energy values for thermal channels and to reflected solar radiation values for visible channels, respectively. For each band there is a value for slope and intercept. The PVs of the thermal bands are converted into energy by:

$$E = S * PV + I \tag{3.7}$$

where E = energy in milliwatts/m²*steradians*cm⁻¹, PV = pixel value, S = slope in milliwatts / m²*steradians*cm⁻¹, I = interference in milliwatts/m²*steradians*cm⁻¹. Then, by Planck's equation, the energy can be converted to a brightness temperature:

$$T = \frac{C_2 \lambda}{\ln\left(\frac{C_1 \lambda^{-5}}{W} + 1\right)} \tag{3.8}$$

where T is in degrees Kelvin. For channels NOAA-AVHRR 3, 4 and 5, coefficients are used to create an image at values of 270–310 K (−3°C to 37°C). Values outside this range are 0. The resulting image has a minimum of 4 (−3°C) and steps of 0.2° per PV. A combination of channels 4 and 5 based on the McClain method is used to calculate SST. The scale is set so that the 10°C corresponds to the 50-pixel value and the increase is 0.1 points per level. This results in temperatures of 1 to 30°. The algorithm calculates gain and intercept for thermal channels to derive the correlation between PV and radiation in a matrix. The radiations are then transformed into heat by Planck's law and then corrected for the nonlinearity of the sensor. Precision temperature can reach 0.50°C and depends on atmospheric conditions, wind, proximity to land, water vapor in the atmosphere and pollution.

3.2.3 ATMOSPHERIC TEMPERATURE

A case study in the region of Thessaly, central Greece, is considered for the calculation of atmospheric temperature. The calculation of air temperature follows certain steps. At first, Land Surface Temperature (LST) is estimated by means of NDVI and BT images (Dalezios et al., 2012). The procedure for the extraction of LST is based on the following empirical Equation (3.9) (Kanellou et al., 2011):

$$T = (\text{image pixel} + 31990) \times 0.005 \tag{3.9}$$

where T is temperature in Kelvin (K), which is converted to values of degrees Celsius (°C) and image pixel is the pixel value from the thermal band. The algorithm of "split window" (Becker and Li, 1990) is used for the elimination of the water vapor effect in infrared radiation and the transmitted radiation from the surface. This method conducts atmospheric correction of the satellite data including water vapor absorption and is given by Equation (3.10):

$$T = 1.274 + (T4 + T5)/2\left[1 + 0.15616\left\{(1-e)/e\right\} - 0.482de/e^2\right]$$
$$+ (T4 - T5)/2\left[6.26 + 3.989\left\{(1-e)/e\right\} + 38.33de/e^2\right] \tag{3.10}$$

where T is the LST in °C, and T4 and T5 are the values of the thermal satellite bands 4 and 5, respectively. The variables e and de of Equation (3.10) are defined by:

$$e = (e4 + e5)/2 \tag{3.11}$$

$$de = e4 - e5 \tag{3.12}$$

where e4 and e5 are the reflection values of bands 4 and 5, respectively, which estimate the transmission of infrared radiation from the surface (Dalezios et al., 2020), which are provided by the following empirical equations (Van de and Owe, 1993):

$$e4 = 1.0094 + 0.047\ln(NDVI) \tag{3.13}$$

$$e5 = e4 + 0.01 \tag{3.14}$$

where NDVI refers to the index values. A regression analysis is conducted between the derived monthly LST values from Equation (3.10) and the corresponding monthly air temperature (T_{air}) values from Larissa station for the 20-year period (1981–2001) (Dalezios et al., 2020). The resulting empirical Equation (3.15) is given below, along with the plotted values (Figure 3.3a) (Dalezios et al., 2012). Finally, from Equation (3.15) monthly air temperature maps (images) of Thessaly can be produced on a pixel basis.

$$T_{air} = 0.6143 \times LST + 7.3674, R2 \approx 0:82 \tag{3.15}$$

The LST images and ground measurements of air temperature derived from Larissa meteorological station are used to produce air temperature maps through regression analysis (number of observations = 500, standard error = 0,4). Moreover, Figure 3.3b refers to the following sub-section, where the explanation is provided.

3.2.4 El Niño/La Niña

El Niño is defined by prolonged differences in Pacific Ocean surface temperatures when compared with the average values. Indeed, El Niño is a weather pattern, which occurs in the Pacific Ocean, when unusual equatorial winds cause warm surface water to move from the equator to the east, toward Central and South America (Figure 3.4a). The beginning of an El Niño is marked in the SST pattern when warm water spreads from the west Pacific and the Indian Ocean to the east Pacific. One characteristic of El Niño is that it can cause more rain than usual in South and Central America and in the USA. Moreover, El Niño takes the rain with it, causing extensive drought in the Western Pacific and rainfall in the normally dry eastern Pacific. El Niño conditions last for several months,

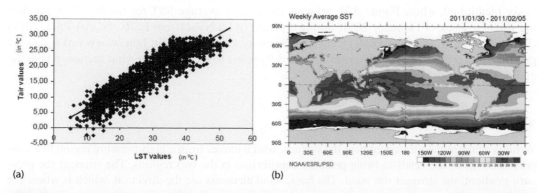

(a)   (b)

FIGURE 3.3 (a) LST and air temperature (T_{air}) for Larissa (0°C) (from Dalezios et al., 2012). (b) Weekly average sea surface temperature for the World Ocean during the first week of February 2011, during a period of La Niña (from ESRL/NOAAPSD, 2011) (Earth System Research Laboratory Physical Sciences Division, credit: http://www.esrl.noaa.gov/psd/map/images/sst/sst.gif).

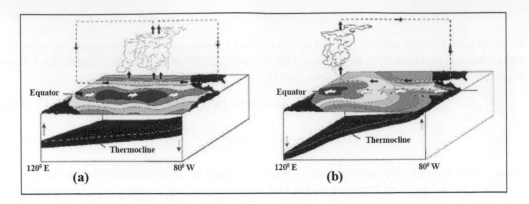

FIGURE 3.4 (a) Circulation during El Niño. (b) Circulation during La Niña.

resulting in extensive ocean warming and the reduction in Easterly Trade winds, which limit upwelling of cold nutrient-rich deep water and have a significant economic impact on local fishing for an international market (Zou et al., 2006). El Niño's warm rush of nutrient-poor tropical water, heated by its eastward passage in the Equatorial Current, replaces the cold, nutrient-rich surface water of the Humboldt Current.

On the other hand, La Niña episodes affect global weather patterns and are associated with less moisture in the air over cooler ocean waters (Figure 3.4b). As a result, there is less rain along the coasts of North and South America and along the equator, and more rain in the far Western Pacific. For illustrative purposes, Figure 3.4a presents the circulation during El Niño and Figure 3.4b presents the circulation during La Niña, respectively.

A warming or cooling of at least 0.5°C averaged over the east-central tropical Pacific Ocean is considered an accepted definition. Indeed, it is classified as El Niño/La Niña "conditions" when this warming or cooling occurs for only seven to nine months, and it is classified as El Niño/La Niña "episodes" when it occurs for longer (Vinnikov et al., 2006). It has been recorded that this weather anomaly occurs at irregular intervals of 2–7 years and its duration ranges from nine months to two years, where the average periodicity is 5 years (UAH, 2017). For illustrative purposes, Figure 3.5a shows the 1997 El Niño and the 1998 La Niña observed by TOPEX/Poseidon altimetry sea surface height data, where the white areas off the tropical coasts of South and North America indicate the pool of warm water (RSS/MSU, 2017). The case of La Niña has been previously mentioned, where Figure 3.3b presents the weekly average SST for the World Ocean during the first week of February 2011, during a period of La Niña (from ESRL/NOAA/PSD, 2011) (Earth System Research Laboratory Physical Sciences Division, credit: http://www.esrl.noaa.gov/psd/map/images/sst/sst.gif). Moreover, Figure 3.5b refers to the following wind section, where the explanation is provided.

3.3 WIND

Wind is a horizontal air current near the surface, which moves from a high-pressure region to a low-pressure region and tends to bring pressure equilibrium in the two regions. The stronger the pressure gradient, the stronger the wind. The basic wind elements are the direction, which is where the wind comes from or originates and expressed in degrees starting from the magnetic north, and the intensity expressing the wind speed. In this section, the Aeolus wind mission (Copernicus) is briefly presented, satellite-based wind applications are considered, such as satellite-based wind fields, the jet stream, as well as assessment of satellite-based local and daily winds.

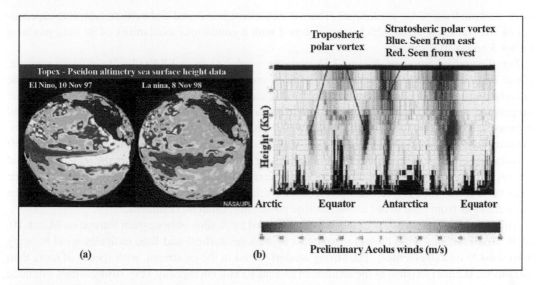

FIGURE 3.5 (a) The 1997 El Niño and the 1998 La Niña observed by TOPEX/Poseidon altimetry sea surface height data. The white areas off the tropical coasts of South and North America indicate the pool of warm water (from RSS/MSU, 2017). (b) First wind data from ESA's Aeolus satellite. These data are from three quarters of one orbit around Earth (image credit: ESA/ECMWF) (from ESA, Aeolus 2019).

3.3.1 Aeolus Wind Mission (Copernicus)

Based on Greek mythology, **Aeolus** is the controller of the winds and ruler of the floating island of Aeolia. In Homer's *Odyssey*, Odysseus received a favorable wind and a bag, where the unfavorable winds were confined. According to the myth, Odysseus' companions opened the bag and the winds escaped and drove them back to the island. Although Aeolus appears as a human in Homer, later he was described as a minor god.

Aeolus is currently an ESA (European Space Agency) Earth Explorer Core Mission, which is a science-oriented mission within its Living Planet Program. The main goal of Aeolus' mission is to provide wind profile measurements for an improved analysis of the global three-dimensional wind field, where the vertical resolution should comply with the accuracy requirements of WMO (World Meteorological Organization). It is significant to understand several aspects of climate research and weather forecasting, such atmospheric dynamics, which involve the global transport of aerosols, energy, chemicals, water and other airborne materials. Specifically, Atmospheric Dynamics Mission (ADM)-Aeolus constitutes a demonstration project for the Global Climate Observing System (GCOS) (Adamson et al., 2010; Straume et al., 2012; ESA, 2019).

The measurement data contribute to the fulfillment of the **primary** goals of Aeolus: (1) development of accurate wind profiles throughout the troposphere and lower stratosphere, which is a major existing deficiency in the Global Observing System. (2) development of data for the global atmospheric circulation and relevant features, such as precipitation systems, stratospheric/tropospheric exchange, the El Niño and the Southern Oscillation phenomena. (3) direct contribution to the Earth's global energy budget. The objectives of the **secondary** mission are related to the development of data sets for model variation and short-term "wind climatologies", which contribute to: (4) improve the understanding of atmospheric dynamics and the global atmospheric transport and cycling of aerosols, energy, chemicals, water and other airborne materials. (5) validate climate models through high-quality wind profiles from a global measurement system. (6) generate several derived products, such as cloud top altitudes, aerosol properties and tropospheric height. The ADM-Aeolus measurements are assimilated in numerical forecasting models, in order to enhance the quality of operational short- and medium-range predictions. There are excellent horizontal and vertical sampling capabilities of the Aeolus system, which are combined with a continuous availability of its data products within 3 hours after sensing.

For illustrative purposes, Figure 3.5b presents wind data from ESA's Aeolus satellite mission. These data are from three quarters of one orbit around the Earth. The image shows large-scale easterly and westerly winds between the Earth's surface and the lower stratosphere, including jet streams. As the satellite orbits from the Arctic toward the Antarctic, it senses strong westerly wind streams, called tropospheric vortices (shown in blue) each side of the equator at mid latitudes. Orbiting further toward the Antarctic, Aeolus senses the strong westerly winds (shown in blue left of Antarctica and in red right of Antarctica) circling the Antarctic continent in the troposphere and stratosphere (Stratospheric Polar Vortex). The overall direction of the wind is the same along the polar vortex, but since the Aeolus wind product is related to the viewing direction of the satellite, the color changes from blue to red as the satellite passes the Antarctic continent.

The image of Figure 3.6a presents winds measured by Aeolus over western Europe on March 10, 2019. Red indicates wind blowing from east to west (easterlies) and blue indicates wind blowing from west to east (westerlies). The strong westerly wind in the jet stream, with speeds of more than 200 km/hr, is clearly visible at the altitude of around 10 km. On this day, very strong winds extended from the jet stream all the way down to the surface and caused problems for traffic and construction. Black areas indicate where the satellite could not measure winds owing to thick cloud layers (from ESA, Aeolus 2019). Moreover, Figure 3.6b shows the wind measured by the Aeolus satellite while crossing the cyclone Idai west of Madagascar on March 11, 2019. Red indicates wind blowing from east to west (easterlies) and blue indicates wind blowing from west to east (westerlies). Since Aeolus measures wind in the cloud-free atmosphere, and within thin clouds and on top of thick clouds, the measurements here are those surrounding Idai. The black patch is the part of the cyclone which was covered by a thick layer of spiral-shaped clouds. The image shows strong easterly winds north of the hurricane (in red on the left of the image), with wind speeds up to 150 km/hr (above 40 m/s). In the upper right corner (altitude of 22–25 km), the tropical stratospheric easterly jet can be seen in red, and lower down on the right (altitude of 10–16 km) the sub-tropical westerly jet in the southern hemisphere is visible in blue (from ESA, Aeolus 2019).

3.3.2 Satellite-based Wind Fields

Wind fields can be delineated by satellites. High-resolution sea surface wind fields can be reconstructed from multi-sensor satellite data. An application is presented in the Grand Banks of Newfoundland of Atlantic Canada (Tang et al., 2014). Indeed, it has long been recognized that Newfoundland, Canada, has a higher average wind speed compared to several other places in the country. The ocean circulation and the marine climate have strong interactions in this region. Specifically, the Grand Banks southeast of Newfoundland is near the intersection of the poleward Gulf Stream and North Atlantic Current and the equatorward Labrador Current (Figure 3.7) (Tang et al., 2014). As a result, the wind field studies in this special geographic region have great significance in weather forecasting, air-sea interactions, atmospheric dynamics and climate.

The RADARSAT-2 satellite operates in a sun-synchronous, circular, near-polar orbit at a mean altitude of 797 km. A case study is considered, where six-hourly ocean wind fields from blended products (BPs) (including multi-satellite measurements) with 0.25° spatial resolution and 226 RADARSAT-2 synthetic aperture radar (SAR) wind fields with 1-km spatial resolution were used to reconstruct new six-hourly wind fields with a resolution of 10 km for the period August 2008–December 2010, with the exception of the period July–November 2009 (Tang et al., 2014). BP combines measurements from six satellites (SSM/I F13, SSM/I F14, TMI, QuikSCAT, SSM/I F15, AMER-E) from the US National Climatic Data Center (Zhang et al., 2006). The BP used in this study from the available multiple resources had been produced to fill data gaps and aliases associated with the subsampling by the individual satellite observations (Tang et al., 2014).

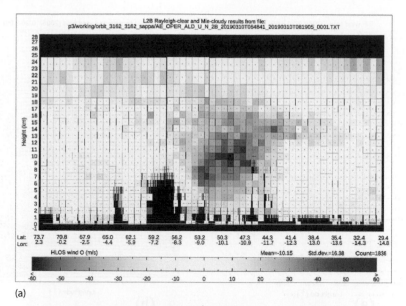

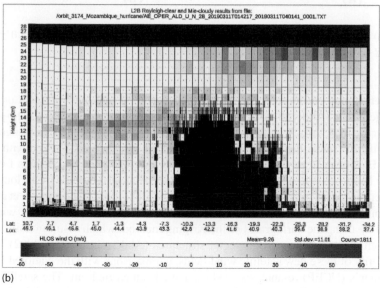

FIGURE 3.6 (a) The image shows winds measured by Aeolus over western Europe on March 10, 2019 (image credit: ECMWF–M. Rennie) (from ESA, Aeolus 2019). (b) Wind measured by the Aeolus satellite while crossing the Cyclone Idai west of Madagascar on March 11, 2019 (image credit: ECMWF–M. Rennie) (from ESA, Aeolus 2019).

In this case study, 226 RADARSAT-2 SAR images were collected from MDA Geospatial Services Inc. (MDA GSI) over the Grand Banks from 2008 to 2010. The reconstruction process is based on the heapsort bucket method with topdown search and the modified Gauss–Markov theorem (Tang et al., 2014). The retrieved wind speeds are compared with buoy-measured wind speeds in order to assess the performance of the proposed approach for reconstructing wind field. The buoy wind measurements are generally reported on the hour and represent 10-min averages. The reconstructed wind is averaged over 10 km in space and is a proxy point measurement in time (Tang et al., 2014). For illustrative purposes, Figure 3.7a presents the BP wind field at 03:00, October 20, 2008 and Figure 3.7b presents the reconstructed wind field at 03:00, October 20, 2008.

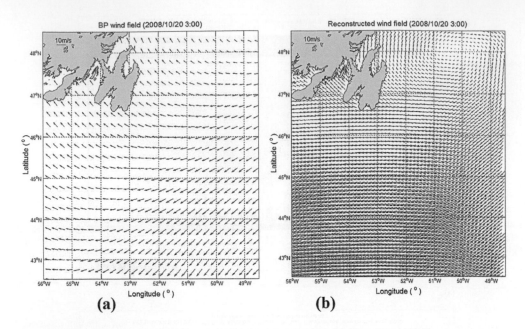

FIGURE 3.7 (a) BP wind field at 03:00, October 20, 2008. (b) Reconstructed wind field at 03:00, October 20, 2008 (from Tang et al., 2014).

Scatterometers and radiometers on several polar-orbiting satellites routinely produce oceanic surface wind field data (Abderrahim et al., 2002). Accurate surface wind speed estimates constitute a significant factor in determining the velocity and magnitude of air-sea gas exchange (Abderrahim et al., 2002). For this study, the NASA scatterometer (NSCAT) data are merged with scatterometer data from the European Remote Sensing (ERS) satellite 2, and the wind speeds from two of the Special Sensor Microwave/Imagers (SSMII) and produced daily 10 latitude by 10 longitude gridded wind fields over the global ocean from September 1996 to June 1997 (Abderrahim et al., 2002). This time period coincides with the lifetime of the NSCAT aboard ADEOS-J. These wind fields have been produced by using the variograms of the Kriging method, which result in daily wind fields by considering both space and time wind vector structures (Abderrahim et al., 2002). To investigate the global patterns of these new satellite wind fields, comparisons with the National Environmental Prediction Center's (NCEP) re-analysis products have been carried out. The satellite data and the NCEP products have a similar statistical error structure, but the merged wind fields provide complete coverage at much higher spatial resolution.

3.3.3 Jet Stream

There are two types of jet stream, namely polar and sub-tropical. Polar jet streams are the more powerful of the two and are typically located 7–12 km above sea level. The weaker sub-tropical jet streams are usually situated 10–16 km above sea level. Jet streams extend for thousands of kilometers around the planet and are usually continuous, but meandering, over these long distances, whereas the width of a jet stream is typically a few hundred kilometers and each one is only a few kilometers deep. They normally take a general west to east path across the Atlantic, however, with different ripples and bulges. The jet steers the areas of low and high pressure in the atmosphere, resulting in different weather types on the ground. The curve in the jet can help develop or destroy areas of low pressure, leading to a decrease in storm severity, which supports forecasters for storm prediction.

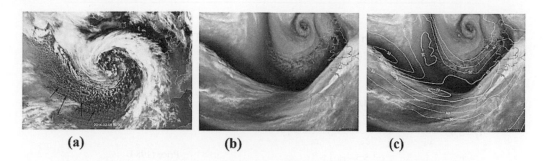

(a)   **(b)**   **(c)**

FIGURE 3.8 (a) The position of the jet stream is shown (February 8, 2014), where there is a color difference, i.e. red changes to purple (from EUMETSAT). (b) The water vapor image from February 1, 2014 shows the location of the jet stream as a strong gradient from high to low humidity (from white to dark colors) on the side of the frontal cloud bands that is facing toward the pole. (c) Isotachs – lines showing points of equal wind speed in knots – make the jet stream more obvious in the water vapor image from February 1, 2014.

The case of Europe is considered, where the jet stream often moves north during the summer, deflecting low pressure systems that have come across the Atlantic, and allowing for more settled conditions. In winter, the jet stream can be expected to move further south, allowing storms from the Atlantic to reach Europe, at first hitting Ireland, the United Kingdom, and France. The case of winter 2014 is considered, where the North Atlantic jet stream was 30% stronger than normal, resulting in several extremely severe storms for the period January 29, 2014–February 11, 2014 in many parts of Europe (EUMETSAT, 2014). Based on airmass RGB animation it was possible to track the jet stream that was reaching Europe around the area of Bay of Biscay (Figure 3.8a). The position of the jet stream is presented in Figure 3.8a (February 8, 2014), showing a color difference, i.e. red changes to purple (EUMETSAT, 2014). Moreover, the location of the jet stream can be identified in water vapor images as a strong gradient from high to low humidity (from white to dark colors) on the side of the frontal cloud bands that is facing the pole (Figure 3.8b, February 1, 2014) (EUMETSAT, 2014). In addition, the isotachs, i.e. lines showing points of equal wind speed, make the jet stream more obvious in the water vapor image from February 1, 2014 (Figure 3.8c) (EUMETSAT, 2014). For aviation, the jet stream can be used for a faster journey from west to east over the Atlantic, although it is also a source of turbulence.

3.4 CLOUDS AND ATMOSPHERIC FEATURES

In this section, remotely sensed cloud identification and classification is presented. Atmospheric features are also considered, including atmospheric sounding, cloud top temperature (CTT), temperature and humidity profile, cloud structure and thickness, as well as fog (Table 3.2).

3.4.1 CLOUD IDENTIFICATION AND CLASSIFICATION

The cloud types identified from satellite observations are called "cloud types", and are distinguished from the so-called "cloud forms", which are the cloud types determined by visual surface observations. In satellite-based cloud type identification, the cloud types are classified into seven groups: Ci (cirrus-high level clouds), Cm (middle level clouds), St (stratus/fog), Cb (cumulonimbus), Cg (cumulus congestus), Cu (cumulus), and Sc (stratocumulus) (Table 3.3). These cloud types are classified as stratiform clouds (Ci, Cm, St) or convective clouds (Cb, Cg, Cu). Moreover, Sc has an intermediate character between stratiform and convective clouds. The satellite-based classification

TABLE 3.2

Synopsis of Algorithms for Split-Windows (T_s T_4, T_5, in °K, and $T_0 =$ 273.15°K, $\varepsilon = (\varepsilon_4 + \varepsilon_5)/2$

Algorithms	Reference
$T_s = 1.274 + PA\dfrac{T_4 + T_5}{2} + MA\dfrac{T_4 + T_5}{2}$	Becker and Li (1990)*
$T_s = 3.45\dfrac{T_4 - T_0}{\varepsilon_4} - 2.45\dfrac{T_5 - T_0}{\varepsilon_5} + 40\dfrac{1 - \varepsilon_4}{\varepsilon_4} + T_0$	Prata and Platt (1991)
$T_s = \left[T_4 + 3.33\left(T_4 - T_5\right)\right]\left[\dfrac{5.5 - \varepsilon_4}{4.5}\right] + 0.75 T_5\left(\varepsilon_4 - \varepsilon_5\right)$	Price (1984)
$T_s = T_4 + 3.00(T_4 - T_5) + 51.57 - 52.45\varepsilon$	Ulivieri et al. (1985)
$T_s = T_4 + 1.8(T_4 - T_5) + 48(1 - \varepsilon) - 75(\varepsilon_4 - \varepsilon_5)$	Ulivieri et al. (1994)
$T_s = T_4 + 1.06(T_4 - T_5) + 0.46(T_4 - T_5)^2 + 53(1 - \varepsilon_4) - 53(\varepsilon_4 - \varepsilon_5)$	Sobrino et al. (1993)
$T_s = T_4 + A(T_4 - T_5) + B$	Sobrino et al. (1993)**
$T_s = 1.0162 T_4 + 2.657(T_4 - T_5) + 0.526(\sec\theta - 1)(T_4 - T_5) - 4.58$	NESDIS (May et al. 1992)

$$*PA = 1 + 0.15616\frac{1 - \varepsilon}{\varepsilon} - 0.482\frac{\varepsilon_4 - \varepsilon_5}{\varepsilon^2}, MA = 6.26 + 3.98\frac{1 - \varepsilon}{\varepsilon} + 38.33\frac{\varepsilon_4 - \varepsilon_5}{\varepsilon^2}$$

$$**A = 0.349W + 1.320 + (1.385W - 0.204)(1 - \varepsilon_4) + (1.056W - 10.532)(\varepsilon_4 - \varepsilon_5)$$

$$B = (1 - \varepsilon_4)/\varepsilon_4 T_4\left[-0.146W + 0.561 + (0.575W - 1.966)(\varepsilon_4 - \varepsilon_5)\right]$$

$$12(1 - \varepsilon_5)/\varepsilon_5 T_5\left[-0.95W + 0.320 + (0.597W - 1.916)(\varepsilon_4 - \varepsilon_5)\right]$$

TABLE 3.3

Cloud type classification by satellite imagery

Cloud Type	Classification	
Ci: High Level Cloud	Stratiform clouds	High level clouds
Cm: Middle Level cloud		Middle level clouds
St: Stratus/fog		Low level clouds
Sc: Stratocumulus		
Cu: Cumulus	Convective clouds	
Cg: Cummulus congestus		
Cb: Cumulonimbus		

leads to high, middle and low-level clouds, which is based on cloud top height. On the other hand, the classification by surface meteorological observation is based on cloud base height.

Computerized cloud type identification uses water vapor and the infrared split window channel imagery besides the visible and infrared imagery. Figure 3.9 presents a cloud type identification diagram, which is based on visible and infrared imageries.

Meteorological satellites have certain advantages, such as high temporal resolution, a wide range of spatial observations, or all-weather observation. Cloud classification is one of the fundamental works of satellite cloud image processing (Bankert et al., 2009). At the present time, there are widely used methods for cloud classification, which are based on thresholds, such as simple threshold methods and histogram-based methods, statistical-based methods and artificial

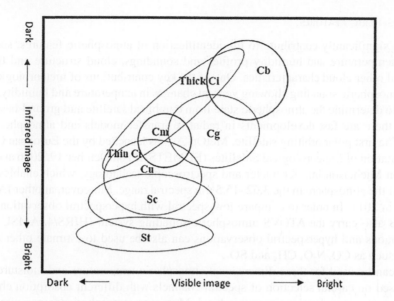

FIGURE 3.9 Cloud type identification diagram.

intelligence-based methods (Tapakis and Charalambides, 2013). Specifically, simple threshold methods (Fiolleau and Roca, 2013) analyze the spectral characteristics of different channels to inverse the brightness temperature for each pixel, then take the brightness temperature and gray value of each pixel with the difference of brightness temperature between channels to determine a series of thresholds as comprehensive criterion. The difficulty lies in the determination of the series of thresholds.

Moreover, the histogram-based method (Kärner, 2000) is considered an improvement of the simple threshold method, since it uses statistical properties of the partial or global histograms of satellite cloud images. However, for complex weather systems, it is still difficult to obtain adaptive thresholds. With regards to the statistical-based methods, such as C-means, fuzzy C-means clustering (Ameur, 2004), or similar methods based on the cluster analysis of the sample set, the pixels are classified into different types of clouds by computing the distance between cloud centers and samples. Nevertheless, the comprehensive data have contributed to several meteorological applications, since different imaging channels reflect atmospheric physics parameters, such as top brightness temperature, albedo and water vapor content in different aspects. Moreover, recent results on the information of these channels have shown that the inversion accuracy of clouds and underlying surface can be significantly improved (Jin et al., 2016).

Global atmospheric heat exchanges, which are highly dependent on the variation of cloud types and amounts, constitute another application of cloud classification. In order to have a better understanding of these exchanges, an appropriate cloud type classification method is required. In a recent study, an alternative approach was considered to the often-used cloud optical and thermodynamic properties-based classifications (Dim and Takamura, 2013). This approach is based on the application of edge detection methods on cloud top temperature (CTT) derived from global satellite maps. Specifically, the edge detection methods are based on the principle that the neighborhood of a pixel contains information about its intensity. Moreover, the variation of this intensity (gradient) allows the decomposition of the image into different cloud morphological features. Specifically, this approach produces the gradient map, which is then used to distinguish various types of clouds. High gradient areas would correspond to cumulus-like clouds, whereas low gradient areas would be associated with stratus-like clouds (Dim and Takamura, 2013).

3.4.2 ATMOSPHERIC FEATURES

Satellites can significantly contribute to the identification of atmospheric features, such as three-dimensional temperature and humidity profiles and soundings, cloud structure and thickness, as well as several other cloud characteristics. One of the key contributions of meteorological satellites consists of atmospheric sounding showing vertical changes in temperature and humidity. Nowadays, it is possible to determine the atmospheric state from combined satellite and ground-based measurements, since there are fast developments in radiative transfer models and atmospheric sounding technology. The first polar orbiting satellite, MetOp-A, was launched by the European Organization for the Exploitation of Meteorological Satellites (EUMETSAT) on October 19, 2006 and carried the IASI based on Michelson interferometer and spectroscopic technology, which enables continuous observation of the atmosphere in the 3.62–15.5 µm spectral range. Moreover, another IASI followed on MetOp-B in 2011. In order to compare low-spectral with hyperspectral observations, MetOp-A and MetOp-B also carry the ATOVS atmospheric sounding system (HIRS/4, AMSU and MHS). IASI's continuous and hyperspectral observations can also be used to estimate other atmospheric constituents, such as CO, N_2O, CH_4 and SO_2.

Satellites can be used for three-dimensional atmospheric temperature and moisture soundings, which are based on careful selection of spectral channels with different absorption characteristics within an atmospheric molecular absorption band. Moreover, atmospheric temperature and humidity can be indirectly measured by satellite IR remote sensing. Indeed, the direct measurements are radiances in different spectral bands or the path delays in a specific FOV at a given time. Moreover, strong absorption channels detect radiation from high in the atmosphere, whereas weak absorption channels detect radiation from low in the atmosphere down to the earth surface. Specifically, the carbon dioxide (CO_2) infrared (IR) absorption bands, with CO_2 mixed almost uniformly in the air, can provide information on an atmospheric temperature profile for any given region around the world and the water vapor (H_2O) IR absorption bands can provide information on the atmospheric humidity profile. A retrieval method is required to obtain the desired parameters, such as concentrations of absorbing gases like CO_2, O_3 and CH_4, the vertical distributions of atmosphere temperature and humidity, and surface (land, ocean) as well as cloud characteristics of any region on the Earth. Indeed, the physical retrieval method is considered reliable in clear and partly cloudy skies. Nevertheless, the estimation difficulties increase and become more complicated in cloudy regions with measured radiances highly affected by absorption, scattering, and emission of clouds and aerosols (Li et al., 2016).

Moreover, satellite-based three-dimensional atmospheric temperature and humidity profiles have been widely used in NWP, warning and monitoring of severe weather, forecasting and monitoring of atmospheric environment, as well as climate prediction and monitoring. However, a significant challenge still exists to explore possible solutions that account for clouds and precipitation in the atmosphere (Li et al., 2016). Indeed, hyperspectral IR combined with advanced microwave measurements are expected to provide improved three-dimensional structure of the atmosphere continuously over the globe. Specifically, the FY-4A hyperspectral atmospheric sounder is the first of several planned missions to monitor the vertical structure of the atmosphere hourly and to capture the atmospheric motions in both horizontal and vertical directions ahead of high-impact weather events. Moreover, with the development of advanced detectors onboard CubeSats, such as NASA's Time-Resolved Observations of Precipitation structure and storm Intensity with a Constellation of Smallsats (TROPICS, http://tropics.ll.mit.edu) with a microwave sounder onboard, it is foreseeable that CubeSats and similar technologies may also offer higher temporal resolution to global atmospheric sounding. Figure 3.10 shows the computed IASI brightness temperatures in clear skies showing: (1) the absorption lines (top) of the carbon dioxide band (from 680 to 770 cm^{-1}); and (2) of the water vapor band (from 1300 to 1500 cm^{-1}) at high spectral resolution. US Standard atmosphere with nadir view is assumed (Menzel et al., 2017).

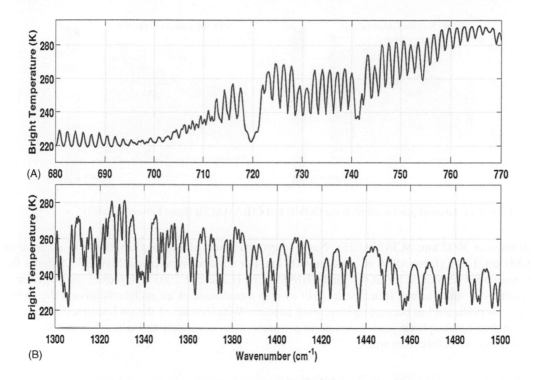

FIGURE 3.10 Calculated IASI brightness temperatures in clear skies revealing the absorption lines (top) of the carbon dioxide band (from 680 to 770 cm⁻¹) and (b) of the water vapor band (from 1300 to 1500 cm⁻¹) at high spectral resolution. US Standard atmosphere with nadir view is assumed (from Menzel et al., 2017).

3.5 ATMOSPHERIC CHEMISTRY

Air pollution constitutes a global environmental health problem, especially for those living in urban areas. National and international agencies have developed directives and recommendations for air pollution and radiation exposure. There are several examples, which include ozone, UV radiation, acidifying compounds and aerosol particles that can carry toxic substances (Lelieveld, 2003). Specifically, the wide range of scales that are used has implications on the spatial and temporal resolution of operational information, which is required to monitor the atmospheric environment.

Moreover, there are the ESA satellite sensors, which include the Global Ozone Monitoring experiment (GOME), the Global Ozone Monitoring by Occultation of Stars (GOMOS), the Michelson Interferometric Passive Atmospheric Sounder (MIPAS) and the Scanning Imaging Absorption Spectrometer for Atmospheric Cartography (SCIAMACHY) (Simon, 2004). Specifically, GOME was based on ERS-2 satellite and was launched in 1995 with nadir viewing mode, having a leading role in retrieval and exploitation of products (e.g. O_3, NO_2, OClO, BrO), by DOAS analyses and evaluation of UV radiation intensity at the Earth's surface. There was also the ENVISAT atmospheric chemistry mission involving GOMOS, MIPAS and SCIAMACHY. Moreover, GOMOS occultation mode involves vertical profiles with 1.7 km sampling and global coverage of self-calibration. There is also the MIPAS viewing mode. For CHIAMACHY, the viewing modes involve at nadir: limb and solar occultation viewing modes; 220-1750 nm + two IR channels; surface: albedo, UV irradiance; troposphere: O_3, NO_2, CO, BrO, CH_4, H_2O, HCHO, SO_2, aerosols; stratosphere: O_3, NO_2, N_2O, CO, CO_2, CH_4, BrO, OClO, aerosols. Observing modes for the above systems include estimations for a tangent point around 30 km altitude: GOME: Nadir $\Rightarrow$ ground pixel resolution; GOMOS: Occultation and limb $\Rightarrow$ air mass at 3000 km; MIPAS: Limb (side and backward) $\Rightarrow$

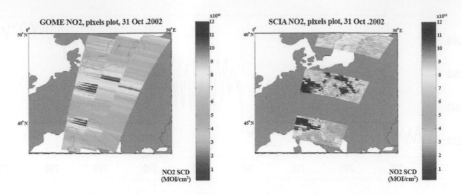

FIGURE 3.11 Ground pixel resolution for GOME and CHIAMACHI (from Lelieveld, 2003).

air mass at 3000 km; SCIAMACHY: Nadir ⇒ ground pixel resolution; Occultation ⇒ air mass at 3000 km; Limb (forward) ⇒ air mass at 3000 km. For illustrative purposes, Figure 3.11 shows the ground pixel resolution for GOME and CHIAMACHI (Lelieveld, 2003). Moreover, the Copernicus Sentinel-5P satellite was launched in 2017 to map a multitude of air pollutants around the globe, since air pollution has become an increasing concern. With its state-of-the-art instrument, Tropomi, it can detect atmospheric gases to image air pollutants more accurately and at a higher spatial resolution than ever before from space.

3.5.1 STRATOSPHERIC OZONE AND UV RADIATION

Stratospheric O_3 depletion enhances solar UV penetration to the earth's surface. Surface UV levels can be derived from satellite measurements of column O_3, clouds and aerosols. Moreover, a subtle balance exists in the atmosphere between solar energy transmission, photochemistry and attenuation of harmful short-wave radiation. Ozone absorbs most UV-B (280–315 nm), and only a few percent reaches the earth's surface. Indeed, UV-B damages biological tissue. On the other hand, UV-A (315–400 nm) is hardly attenuated by O_3, and it also affects human health (e.g. sunburn). Specifically, about 90% of atmospheric O_3 is in the stratosphere, and as a result its depletion by nitrogen oxides and halocarbons constitutes serious concern about the trend in surface UV-B irradiance.

Polar stratospheric clouds (PSC) can be formed as a result of the exceptional cold conditions in the Antarctic stratosphere, where halogen compounds are activated from inert reservoir species. Indeed, due to the optimum between low temperatures and the availability of sunlight, these halogens, notably chlorine from chlorofluorocarbons (CFCs), cause dramatic O_3 loss during the Antarctic spring, resulting into the "ozone hole". Moreover, although column ozone changes are of prime importance, surface UV is also controlled by several other parameters that affect radiation transfer through scattering and absorption. Clouds generally reduce UV depending on their optical thickness. Scattered or thin clouds over a highly reflective surface enhance surface UV, whereas thicker clouds cause a reduction. Figure 3.12a shows the **stratospheric ozone** hole over Antarctica on September 4, 2018. Strong winds, called the Stratospheric Polar Vortex, around the South Pole play an important role in the depletion the ozone at this time of the year. Low ozone is shown in blue and high in pink (image credit: KNMI–Temis, released on September 12, 2018).

3.5.2 GREENHOUSE GASES, AEROSOLS AND TRANSBOUNDARY AIR POLLUTION

Greenhous gases, such as carbon dioxide (CO_2), methane (CH_4), tropospheric O_3 and several minor greenhouse gases, have increased significantly in the atmosphere and contribute to climate change. The long-lived greenhouse gases (e.g. CO_2 and CH_4) are regulated through the Kyoto Protocol. Specifically, future amendments of the Kyoto Protocol are expected to additionally address short-lived

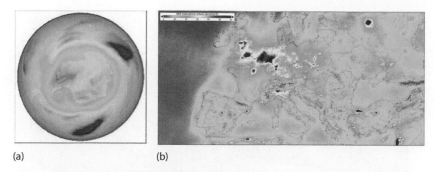

(a)   (b)

FIGURE 3.12 (a) Ozone hole over Antarctica on September 4, 2018. Strong winds, called the Stratospheric Polar Vortex, around the South Pole play an important role in the depletion of the ozone at this time of year. Low ozone is shown in blue and high in pink (image credit: KNMI–Temis, released on September 12, 2018). (b) Sentinel-5P map of nitrogen dioxide over Europe (from ESA, March 12, 2019).

greenhouse gases (e.g. tropospheric O_3) and aerosols (e.g. black carbon). Moreover, in addition to regional trend monitoring of these species, geostationary satellite observations of CO_2 and CH_4 can be used to estimate emissions, provided that the measurements are performed with high accuracy.

The European Convention on Long-Range **Transboundary Air Pollution** (LRTAP) addresses emissions and transport of pollutants to control air quality and reduce acidification and eutrophication, e.g. by nitrogen oxides and sulphur species. Moreover, unpredictable pollution sources, such as volcanoes, forest fires and other catastrophic events can affect human health and give rise to safety risks, e.g. for air traffic. Indeed, geostationary satellite measurements of several reactive gases, e.g. CO, SO_2, O_3 and aerosols, can trace pollutant plumes. In addition, these measurements can provide proxies of air movement that are useful for numerical weather prediction (NWP). Moreover, nitrogen dioxide is the pollutant with the strongest evidence of health effects. In the EU, the largest contributor to nitrogen dioxide emissions is the road transport sector, as well as pollution produced from industrial activities, and residential combustion.

Aerosols affect the atmospheric radiation budget and the microphysical properties of clouds, thus, directly and indirectly influence the earth's climate. Pollutant aerosols over water masses cause solar radiation attenuation, resulting in the reduction of evaporation and consequently precipitation. Moreover, aerosols cause solar radiation absorption, which affects heating rates in the lower troposphere, influencing convection and moisture transports. In addition, direct and indirect aerosol effects on the hydrological cycle are considered substantial, although these effects are as yet poorly quantified. Indeed, aerosol optical thickness, particle size and aerosol absorption of solar radiation are parameters which can be detected by satellite measurements.

Air quality is a serious concern. The Copernicus Sentinel-5P satellite was launched in October 2017 to map a multitude of air pollutants around the globe. The satellite carries the most advanced sensor of its type to date: Tropomi. This state-of-the-art instrument detects the unique fingerprint of atmospheric gases to image air pollutants with high accuracy and at a higher spatial resolution (3.5×7 km) than ever before. A case is presented, where measurements gathered between April and September 2018 have been averaged to show exactly where **nitrogen dioxide** is polluting the air. This kind of pollution results from traffic and the combustion of fossil fuel in industrial processes. It can cause significant health issues by creating respiratory problems and irritating the lungs. The map shows emissions from major cities, but also medium-size towns. With Copernicus Sentinel-5P's Tropomi instrument, pollution can be observed from individual power plants and other industrial complexes, major highways, and many more ship tracks can be identified than before. On October 17, **sulphur dioxide** and formaldehyde joined the list of air pollutants routinely available for services, such as air-quality forecasting and volcanic ash monitoring. Figure 3.12b shows a nitrogen dioxide map of Europe between April and September 2018.

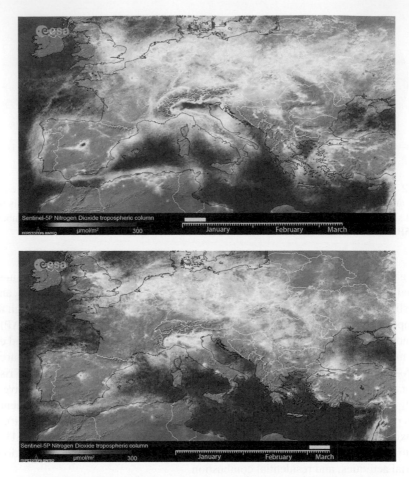

FIGURE 3.13 (a) shows the nitrogen dioxide emissions across Europe as at January 1, 2020; and (b) shows the corresponding nitrogen emissions as at March 11, 2020 (from ESA, March 12, 2020b).

The Copernicus Sentinel-5P satellite has also revealed the decline of air pollution, specifically nitrogen dioxide emissions, over Italy and China, respectively, due to coronavirus (COVID-19). This reduction is particularly visible in northern Italy, which coincides with its nationwide lockdown to prevent the spread of coronavirus (Figure 3.13a,b). Specifically, Figure 3.13a shows the nitrogen dioxide emissions across Europe as at January 1, 2020 and Figure 3.13b shows the corresponding nitrogen emissions as at March 11, 2020, using a ten-day moving average based on the Tropomi instrument onboard the Copernicus Sentinel-5P satellite. Similarly, Figure 3.14a shows the nitrogen dioxide emissions over China as at January 1, 2020 and Figure 3.14b shows the corresponding nitrogen emissions as at March 20, 2020, using a ten-day moving average based on the Tropomi instrument onboard the Copernicus Sentinel-5P satellite.

Moreover, Figure 3.15a shows the sulphur dioxide emissions for the period November 2017–July 2018. The Copernicus Sentinel-5P mission has also been used to produce global maps of two atmospheric gases responsible for making our world warmer: **methane**, which is a potential greenhouse gas and tropospheric **ozone**, which is a greenhouse gas and a pollutant in the lower part of the atmosphere. The maps offer an insight into the source of these gases. Both products are important for monitoring climate change. Although carbon dioxide is more abundant in the atmosphere and therefore more commonly associated with global warming, methane is about 30 times more potent as a heat-trapping gas. It enters the atmosphere mainly from the fossil fuel industry, landfill sites,

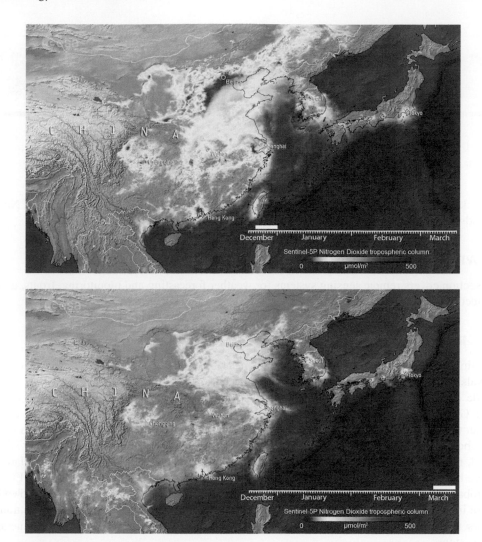

FIGURE 3.14 (a) shows the nitrogen dioxide emissions over China as at January 1, 2020; and (b) shows the corresponding nitrogen emissions as at March 20, 2020 (from ESA, March 20, 2020a).

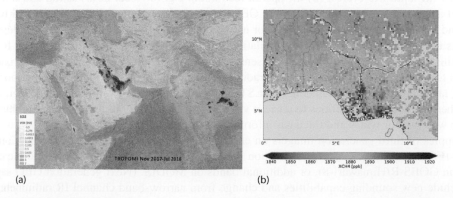

FIGURE 3.15 (a) Sentinel-5P map of sulphur dioxide for the period November 2017–July 2018 (from ESA, October 24, 2018). (b) Sentinel-5P map of Methane over wetlands in Nigeria (from ESA, March 4, 2019).

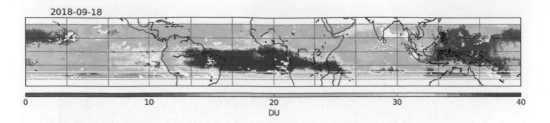

FIGURE 3.16 Sentinel-5P map of Global tropospheric ozone (from ESA, March 4, 2019).

livestock farming, rice agriculture and wetlands. Figure 3.15b shows the methane map of Nigeria wetlands and Figure 3.16 shows the global map of the tropospheric ozone.

3.6 WEATHER SYSTEMS

This section covers remote sensing systems for weather forecasting and monitoring, fronts, as well as detection and tracking of storms, cyclones, hurricanes, tornadoes and lightning.

3.6.1 REMOTE SENSING SYSTEMS FOR WEATHER FORECASTING AND MONITORING

Operational satellite systems provide invaluable measurements of atmospheric parameters at regular intervals on a global scale. Geostationary (GEO) and low-Earth-orbiting (LEO) satellite systems provide data in different spatial and temporal resolutions (Kidd et al., 2009; Thies and Bendix, 2011). GEO satellite systems circulate the Earth at an altitude of circa 36,000 km above the equator with an orbital period of 24 h. As a result, GEO satellites appear to be stationary above a certain point above the equator, which enable observations with a high temporal resolution of 15 or 30 min. Indeed, meteorological now-casting applications are based on operational geostationary platforms and are becoming increasingly significant for climate research, since longtime series of GEO data are globally available. Moreover, operational polar orbiting platforms (LEO) complement the GEO system in the global weather satellite system. Indeed, LEO satellites are frequently used to transfer sensors from experimental missions into operational use, thus, contribute to the next generation GEO systems.

Current (next generation) **GEO systems** primarily emphasize NWP and climate research in order to improve the systems, while at the same time to warranty data and product continuity. There are several improvements, which emphasize : (1) the temporal resolution to observe atmospheric phenomena with rapid life cycles; (2) the spatial resolution, e.g. for local now-casting and to capture subpixel atmospheric phenomena from the GEO orbit; (3) the radiometric, e.g. the signal to noise ratio; and (4) spectral resolution to allow for new products and to boost the accuracy of developed products to the requirements of data assimilation in NWP models. Recent GEO missions focus on passive instruments and mainly two proven sensor families: (1) multispectral narrow and broad band imagers to retrieve data on clouds, winds, radiation balance, aerosol, cloud and water vapor (WV) motion, land and sea surface temperatures (LST/SST), as well as Earth's snow and ice cover, and (2) atmospheric sounding capabilities to retrieve vertical profiles, mainly of temperature and humidity, but also water vapor winds and trace gas information.

Examples of third generation imagers on current GEO systems include the Flexible Combined Imager (FCI) on Meteosat Third Generation (MTG), partly also the Advanced Baseline Imager (ABI) on GOES-R/Himawari-8), or additional bands on MODIS. (Next generation GEO satellites also include new sounding capabilities and change from narrow-band channel IR radiometer solutions to more complex systems as spectrometer technology to improve accuracy and vertical resolution of the developed products. Examples include an IASI-like (Infrared Atmospheric Sounding

Interferometer) on LEO MetOP system, the Interferometric Infrared Sounder (IIS) on board the Chinese FY-4 GEO mission and the 10-channel Michelson Fourier Transform spectrometer on MTG, which provide hyper-spectral sounding information in two bands, new sensor type, which is added to Meteosat and GOES third generation (MTGLI (Lightning Imager)/GOES-R – GLM (Geostationary Lightning Mapper)) and to the Chinese second generation FY-4 satellites, which is devoted to the detection and now-casting of lightning.

Next generation **LEO missions** also continue to improve the spatial, spectral and radiometric resolution and thus to improve the operational products for now-casting and climate research, but also the implementation of new operational products. Most LEO programs have been recently updated, such as, e.g. the FY-3 mission of CMA, the Russian Meteor-MN1 (http://planet.iitp.ru/; accessed March 24, 2011) and the USA second generation LEO Joint Polar Satellite System (JPSS). This multi-purpose imager represents a subset of the successful MODIS sensor onboard Terra/Aqua going into operational uses. It has the high potential to observe cloud and retrieve their properties with higher accuracy. The key instrument to construct vertical profiles of atmospheric temperature and moisture is an interferometric sounding sensor, the Cross-track Infrared Sounder (CrIS) in combination with the Advanced Technology Microwave Sounder (ATMS), which can significantly improve temperature and moisture retrievals in comparison to current in-orbit sounder technology (Kleespies, 2007).

It is worth mentioning the long-standing partnership between ESA and EUMETSAT and, specifically, the Meteosat Third Generation (MTG), which will soon take over the reins of providing critical data for weather forecasting. While this third generation of weather satellites guarantees the continuity of data for weather forecasting from geostationary orbit, the combination of its higher resolution images and state-of-the-art sounding products is set to take weather forecasting to the next level, particularly in the challenging task of "now-casting". Now-casting relates to the monitoring and prediction of rapidly evolving, and potentially damaging, weather phenomena, such as severe thunderstorms. Earlier detection of such phenomena is expected to increase the reaction time for issuing severe weather warnings and implementing measures to avoid potential catastrophic consequences. To ensure the continuity of data for at least the next 20 years and to provide the wide range of observations needed by meteorologists, the mission is based on a series of two types of satellite: four MTG-Imagers and two MTG-Sounders (ESA, 2019).

3.6.2 Remote Sensing for Storm Detection and Tracking

Satellite systems are used for detection and tracking of storms, cyclones, hurricanes, tornadoes and lightning.

Thunderstorm is defined as a towering cumulus cloud with strong updrafts and downdrafts, accompanied by lightning and thunder (Dalezios, 2017). It is driven by convection, in which moist warm air rises aloft within a cooler environment. Thunderstorms may be accompanied by various conditions including high winds, heavy rain, hail, and at times tornadoes. The **tornado** remains one of the most feared and destructive atmospheric phenomena on the planet. It is characterized by a relatively small-scale (Orlanski 1975) columnar vortex circulation, which is often, but not always, associated with a funnel cloud. Tornadoes can range in size from a few meters to 3 km in diameter, as defined by maximum funnel diameter just below cloud base or damage-path width. Most tornadoes and waterspouts rotate counterclockwise (cyclonically), although a few intense anticyclonic tornadoes have been documented.

Lightning is a transient, high-current discharge, whose path length is measured in kilometers. There are two types of events, namely intra-cloud (IC), when discharges occur within the cloud, and cloud-to-ground (CG) events. In most storms, ICs greatly outnumber CGs, the ratio approaching 100 or more in some severe storms. Lightning generally is associated with thunderstorms. Conditions conducive to lightning can occasionally occur within tropical storms and hurricanes and winter snowstorms in which convective activity becomes locally intense. Thus, an association between the number of thunderstorms and lightning occurrences was sought. Satellite cloud-top climatologies have also been used as a surrogate for lightning inputs in global climatological models.

A recent analysis investigated the spatiotemporal variability of the convective parameters and associated lightning flash rates during the period 1997–2013 including the El Niño and La Niña episodes (Upal et al., 2017). The analysis has indicated that, although the occurrence of convective activities during the El Niño (La Niña) is reduced (increased), the occurrence of lightning flashes is enhanced (diminished) during this period, which may result in the direct warming of the atmosphere in relation to changing patterns of regional climate. In another study, wildland fires originated by lightning in Catalonia (NE Spain) are systematically examined through lightning and precipitation data. The results suggest that there is no apparent link between the amount of precipitation and the holdover duration, indicating that the survival phase of lightning-ignitions is mainly driven by the daily cycle of solar heating (Pineda and Rigo, 2017). Moreover, in a similar research, the forest fires trends caused by lightning in climate change scenarios were investigated (Arif, 2012). The objective of this study was to find a reliable link between the lightning frequency and the main prognostic variables in climate models, and the possible feedback between the Climate change system and the Cloud-to-Ground lightning frequency was also considered. The study has indicated the interaction between carbon components changes, the vegetation changes and aerosol emission from forest fires and lightning activity.

Tropical cyclones are a special class of large-scale rotating wind systems, which occur over sizable portions of the tropical and subtropical oceans. Tropical systems derive their energy from clusters of thunderstorms. Tropical systems include loosely organized thunderstorm systems, namely mesocyclone systems (MCSs), and tightly structured features, such as tropical depressions, tropical storms and hurricanes. Important basic features include: (1) a central warm core; (2) an environment, where only small changes of wind direction and wind speed (shear) occur through the vertical extent of the troposphere and the tropical cyclone's center has comparatively little slope with height. The main precursor, which is required to maintain a tropical cyclone, is to have an ocean temperature of at least 26.5°C spanning through at a minimum depth of 50-meters (Thompson and Solomon, 2009).

Figure 3.17a, captured by the Copernicus Sentinel-3 mission on March 13, 2019, shows cyclone Idai west of Madagascar and heading for Mozambique. Cyclone Idai was also mentioned in Figure 3.6b, with reference to the Aeolus mission. Part of the width of the storm Idai is around 800–1000 km. The outcome of the storm Idai was widespread devastating disasters in several countries, such as Mozambique, Malawi and Zimbabwe, including thousands of human lives, houses, roads and croplands being submerged. The International Charter Space and Major Disasters and the

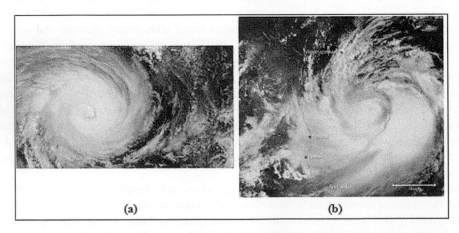

(a)   (b)

FIGURE 3.17 (a) Cyclone Idai west of Madagascar. Captured by the Copernicus Sentinel-3 mission, this image shows Cyclone Idai on March 13, 2019 west of Madagascar and heading for Mozambique (from ESA, March 13, 2019). (b) Sentinel-3 image of Cyclone Fani, India, May 1, 2019 (from ESA, May 2, 2019).

Copernicus Emergency Mapping Service provided maps of flooded areas based on satellite data to support emergency response efforts (image credit: ESA, the image contains modified Copernicus Sentinel data (2019), processed by ESA, CC BY-SA 3.0 IGO).

Similarly, Figure 3.17b, captured by Copernicus Sentinel-3 image on May 1, 2019, shows cyclone Fani in the Bay of Bengal, which was heading westwards and making landfall on India's east coast. Heavy rainfall and flooding occurred along the Odisha coast with wind speeds of up to 200 km per hour, which caused the evacuation of around 800,000 people from the nearby low-lying areas. In Figure 3.17b the storm width is estimated to be around 700–800 km. Once cyclone Fani made land-fall, it moved north-east, hitting Bangladesh and Bhutan (released May 2, 2019 at 4:45 p.m. Copyright contains modified Copernicus Sentinel data (2019), processed by ESA, CC BY-SA 3.0 IGO).

Tropical cyclones create surface temperature signatures, which can be detected by remotely sensed SST. Indeed, the result of a hurricane passage is mainly mixed layer deepening and surface heat losses in terms of SST cooling (Fu and Johanson, 2004). Specifically, SSTs are reduced 0.2°C to 0.4°C, due to Saharan dust outbreaks across the adjacent northern Atlantic Ocean lasting for sev-eral days. Moreover, extratropical cyclones, concentrated phytoplankton blooms caused by seasonal cycles or agricultural runoff and rapid influxes of glacial fresh water constitute additional sources of short-term SST fluctuation (Mears and Wentz, 2005). As already mentioned, SSTs play a key role in tropical cyclogenesis. Similarly, SSTs are equally significant in determining the formation of sea fog and sea breezes. Indeed, an air mass can be significantly affected by heat from underlying warmer waters over distances of 35 to 40 kilometers. In summary, the initiation of SSTs into atmospheric models is considered significant, since SSTs affect the behavior of the Earth's atmosphere above.

Storm tracking. Increased climate variability and change may cause an increase in the frequency and severity of extreme weather events. It is, thus, significant to develop and implement accurate forecasting and tracking of storm events. Indeed, cyclones Fani, Idai and Kenneth have recently brought devastation to millions (ESA, 2019). Specifically, ESA's satellite SMOS (Soil Moisture and Ocean Salinity) was built to measure soil moisture and ocean salinity to better understand the water cycle. In addition, the SMOS can support monitoring and improving forecasting of large storms. Specifically, SMOS carries a microwave radiometer to capture images of brightness temperature. Then, soil moisture and ocean salinity can be retrieved through these measurements, which cor-respond to radiation emitted from the surface of the Earth. Indeed, microwave emission from the surface is affected by strong winds over the oceans, which create waves. In other words, the changes in radiation can be linked directly to the strength of the wind over the sea. Figure 3.18 shows surface winds under cyclone Idai from SMOS (ESA, May 14, 2019).

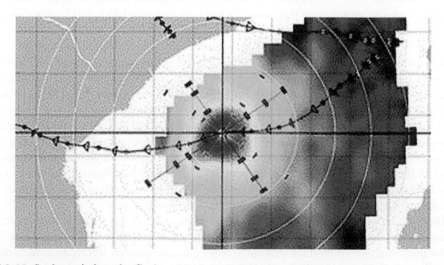

FIGURE 3.18 Surface winds under Cyclone Idai from SMOS (from ESA, May 14, 2019).

Severe Weather Warning. Measurements of relatively weak water vapor and carbon dioxide absorption lines can provide key information for monitoring the evolution of the lower-tropospheric thermodynamic state. This is because advanced IR sounders within the IR window region can isolate these weak absorption lines. Moreover, hyperspectral IR sounders produce vertical profile information, which can be very important for severe storm now-casting and short-term forecasting in the mesoscale environment. Typical examples include the indices related to storms, atmospheric stability and the structure of the boundary layer.

3.7 CLIMATIC ANALYSIS AND CLIMATE CHANGE

Greenhouse gases, aerosols or feedback mechanisms of moisture and clouds constitute significant factors affecting climate change. Remote sensing potential in terms of hyperspectral IR instruments, such as AIRS, IASI and CrIS, are considered significant data sources in relevant studies. Specifically, important greenhouse gases in the atmosphere, such as the upper tropospheric water vapor (UTWV), contribute to global warming. However, more information is required for UTWV in current reanalysis data and climate models. Indeed, UTWV has been recently analyzed through hyperspectral observation data from AIRS, IASI and CrIS. In addition, recent studies of seasonal variance in monsoon regions, model assessment and water vapor feedback mechanisms have used UTWV products from hyperspectral sounders.

Moreover, clouds constitute a considerable factor in the energy balance and the hydrological cycle in the earth atmosphere system, in studies of climate variability and change. However, the various cloud types, non-isotropic radiance distribution, complex structures, as well as the variety of cloud depictions in climate models, lead to large uncertainties in the cloud feedback mechanisms. As a result, the study of clouds remains an important component of climate change research, where satellite observations can provide valuable cloud information. Specifically, cloud profiles and cloud phase information have been produced by CloudSat and CALIPSO, as well as by short wavelength IR measurements of clouds through MODIS, GOES and MISR (Multi-angle Imaging SpectroRadiometer).

Ocean temperature is related to ocean heat content, which constitutes a significant component of global warming. Figure 3.19 presents the annual mean temperature change for the land and the ocean (1950–2020) (from NASA GISS, 2018). In Figure 3.19, surface temperature data indicated substantial global average warming, which was not shown in early versions of satellite and radiosonde data. However, there is still a significant difference between what climate models predict and what the satellite data show for warming of the lower troposphere. Specifically, the climate models predict significantly more warming than what the satellites measure (RSS, 2017). The climate models also overpredict the results of the radiosonde measurements.

In the tropics, this significant global average warming in the troposphere compared to the surface becomes more obvious, which is present in the models but not in observed data. Specifically, in the tropics, surface temperature changes are amplified in the free troposphere. Indeed, tropospheric amplification of surface temperature anomalies is caused by the release of latent heat by moist, rising air under convective conditions (Santer et al., 2006). There are changes with the increase of the time horizon, since models and observations show similar amplification behavior for monthly and inter-annual temperature variations, but not for decadal temperature changes. Specifically, the UAH (University of Alabama at Huntsville) data set has shown an overall warming trend since 1998, which is less than the RSS version. In summary, the troposphere has warmed and the stratosphere has cooled during the past 40 years based on satellite data sets, and both of these trends are consistent with the increase of atmospheric concentrations of greenhouse gases.

Satellites, which observe the Earth, provide a clear picture of changes across the entire planet. They measure and monitor the vast oceans, land, atmosphere and areas that are difficult to reach, such as the polar regions. To produce data suitable for climate research, information from many different satellite missions needs to be combined to produce data sets that span decades. The Climate

Annual Mean Temperature Change for Land and for Ocean

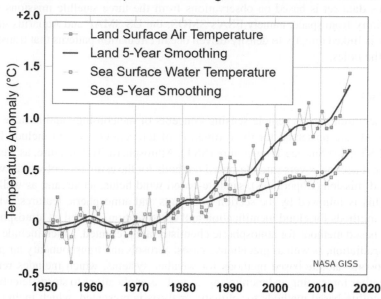

FIGURE 3.19 Annual (thin lines) and five-year lowest smooth (thick lines) for the temperature anomalies averaged over the Earth's land area and sea surface temperature anomalies (blue line) averaged over the part of the ocean that is free of ice at all times (open ocean) (from NASA GISS, 2018).

Change Initiative of ESA (ESA, 2019) supports the long-term generation of data on climate variables for more accurate carbon modeling. These data include measuring concentrations of chlorophyll in the oceans and carbon dioxide in the atmosphere, as well as mapping the amount of vegetation on land, among others. Moreover, the global ocean circulation relies primarily on polar regions. Indeed, Satellite data used in the Climate Change Initiative uses satellite data, which provide predictions of ice retreat and are considered essential for monitoring rapid changes in the cryosphere and for the long-term tracking of SST.

Some examples are presented. The Danish Meteorological Institute recently reported that heat-wave conditions had occurred in Greenland during an early Arctic summer in June, causing wide-spread melting across its icesheet surface (Figure 3.20a) (ESA, July 5, 2019). Regional changes are also linked to periodic inter-annual climate events, such as El Niño. Specifically, salinity plays a role in the intensification of the global water cycle. Measurements of sea-surface salinity and SST, which determine the thickness of the surface mixed layer, have the potential to help understand the development of extreme weather events, such as cyclones (Figure 3.20b) (ESA, May 14, 2019).

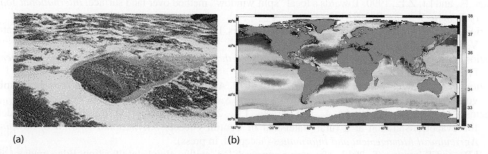

(a)   (b)

FIGURE 3.20 (a) Melting ice over Greenland (from ESA, July 5, 2019). (b) Global salty water (from ESA, May 14, 2019).

Moreover, the thermohaline ocean circulation depends on sea-surface salinity with a data set of nine years. This data set is based on observations from the three satellite missions that measure sea-surface salinity from space, namely ESA's SMOS, the US SMAP and Aquarius missions. Sea-surface salinity is linked directly to density-driven ocean circulation patterns that transfer heat from the Tropics to the poles.

3.8 SUMMARY

The initial presentation covers remote sensing concepts of meteorology and climate. Then, satellite-based methods are presented for the estimation of temperature, which include Land Surface Temperature (LST), Sea Surface Temperature (SST), Atmospheric Temperature, as well as El Niño/ La Niña. The next section covers satellite-based methods for the estimation of winds, which include the Aeolus wind mission (Copernicus), satellite-based wind fields, jet stream, as well as local and daily winds. This is followed by the section on clouds and atmospheric features, which involves satellite-based methods for cloud identification and classification, as well as atmospheric features. Then, satellite-based methods for atmospheric chemistry are presented, which include stratospheric ozone and UV radiation, as well as greenhouse gases, aerosols and transboundary air pollution. The next section covers satellite-based methods for weather systems, which includes remote sensing systems for weather forecasting and monitoring, as well as applications to storm detection and tracking. Finally, satellite-based methods for climatic analysis is presented, which includes a satellite-based methodology for agroclimatic classification and zoning, as well as satellite-based methods to address climate change issues.

REFERENCES

Abderrahim, B., Katsaros, K.B., Drenman, W.M., and Forde, E.B., 2002. Daily surface wind fields produced by merged satellite data. *Geophysical Monograph*, 127, 343–349.

Adamson, K., Bargellini, P., Nogueira, T., Nett, H., and Caspar, C., 2010. *ADM-Aeolus: Mission Planning, Re-use Autonomy and Automation, Proceedings of the SpaceOps 2010 Conference*, Huntsville, ALA, USA, April 25–30, 2010.

Ameur, S., 2004. Cloud classification using the textural features of Meteosat images. *International Journal of Remote Sensing*, 25, 4491–4503, 10.1080/01431160410001735120

Arif, A., 2012. The Cloud-to-Ground Lightning activity and climate change. Ph.D. thesis, March 2012, 127.

Babita, K., Mohammad, T., Shahfahad, S., Javed, M., Mohd Firoz, K., and Atiqur, R., 2018. Satellite-driven land surface temperature (LST) using landsat 5, 7 (TM/ETM+ SLC) and Landsat 8 (OLI/TIRS) data and its association with built-up and green cover over Urban Delhi, India. *Remote Sensing in Earth Systems Sciences*, 1–16, 10.1007/s41976-018-0004-2

Bankert, R.L., Mitrescu, C., Miller, S.D., and Wade, R.H., 2009. Comparison of GOES cloud classification algorithms employing explicit and implicit physics. *Journal of Applied Meteorology and Climatology*, 48, 1411–1421.

Becker, F., and Li, Z.L., 1990. Towards a local "split window" method over land surface. *International Journal of Remote Sensing*, 3, 369–393.

Dalezios, N.R., 2011. Remote Sensing in Water Resources. Lecture notes, University of Thessaly Publications, Version 2, 248p.

Dalezios, N.R. (ed.), 2017. *Environmental Hazards Methodologies for Risk Assessment and Management*. IWA, London, UK, 534.

Dalezios, N.R., Blanta, A., and Spyropoulos, N.V., 2012. Assessment of remotely sensed drought features in vulnerable agriculture. *Natural Hazards and Earth System Sciences*, 12, 3139–3150.

Dalezios, N.R., Dercas, N., Blanta, A., and Faraslis, I.N., 2020. Remote sensing in water balance modelling for evapotranspiration at a rural watershed in Central Greece. *International Journal of Sustainable Agricultural Management and Informatics* (accepted, in press).

Dim, J.R., and Takamura, T., 2013. Alternative approach for satellite cloud classification: Edge gradient application. *Advances in Meteorology*, 2013, Article ID 584816, 8, doi:10.1155/2013/584816

ESA, 2018. Suphur dioxide, October 2018, https://www.esa.int/ESA_Multimedia/Images/2018/10/Sulphur_dioxide#.X_6wAr6HauE.link

ESA, 2019. ADM-Aeolus Science Report, ESA, SP-1311, April 2008, http://esamultimedia.esa.int/docs/SP-1311_ADM-Aeolus_FINAL_low-res.pdf.

ESA, 2020a. Coronavirus: nitrogen dioxide emissions drop over Italy, March 2020, https://www.esa.int/ESA_Multimedia/Videos/2020/03/Coronavirus_nitrogen_dioxide_emissions_drop_over_Italy#.X_6uFgYNljk.link

ESA, 2020b. COVID-19: nitrogen dioxide over China, March 2020, http://www.esa.int/Applications/Observing_the_Earth/Copernicus/Sentinel-5P/COVID-19_nitrogen_dioxide_over_China#.X_6ub4s_BTM.link

ESRL/NOAA/PSD, 2011. Earth System Research Laboratory Physical Sciences Division, credit: http://www.esrl.noaa.gov/psd/map/images/sst/sst.gif

EUMETSAT, 2014. Jet stream tracking from January 29, 2014 to February 11, 2014

Fiolleau, T., and Roca, R., 2013. An algorithm for the detection and tracking of tropical mesoscale convective systems using infrared images from geostationary satellite. *IEEE Transactions on Geoscience and Remote Sensing*, 7, 4302–4315.

Fu, Q., and Johanson, C.M., 2004. Stratospheric influences on msu-derived tropospheric temperature trends: a direct error analysis. *Journal of Climate*, 17, 24, 4636–4640.

Jin, W., Gong, F., Zeng, X., and Fu, R., 2016. Classification of clouds in satellite imagery using adaptive fuzzy sparse representation. *Sensors*, 16, 2153.

Kanellou, E., Spyropoulos, N., and Dalezios, N.R., 2011. Geoinformatic intelligence methodologies for drought spatiotemporal variability in Greece. *Water Resources Management*, 26, 1089–1106.

Kärner, O., 2000. A multi-dimensional histogram technique for cloud classification. *International Journal of Remote Sensors*, 21, 2463–2478. doi: 10.1080/01431160050030565

Kidd, C., Levizzani, V., and Bauer, P., 2009. A review of satellite meteorology and climatology at the start of the twenty-first century. *Progress in Physical Geography*, 33, 474–489.

Kleespies, T.J., 2007. Relative information content of the advanced technology microwave sounder and the combination of the advanced microwave sounding unit and the microwave humidity sounder. *IEEE Transactions on Geoscience and Remote Sensing*, 45, 2224–2227.

Krien, Y., Gaël, A., Raphaël, C., Chris, R., Ali, B., Jamal, K., Didier, B., Islam, A.K.M.S., Durand, F., Laurent, T., Philippe, P., and Zahibo, N., 2018. Can we improve parametric cyclonic wind fields using recent satellite remote sensing data? *Remote Sensing*, 10, 1963

Lelieveld, J., 2003. Geostationary Satellite Observations for Monitoring Atmospheric Composition and Chemistry Applications. Report, 64.

Li, X., Chen, M., Le, H.P., Wang, F., Guo, Z., Iinuma, Y., Chen, J., and Herrmann, H., 2016. Atmospheric outflow of PM2.5 saccharides from megacity Shanghai to East China Sea: Impact of biological and biomass burning sources. *Atmospheric Environment*, 143, 1–14.

May, D.A., Stowe, L.L., Hawkins, J.D., and McClain, E.P., 1992. A correction for Saharan dust effects on satellite sea surface temperature measurements. *Journal of Geophysic Research*, 97, C3, 3611–3619.

Mears, C.A., and Wentz, F.J., 2005. The effect of diurnal correction on satellite-derived lower tropospheric temperature. *Science*, 309, 5740, 1548–1551

Menzel, W., Schmit, T., Zhang, P., and Li, J., 2017. Satellite based atmospheric infrared sounder development and applications. *The Bulletin of the American Meteorological Society*, 10.1175/BAMS-D-16-0293.1.

NASA GISS, 2018. https://www.giss.nasa.gov/research/2018.html

NOAA/ESRL/PSD, 2011. Earth System Research Laboratory Physical Sciences Division. Credit: http://www.esrl.noaa.gov/psd/map/images/sst/sst.gif

Obiefuna, J.N., Nwilo, P.C., Okolie, C.J., Emmanuel, E.I., and Daramola, O.E., 2018. Dynamics of land surface temperature in response to land cover changes in Lagos Metropolis. *Nigerian Journal of Environmental Sciences and Technology (NIJEST)*, 2, 2, 148–159.

Orlanski, I., 1975. A rational subdivision of scales for atmospheric processes. *Bulletin of the American Meteorological Society, AMS*, 56, 5, 527–530.

Pineda, N., and Rigo, T., 2017. The rainfall factor in lightning-ignited wildfires in Catalonia. *Agricultural and Forest Meteorology*, 239, 2017, 249–263.

Prata, A.J., and Platt, C.M.R., 1991. *Land surface temperature measurements from the AVHRR. Proceedings of 5th A VHRR Data Users Conference*, Tromso, Norway, 433–438.

Price, J.C., 1984. Land surface temperature measurements from the split window channels of the NOAA 7 Advanced Very High-Resolution Radiometer. *Journal of Geophysic Research*, 89, D5, 7231–7237.

Remote Sensing Systems (RSS), 2017. Climate Analysis: Tropospheric Temperature. remss.com. Retrieved February 3, 2017.

Remote Sensing Systems (RSS)/MSU and AMSU Data/Description, 2017. Retrieved February 6, 2017.

Rosenzweig, C., and Hillel, D., 2008. *Climate Variability and the Global Harvest: Impacts of El Niño and Other Oscillations on Agroecosystems.* Oxford University Press, United States, 31

Santer, B.D., Penner, J.E., Thorne, P.W., Collins, W., Dixon, K., Delworth, T.L., Doutriaux, C., Folland, C.K., Forest, C.E., Hansen, J.E., Lanzante, J.R., Meehl, G.A., Ramaswamy, V., Seidel, D.J., Wehner, M.F., and Wigley, T.M.L., 2006. How Well Can the Observed Vertical Temperature Changes be Reconciled with our Understanding of the Causes of these Changes?, pp. 89–118, in: *Temperature Trends in the Lower Atmosphere: Steps for Understanding and Reconciling Differences.* S.J. Hassol, C.D. Miller et al.

Simon, P., 2004. Atmosphere remote sensing: Validation of satellite products. *Dragon Advanced Training Course in Atmosphere Remote Sensing,* 43.

Sobrino, J.A., Caselles, V., and Coll, C., 1993. Theoretical split window algorithms for determining the actual surface temperature. *Il Nuovo Cimento,* 16, 3, 219–236.

Straume, A.G., Ingmann, P., and Wehr, T., 2012. *ESA's spaceborne lidar missions: Candidate and selected concepts for wind, aerosols and CO2 monitoring.* Proceedings of the 26th International Laser Radar Conference (ILRC 26), Porto Heli, Peloponnesus, Greece, June 25–29, 2012.

Tang, R., Deyou, L., Guoqi, H., Zhimin, M., and de Young, B., 2014. Reconstructed wind fields from multi-satellite observations. *Remote Sensing,* 6, 2898–2911, doi:10.3390/rs6042898

Tapakis R., and Charalambides, A.G., 2013. Equipment and methodologies for cloud detection and classification: A review. *Solar Energy,* 95, 392–430

Thies, B., and Bendix, J., 2011. Review: Satellite based remote sensing of weather and climate: recent achievements and future perspectives. *Meteorological Applications,* 18, 262–295.

Thompson, D.W.J., and Solomon, S., 2009. Understanding recent stratospheric climate change (PDF). *Journal of Climate,* 22, 8, 1934.

UAH 6.0 TMT data (trend data at bottom of file). nsstc.uah.edu. The National Space Science and Technology Center. January 2017. Retrieved February 3, 2017.

Ulivieri, C., and Cannizzaro, G., 1985. Land surface temperature retrievals from satellite measurements. *Acta Astronautica,* 12, 12, 977–985.

Ulivieri, C., Castronuovo, M.M., Francioni, R., and Cardillo, A., 1994. A split window algorithm for estimating land surface temperature from satellites. *Advances in Space Research,* 14, 3, 59–65.

Upal, S., Devendraa, S., Midya, S.K., Singh, R.P., Singh, A.K., and Kumar, S., 2017. Spatio-temporal variability of lightning and convective activity over South/ South-East Asia with an emphasis during El Niño and La Niña. *Atmospheric Research,* 197, 150–166.

US National Research Council, 2000. *Committee on Earth Studies. Atmospheric Soundings. Issues in the Integration of Research and Operational Satellite Systems for Climate Research: Part I. Science and Design.* Washington, DC: National Academy Press, 17–24

Van de Griend, A.A., and Owe, M., 1993. On the relationship between thermal emissivity and the normalized difference vegetation index for natural surfaces. *International Journal of Remote Sensors,* 14, 1119–1137.

Vinnikov, K.Y., Grody, N.C., Robock, A., Stouffer, R.J., Jones, P.D., and Goldberg, M.D., 2006. Temperature trends at the surface and in the troposphere. *Journal of Geophysical Research,* 111, D3, D03106

Weng, Q., 2003. Fractal Analysis of Satellite-Detected Urban Heat Island Effect (PDF). *Photogrammetric Engineering and Remote Sensing,* 555–566.

Weng, Q., Lu, D., and Schubring, J., 2004. Estimation of land surface temperature – vegetation abundance relationship for urban heat island studies. *Remote Sensing of Environment,* 89, 4, 467–483.

Zareie, S., Khosravi, H., and Nasiri, A., 2016. Derivation of Land Surface Temperature from Landsat Thematic Mapper (TM) sensor data and analysing relation between Land Use changes and Surface Temperature. *Solid Earth Discuss.* doi:10.5194/se-2016-22

Zhang, Y., Balzter, H., Liu, B., and Chen, Y., 2016. Analyzing the impacts of urbanization and seasonal variation on land surface temperature based on subpixel fractional covers using landsat images. *IEEE Journal of Selected Topics in Applied Earth Observations and Remote Sensing,* 10, 4, 1344–1356.

Zhang, H.-M., Bates, J.J., and Reynolds, R.W., 2006. Assessment of composite global sampling: Sea surface wind speed. *Geophysical Research Letters,* doi:10.1029/2006GL027086

Zou, C.-Z., Goldberg, M.D., Cheng, Z., Grody, N.C., Sullivan, J.T., Cao, C., and Tarpley, D., 2006. Recalibration of microwave sounding unit for climate studies using simultaneous nadir overpasses. *Journal of Geophysical Research,* 111, D19, D19114.

4 Hydrology and Water Resources

4.1 INTRODUCTION: REMOTE SENSING CONCEPTS OF HYDROLOGY AND WATER RESOURCES

There is an increasing trend for the use of remote sensing in hydrology and water resources. Indeed, the reliability and efficiency of remote sensing data and methods have gradually increased (Dalezios et al., 2012; Dalezios, 2014; Dalezios et al., 2017; Dalezios et al., 2020a). Specifically, current satellite systems provide continuous data in space and time over the globe and, thus, they are potentially better and relatively inexpensive tools for regional applications in hydrology and water resources (Thenkabail et al., 2004). At the present time, operational users can rely on a continued supply of remote sensing data for hydrological applications, such as operation of hydrological systems and water resources management or monitoring components of the hydrological cycle than using conventional meteorological data (Dalezios et al., 2020a). Figure 4.1 shows an illustration of the hydrological cycle. In hydrology and water resources, appropriate remote sensing systems are mainly considered as weather radars and low-Earth-orbiting (LEO) satellite systems, since the requirements are for daily data acquisition and coverage. Satellite-based Earth-atmosphere observations exploit geostationary Earth-orbiting (GEO) and LEO satellite systems, which provide low spatial and high temporal resolution data. The series of geosynchronous, polar-orbiting meteorological satellites, such as the National Oceanic and Atmospheric Administration/Advanced Very High-Resolution Radiometer (NOAA/AVHRR) or Copernicus, fulfill the above requirements and there are already long series of databases (Copernicus, 2015; Eumetsat, 2015; NOAA KLM, 2015; Dalezios et al., 2020a).

In Chapter 5, remote sensing data and methods for hydrology and water resources are considered. At first, remote sensing of the hydrological cycle is presented, including components, such as evapotranspiration, interception, soil moisture and groundwater. This is followed by remote sensing of water quality, remote sensing modeling of hydrological systems, remote sensing of snow, remote sensing for agrohydrology and remote sensing of urban hydrology.

4.2 REMOTE SENSING OF THE HYDROLOGICAL CYCLE

This section covers remote sensing methods for measuring and estimating components of the hydrological cycle, such as evapotranspiration, soil moisture, interception and groundwater.

4.2.1 REMOTE SENSING OF EVAPOTRANSPIRATION (ET)

There is recently significant progress in the detection of several parameters through remote sensing, such as surface temperature, surface soil moisture, surface albedo, incoming solar radiation and vegetative cover. Evapotranspiration (ET) over a range of temporal and spatial scales can be estimated indirectly by combining remote sensing observations with ancillary meteorological data (Schultz and Engman, 2000).

Remote sensing of evaporation variables. Evaporation of lakes or reservoirs or at upwind shore stations can be computed from measurements of air temperature and radiation made at the same locations, which allows the recording of several parameters in sequence on a single multichannel recorder for a typical time period of ten days or two weeks. Moreover, remotely sensed estimation

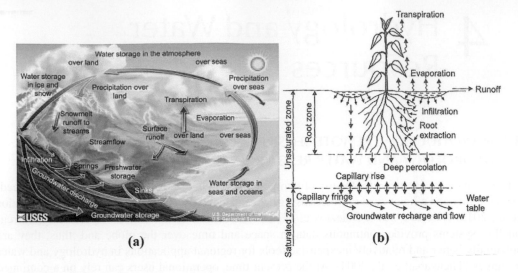

FIGURE 4.1 The hydrological cycle (a) global scale (from USGS), (b) micro-scale.

of evaporation is conducted through several parameters by measuring the electromagnetic radiation in a specific waveband reflected or emitted from the Earth's surface. Specifically, the surface temperature can be estimated from Multispectral Scanner (MSS) measurements at thermal infrared (IR) wavelengths of the emitted radiant flux (Engman and Gurney, 1991). Similarly, the surface albedo can be estimated from measurements covering the entire visible and near-infrared waveband under clear-sky conditions (Brest and Goward, 1987). Moreover, the incoming solar radiation can be estimated from satellite observations of cloud cover primarily from geosynchronous LEO satellite systems using MSS in the visible, near-infrared and thermal infrared parts of EMS (Gautier et al., 1980).

Remote sensing of evapotranspiration (ET) variables. Remotely sensed direct estimation of ET is affected by atmospheric parameters, such as near-surface air temperature, near-surface water vapor gradients and near-surface winds, for which there has been little progress. Moreover, with remote sensing areal coverage can be achieved in the spatial extrapolation process of ET. Recently, researchers have begun using satellite data (Granger, 1997) to estimate regional actual evapotranspiration by measuring the electromagnetic radiation and estimating surface temperature, surface albedo and incoming solar radiation as for evaporation described earlier. In addition, the soil moisture may be estimated using the measurement of microwave properties of the soil, i.e. microwave emission and reflection or backscatter from soil. Furthermore, the EOS program should provide the necessary data for evaluating ET on local, regional and global scales. Indeed, sensible heat flux components can be measured by the EOS instruments, whereas the latent heat flux cannot be measured directly, however, EOS instruments can provide some sampling capability.

There have been recent advancements in the availability and quality of satellite-based estimation of all the main variables of the hydrological cycle (McCabe et al., 2008), leading to feasible water balance assessment in areas with poor ground information. Moreover, the application of remote sensing to evapotranspiration, water-use efficiency and irrigation water requirements is growing rapidly, mainly due to the increasing availability of new satellite systems along with their capabilities (Dalezios et al., 2019b). Satellite-based earth-atmosphere observations exploit GEO systems, such as the recently available evapotranspiration product from the Moderate Resolution Imaging Spectroradiometer (MODIS) Global Evapotranspiration Project (Mu et al., 2011). Similarly, LEO satellite systems have also developed a new evapotranspiration product, which is available from Landsat-8. The recent Sentinel system, which is part of the Copernicus program from the European

Space Agency (ESA), has developed remote sensing applications in evapotranspiration analysis (Malenovský et al., 2012; ESA, 2014; Dalezios et al., 2018b). There is a very recent trend, which consists of increasing the number of available bands in new satellites, such as Quickbird, IKONOS, WV-2, WV-3 or WV-4. Similarly, there is a tendency to further improve the spatial resolution of these satellites, reaching the level of micro-remote sensing with resolution of one meter or less resulting in new and valuable information for environmental applications (Dalezios et al., 2018b).

The existing limitations in direct measurements of evaporative fluxes from space suggest the use of simulation methods along with remote sensing observations of related variables. Indeed, the assessment and evaluation of the large-scale performance of these evaporation models is considered extremely useful and can contribute to a better understanding of the Earth energy and water cycles. Specifically, the recent Soil Moisture and Ocean Salinity (SMOS) system for evapotranspiration focuses on the investigation of the potential advantages of using SMOS soil moisture and vegetation optical depth (VOD) to reduce uncertainty in satellite-based evaporation estimates. Moreover, the Amsterdam methodology (GLEAM: Global Land Evaporation: the Amsterdam Methodology) has been applied to simulate evaporation fields over continental Australia (Miralles et al., 2011). Furthermore, in-situ networks of eddy-covariance towers and soil moisture probes have been used to validate model estimates of terrestrial evaporation and root-zone soil moisture. Indeed, results from this research effort indicate that it is possible to improve the soil moisture and evaporation estimates of GLEAM by using the assimilation of SMOS soil moisture, since both ascending and descending SMOS soil moisture have a positive impact on the model simulations. Specifically, the descending soil moisture retrievals from AMSR-E have a sufficiently high quality to improve the model simulations, where the positive impact is approximately 5% lower than when SMOS soil moisture is assimilated. On the other hand, the quality of the ascending AMSR-E soil moisture is found to be too low to improve the model simulations at the examined study areas.

There is a comprehensive survey of published methods to estimate ET through remote sensing, which highlights major issues and challenges (Kalma et al., 2008; Mateos et al., 2013; Calera et al., 2017; Campos et al., 2017). The commonly applied remotely sensed ET methods can be classified into two classes: analytical methods and semi-empirical methods. In particular, the analytical methods emphasize the physical processes at the appropriate scale with varying complexity and use a variety of direct and indirect data sources from remote sensing and ground-based measurements (Li et al., 2009). On the other hand, semi-empirical methods are usually based on empirical relationships and use data mainly derived from remote sensing observations with minimum ground-based measurements.

As already mentioned, most of the crop ET estimation methodologies from remote sensing are based either on the leaf area model (Monteith and Unsworth, 1990), or on developments and adjustments of the Penman–Monteith equation (Dalezios et al., 2020a). Specifically, meteorological conditions and several agronomic and crop features are required, such as Leaf Area Index (LAI), surface albedo, crop height and soil water status. Indeed, the Penman–Monteith equation can be applied to estimate ET based on the determination of surface and aerodynamic resistances of a crop cover with known characteristics. The Penman–Monteith equation is considered the standard procedure to estimate crop water requirements under standard conditions in irrigation water management, known as the FAO-56 method (Allen et al., 1998). Moreover, for the surface energy balance method or for the computation of soil water balance, the estimation of crop ET requires additional data under non-standard conditions of water-stressed crops (Dalezios et al., 2020a). The FAO-56 method is widely used in irrigation practice and is based on a two-step procedure, since ET is estimated as a product of two factors (Doorenbos and Pruitt, 1977; Allen et al., 1998).

There are three main remote sensing approaches for the estimation of ET: (1) remote sensing measurements directly incorporated into the Penman–Monteith equation (RS-PM); (2) reflectance-based crop coefficient or reflectance-based K_{cb}; and (3) remotely sensed surface energy balance (RSEB) method (Dalezios et al., 2020a). Figure 4.2 shows a schematic representation of the

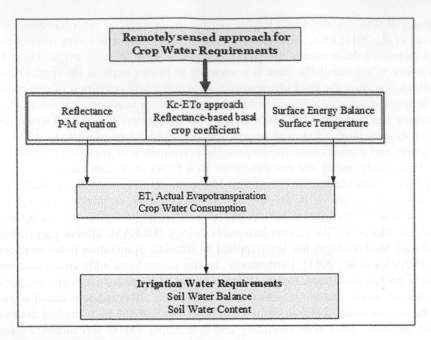

FIGURE 4.2 Remotely sensed approaches for evapotranspiration estimation.

framework for the integration of the different models for the assessment of crop water requirements (CWR) and net irrigation water requirements (NIWR) (Calera et al., 2017; Dalezios et al., 2019b). A brief presentation of the three methods follows.

4.2.1.1 Remotely Sensed Penman–Monteith (RS-PM) Direct Approaches

The direct estimation of the Penman–Monteith (RS-PM) equation is based on the canopy parameters related to the surface properties and specifically the surface and canopy resistances, namely r_s and r_c, respectively, and the net radiation (R_n) (Calera et al., 2017; Dalezios et al., 2020a). The Penman–Monteith (RS-PM) equation can be used to estimate the maximum fluxes of evaporation (E) from soil and transpiration (T) from plant leaves (Calera et al., 2017). The aforementioned three parameters depend on three parameters, which are derived from RS data. Specifically, the canopy resistance r_c is inversely related to the active LAI and depends on the maximum crop-specific resistance of a single leaf. Moreover, the active LAI is the index of the leaf area, which contributes to the surface heat and vapor transfer and it generally covers the upper, sunlit portion of a dense canopy (Allen et al., 1998).

Another parameter derived from RS data is the canopy height, which is considered the canopy architecture parameter used in the estimation of r_c. Moreover, the radiative component of the Penman–Monteith (RS-PM) equation prevails over the aerodynamic term in irrigated fields, which allows a fixed value for crop height to be considered without loss of accuracy. It is also mentioned that the RS-PM equation can be considered stable when applied over agricultural fields, since stability is affected by the meteorological conditions. The computation of the other two remotely sensed derived parameters, such as LAI and surface albedo from VIS-NIR observations, is based either on empirical relationships with different VIs or physically-based methods, such as radiative transfer models (Shi et al., 2016), which allows for a validation based on the estimated accuracy of albedo and LAI (Calera et al., 2017). An analogous application could be considered the MOD16 global ET product (Mu et al., 2011), which has also been applied in regional coverage of natural vegetation and is currently the basis of an operational irrigation advisory service in Austria, Italy and Australia (Vuolo et al., 2015).

4.2.1.2 The Reflectance-based Basal Crop Coefficient (Kcb)

This is the $K_c - ET_o$ procedure, which is a two-step procedure, either single or dual coefficient, based on the following P-M equation:

$$ET = K_s\,K_c\,ET_o = \left(K_s\,K_{cb} + K_e\right)ET_o \tag{4.1}$$

where K_c is the crop coefficient, K_{cb} is the basal crop coefficient, which is analogous to a transpiration coefficient with tabulated values adapted to local conditions, K_e is the evaporative component of the bare soil fraction (Wright, 1982) calculated on the basis of soil water balance in the upper soil layer, and K_s is the water stress coefficient, which refers to the soil water content through the water balance in the root soil layer (Calera et al., 2017). The crop coefficient K_c includes three parameters, namely K_{cb}, K_e and K_s. The basal crop coefficient (K_{cb}) is defined as the ratio between plant transpiration in the absence of water stress and reference evapotranspiration (ET_o) and the evolution of K_{cb} over time can be represented by a smooth continuous function. The term $K_{cb}\,ET_o$ represents the maximum transpiration of an unstressed canopy, the term $K_s\,K_{cb}\,ET_o$ represents the actual transpiration of a canopy and the term $K_e\,ET_o$ is the evaporation from bare soil fraction.

There is a direct relationship between K_{cb} and the fraction of photosynthetic active radiation (fPAR), as well as a relationship with Vegetation Indices (VIs), which are derived from multispectral satellite images (Bausch and Neale, 1987). Several researchers (Choudhury et al., 1994; Allen et al., 2005; Irmak et al., 2011) also have shown that vegetation indices from remote sensing could be used to assess "basal" crop coefficients for agricultural crops. Moreover, the development of different measurement methods for crop ET, such as lysimeters, eddy covariance and Bowen ratio methods, has contributed with additional data for the development of empirical Kcb-VI relationships, increasing, thus, the number of crops to be considered for ET estimation (Calera et al., 2017).

All VIs are based on high-quality reflectance measurements from any particular sensor, whereas the Normalized Difference Vegetation Index (NDVI), being still one of the most widely used VI today, can be calculated from a data set utilizing only red and NIR bands. Moreover, the computation of NDVI constitutes most common remote sensing approach for the assessment of K_c, which is called the $K_c - NDVI$ approach (Bhavsar and Patel, 2016). This falls into the category of assessing the relationship between spectral indices and crop coefficients, which is highly recommended today (Kamble et al., 2013). The advantage of using Kcb-VI is recognized for almost every crop and especially for fruit trees (Calera et al., 2017). The improvements in farm management practices have a significant effect on the actual value of the crop coefficient, and K_{cb} derived from VI can capture these variations. Thermal-based remote sensing models (Irmak et al., 2011) allow for the determination of latent heat fluxes, leading, thus, to the estimation of actual ET of crops. When these methods are applied over irrigated areas, where the conditions are usually standard, they result in the calibration of the single Kc-VIs relationships without expensive field campaigns measuring ET (Calera et al., 2017).

4.2.1.3 Remote Sensing Surface Energy Balance (SEB)

At present, the remote sensing surface energy balance (SEB) approaches and algorithms and more specifically the residual methods constitute a significant group of methodologies (Calera et al., 2017). Remote sensing based SEB algorithms convert satellite sensed radiances into land surface features, such as vegetation indices, leaf area index, albedo, surface roughness, surface emissivity and surface temperature to estimate ET as a "residual" of the land surface energy balance equation (Gowda et al., 2007):

$$LE = Rn - H - G \tag{4.2}$$

where LE is the latent heat flux (W/m^2), R_n is net radiation (W/m^2), G is soil heat flux (W/m^2) and H is sensible heat flux (W/m^2). (W/m^2) in latent heat flux can easily later converted to mm.

The estimation of H differs in most recent SEB models (Gowda et al., 2007). Indeed, these models include: the Surface Energy Balance Algorithm for Land (SEBAL; Bastiaanssen et al., 2005) and Mapping Evapotranspiration with Internalized Calibration (METRIC; Allen et al., 2011) that uses hot and cold pixels within the satellite images to develop an empirical temperature difference equation; the Two Source Model (TSM; Kustas and Norman, 1996), where the energy balance of vegetation and soil are modeled separately and then combined to estimate total LE; and the Surface Energy Balance Index (SEBI; Menenti and Choudhury, 1993) based on the contrast between wet and dry areas. Other variations of SEBI include the Simplified Surface Energy Balance Index (S-SEBI; Roerink et al., 2000), and the Surface Energy Balance System (SEBS; Su, 2002). Other EB models are: the excess resistance (kB^{-1}; Su, 2002), the aerodynamic temperature parameterization models (Chávez et al., 2005), the beta (β) approach (Chehbouni et al., 1996), and most recently the ET Mapping Algorithm (ETMA; Gowda et al., 2007). Finally, two further methods should be mentioned, namely the Triangle Method derived from the Crop Water Stress Index (CWSI), and the Trapezoid Method after introducing a Water Deficit Index (WDI) for ET estimation based on the Vegetation Index/Temperature trapezoid to extend the application of CWSI over fully to partially vegetated surface areas (Li et al., 2009). Soil-Vegetation-Atmosphere-Transfer (SVAT) models are also very effective, but very difficult to apply.

Similarly, there are remotely sensed algorithms for the estimation of daily evapotranspiration, which essentially calculate the sensible heat flux in order to obtain the latent heat flux (Kustas and Norman, 1996; Psilovikos and Elhag, 2013). However, the validity of these algorithms is restricted to small-scale regions, since at large scale there are difficulties with surface geometry and thermal variability, as well as lack of meteorological consistency (Wu et al., 2010). Recently, an adequate algorithm has been developed for daily evapotranspiration estimation for agriculture lands, namely the Surface Energy Balance System (SEBS) (Su, 2002), which has been previously mentioned and successfully implemented over the Nile Delta region (Psilovikos and Elhag, 2013). Specifically, SEBS considers different land surface physical and biological parameters that are derived from both the Advance Along Track Scanning Radiometer (AATSR) and the Medium Spectral Resolution Imaging Spectrometer (MERIS) imagery (Psilovikos and Elhag, 2013).

4.2.2 Remote Sensing of Soil Moisture

Remote sensing of soil moisture is becoming increasingly significant for measurement and estimation of electromagnetic energy that has either been reflected or emitted from the soil surface (Schmugge et al., 1980). In order to progress to operational soil moisture monitoring by remote sensing methods, multi-frequency and multi-polarization satellite data are required. It has also been illustrated that the synergistic use of optical and microwave data is advantageous. The relationship between the soil's liquid water content and its permittivity constitutes the basic concept for retrieving soil moisture estimates from satellite observations (Psilovikos and Elhag, 2013). Indeed, satellite-based soil moisture estimates typically provide information only about the top-soil layer, i.e. around 5 cm of soil depth (Njoku et al., 2003; Entekhabi et al., 2010). Visible, infrared (near and thermal), microwave and gamma data can be used for remote sensing of soil moisture, whereas the most promising techniques are based on active and passive microwave data (Schultz and Engman, 2000). Remotely sensed methods for measurement and estimation of soil moisture (SM) are classified into **direct** methods, which cover nearly all portions of the electromagnetic spectrum and include methods based on radiation reflection and emission, and **indirect** methods, which include radiation transmission.

4.2.2.1 Direct Methods

1. **Reflection.** Reflected solar radiation energy is not a very promising method for soil moisture detection, since the soil spectral reflectance as a function of moisture content depends on several other variables, such as surface roughness, soil texture, spectral reflectance of dry soil, geometry of illumination and organic matter. Moreover, the visible and near-infrared methods

are based on the reflected solar radiation and are not considered viable, since there are too many noise elements. The aforementioned factors, along with the fact that observations represent only a very thin surface layer, limit the utility of solar reflectance measurements for soil-moisture determination.

2. **Emission.** In emitted radiation, water has a strong influence on the thermal and dielectric properties of soils that make the remote sensing of soil moisture possible through measurements at the thermal infrared (10 m) and microwave (1–50 cm) wavelengths.

Thermal radiation region (3–15μm). In the thermal infrared measurements (8–12μm), the effects of the atmosphere are minimized, but not negligible, and the emitted energy is near the maximum for terrestrial surfaces. The time rate of change of soil temperature is a function of internal and external factors:

The internal factors are thermal conductivity (K) and heat capacity (C), which can be combined to obtain the thermal inertia: $P = (KC)^{\frac{1}{2}}$. Since K and C are strong functions of soil moisture, the thermal inertia and resultant soil temperatures will be dependent on soil moisture. **The external factors** are primary meteorological that affect the energy balance at the surface of the earth, such as incoming solar radiation, ambient air temperature, relative humidity, wind and cloudiness. The combined effect of these external factors justifies the diurnal variation of surface temperature. If the variability in these external factors can be held reasonably constant or accounted for, then the time rate of change of remote estimates of soil temperature or differences in surface temperature are related to soil moisture. Moreover, thermal inertia is an indication of the soil's resistance to this driving force and since both K and C of a soil increase with an increase of soil moisture, the resulting thermal inertia increases. Indeed, the effect of surface evapotranspiration (ET) in reducing the net energy input to the soil from the sun is a complicating factor. Evaporation has an impact on soil water by reducing the amplitude of the surface temperature diurnal range. As a result, the combination of soil moisture and surface evaporation justifies the day-night temperature difference.

Thermal infrared methods have been successfully employed to measure the few surface centimeters of soil moisture. However, there is a limitation to the thermal approach when the surface is covered with vegetation. Evaluation of the soil moisture has been attempted through observation of the Apparent Thermal Inertia using both AVHRR data from SPOT and Landsat, as well as geostationary images (WMO, 1993). Moreover, in the visible and near-infrared parts of the electromagnetic spectrum (EMS), the reflection from bare soil can only be used for the estimation of soil moisture under limited conditions. Indeed, for soil moisture, more spectral bands and a much higher geometric resolution in the visible/near-infrared (VIS/NIR) range are required, than that available from NOAA, Landsat and SPOT satellites. Nevertheless, one of the most accurate remote sensing methods, which has been developed for soil moisture measurement, is the use of gamma radiation. Indeed, the attenuation of gamma radiation can be used to determine changes in soil moisture in the top 20–30 cm of the ground. This method requires some field measurements of soil moisture during the measurement flight, since the absolute values of soil moisture cannot be provided (WMO, 1992b). Furthermore, soil moisture has also been estimated by using precipitation indices and operational applications have been developed by FAO using geostationary imagery over intertropical regions (WMO, 1993). With the advent of the International Geosphere–Biosphere Program (IGBP) the need for high-resolution data is gradually increasing.

Emitted microwave radiation (1–52cm). The dielectric properties of water strongly influence the dielectric properties of soils. This is an indication that both microwave reflection and emission may be related to soil moisture content, which can be measured in the microwave (MW) region of the spectrum by radiometric (passive) and radar (active) methods (Schmugge et al., 1980).

Microwave measurements are also sensitive to polarization of the emitted or reflected energy, roughness of the surface, the observation or incidence angle of emitted or reflected energy and the amount of vegetation. There are soil moisture retrieval methods that combine information from multiple sensors to improve both the spatial coverage/resolution and the quality of estimates (Liu et al., 2011; Psilovikos and Elhag, 2013). There are also empirical relationships between active microwave backscatter and soil moisture (Takada et al., 2009). Moreover, similar relationships can be used to link passive microwave information to soil's volumetric water content (Njoku et al., 2003; Psilovikos and Elhag, 2013).

Passive microwave. The passive microwave region has been most exploited, since there is a direct physical relationship between soil moisture and the reflection or emission of radiation. The thermal emission from the surface is measured by a microwave radiometer, where the intensity of the observed emission is essentially proportional to the product of the temperature and emissivity of the surface. The product is commonly referred to as brightness temperature T_B, the value of which at a height above the ground is given by:

$$T_B = \tau \left(r\,T_{sky} + (1-r)\,T_{soil} \right) + T_{atm} \tag{4.3}$$

where τ is the atmospheric transmission, r is the surface reflectivity, (1-r) = e is the emissivity, the term $(\tau\,r\,T_{sky})$ is the reflected sky brightness temperature, the term $(1-r)T_{soil}$ is the emission from the soil and T_{atm} is the contribution from the atmosphere between the surface and the receiver. The amount of energy generated at any point within the soil volume depends on the soil dielectric properties, i.e. soil moisture, and the soil temperature at that point (Schmugge et al., 1980). The thermal microwave emission from the soil surface can be expressed by:

$$T_B = e \int T(z)\alpha(z)\exp\left[-\int \alpha(z)dz \right]dz \tag{4.4}$$

where T(z) is the temperature profile and $\alpha(z)$ is the absorptivity as a function of depth, depending on moisture content. This equation can be solved numerically. Prediction of T_B is satisfactory for smooth surfaces and the effective sampling depth is of the order of a few tenths of a wavelength. Thus, for a 21-cm wavelength radiometer this depth is 2–5cm. The range of the dielectric constant produces a change in emissivity from greater than 0.9 for dry soil to less than 0.6 for a wet soil, assuming an isotropic soil with a smooth surface (Schmugge et al., 1980). The potential of passive microwave sensors for estimating soil moisture on an operational basis must be performed with aircraft and spacecraft sensors that integrate large areas of natural terrain, e.g. Skylab and Nimbus.

Active microwave. The active microwave band offers the greatest potential in terms of penetration through vegetation cover for sensing the moisture of the underlying soil medium (Ulaby et al., 1982). Among active (radar) and passive (radiometer) microwave sensors, only radar can provide the required information at a spatial resolution compatible with the needs of soil moisture estimation. Synthetic aperture radar (SAR) can effectively be used to obtain quantitative measurements, since **radar backscatter response** is affected by soil moisture, among others. Multitemporal radar images can detect soil moisture changes over time, since radar is sensitive to the soil's dielectric constant, which is a function of the amount of soil water.

The **backscattering coefficient** σ_o describes the scattering properties of terrain, which is driven by the geometrical and dielectric properties of the surface (or volume) relative to wave properties. The σ_o is a function of surface (or volume) roughness, vegetation and snow cover (if not bare). Thus, σ_o of terrain depends on the soil moisture content of an effective surface layer, whose thickness is governed by the penetration properties of the terrain at this wavelength. The variation of σ_o with soil moisture, surface roughness, incidence angle and observation frequency have been studied in the past in ground-based experiments using a truck-mounted 30- to 1.6-cm wavelengths active microwave system (Ulaby, 1974; Schmugge et al., 1980). The results have indicated that coefficient σ_o may extend in depth from about one wavelength for relativity dry soil conditions to about a tenth of a wavelength for very wet soil conditions (Ulaby et al., 1982). As a result, the soil dielectric constant strongly depends on the moisture content. In addition to surface roughness, another soil variable with an influence on the σ_o response to moisture is **soil texture**. To incorporate soil texture in the microwave response to soil moisture, the latter can be expressed in terms of percent of field capacity (FC) (Schmugge et al., 1980). C-band observations are superior to other wavelengths, since soil moisture information at a depth of several meters can be obtained from methods based on short pulse radar (wavelengths of 5–10 cm).

Natural **terrestrial gamma radiation** can be used for soil moisture measurements, since gamma radiation is strongly attenuated by water. Specifically, the gamma radiation attenuation technique is based on the difference between the natural terrestrial gamma-radiation flux measured from dry and wet soils (Jones and Carrol, 1983). There is a principle that the presence of moisture in the soil causes an effective increase in soil density resulting in an increased attenuation of the gamma flux for relatively wet soil and a correspondingly lower flux at the ground surface. Reliable real-time mean areal soil-moisture measurements can be made for the upper layer of soil if both background and current uncollided terrestrial gamma-count rates and background soil-moisture data are available (Jones and Carrol, 1983).

4.2.2.2 Indirect Estimation of Soil Moisture

The condition of the vegetation may be observed and may possibly serve as an indicator of soil moisture status in the root zone. The Crop Water Stress Index (CWSI) is derived from energy balance considerations by the equation:

$$T_C - T_A = \frac{\gamma\left(1 + r_c / r_a\right)}{\Delta + \gamma\left(1 + r_c / r_a\right)} * \frac{r_a R_n}{\rho c_p} - \frac{e_s - e}{\Delta + \gamma\left(1 + r_c / r_a\right)} \tag{4.5}$$

where γ = psychometric constant (Pascals C^{-1}), Δ = slope of saturated vapor pressure-temperature relation (Pascals C^{-1}), r_c = crop resistance (sm^{-1}), r_a = aerodynamic resistance (sm^{-1}), R_n = net radiation (Wm^{-2}), ρ = density of air (kgm^{-3}), C_p = specific heat of air ($Jkg^{-1}C^{-1}$), e_s = saturated vapor pressure of air (Pascals) at temperature T_A, e= actual vapor pressure of the air (Pascals) and T_C = canopy temperature. This equation shows that $T_C - T_A$ depends on the vapor pressure deficit and the net radiation that can be estimated from incoming solar radiation data. The remaining term is to be evaluated is r_a. One way is to measure $T_C - T_A$ when the plant is no longer transpiring, where $r_c \rightarrow \infty$. Then:

$$T_C - T_A = r_a R_n / \rho c_p \tag{4.6}$$

from which r_a can be calculated. Then solve for the ratio r_c/r_a that can be used in:

$$TR / ET_p = \frac{\Delta + \gamma}{\Delta + \gamma\left(1 + r_c / r_a\right)} \tag{4.7}$$

where TR = transpiration, ETp= potential evapotranspiration. The ratio TR/ET_p or its complement $1 - TR/ET_p$ is defined as the crop water stress index. The index is calculated from a one-time measurement of surface temperature, air temperature (wet and dry bulb), with an estimate of net radiation. Under uniform environmental conditions, it may be possible to relate the ratio to the average daily evapotranspiration. An airborne or spaceborne infrared scanner may effectively "map" ground temperatures. To be useful as a soil – moisture status estimation technique, imagery at frequent intervals is needed, perhaps several times per week during periods of high evapotranspiration. Satellite systems are capable of measuring surface temperatures from space. Equation (4.6) serves as a guide as to when temperature differences due to irrigation or soil-moisture differences might be detectable. Recent studies on the interrelationships of soil water on remotely measured surfaces temperatures, including plant canopy temperatures, seem to relate well to soil moisture in the upper layers.

In summary, remote sensing methods of soil moisture offer rapid data collection over large areas on a repetitive basis. The major problems are spatial resolution, penetration depth and cost.

4.2.2.3 SMOS and SMAP Missions

Soil Moisture and Ocean Salinity (SMOS) mission. The Soil Moisture and Ocean Salinity (SMOS) mission of ESA was launched in 2009. Since then it has been providing global observations of emissions from Earth's surface, particularly soil moisture and ocean salinity, which are two significant variables in the hydrological cycle (Crapolicchio et al., 2010). SMOS carries the passive L-Band 2D Microwave Imaging Radiometer with Aperture Synthesis (MIRAS) instrument, receiving in the 1400-1427-MHz protected wavelengths. SMOS produces global maps every three days of surface soil moisture (SSM) content at a 50 km spatial resolution (Kerr et al., 2010). Moreover, SMOS has the ability to relate emitted longwave radiation to moisture content of the first few centimeters of the top soil. In Figure 4.3, global observations of soil moisture content at a maximum depth of 5 cm from the surface are presented, where white areas show retrievals related to dense vegetation or snow and ice cover. Indeed, microwave response is strongly dominated by SSM and vegetation water content (Kerr et al., 2010). Furthermore, the synthetic aperture image provides observations of long wavelengths with an acceptable spatial resolution and is produced from 69 antennas positioned along three arms. The following topics could be considered as some of the potential applications of SMOS: early crop failure, crop yield indicator, warning of flood vulnerability, input to energy/water cycle, information on weather conditions, assimilation into numerical weather prediction, information for tracking hurricanes and measuring thin sea ice.

SMOS for hydrological modeling. As already mentioned, the SMOS mission provides new soil moisture measurements, which are considered novel accurate data with a high acquisition frequency at global scales. Hydrological models can accept satellite information, however, the low resolution of SMOS data causes some difficulties (Dalezios, 2014). ESA has undertaken a study to incorporate SMOS into hydrological models, the targets of which are the following: (1) the exploration of the potential and limits of SMOS data including both L1 brightness temperatures (TB) and L2 soil moisture to be used in hydrological modeling over large watersheds and at continental scales; (2) the development and validation of suitable robust end-to-end methodologies, i.e. from data acquisition to data assimilation approaches, which enable the effective exploitation of SMOS data including L1 brightness temperatures and L2 soil moisture into hydrological models; (3) the implementation of a few case studies of the developed methodologies and the demonstration of the benefits of the

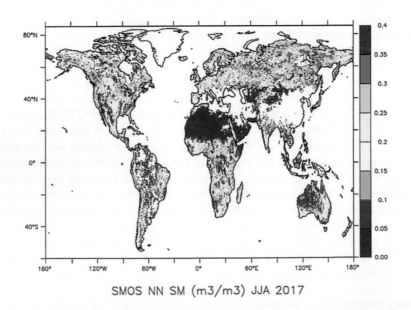

SMOS NN SM (m3/m3) JJA 2017

FIGURE 4.3 Global SMOS for three summer months (JJA) in 2017.

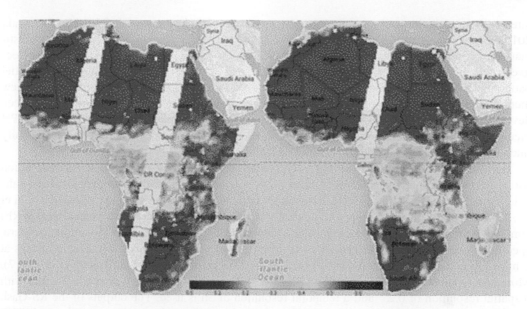

FIGURE 4.4 3-day satellite-based soil moisture (m3/m3) information from the SMAP mission on (left) June 4, 2016, and (right) March 4, 2016 (Source: Princeton University's African Flood and Drought Monitor, Multi-Sensor Remote Sensing of Drought from Space – Scientific Figure on ResearchGate. Available at: https://www.researchgate.nct/figure/day-satellite-based-soil-moisture-m-3-m-3-information-the-SMAP-mission-on-left_fig4_322518633 [accessed January, 13, 2021].

SMOS data for hydrological modeling with emphasis on floods. Specifically, the study objectives are addressed using the Variable Infiltration Capacity (VIC) model (Hamman ct al., 2018), SMOS observations, and the corresponding forcing and validation data over the Upper Mississippi and the Murray Darling Basins in the USA and Australia, respectively.

Soil Moisture Active Passive (SMAP) mission. On January 31, 2015, NASA launched the Soil Moisture Active Passive (SMAP) satellite. The SMAP mission is designed to measure soil moisture and freeze/thaw state for non-liquid water on Earth surfaces. These measurements have applications in several fields of environmental science disciplines, such as climate change, hydrology, carbon-energy cycle, weather forecasts, crop-yield predictions. Moreover, thc SMAP mission data can help climate models to estimate future trends in water resource availability. The SMAP mission is equipped with two instruments, an L-band radar (active) sensor and highly sensitive radiometer (passive) operating at 1.41GHz. Figure 4.4 presents 3-day satellite-based soil moisture (m3/m3) information from the SMAP mission on June 4, 2016 (left), and March 4, 2016 (right).

4.2.3 Remote Sensing of Interception

Interception is the portion of the precipitation which may be stored or collected by vegetation cover and eventually evaporated. Interception loss is called the volume of lost water. The interception loss could be a very significant factor in water balance studies, however, in major storm events and floods is generally neglected. Usually about 10 to 20 percent of the precipitation falling during the growing season is intercepted and returned to the hydrological cycle through evaporation. Several methods are used to measure rain interception, i.e. water stored in the canopy, time of leaf wetness (Lundberg, 1993), canopy interception storage capacity (Klaassen et al., 1998), canopy evapotranspiration, interception evaporation (Sharma, 1985) and throughfall.

The development of satellite earth observation systems could be used to estimate the global rainfall interception loss and its spatiotemporal variability. In a modeling effort, an analytical satellite-based model, namely the Gash model, has been revised and applied for global rainfall interception estimation using mainly precipitation, leaf area index, canopy height (Miralles et al., 2010). Moreover, gridded Leaf Area Index (LAI) and fractional vegetation cover (FVC) are estimated from images using spectral vegetation indices and based on spectral mixture analysis, respectively, which has proven to be a valuable and practical approach to quantitatively assess spatial patterns of interception loss for given rainfall events (de Jong and Jetten, 2007). Furthermore, a new methodology for estimating forest rainfall interception from multi-satellite observations has been developed, namely the Climate Prediction Center morphing technique (CMORPH). This CMORPH precipitation product is used as driving data and is applied to Gash's analytical model to derive daily interception rates at global scale (Miralles et al., 2010).

The remote sensed Gash model has been recently developed for interception loss estimation using forcing data and remote sensing observations at regional scale and has been applied and validated in the upper reach of the Heihe River Basin of China for different types of vegetation (Cui et al., 2017). Specifically, the hourly forcing data from the Weather Research and Forecasting (WRF) simulation model is used to drive the RS-Gash model and the forcing data consist of air temperature, specific humidity, air pressure, as well as downward short-wave and long-wave radiation fluxes at 2 m height and wind speed at 10 m height with a spatial resolution of 5 km (Calera et al., 2017). Moreover, tropical Rainfall Measurement Mission (TRMM) 3B42 Version 7 precipitation data with a spatial resolution of 0.25° and a temporal resolution of 3h is used. The outputs of the RS-Gash model are mean rainfall rate, mean evapotranspiration rate, vegetation storage capacity and FVC. Figure 4.5 shows the flowchart of the RS-Gash model. The resolution of 5 km is downloaded to 30 m for all the variables by using the bilinear resampling method. To eliminate the effect of mixed pixels and the scale error, the RS-Gash model is applied at a fine scale of 30 m with the high-resolution vegetation area index retrieved by using the unified model of bidirectional reflectance distribution function (BRDF-U) for the vegetation canopy.

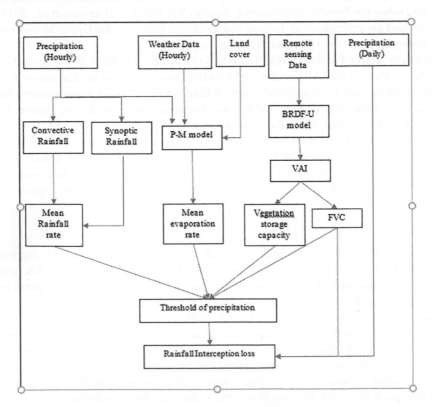

FIGURE 4.5 Flow chart of the RS-Gash model (from Cui et al., 2017).

4.2.4 Remote Sensing of Groundwater

Remote sensing methods used to map areas of groundwater include aerial and satellite imagery in the visible, infrared and microwave regions of EMS. Remote sensing data require substantial interpretation, which means that they cannot be used directly for groundwater, leading to surface features for the delineation of aquifers. Specifically, topography, morphology and vegetation are among the required surface features. Satellite imagery detects large areas, which cannot be viewed from ground surveys or even low-level aerial photography. Indeed, vegetation characteristics, drainage patterns, landforms, linear and curvilinear features, land-use patterns, image tones and textures can be used to extract groundwater information. Sedimentary strata or certain rock outcrops may indicate potential aquifers. Moreover, structural features, such as fracture traces, faults and other linear features can delineate the possible presence of groundwater. Soil moisture measurements and changes in vegetation types and temperature may be used as inference for shallow groundwater. Furthermore, soils, vegetation and shallow or perched groundwater may be used as inference for groundwater recharge and discharge areas within drainage basins (Engman and Gurney, 1991). A relatively recent comprehensive state-of-the-art review of remote sensing applications to groundwater can be found (Meijerink in Schultz and Engman, 2000).

Radar, as an active remote sensor, has an all-weather capability and can be used to detect subtle geomorphic features. Radar can also penetrate dry sand sheets to disclose abandoned drainage channels and to successfully reveal networks of valleys and smaller channels buried by desert sands. Radar can also provide information on soil moisture, as described earlier. An increase in soil moisture near the surface in arid areas may be an indicator of water, which can be detected by radar several decimeters below the ground surface. Moreover, near-viewing short pulse radars installed on mobile ground or aircraft platforms provide information on the depth to a shallow water table down to 5–50 m (WMO, 2009). In addition, radar imagery has the potential to penetrate rainfall and dense tropical rainforest for the creation of a groundwater exploration geological map (Engman and Gurney, 1991).

Electromagnetic sensors, which are developed for the mineral industry, can be used for airborne exploration of groundwater (Engman and Gurney, 1991). Aquifers at depths greater than 200 m have been mapped by this type of equipment (Paterson and Bosschart, 1987). Moreover, Landsat, SPOT or Copernicus satellite data, supported by aerial photography, are extensively used for groundwater inventories, essentially for locating potential sources of groundwater, leading to inferences about rock types, structure and stratigraphy. Moreover, IR and thermal imagery are effectively used to detect springs, as well as underwater springs. In addition, IR images can be used for mapping soil type and vegetation, which, in turn, can be used in groundwater exploration. Thermal IR imagery can also extract information on subsurface moisture and perched water tables at shallow depths based on temperature differences (van de Griend et al., 1985). Moreover, shallow groundwater tables have been measured by passive microwave radiometry.

4.3 REMOTE SENSING OF WATER QUALITY

Remote sensing data of water quality are considered very useful to resource managers, since they provide a synoptic view, which is unmatched by surface data collection methods. Mainly visible and near-infrared electromagnetic radiation through light can collect water quality information from a remote location (Alfoldi and Munday, 1978; WMO, 2009). Specifically, the variation in the intensity and color of light radiation through water lead to the nature and concentration of the water constituents responsible for the attenuation of the light. Figure 4.6 presents the various contributions to the total signal received by a remote sensor over water, where only the volume reflectance contains water quality information. Specifically, the volume reflectance is distorted by attenuation of the original radiance, further attenuation of the subsequent water reflectance and the addition of other sources of radiance, such as the atmospheric (path) radiance. This section briefly presents remote

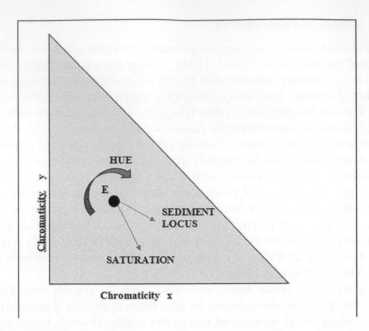

FIGURE 4.6 Remote sensing process for water quality information: Landsat Chromaticity Space. "E" at (x,y) = (0.333, 0.333) is the achromatic point. From "E", azimuthal variation defines "hue", while radial distance represents "saturation".

sensing data and methods for suspended sediments, turbidity and salinity, vegetation reflectance and chlorophyll, eutrophication and trophic state, as well as waves and oil spills.

4.3.1 SUSPENDED SEDIMENT

The reflectance of radiation in the visible and IR parts of EMS is used for the determination of the amount of sediment in water (WMO, 1993). In general, reflection is a non-linear function of the concentration of suspended sediments, where the maximum reflectance depends on wavelength and suspended sediment concentration. The relationship between reflectance and suspended solid concentration has a significant impact on the accuracy of remote measurements of suspended solids. Specifically, at high suspended solid concentrations, the nonlinearity becomes significant, leading to a noticeable decrease of the radiance resolution with increasing concentration. As a result, the accuracy decreases with increasing suspended solid concentrations. Moreover, another serious effect of the nonlinear reflectance versus suspended solid concentration relationship is the nonhomogeneous distribution of suspended solid concentration within a single pixel, which may lead to a remarkable underestimation of suspended solid concentration of the order of 30 percent (Alfoldi and Munday, 1978). There are two characteristics to mention: reflectance increases with increasing suspended solid concentration; and peak reflectance shifts to higher wavelengths with increasing concentrations, which permits the use of proportional brightness or "color" as a measure of suspended solid concentration. In the near-IR, suspended matter and a shallow bottom may affect water radiometric response (Chuvieco, 2000). For shallow waters, reflectance is high and absorption is low due to high bottom reflectance. However, this effect is complex, since soil radiometric behavior is affected by its chemical composition, structure, texture and humidity.

Since suspended sediments and turbidity are closely linked in most water bodies, estimates of turbidity can also be considered. However, this method is limited by the requirement to collect field data for the calibration of the relationship between suspended sediments and reflectance. Moreover, mapping of suspended sediment concentrations can be conducted in river plumes using scanner data without calibration data, leading to inferences about sediment deposition patterns in lakes and estuaries. A case is presented, where a system for automated multidate Landsat MSS measurement

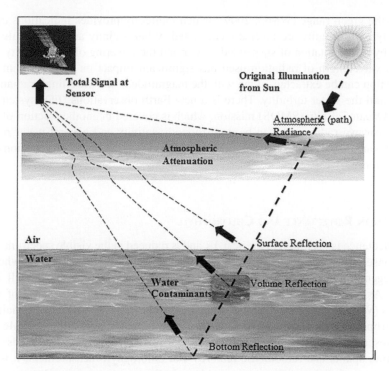

FIGURE 4.7 Components of the total signal received by a remote sensor over water.

of suspended sediment concentration (S) has been implemented and verified on nine sets of data from the Bay of Fundy, Nova Scotia, Canada (Munday et al., 1979). The system uses a classical approach, namely the chromaticity analysis, to provide pixel-by-pixel adjustment of atmospheric variations, allowing reference calibration data from one or several dates to be spatially and temporally extrapolated to other regions and to other dates. In this study, the correlation between a Landsat "chromaticity coefficient" and $\log_e S$ was $r = 0.965$. Effects of sediment type and size were negligible. The system can be used to measure chlorophyll in the absence of atmospheric variations and turbidity. In this study, S contour maps were used to initialize and calibrate a numerical model and were also used to identify sediment transport paths and hydrodynamic flow (Munday et al., 1979). Figure 4.7 presents the Landsat chromaticity diagram. Specifically, by plotting the chromaticity values of Landsat data, which represent a range of suspended sediment concentrations (S) values, the atmospheric homogeneity of the scene in question may be determined. In Figure 4.7, point E is the "achromatic" point and from "E" azimuthal variation defines "hue", whereas radial distance represents "saturation".

4.3.2 SALINITY AND TURBIDITY

Remote sensing can be used to measure turbidity, since turbidity of a water body is defined as its optical transparency or opacity. From the visible domain to the near-IR, a black body absorbing the whole incident radiation signifies the radiometric response of pure water. This well-known property is used to easily identify the presence of water on a satellite image. Water radiometric response can be inversely affected by several factors, such as water salinity and turbidity, soil composition or vegetation presence. Since turbidity causes light attenuation by both organic and inorganic water constituents, the relationship between reflected light and turbidity can be expected to vary among different types of water bodies. The best positive correlation between radiometric response and turbidity is in the green range (WMO, 2009), which indirectly provides indications on salinity. In general, salinity and turbidity are inversely correlated. When salinity arises, it results in flocculation followed by sedimentation of suspended matter and the lowering of the turbidity of the water. Moreover, the wavelength of radiation used has significant impact on the maximum depth from which information can be extracted, along with the magnitude of downwelling irradiance below the water surface and the water turbidity. There is a new Earth observations (EO) system, called Soil Moisture and Ocean Salinity (SMOS) mission, which is covered in another section of this chapter. Moreover, microwave radiometry can be used to study salinity and general water mineralization (WMO, 2009), since microwave emissivity is sensitive to water conductivity variations, and thus to water composition.

4.3.3 VEGETATION REFLECTANCE AND CHLOROPHYLL

Vegetation reflectance is very high in the near-IR and is reduced in the visible spectrum. Specifically, the high response in the near-IR is caused by the internal cell structure of leaves, whereas the low response of vegetation in the visible range is due to strong absorption of chlorophyll, especially in the red range. As a result, it is better to use optical images to study the presence of vegetation in shallow waters (WMO, 2009). Detecting and measuring chlorophyll is significant mainly due to its indication of trophic status and possibly of the presence of man-made pollutants. Measurement of chlorophyll presents more difficulties than suspended solids, since increased demands are put on the sensor's spectral, spatial and radiometric features. For chlorophyll measurement, it is desirable to claim the maximum of all three. The nature of chlorophyll is organic, thus, tends to be more spatially variant than inorganic sediment, which requires high spatial resolution. Moreover, narrow spectral bands in the absorption and/or reflectance bands must be used to detect these chlorophyll-specific characteristics. The problem of within-pixel non-homogeneity also exists in chlorophyll, since there is high spatial variability and reflectance is nonlinear with chlorophyll concentrations. Furthermore, the combined chlorophyll/sediment feature can show an increased, then decreased, reflectivity at certain wavelengths when concentrations are increased, which makes remote measurements more complex. The Coastal Zone Color Scanner (CZCS) or AVHRR images have been mainly used for the assessment of chlorophyll quantity in the ocean and estuaries (WMO, 1993). This assessment is limited to cases where suspended matter concentration is low enough not to mask the reflectance corresponding to that of chlorophyll, which is valid in macrophytes and aquatic vegetation studies (WMO, 2009).

4.3.4 EUTROPHICATION AND TROPHIC STATE

In the process of eutrophication, the natural aging process of a water body, an aquatic ecosystem changes to a terrestrial one. Moreover, trophic state identifies a body of water by its "relative level of nutrient richness and nutrient utilization in organic production" (Schmugge et al., 1980). Indeed, one of the appealing possibilities for use of satellite imagery is that of remote detection of trophic states of shallow coastal lakes. Since such determination is feasible, then it is possible to develop a monitoring methodology for such lakes. Indeed, the feasibility of differentiating between different lakes and determining their respective trophic states is based on Landsat MSS imagery. Specifically, the application of color-additive imagery enhancement methods can be used to monitor the trophic states of different lakes on a seasonal basis. The key to the comparison of data from different images is to have essentially constant reflectance characteristics. The lake reflectance can be related to their trophic states through comparison of the false-color renditions in the viewer screen with a standard interference color chart in combination with brightness measurements made directly from the viewer screen.

4.3.5 Waves and Oil Spills

The presence of waves can cause roughness, which can be detected by a radar image. At the same time, surface anomalies can be detected, such as those due to indiscriminate oil discharge (for further details see Chapter 6). The microwave domain permits a certain penetration within water. Indeed, it is possible to distinguish between a rough and a smooth surface by a lambertian or symmetric response, respectively. Moreover, remote sensing in the thermal IR domains and microwave radiation can be used to assess surface water temperature (Engman and Gurney, 1991). Microwave radiation is less sensitive to atmospheric conditions, thus, it is expected to be used more often, although its resolution is rough compared with that of the IR.

4.4 REMOTE SENSING MODELING OF HYDROLOGICAL SYSTEMS

This section covers the hydrological remote sensing potential in terms of data and methods within existing hydrological models, which represent the land phase of the hydrological cycle. Figure 4.8 illustrates a hydrological system. System operation depends on the nature of the system and follows natural laws. The system equations represent the natural processes of conversion of rainfall into runoff, which is the output of the system.

4.4.1 Hydrological Models and Remote Sensing

4.4.1.1 Remote Sensing Potential in Hydrological Models

During the last 20 years the technological and scientific advances in the field of remote sensing have resulted in the gradual improvement of the level of accuracy in quantitative assessment of several hydrological parameters and variables (Beven and Young, 2013; Dalezios, 2014). At the present time, remote sensing data and methods provide direct measurements of land characteristics, vegetative cover and the states of water in the hydrological cycle (Dalezios et al., 2020a). Moreover, direct estimation and assessment of hydrological parameters is feasible, such as precipitation, temperature, evapotranspiration, soil moisture, snow cover and snow depth, as well as water and energy balance. Moreover, meteorological satellites provide information on weather monitoring and prediction models, ocean temperatures and moisture locations, climate research, cryosphere (ice, snow, glaciers) detection and extent, land temperatures, crop conditions and hazard detection (Dalezios et al., 2018b). The capability of meteorological satellites, along with the increasing remote sensing reliability and the continuous technological and scientific advances, enables tracking of the atmosphere, ensuring real-time coverage of short-term dynamic events, such as local storms and

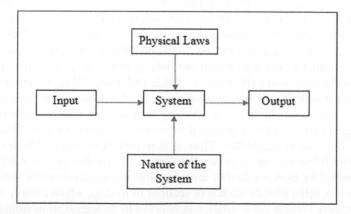

FIGURE 4.8 Schematic presentation of the hydrological system.

mesoscale convective systems, and detecting meteorological parameters quantitatively, such as precipitation, temperature, humidity and wind, among others (Dalezios, 2017; Dalezios et al., 2018b). Similarly, it is worth mentioning that existing successful uses of environmental satellite data in monitoring and management of water resources include, among others, flood area mapping, snow-cover measurements for runoff estimation, determination of the areal extent of surface water and wetlands and detection of water pollution.

4.4.1.2 Suitability of Hydrological Models to Remote Sensing

Most of the existing hydrological models have been originally developed for use with ground measurements (Peck et al., 1982; Fortin et al., 2001; Andersen, 2008; Dalezios, 2014.). Modeling is considered in the three levels of spatial scaling, in temporal scaling and within the structure of the models. Indeed, a major feature of any hydrological model is the scale of the model, namely the horizontal scale (basin size), the vertical scale (soil) and the time scale (time step). It is desirable to match the scale of the model to the scale of the remote sensing data and observations (Dalezios, 2014; Dalezios et al., 2020a). An inherent trade-off exists between the timelessness and accuracy of observations and the complexity of the model. For instance, an alternative approach for the horizontal scale is to match the horizontal scale of the model to the observations. Similarly, the vertical scale of the model should match the scale of observations. Moreover, the time step of the model must be short enough to identify significant variations in observed data.

Precipitation is an input to all hydrological models and no modification to a model would be required for its use. An improved estimation of the amount and the spatiotemporal distribution of precipitation could lead to better prediction and evaluation of the water-yield potential, thereby enabling better management and conservation of water resources. The subject of precipitation has been extensively covered in Chapter 3. Similarly, evapotranspiration receives special reference, due to its increasing significance in hydrological simulation, as well as its advancement in accuracy as a result of improved remote sensing technology and methods (Dalezios et al., 2019b; Dalezios et al., 2020a). Several other state variables and parameters have also been considered with remote sensing capabilities. Some variables, such as areal extent of snow cover, land cover and impervious area, can be considered to have operational measurement techniques at the present time and may be obtained through analysis of Landsat images (Schmugge et al., 1980; Dalezios, 2014). Other variables, such as soil moisture, areal extent of frozen ground and water equivalent of snow cover, although they can currently be measured effectively from satellites, require further research on combined active/passive microwave techniques (Schmugge et al., 1980; Dalezios, 2014).

At present, there is limited use of remote sensing data and methods with the current form of hydrological models. There are several strategies for employing remote sensing data and methods in hydrological models. The first strategy remains to estimate the inputs to the models. A second strategy is to update the state of the model to be consistent with the observed data. A third strategy is to calibrate the parameters of the models (Dalezios, 2014; Dalezios et al., 2020a). Nevertheless, extensive reviews of hydrological models have indicated that in their present configurations there is no significant potential to use remote sensing data as inputs, to update the state of the model or to calibrate model parameters (Peck et al., 1982; Dalezios, 2014). In summary, the greatest benefit in operational hydrology from the incorporation of remote sensing data and methods can be realized after major modifications of the existing hydrological models. However, some models can take advantage of the significant potential of remote sensing even with minor modifications (Dalezios, 2014; Dalezios et al., 2020a). Thus, it is important to consider both the requirements of the measurement frequency and accuracy. Moreover, the greatest potential for remote sensing could be implemented by high resolution, active/passive microwave satellite sensors. Finally, the development of a new generation of models or sections of models, which comply with the continuous advances of remote sensing capabilities, can improve hydrological simulation (Dalezios, 2014; Dalezios et al., 2020a).

4.4.2 REMOTE SENSING IN SCALING LEVELS OF HYDROLOGICAL SYSTEMS

Modeling efforts are investigated in water resources management tasks in three levels of temporal and spatial scaling, namely large-scale hydrological systems or water supply systems, meso-scale hydrological systems for flood protection and forecasting and operational hydrological systems (Schultz and Engman, 2000) (Table 4.1). A brief description follows.

4.4.2.1 Large-scale Hydrological Systems

Large-scale hydrological systems include water supply systems referring to large areas. This type of modeling requires long-term time series data records of hydrological variables with the minimum time step being the month and refers to large-scale watersheds and regions. For instance, the conventional data source could be observed or generated runoff data (Dalezios, 2017; Dalezios et al., 2020b). Moreover, the potential remote sensing data source could be infrared (IR) data from polar orbiting meteorological satellites, such as NOAA and TIROS N. Finally, the employed hydrological model could be a transfer function model in convolution integral, which is a stochastic black-box model based on the theory of linear systems (Dalezios, 2017; Dalezios et al., 2020b).

The HyMeX (Hydrometeorological Mediterranean eXperiment) is considered a case study in this category, since the Mediterranean region is one of the climate hotspots, where the climate change impacts are both pronounced and documented (Figure 4.9) (Pellet et al., 2019). The objective of HyMeX is to improve our understanding of the hydrological cycle from the meteorological to climate scales, which remains a challenge with the incorporation of Earth observations (EO). Specifically, EO-based precipitation and evaporation, arc considered multiple with large uncertainties and incoherencies among these products. As a result, merging and/or integration techniques have been developed to reduce these issues (Pellet et al., 2019).

An improved methodology is implemented that closes not only the terrestrial, but also the atmospheric and ocean budgets (Pellet et al., 2019). The new scheme imposes a spatial and temporal multiscale budget closure constraint. Moreover, the budget closure constraint is used simultaneously at different spatial (basin and sub-basin) and temporal (monthly and annual) scales (at the resolution of 0.25°). In addition, a new spatial interpolation approach is introduced to downscale the basin-scale closure constraint to the pixel scale (at the resolution of 0.25°). This new framework is applied to the Mediterranean basin to provide an updated water cycle (WC) budget. The methodology includes single weighting (SW), post-filtering (PF) and integration (INT) (Pellet et al., 2019). The provided Mediterranean WC budget is based mostly on observations. Figure 4.10 shows the mean annual fluxes (km3 yr $^{-1}$) of the Mediterranean WC and associated uncertainties in SW (small font) and INT (large font) during the 2004–2009 period (Pellet et al., 2019).

4.4.2.2 Meso-scale Hydrological Systems

Meso-scale hydrological systems include rainfall-runoff modeling, rural and urban watershed modeling, water balance and involve flood protection and forecasting, design flood, flood routing or flood mapping (Schultz and Engman, 2000). Short-term extreme hydrological events with a time step of days, hours or even 10 minutes, for example, in case of urban systems, are used in this type of modeling (Dalezios, 2017; Dalezios et al., 2020b). Moreover, the conventional data source could be observed or extrapolated runoff data. In addition, the potential remote sensing data source could be IR data from geostationary meteorological satellites, such as GOES, METEOSAT, GMS (Dalezios et al., 2020a). Finally, hydrological modeling could include a rainfall model in association with a rainfall-runoff model of distributed or lumped system type, e.g. unit hydrograph (Dalezios, 2017; Dalezios et al., 2020b).

A flood mapping case is presented, which integrates information produced from multiple sources to support the estimation of flood inundation extent (Rosser et al., 2017). Both SAR and multispectral satellite remote sensing is commonly used for assessing and mapping flood inundation extent (Psilovikos and Elhag, 2013). In this case, medium-scale optical remote sensing and a high-resolution

TABLE 4.1

Potential use of remote sensing information of water resources management tasks (from Schultz and Engman, 2000)

Water Resources Management Tasks	Data Requirements			Potential Remote Sensing Data Source (as example)	Hydrological Model (as example)	Example Given
	Length of Time Series	Time Intervals	Data Source (conventional)			
Large-scale hydrological systems	Long-term (continuous for many years)	Months	Observed or generated runoff data	IR data from polar orbiting satellites (e.g. NOAA, TIROS)	Transfer function in convolution integral (theory of linear systems)	River system in Southern France
Meso-scale hydrological systems	Short-term (extreme events of many years)	Hours, days, 10 minutes (urban systems)	Observed or extrapolated runoff data	IR data from geostationary satellites (e.g. GOES, METEOSAT, GMS)	Rainfall model plus rainfall-runoff model of distributed or lumped system type (e.g. unit hydrograph)	General
Operational hydrological systems	Short-term	10 minutes (urban), hours, days	Observed in real-time forecast rainfall + runoff	Ground-based weather radar, IR data from geostationary satellites	Rainfall model + rainfall-runoff model (distributed system type)	River system in Southern Germany

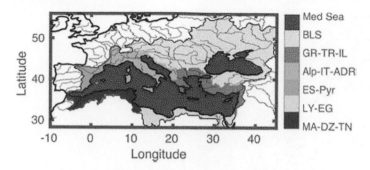

FIGURE 4.9 Region of interest. Sub-basins have been computed using a hydrological model (Wu et al., 2011), and rivers are from HydroShed (http://www.hydrosheds.org/, last accessed: January 24, 2019).

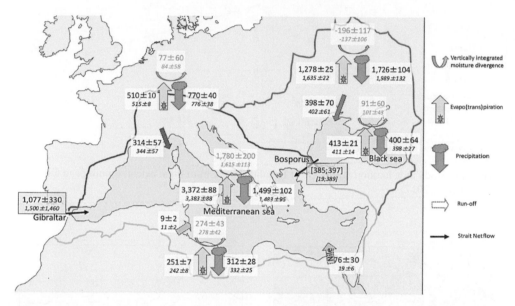

FIGURE 4.10 Mean annual fluxes (km3/yr) of the Mediterranean WC and associated uncertainties in SW (small font) and INT (large font) during the 2004–2009 period.

terrain model are combined with observations made by in situ crowd users (Figure 4.11). The developed methodology consists of integration of near real-time Landsat-8 imagery, along with ancillary topographic data (derived from a LIDAR DTM). These data sets are processed and classified, and the results are then integrated using the fusion model (Psilovikos and Elhag, 2013). Specifically, the operational land imager (OLI) instrument of Landsat-8 provides multispectral data at 30-m spatial resolution. The Modified NDWI (MNDWI) index is computed, which constitutes a frequently applied approach to detect water given spectral response at appropriate infrared wavelengths (Psilovikos and Elhag, 2013). Indeed, MNDWI has the advantage of suppressing the response of both vegetation and built-up areas leading to enhanced water detection for these areas (Xu, 2006). This is applicable to the landscape found in the Oxford study area (Figure 4.12). MNDWI can be expressed as:

$$MNDWI = (Green - MIR)/(Green + MIR) \qquad (4.8)$$

where MIR corresponds to band 7 (SWIR2, 2.11–2.29 μ) of the L8-OLI instrument.

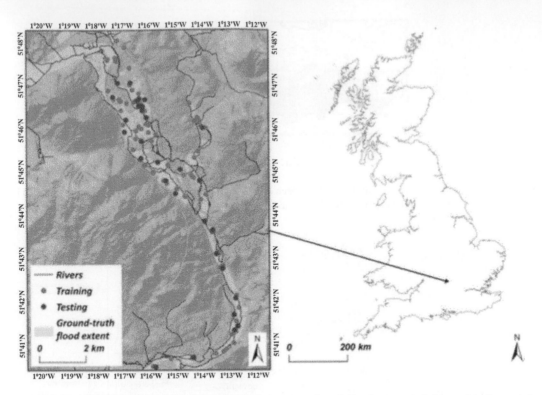

FIGURE 4.11 Study area location within Great Britain, ground-truth flood extent, training and testing points (from Rosser et al. 2017).

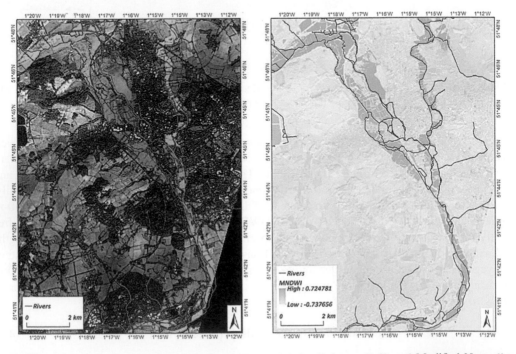

FIGURE 4.12 False color composite (RGB = 632) Landsat-8 reflectance (left) and Modified Normalized Difference Water Index (MNDWI, right) (from Rosser et al., 2017).

FIGURE 4.13 Flood mapping by Sentinel-1 caused by cyclone Idai (ESA, March 20, 2019) (from modified Copernicus Sentinel data (2019), processed by ESA, CC BY-SA 3.0 IGO).

Another case is the floods caused by cyclone Idai (Figure 4.13), which affected millions of people in Mozambique, Malawi and Zimbabwe, and was considered one of the southern hemisphere's worst storm. The image of Figure 4.13 is from Copernicus Sentinel-1 and shows the extent of the flooding, depicted in red, around the port town of Beira in Mozambique on March 19, 2019. This mission also supplies imagery through the Copernicus Emergency Mapping Service to aid relief efforts (ESA, March 20, 2019).

4.4.2.3 Operational Hydrological Systems

These systems include real-time hydrological forecasting-now-casting, flash floods, or reservoir operation and involve mainly the use of weather radar. This type of modeling requires short-term or even real-time data with a daily time step or hours or even 10 minutes (Schultz and Engman, 2000). Real-time rainfall observations, rainfall forecast, as well as observed runoff constitute the conventional data source (Dalezios, 2017; Dalezios et al., 2020b). Moreover, the potential remote sensing data source consists of ground-based weather radar and IR data from geostationary meteorological satellites. Finally, the hydrological modeling system to be used should include a rainfall model and a rainfall-runoff model, preferably of distributed system type, in order to conduct now-casting of extreme events in real-time or semi real-time (Dalezios, 2017; Dalezios et al., 2020b).

Flash floods are rapidly rising flood waters that are the result of excessive rainfall or dam break events (WMO, 2009). Specifically, rain-induced flash floods are excessive water flow events that develop within a few hours of the causative rainfall event, usually in mountainous areas or in areas with extensive impervious surfaces, such as urban areas (WMO, 2009; Sene, 2016). Most of the flash floods observed are rain induced. Excessive volumes of stored water can be released in a short period of time from breaks of natural or human-made dams with catastrophic consequences downstream. Examples are the break of ice jams or temporary debris dams (Sene, 2016).

A case is presented, which was a typical autumn event (September 2006) during which several mesoscale convective systems crossed Catalonia from southeast to northwest (Alfieri et al., 2011).

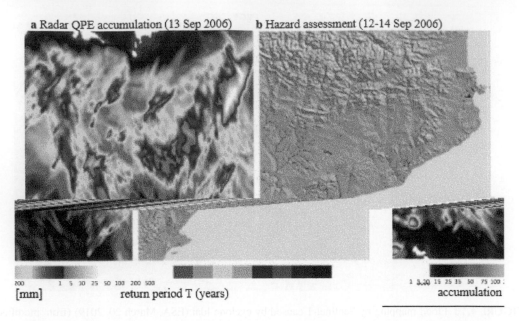

FIGURE 4.14 Results obtained with the radar-based hazard assessment system during the flash flood event of September 12–14, 2006 in Catalonia, Spain. (a) radar QPE accumulation for September 13, 2006; (b) hazard assessment for the entire event (the red dashed ellipses indicate the areas where floods were reported) (from Alfieri et al., 2011).

The maximum rain gauge accumulations were reported in Constantí (near Tarragona) with 267 mm and in the area of the Gulf of Roses (north of Girona) with 256 mm, 216 mm in 24 h (Figure 4.14). This event caused significant material losses (intense flooding in urban areas in the regions of Tarragona and Barcelona and failure of the road and railway networks due to flooding and landslides) and one casualty (Alfieri et al., 2011). The intense rainfall produced flash floods in ephemeral torrents near the coast (Psilovikos and Elhag, 2013) and in some sub-basins of the main rivers as, for example, in the lower part of the rivers Llobregat (5000 km^2) and La Muga (850 km^2). The estimated rainfall accumulation map for September 13, 2006 is shown in Figure 4.14a by combining radar QPE and the available rain gauge records, which show a reasonable agreement, except for the largest accumulations (probably more affected by attenuation of the radar signal due to intense rain). Figure 4.14b shows the summary of the maximum hazard level estimated at each point of the drainage network throughout the event and how the system was able to successfully detect the importance of the event and identify the areas most affected by intense rainfall and flash floods (Alfieri et al., 2011).

Another case is presented (Destro et al., 2018) using data from an extreme rainstorm that impacted the 140 km^2 Vizze basin in the Eastern Italian Alps on August 4–5, 2012 (Figure 4.15). Rain gauge observations and radar recording were used to derive distributed rainfall information for the storm event. Five-minute rainfall data were collected by 14 rain gauges inside and close to the basin (Destro et al., 2018). The single polarization, C-band, Doppler radar, located at Mt. Macaion, about 60 km from the basin at an altitude of 1860 m a.s.l., provided rainfall estimates at spatial and temporal resolutions of 1 km^2 and 5 min, respectively. A reflectivity-to-rain rate (Z-R) relationship derived from measurement of convective events in the study region was used to obtain rain rate estimates, where the residual mean field bias with respect to the rain gauge data was also adjusted (Marra et al., 2014; Destro et al., 2018).

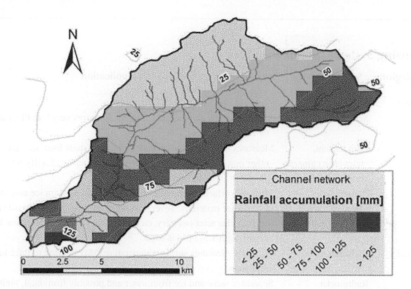

FIGURE 4.15 Radar rainfall spatial distribution of the August 4–5, 2012 rainstorm over the Vizze Basin, north Italy: event-rainfall accumulation (Destro et al., 2018).

4.5 REMOTE SENSING OF SNOW

Remote sensing of snow can be accomplished using gamma rays, visible and near-IR, thermal IR and microwaves. Indeed, the microwave band has the greatest overall potential followed by the visible and near-infrared band (WMO, 2009). Specifically, thermal infrared is limited in potential, but it can be used from space in night-time situations. Moreover, the gamma ray portion is extremely limited by the fact that the sensing must be carried out with low altitude aircraft, and to a lesser extent that it is only sensitive to a finite snow-water equivalent (WMO, 2009).

Snow parameters and properties can be determined by using different approaches. The availability of data from existing satellites or from experimental aircraft and truck programs mostly constitutes the driving force for these approaches (WMO, 2009). Specifically, remote sensing data are currently being used operationally in snow cover and snow-water equivalent assessments, as well as seasonal snowmelt runoff forecasts. At the present time, the potential of satellites to provide usable information on snowpack dynamics is now widely recognized, and many schemes exist that employ satellite-derived snow measurements for snow melt runoff prediction (WMO, 2009).

4.5.1 Snow Parameters

In snow hydrology, to adequately monitor the snow resource of a watershed and to simulate and predict snowmelt runoff, the following parameters are critical: snow covered area (SCA), snow depth and density and snow water equivalent (SWE) (Pellet et al., 2019). Quantitative or qualitative snow and ice data constitute significant validation data source for a wide range of other hydrological variables. For example, background data indicating the nature and extent of ice conditions may explain and possibly correct anomalous river-stage data during the winter months. Table 4.2 presents remote sensing systems useful for snow and ice studies.

4.5.1.1 Snow Cover Area (SCA) Analysis

A brief description of existing satellite systems for SCA analysis is presented, which is followed by methods for SCA analysis.

TABLE 4.2

Remote Sensing Systems for Snow and Ice

Wavelength Region	Sensing Device	Application
Passive sensors		
Gamma rays	Scintillometer	Measures snow and ice mass directly. Requires very low flight altitudes ($\approx$ 100 m above terrain).
Visible	Film camera, television camera, or photomultiplier	Measures surface reflectance, distinguishes snow from ice and snow/ice from other materials. Several wavelengths may permit classifying snow surface character (wet/dry, etc.).
Near-infrared: red-edge	Camera or photomultiplier	Measure surface reflectance, distinguishes snow from ice and snow/ice from other materials. Several wavelengths may permit classifying snow surface character (wet/dry, etc.). Also, may differentiate snow from cloud if wavelength 1.5-3 µm is used.
Thermal infrared	Radiometer	Primarily measures surface temperature. May be used to infer lake or sea ice thickness.
Microwave	Radiometer	Separates snow and ice from water and possibly from land, multi-year sea ice from new sea ice. Several wavelengths may distinguish snow wetness, may possibly permit measurement of grain size, density, mass and layering.
Active sensors		
Visible	Laser altimeter	Measures surface roughness, altitude.
Microwave	Radar	Measures surface altitude, form, roughness, may possibly measure subsurface properties if several wavelengths used.
Microwave	Altimeter	Measures variations in altitude of surface to within ± 10 cm.

Satellite systems. Only satellites enable seasonal snow cover to be monitored periodically and efficiently and on a sufficiently large scale. Operational snow mapping can be conducted through significant remote sensing data from available meteorological and environmental satellites, such as Satellites pour l'observation de la terre (SPOT), Landsat, National Oceanic and Atmospheric Administration (NOAA), Geostationary Operational Environmental Satellite (GOES), Earth Observation Satellites (EOS) and Defense Meteorological Satellite Program (DMSP). The areal extent of the snow cover is mapped operationally in many countries using data from meteorological satellites. Operational snow mapping schemes are usually not required for small areas (WMO, 2009). Indeed, the spatial resolution of the sensors involved drives the accuracy of the snowpack delineation and snow cover area. Nevertheless, the choice of satellite for snow mapping depends upon the smallest partial area of the region to be monitored.

The visible and the near-infrared regions of the electromagnetic spectrum (EMS) are mostly used for detecting and monitoring snow cover, since the reflectivity (albedo) depends on snow properties, such as the water content, surface roughness, grain size and shape, or depth and presence of impurities. Specifically, the Landsat satellite through its multispectral scanner (MSS) in the **visible red band** (0.6–0.7 m) has been widely used for snow cover mapping, due to its strong contrast with snow-free areas. However, besides spatial resolution, the snow mapping capabilities of Landsat and SPOT are hindered by their inadequate frequency of coverage. This leads to the use of NOAA polar-orbiting satellites with the Advanced Very High-Resolution Radiometer (AVHRR). Indeed, the frequency of coverage for NOAA is every 12 hours, which is much higher than Landsat with a frequency every 16 to 18 days. However, the problem with the NOAA-AVHRR data is that the resolution of 1 km (in the visible red band (0.58–0.68 m)) may be insufficient for snow mapping on small basins. Nevertheless, there is an operational alternative with EOS AM and PM satellites, which carry the Moderate Resolution Imaging Spectroradiometer (MODIS) instrument, providing daily data at reasonably high spatial resolutions.

In the **near-infrared** band snow can be detected, however, the contrast between a snow and a no-snow area is considerably lower than with the visible region of EMS. Indeed, the near-infrared band, when available, serves as a useful discriminator between clouds and snow, since the contrast between clouds and snow is greater in the Landsat TM Band 5 (1.57–1.78 m). Visible/near-infrared difference data from NOAA imagery can be used to locate areas of complete or partial snow cover and identify melt and accumulation zones. Daily snow area maps can be produced and subsequently can be composited to generate weekly estimates of snow distribution. This technique can be considered for operational in several parts of the world (WMO, 2009).

Cloud cover hinders **thermal infrared** data, which have, thus, limited importance for snow mapping and measuring snow properties. Moreover, the surface temperature of snow is not always that much different from the surface temperatures of other adjacent areas with different cover, such as rock or grass (WMO, 2009). However, thermal infrared data can be useful to help identify snow/no snow boundaries, and for discriminating between clouds and snow with AVHRR data. Nevertheless, the best result in snowline mapping can be achieved by combining the information of the thermal emission and the reflectance in the visible part of EMS (WMO, 1992b).

A major advantage of the **microwave** approach is the ability to penetrate cloud cover and map snow extent. Indeed, the microwave wavelength at about 1 cm has the greatest overall potential for snow mapping, due to its cloud penetration or all-weather capability. However, the poor passive microwave resolutions from space (about 25 km) constitutes a current major drawback. As a result, only very large areas of snow cover can be detected. Moreover, there are data processing geophysical algorithms for satellite microwave radiometers, such as the Special Sensor Microwave Imager (DMSP) SSM/I, which can be used to produce large-scale snow cover extent maps. It is clear that these maps are considered reliable over large flat regions with little or low-lying vegetation when snow is dry. Indeed, high-resolution active microwave sensors can be used to potentially solve the resolution problem.

New satellites systems, such as Copernicus Sentinel-1, Sentinel-2 and ESA's CryoSat, can also be used to monitor glacial change, which has a real impact on water supplies downstream. For example, part of the Himalayas, known as "the third pole", since these high-altitude ice fields contain the largest reserve of freshwater outside the polar regions, provides freshwater for over 1.3 billion people in Asia, nearly 20% of the world's population.

Remote sensing methods of SCA analysis. There are several remote sensing methods for SCA analysis (Rango (ed.), 1975), such as photointerpretation and image interpretation, which is further classified into: (1) zoom transfer scope method; (2) grid analysis method; and (3) digital processing methods.

1. **The zoom transfer scope method** is used to precisely register the satellite imagery and the watershed boundaries. This instrument superimposes the imagery on base-map overlays at various scales through a system of mirrors, lenses and scale adjustments with the additional capability of removing image distortion. The snow cover on the Landsat scene at 1:1,000,000 scale is mapped on the watershed overlay at 1:250,000 scale. This is conducted by tracing the line separating snow from non-snow-covered areas for the entire basin. Using several elevation zones, this snowline will cut through various zones. A snow-cover depletion-curve for each zone can be obtained. These curves can be used for comparing snow cover retreat in other years, for obtaining daily zonal snow-cover values for use in modeling (Rango et al., 1979) and for graphically predicting snowmelt runoff. Difficulties in the identification of snow cover and snowline in satellite images are due to clouds, forest cover, bare rock and mountain shadows (Rango, 1975). Remotely sensed snow-water-equivalent and temperature data have been used to construct modified snow cover depletion curves for use in the snowmelt-runoff model for snowmelt forecasts in the Rio Grande basin and elsewhere (Rango and van, 1990; WMO, 2009).

2. **Grid Analysis Method.** More detailed snow-cover analysis, using photo interpretation, results when a snow-interpretation grid system is used. The grid system permits decisions for discrete parts of the basin as to the areal extent of snow, density and type of vegetation cover, elevation, aspect, bare rock and shadows. These grid-unit techniques are also well suited for special snow interpretations, such as inference of snow water-equivalent.

3. **Digital Processing Methods.** Snow area is an easily identifiable parameter. When analyzing numerous basins digital techniques are preferred. Supervised classifications of snow are easily accomplished: delineating and separating snow in trees, in shadows, from bare rock and from clouds. For determining snow area, the two major subclasses of hydrological importance are dry and wet snows. For example, Landsat classifications tend to be too detailed, sometimes providing 20 or more categories of dry snow. The advantage is the capability to merge the snow data with conventional topographic data, which permits classification of snow not only by area, but also with elevation. Another capability is the automatic discrimination between snow and clouds if a band at 1.55–1.75µm is available.

4.5.1.2 Snow Depth Analysis and Snow Water Equivalent (SWE)

Digital data from microwave scanners (Nimbus 5-7) can be computer processed to yield microwave-brightness temperature data averaged over various size-sub-divided study areas. Over large homogeneous areas, a significant regression relationship between snow depth (criterion variable) and microwave brightness-temperature (predictor variable) can be developed (Rango et al., 1979; Schmugge et al., 1980). The estimation of snow depth under dry-snow conditions is a possibility. Remote sensing methods to estimate snow depth or SWE also contribute considerably to already existing SCA capability (WMO, 2009). Such a method provides detailed knowledge of the snowpack in both mountains and flatland basins for runoff prediction purposes. Then, it is possible to estimate large areas of total snow volume, if the method is combined with SCA (Schmugge et al., 1980). This is very useful, since such an estimate would specify the maximum potential of snow water production and locate that water in relation to various elevation zones or sub-basins of the watershed.

Radiometers sensitive to microwave energy emitted by the earth's surface and other natural objects as snow can be used to measure snow depth or SWE. In passive microwave radiometers, the emitted microwave radiation is measured, expressed as brightness temperatures, which is affected by the depth of snow and infers the characteristics of the snowpack. Several radiometers are used at different wavelengths, which seems to be necessary to separate the emission of the snow from the soil, since the observed radiation can come from the underlying soil, snow and atmosphere. Moreover, radiation from dry snow is strongly influenced by crystal or grain size, since the radiation emitted by the snowpack is scattered on its way to the surface by the snow grains. Specifically, larger grain sizes cause more scattering resulting in a lower emissivity. The snow emissivity is affected by the onset of melting, which produces liquid water in the snowpack. Specifically, a significant increase in internal absorption of the microwave radiation is observed, since the water coats the snow grains, resulting in a decrease in the snow emissivity. This liquid water in the snowpack is called free water equivalent.

There is a significant relationship between snow depth and brightness temperature, which could be used for snow surveys over large areas (Rango et al., 1979; Schmugge et al., 1980). Currently the use of such relationships is restricted to large homogeneous regions, such as plains, due to poor sensor resolution. Drawbacks for hydrology are poor spatial resolution (25 km) and limited spatial coverage. However, enough representative data sets should be collected for reliable runoff prediction during rapid snowmelt. Moreover, with resolution improvement, applications should be found in other zones, such as inter mountain valleys and large mountain open areas (Schmugge et al., 1980). Such capabilities would be more directly applicable to seasonal and short-term runoff forecasts.

Snow water equivalent (SWE). The Special Sensor Microwave/Imager (SSM/I) data are being used operationally to produce snow-water equivalent maps of several parts of the world, such as the Canadian prairies (WMO, 2009). The active microwave region has a similar potential as the passive microwave region, and Synthetic Aperture Radar (SAR) can provide high-resolution data. Single-frequency systems, such as ERS-1, are likely to be limited to the recognition of the onset of melt and the delineation of wet snow extent, however, multifrequency and multi-polarization SAR measurements can overcome these problems.

4.5.2 Snowmelt Runoff Analysis

Empirical or deterministic modeling approaches are considered in snowmelt runoff procedures. The choice of approach depends on both the availability of data to quantify the snowpack and the required detail of the output. Specifically, the required quantitative information includes the areal extent of the snow (S), the snow water equivalent (SWE) and the condition or properties of the snow, such as depth, density, grain size and presence of liquid water (Engman and Gurney, 1991). Moreover, another characteristic feature of the seasonal snow cover is the gradual decrease of the areal extent. Regardless of the approach used to conduct day-to-day simulations of snowmelt runoff, the daily snow-covered area in the basin should be known (WMO, 2009). For many basins, there is a very good relationship between runoff and snow cover area (Engman and Gurney, 1991). For operational runoff forecasts the water equivalent must be determined (WMO, 1994). Many large hydropower companies use snow cover extent maps from NOAA/AVHRR, advanced very high-resolution radiometer, on an operational basis as input to their hydropower production planning (Andersen, 1991; WMO, 2009).

Remote sensing is very successful in mountain regions, especially in snow cover mapping, except only in regions with very dense forest cover. The estimation of snow cover features, such as grain size, albedo, layering, surface temperature and snowpack temperature, can be achieved through remote sensing methods using appropriate wavelength bands. This, in turn, leads to reliable estimates of the rate and time of melt water from the surface to lower layers, which eventually produces runoff at the base of the snowpack (Rango and van, 1990). Moreover, the merging of remote sensing data with digital elevation modeling and geographical information systems (GIS) enables different types of data to be combined objectively (Engman and Gurney, 1991). Specifically, a traditional approach uses digital elevation modeling to normalize imagery by using the elevation of the sun and the slope, aspect and elevation of the terrain (Baumgartner, 1988). Similarly, GIS are helpful in combining vegetation masks with satellite imagery. In summary, new models developed to use remote sensing data are expected to further improve snow hydrology predictions (WMO, 2009).

4.6 REMOTE SENSING FOR AGROHYDROLOGY

4.6.1 Remotely Sensed Irrigation and Drainage Water Management

Remote sensing has been traditionally used to identify land use and areas that are cropped, irrigated, waterlogged or flooded (WMO, 2009). Moreover, it can yield information on soil salinity, crop water needs and stress, and crop yields (Dalezios et al., 2019a; Dalezios et al., 2019b). Specifically, several satellite systems, such as the European Copernicus, combined with Landsat Thematic Mapper (TM), SPOT Multispectral Scanner (MSS) or SAR data can be used for obtaining information on land use and crop areas (WMO, 2009). Moreover, the temporal normalized differential vegetation index (NDVI) has be used extensively and is still used to monitor vegetal cover and crop growth (Dalezios, 2015). Similarly, the low-resolution advanced very high-resolution radiometer (AVHRR) satellite imagery has been used operationally to estimate annual crop area, derive 10-day yield indicators and derive quantitative estimates of crop condition and production (WMO, 2009).

Recent developments in remote sensing technology, linked to GIS, are proving valuable in the planning and monitoring of irrigation and drainage systems (WMO, 2009). There is significant

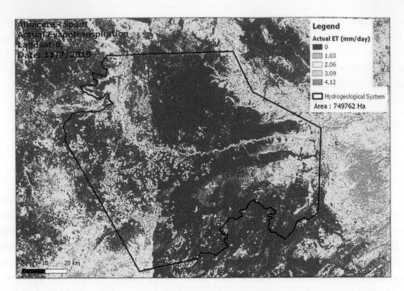

FIGURE 4.16 Actual evapotranspiration of July 11, 2019 from Landsat-8 (30 m resolution) at Albacete, Spain using the Surface Energy Balance (SEB) method (METRIC) used in precision agriculture.

literature on satellite remote sensing applications in irrigation management (Kustas and Norman, 1996; Allen et al., 2005; Dalezios, 2015; Calera et al., 2017). Irrigation has always been significant for cultivation yield in semi-arid regions, such as the Mediterranean basin, mainly due to the high evapotranspiration rates and limited precipitation inputs (Psilovikos and Elhag, 2013; Dalezios et al., 2019b). Precision agriculture based on high-resolution satellite data composes a recent sophisticated approach leading to management of water availability and the need for efficient crop monitoring in vulnerable agroecosystems (Dalezios et al., 2012; Dalezios et al., 2019a).

Precision agriculture methodology for the estimation of the water needs requires a combination of field observations, such as crop characteristics and water requirements, along with meteorological data, in order to assess reference evapotranspiration (ETo) (Psilovikos and Elhag, 2013). High-resolution satellite images are also utilized to evaluate the variability and spatial distribution of crop coefficient (Kc) and crop evapotranspiration (ETc) (Dalezios et al., 2019a) and contribute through processing and analysis to support decision at field level. Timely and easy-to-use maps about water requirements are the natural way in which the information is valuable for precision farming application (Psilovikos and Elhag, 2013). Indeed, the implementation of EO methodology for calculating crop water requirements and mapping irrigated is considered internationally mature (Calera et al. 2017; Dalezios et al., 2019a). Valuable information on spatial variability is added by using time series of high and very high spatial resolution images, such as WV-2 at resolution 0.5 m, to describe, map and assess water stress (Psilovikos and Elhag, 2013). Figure 4.16 shows the actual evapotranspiration of July 11, 2019 from Landsat-8 (30 m resolution) at Albacete, Spain, using the Surface Energy Balance (SEB) method (METRIC), which is used in precision agriculture for crop water requirements.

4.6.2 REMOTELY SENSED WATER SCARCITY MANAGEMENT

At the present time, more than two billion people live without safe water and around four billion people suffer severe water scarcity for a least one month a year, which means that achieving *water for all* is a huge challenge (WMO, 2009). Moreover, climate change is expected to have a significant impact on a growing global population. With remote sensing technology, it is now possible to understand and measure the processes driving the water cycle and the impact of climate change and

FIGURE 4.17 Water scarcity case of Aral Sea. Image of June 15, 2018 from Proba-V, a miniaturized ESA satellite (Copyright from ESA/Belspo – produced by VITO).

human activity (WMO, 2009; Dalezios et al., 2018a; Dalezios et al., 2018b). Moreover, it is also possible to measure and monitor, for example, the changing shape of lakes, reservoirs and rivers through remote sensing, leading to mitigation strategies (Dalezios et al., 2018a). For example, data from the Sentinel-2 mission contribute to the Copernicus Land Monitoring Service, which provides geographical information on land cover and its changes, land use, vegetation state, water cycle and surface-energy variables for a broad range of users across the world (Copernicus, 2015).

The main subject of a sustainable water resources management framework involves technical and scientific analyses of drought and water scarcity management in regions and countries vulnerable to drought and water scarcity (Dalezios et al., 2018a). In every stage of such framework, remote sensing data and methods can significantly contribute to valuable solutions leading to integrated and effective sustainable water resource management. These analyses are expected to explore alternative water resource solutions, such as non-conventional measures related to marginal waters, including rain enhancement, desalination, water harvesting and water treatment and reuse potential, trends and practices under drought and water scarcity conditions (Dalezios et al., 2018a; Dalezios et al., 2018b).

The case of Aral Sea is presented as a typical example of water scarcity (Figure 4.17). Aral Sea was one of the four largest lakes in the world and is now one of the world's major ecological disaster areas. A recent image of Proba-V satellite shows the current situation of Aral Sea, which has shrunk into separate lakes, surrounded by Earth's youngest desert (Figure 4.17). Indeed, Proba-V is a miniaturized ESA satellite, which was launched on May 7, 2013 and tasked to map land cover and vegetation growth across the entire planet every two days. This 100 m resolution image of Figure 4.17 was acquired on June 15, 2018.

4.7 REMOTE SENSING FOR URBAN HYDROLOGY

Natural flood plains have been continuously urbanized, thus contributing to a sharp increase in loss of life and damage to property. Moreover, potential climate change, which is characterized by increased climate variability and uncertainty and increasing severity and frequency of extreme events, makes urban water management worldwide an urgent issue (Dabberdt et al., 2000; WMO, 2009; Dalezios, 2017). There are two types of urban flooding. First, fluvial flooding, where urban areas can be inundated by rivers overflowing their banks. Areas of inundation are forecast from the specific river-stage forecasts. Second, urban flooding can occur in local drainage as a special case of flash flooding. There is extensive literature on urban hydrology and water management based on remote sensing (Dabberdt et al., 2000; WMO, 2009; Dalezios, 2017).

Urban flooding. Typical features of urban watersheds are large areas of impervious or near impervious areas and the combined operation of natural and technological drainage systems, such as sewers, levees, pumps, detention basins and similar (WMO, 2009). Thus, surface runoff as produced from rainfall is highly variable and non-homogeneous, resulting in accelerated flow of water and contaminants leading to higher peaks of hydrographs at the outlet of the basin (WMO, 2009). Moreover, in urban watersheds high spatiotemporal rainfall variability translates into high spatio-temporal runoff variability. The technological drainage and improvements to the natural drainage may lead to earlier and higher peak flows. It is, thus, recognized that the problem of flood prediction and control becomes severe for flood events with return period of 5 to 100 years (WMO, 2009). In addition, the water quality problem can be acute with storms occurring with short return periods of even less than two years. It is evident therefore that it is necessary to use very high spatial and temporal resolution data, models and controls over large urban areas in order to ensure effective water resources management (Dabberdt et al., 2000; Dalezios, 2017; Destro et al., 2018;). Thus, digital terrain elevation data, distributed hydrological models and weather radar data, combined with in situ automated raingauge data and GIS, can be used to develop urban runoff forecast-now-cast and management systems (Cluckie and Collier, 1991; Braga and Massambani, 1997; Georgakakos and Krajewski, 2000; Destro et al., 2018). Similarly, in urban areas with mountainous terrain and convective weather regimes, there is a need to develop urban water resources management systems capable of very high resolution over large urban areas.

Flooding from local drainage. Intense rainfall over an urban area may cause flash flooding of streets and property in low-lying areas or built-up areas in old waterways, underpasses and depressions in highways (WMO, 2009; Destro et al., 2018). Such floods arise primarily from inadequate storm-drainage facilities and are invariably aggravated by debris clogging inlets to pipes and channels or outlets of retention basins. Flood warning schemes can be employed (WMO, 2009). These usually consist of local automated flash flood warning systems or generalized warnings that are based on national flash flood guidance operations (WMO, 2009). Urban flooding usually affects sewer systems, even when wastewater and storm sewerage are piped separately. Forecasts of urban runoff can be helpful in the treatment of sewage and the handling of polluted flood water in combined systems.

4.8 SUMMARY

Remote sensing data and methods for hydrology and water resources have been considered. Specifically, remote sensing of the components of the hydrological cycle has been presented, such as evapotranspiration, interception, soil moisture and groundwater. This is followed by remote sensing of water quality, remote sensing modeling of hydrological systems, remote sensing of snow, remote sensing for agrohydrology and remote sensing of urban hydrology. There is steadily increasing reliability and utilization of remote sensing data and methods in hydrology and water resources year by year, which is expected to continue.

REFERENCES

Alfieri, L., Velasco, D., and Thielen, J., 2011. Flash flood detection through a multi-stage probabilistic warning system for heavy precipitation events. *Advances in Geoscience*, 29, 69–75, https://doi.org/10.5194/adgeo-29-69-2011

Alfoldi, T.T., and Munday, J.C., 1978. Water quality analysis by digital chromaticity mapping of Landsat data. *Canadian Journal of Remote Sensing*, 4, 2, 108–126.

Allen, R.G., Pereira, L.S., Raes, D., and Smith, M., 1998. *Crop evapotranspiration: Guidelines for computing crop requirements. Irrigation and Drainage Paper No. 56*, FAO, Rome, Italy, 300p.

Allen, R.G., Clemmens, A.J., Burt, C.M., Solomon, K., and O'Halloran, T., 2005. Prediction accuracy for project wide evapotranspiration using crop coefficients and reference evapotranspiration. *Journal of Irrigation and Drainage Engineering*, 131, 24–36.

Allen, R.G., Irmak, A., Trezza, R., Hendrickx, J.M.H., Bastiaanssen, W., and Kjaersgaard, J., 2011. Satellite-based ET estimation in agriculture using SEBAL and METRIC. *Hydrological Processes*, 25, 26, December 30. 4011–4027.

Andersen, T., 1991. *AVHRR data for snow mapping in Norway, Proceedings of the 5th AVHRR Data Users Meeting*, Tromsoe, Norway.

Andersen, F.H., 2008. *Hydrological Modeling in a semi-arid area using remote sensing data*. Ph.D. Thesis, Dept. of Geography and Geology, U. of Copenhagen, Denmark, 110p.

Bastiaanssen, W.G.M., Noordman, E.J.M., Pelgrum, H., Davids, G., Thoreson, B.P., and Allen, R.G., 2005. SEBAL model with remotely sensed data to improve water resources management under actual field conditions. *ASCE Journal of Irrigation and Drainage Engineering*, 131, 85–93.

Baumgartner, M.F., 1988. Snowmelt runoff simulation based on snow cover mapping using digital Landsat-MSS and NOAA/AVHRR data, USDA-ARS, Hydrology Lab. Tech. Rep.

Bausch, W., and Neale, C., 1987. Crop coefficients derived from reflected canopy radiation: A concept. *Transactions of the ASAE*, 30, 0703–0709

Beven, K., and Young, P., 2013. A guide to good practice in modeling semantics for authors and referees. *Water Resources Research*, 49, 5092–5098.

Bhavsar, P.N., and Patel, J.N., 2016. Development of relationship between crop coefficient and NDVI using geospatial technology. *Journal of Agrometeorology*, 18, 261–264.

Braga, B., Jr., and Massambani, O. (eds.), 1997. *Weather Radar Technology for Water Resources Management*. UNESCO Press, Montevideo, 516.

Brest, C.L., and Goward, S.N., 1987. Deriving surface albedo measurements from narrow band satellite data. *International Journal of Remote Sensing*, 8, 351–367.

Calera, A., Campos, I., Osann, A., D'Urso, G., and Menenti, M., 2017. Remote sensing for crop water management: From ET modelling to services for the end users. *Sensors (Switzerland)* 17, 1104, doi:10.3390/s17051104

Campos, I., Neale, C.M.U., Suyker, A.E., Arkebauer, T.J., and Gonçalves, I.Z. 2017. Reflectance-based crop coefficients REDUX: For operational evapotranspiration estimates in the age of high producing hybrid varieties, *Agricultural Water Management*, 187, 140–153.

Chehbouni, A., Lo Seen, D., Njoku, E.G., and Monteny, B., 1996. Examination of the difference between radiative and aerodynamic surface temperatures over sparsely vegetated surfaces. *Remote Sensing of Environment*, 58, 2, 176–186.

Choudhury, B.J., Ahmed, N.U., Idso, S.B., Reginato, R.J., and Daughtry, C.S.T., 1994. Relations between evaporation coefficients and vegetation indices studied by model simulations. *Remote Sensing Environment*, 50, 1–17.

Chuvieco, E., 2000. *Fundamentos de la Teledetección Espacial*, 3rd edn, Ediciones RIALP, Madrid.

Cluckie, I.D., and Collier, C.G. (eds.), 1991. *Hydrological Applications of Weather Radar, Environmental Management, Science and Technology Series*. Ellis Horwood, Chichester, 644.

Copernicus, Earth Observation Satellites, 2015. http://www.copernicus.eu/main/satellites [accessed: March 31, 2015].

Crapolicchio, R., Ferrazzoli, P., Meloni, M., Pinori, S., and Rahmoune, R., 2010. Soil Moisture and Ocean Salinity (SMOS) mission: System overview and contribution to vicarious calibration monitoring. *Italian Journal of Remote Sensing*, 42, 1, 37–50.

Cui, Y., Zhao, P., Yan, B., Xie, H., Yu, P., Wan, W., Fan, W., and Hong, Y., 2017. Developing the remote sensing-gash analytical model for estimating vegetation rainfall interception at very high resolution: A case study in the Heihe River Basin. *Remote Sensing*, 9, 661, doi:10.3390/rs9070661

Dabberdt, W.F., Hales, J., Zubrick, S., Crook, A., Krajewski, W., Christopher Doran, J., Mueller, C., King, C., Keener, R.N., Bornstein, R., Rodenhuis, D., Kocin, P., Rossetti, M.A., Sharrocks, F., and Stanley, E.M., Sr., 2000. Forecast issues in the urban zone, Report of the 10th Prospectus Development team of the U.S. Weather Research Program. *Bulletin of the American Meteorological Society*, 81, 9, 2047–2064.

Dalezios, N.R., 2014. "Remote Sensing Potential in Hydrological Simulation," Invited paper in 20-year Anniversary Special Volume of Dept. of Civil Engineering, Univ. of Thessaly, Volos, Greece entitled Advances in Civil Engineering Research, by A. Liakopoulos, E. Mistaskidis and A.Giannakopoulos (eds.), 253–264.

Dalezios, N.R., 2015. *Agrometeorology: Analysis and Simulation (in Greek)*. Libraries of Hellenic Universities (also e-book), Kallipos, 481

Dalezios, N.R. (ed.), 2017. *Environmental Hazards Methodologies for Risk Assessment and Management.* IWA, London UK, 534

Dalezios, N.R., Blanta, A., and Spyropoulos, N.V., 2012. Assessment of remotely sensed drought features in vulnerable agriculture. *Natural Hazards and Earth System Sciences*, 12, 3139-3150.

Dalezios, N.R., Spyropoulos, N.V., and Eslamian, S., 2017. Remote Sensing in Drought Quantification and Assessment, pp. 375-394, in *Vol. 1 of 3-Volume Handbook of Drought and Water Scarcity (HDWS).* Editor: S. Eslamian. Taylor and Francis.

Dalezios, N.R., Angelakis, A.N., and Eslamian, S., 2018a. Water scarcity management: Part 1: Methodological framework. *International Journal of Global Environmental Issues*, 17, 1, 1–40.

Dalezios, N.R., Dercas, N., and Eslamian, S., 2018b. Water scarcity management: Part 2: Satellite-based composite drought analysis. *International Journal of Global Environmental Issues*, 172/3, 267–295.

Dalezios, N.R., Dercas, N., Spyropoulos, N.V., and Psomiadis, M., 2019a. Remotely sensed methodologies for water availability and requirements in precision farming of vulnerable agriculture. *Water Resources Management*, 33, 1499–1519.

Dalezios, N.R., Blanta, A., Loukas, A., Spiliotopoulos, M., Faraslis, I.N., and Dercas, N., 2019b. Satellite methodologies for rationalizing crop water requirements in vulnerable agroecosystems. *International Journal of Sustainable Agricultural Management and Informatics*, 5, 1, 37–58.

Dalezios, N.R., Dercas, N., Blanta, A., and Faraslis, I.N., 2020a. Remote sensing in water balance modelling for evapotranspiration at a rural watershed in Central Greece. *International Journal of Sustainable Agricultural Management and Informatics*(accepted, in press).

Dalezios, N.R., Petropoulos, G.P., and Faraslis, I.N., 2020b. Concepts and Methodologies of Environmental Hazards and Disasters, pp. 3–22, in *Techniques for Disaster Risk Management and Mitigation.* Editors: P.K. Srivastava, S.K. Singh, U.C. Mohanty and T. Murty. AGU-Wiley, 352.

de Jong, S.M., and Jetten, V.G., 2007. Estimating spatial patterns of rainfall interception from remotely sensed vegetation indices and spectral mixture analysis. *International Journal of Geographical Information Science*, 21, 5, 529–545.

Destro, E., Amponsah, W., Nikolopoulos, E.I., Marchi, L., Marra, F., Zoccatelli, D., and Borga, M., 2018. Coupled prediction of flash flood response and debris flow occurrence: Application on an alpine extreme flood event. *Journal of Hydrology*, https://doi.org/10.1016/j.jhydrol.2018.01.021

Doorenbos, J., and Pruitt, W.O., 1977. *Guidelines for Predicting Crop Water Requirements; FAO Irrigation and Drainage Paper No. 24*, FAO, Rome, Italy.

Engman, E.T., and Gurney, R.J., 1991. *Remote Sensing in Hydrology.* Chapman and Hall, London.

Entekhabi, D., Njoku, E.G., O'Neill, P.E., Kellogg, K.H., Crow, W.T., Edelstein, W.N., Entin, J.K., Goodman, S.D., Jackson, T.J., Johnson, J., et al., 2010. The soil moisture active passive (SMAP) mission. *Proceedings of the IEEE*, 98, 5,704–716.

ESA, 2014. Sentinel, Earth online - ESA, https://earth.esa.int/web/guest/missions/esa-future-missions#_56_INSTANCE_hH2r_matmp

ESA, 2019. Floods imaged by Copernicus Sentinel-1, https://www.esa.int/ESA_Multimedia/Images/2019/03/Floods_imaged_by_Copernicus_Sentinel-1#.X_615_yv5eA.link

Eumetsat, Future Satellites, 2015 http://www.eumetsat.int/website/home/Satellites/FutureSatellites/index.html [accessed: April 17, 2015].

Fortin, J., Turcotte, R., Massicotte, S., Moussa, R., Fitzback, J., and Villeneuve, J., 2001. Distributed watershed model compatible with remote sensing and GIS data. I: Description of model. *Journal of Hydrologic Engineering*, 6, 2, 91–99.

Gautier, C., Diak, G., and Masse, S., 1980. A simple physical model to estimate incident solar radiation at the surface from GOES satellite data. *Journal of Applied Meteorology*, 19, 1005–1012.

Georgakakos, K.P., and Krajewski, W.F., (eds.), 2000. Hydrologic applications of weather radar. *Special Issue, Journal of Geophysical Research – Atmospheres*, 105, D2, 2213–2313.

Gowda, P.H., Chávez, J.L., Colaizzi, P.D., Evett, S.R., Howell, T.A., and Tolk, J.A., 2007. Remote sensing-based energy balance algorithms for mapping ET: Current status and future challenges. *Transactions of the ASABE*, 50, 5, 1639–1644.

Granger, R.J., 1997. Comparison of Surface and Satellite Derived Estimates of Evapotranspiration Using a Feedback Algorithm, pp. 71–81, in: *Applications of Remote Sensing in Hydrology: Proceedings of Symposium No. 17.* Editors: G.W. Kite, A. Pietroniro and T. Pultz. National Hydrology Research Institute, Saskatoon.

Hamman, J.J., Nijssen, B., Bohn, T.J., Gergel, D.R., and Mao, Y., 2018. The variable infiltration capacity model, Version 5 (VIC-5): Infrastructure improvements for new applications and reproducibility. *Geoscientific Model Development: Discussions*, https://doi.org/10.5194/gmd-2018-36

Irmak, A., Ratcliffe, I., Ranade, P., Hubbard, K., Singh, R.K., Kamble, B., and Kjaersgaard, J., 2011. Estimation of land surface evapotranspiration with a satellite remote sensing procedure. *Great Plains Research*, 21, 73–88.

Jones, W.K., and Carrol, T.R., 1983. Error analysis of airborne gamma radiation soil moisture measurements. *Agricultural Meteorology*, 28, 19–30.

Kalma, J.D., McVicar, T.R., and McCabe, M.F., 2008. Estimating land surface evaporation: A review of methods using remotely sensed surface temperature data. *Surveys in Geophysics*, 29, 4-5, 421–469.

Kamble, B., Kilic, A., and Hubbard, K., 2013. Estimating crop coefficients using remote sensing-based vegetation index. *Remote Sensing*, 5, 4, 1588–1602.

Kerr, Y., et al., 2010. The SMOS Mission: New tool for monitoring key elements of the global water cycle. *Proceedings of the IEEE*, 98, 666–687.

Klaassen, W., Boseveld, F., and de Water, E., 1998. Water storage and evaporation as constituents of rainfall interception, *Journal of Hydrology*, 212–213, 36–50.

Kustas, W.P., and Norman, J.M., 1996. Use of remote sensing for evapotranspiration monitoring over land surfaces. Hydrological. *Sciences Journal*, 41, 4, 495–516.

Li, Z.-L., Tang, R., Wan, Z., Bi, Y., Zhou, C., Tang, B., Yan, G., and Zhang, X.A., 2009. A review of current methodologies for regional evapotranspiration estimation from remotely sensed data. *Sensors*, 9, 5, 3801–3853.

Liu, Y., Parinussa, R., Dorigo, W., Jeu, R.D., Wagner, W., Dijk, A.V., McCabe, M., and Evans, J., 2011. Developing an improved soil moisture dataset by blending passive and active microwave satellite-based retrievals. *Hydrology and Earth System Sciences*, 15, 2, 425–436

Lundberg, A., 1993. Evaporation of intercepted snow: Review of existing and new measurement methods. *Journal of Hydrology*, 151, 267–290.

Malenovský, Z., Rott, H., Cihlar, J., Schaepman, M.E., García-Santos, G., Fernandes, R., and Berger, M., 2012. Sentinels for science: Potential of Sentinel-1, -2, and -3 missions for scientific observations of ocean, cryosphere, and land. *Remote Sensing of Environment*, 120, 91–101.

Marra, F., Nikolopoulos, E.I., Creutin, J.D., and Borga, M., 2014. Radar rainfall estimation for the identification of debris-flow occurrence thresholds. *Journal of Hydrology*, 519, 1607–1619.

Mateos, L., González-Dugo, M.P., Testi, L., and Villalobos, F.J., 2013. Monitoring evapotranspiration of irrigated crops using crop coefficients derived from time series of satellite images, I: method validation. *Agricultural Water Management*, 125,81–91.

McCabe, M.F., Wood, E.F., Wójcik, R., Pan, M., Sheffield, J., Gao, H., and Su, H., 2008. Hydrological consistency using multi-sensor remote sensing data for water and energy cycle studies. *Remote Sensing of Environment*, 112, 430–444.

Menenti, M., and Choudhury, B.J., 1993. *Parameterization of Land Surface Evaporation by Means of Location Dependent Potential Evaporation and Surface Temperature Range*, FAO, Rome, Italy.

Miralles, D.G., Gash, J.H., Holmes, T.R.H., de Jeu, R.A.M., and Dolman, A.J., 2010. Global canopy interception from satellite observations. *Journal of Geophysical Research*, 115, D16122, doi:10.1029/2009JD013530

Miralles, D.G., Holmes, T.R.H., de Jeu, R.A.M., Gash, J.H., Meesters, A.G.C.A., and Dolman, A.J., 2011. Global land-surface evaporation estimated from satellite-based observations. *Hydrology and Earth System Sciences*, 15, 453–469

Monteith, J.L., and Unsworth, M., 1990. *Principles of Environmental Physics*. Academic Press, Burlington, VT, USA.

Mu, Q.Z., Zhao, M., and Running, S.W., 2011. Improvements to a MODIS global terrestrial evapotranspiration algorithm. *Remote Sensing of Environment*, 115, 1781–1800.

Munday, Jr. J.C., Alfoldi, T.T., and Amos, C.L., 1979. Bay of Fundy Verification of a System of Multidate Landsat Measurement of Suspended Sediment, pp. 622–640, in *Satellite Hydrology*. Editors: M. Deutsch, D.R. Wiesnet and A. Rango. American Water Resources Association (AWRA).

Njoku, E.G., Jackson, T.J., Lakshmi, V., Chan, T.K., and Nghiem, S.V., 2003. Soil moisture retrieval from AMSR-E. *IEEE Transactions on Geoscience and Remote Sensing*, 41, 2, 215–229.

Paterson, N.R., and Bosschart, R.A., 1987. Airborne geophysical exploration for ground water. *Groundwater*, 25, 41–50

Peck, E.L., Keefer, T.N., and Johnson, E.R., 1982. *Suitability of Remote Sensing Capabilities for use in Hydrologic Models. Proceedings, International Symposium on Hydrometeorolgy*, AWRA, Denver, CO, 59–63.

Pellet, V., Aires, F., Munier, S., Prieto, D.F., Jordá, G., Dorigo, W.A., Polcher, J., and Brocca, L., 2019. Integrating multiple satellite observations into a coherent dataset to monitor the full water cycle – application to the Mediterranean region. *Hydrology and Earth System Sciences*, 23, 465–491

Psilovikos, A., and Elhag, M., 2013. Forecasting of remotely sensed daily evapotranspiration data over Nile Delta Region, Egypt. *Water Resources Management*, 27, 4115–4130

Rango, A., (ed.), 1975. *Operational Applications of Satellite Snowcover Observations*. National Aeronautics and Space Administration, NASA Special Publication, SP-391, Washington, DC, 430.

Rango, A., and van Katwijk, V., 1990. Development and testing of a snowmelt-runoff forecasting technique. *Water Resources Bulletin*, 26, 135–144.

Rango, A., Chang, A.T.C., and Foster, J.L., 1979. The utilization of spaceborne microwave radiometers for monitoring snowpack properties. *Nordic Hydrology*, 10, 25–37.

Roerink, G.J., Su, B., and Menenti, M., 2000. S-SEBI: A simple remote sensing algorithm to estimate the surface energy balance. *Physics and Chemistry of the Earth, Part B*, 25, 2, 147–157.

Rosser, J.F., Leibovici, D.G., and Jackson, M.J., 2017. Rapid flood inundation mapping using social media, remote sensing and topographic data. *Natural Hazards*, 87, 103–120

Schmugge, T.J., Jackson, T.J., and McKlim, H.L., 1980. Survey of methods for soil moisture determinations. *Water Resources Research*, 18, 6, 961–979.

Schultz, G.A., and Engman, E.T., (eds.), 2000. *Remote Sensing in Hydrology and Water Management*. Springer, 483.

Sene, K., 2016. *Hydrometeorology*. 2nd ed. Springer, New York, 427.

Sharma, L.M., 1985. Estimating Evapotranspiration, pp. 213–282, in *Advances in Irrigation, Volume 3*. Editor: D. Hillel Academic Press, New York.

Shi, H., Xiao, Z., Liang, S., and Zhang, X., 2016. Consistent estimation of multiple parameters from MODIS top of atmosphere reflectance data using a coupled soil-canopy-atmosphere radiative transfer model. *Remote Sensing of Environment*, 184, 40–57.

Su, Z., 2002. The surface energy balance system (SEBS) for estimation of turbulent fluxes. *Hydrology And Earth Systems Science*, 6 1, 85–99.

Takada, M., Mishima, Y., and Natsume, S., 2009. Estimation of surface soil properties in peatland using ALOS/PALSAR. *Landscape and Ecological Engineering*, 5, 1, 45–58.

Thenkabail, P.S., Gamage, M.S.D.M., and Smakhtin, V.U., 2004. The Use of Remote Sensing Data for Drought Assessment and Monitoring in Southwest Asia, Research Report, *International Water Management Institute*, 85, 1–25.

Ulaby, F.T., 1974. Radar measurement of soil moisture content. *IEEE Transactions on Antennas and Propagation*, AP-22, 257–265.

Ulaby, F.T., Moore, R.K., and Fung, A.K., 1982. *Microwave Remote Sensing: Active and Passive, Vol. II—Radar Remote Sensing and Surface Scattering and Emission Theory*. Addison-Wesley, Advanced Book Program, Reading, MA, 609

van de Griend, A.A., Camillo, P.J., and Gurney, R.J., 1985. Discrimination of soil physical parameters, thermal inertia and soil moisture from diurnal surface temperature fluctuations. *Water Resources Research*, 21, 997–1009.

Vuolo, F., D'Urso, G., De Michele, C., Bianchi, B., and Cutting, M., 2015. Satellite-based irrigation advisory services: A common tool for different experiences from Europe to Australia. *Agricultural Water Management*, 147, 82–95.

WMO, 1992a. Remote Sensing for Hydrology: Progress and Prospects (R. Kuittinen). Operational Hydrological Report No. 36, WMO-No. 773, Geneva.

WMO, 1992b. Snow Cover Measurements and Areal Assessment of Precipitation and Soil Moisture. Operational Hydrology Report No. 35, WMO No. 749, Geneva.

WMO, 1993. Executive Council Panel of Experts on Satellites: Final Report. March 9–10, Geneva.

WMO, 1994. Applications of remote sensing by satellite, radar and other methods to hydrology, Operational Hydrology Report No. 39, WMO-No. 804, Geneva.

WMO, 2009, Guide to Hydrological Practices, Vol. I and II. WMO-No. 168, 6th edition, 598p.

Wright, J.L., 1982. New evapotranspiration crop coefficients. *Journal of the Irrigation and Drainage Division*, 108, 57–74.

Wu, C.D., Lo, H.C., Cheng, C.C., and Chen, Y.K., 2010. Application of SEBAL and Markov models for future stream flow simulation through remote sensing. *Water Resources Management*, 24, 3773–3797.

Wu, H., Kimball, J.S., Mantua, N., and Stanford, J., 2011. Automated upscaling of river networks for macroscale hydrological modeling. *Water Resources Research*, 47, W03517

Xu, H., 2006. Modification of normalised difference water index (NDWI) to enhance open water features in remotely sensed imagery. *International Journal of Remote Sensing*, 27, 3025–3033

NOAA KLM, 2015. User's Guide, Section 3.1, at: https://www1.ncdc.noaa.gov/pub/data/satellite/publications/podguides/N-15%20thru%20N-19/pdf/0.0%20NOAA%20KLM%20Users%20Guide.pdf   [accessed: April 17, 2015].

5 Marine and Coastal Ecosystems

5.1 INTRODUCTION: REMOTE SENSING CONCEPTS OF MARINE AND COASTAL ECOSYSTEMS

The oceans and, in general, the marine environment, provide valuable biophysical resources and contribute to the food chain. Moreover, they make a significant contribution to the evolution of weather systems and CO_2 storage, as well as to the earth's hydrological balance. In addition, oceans serve as transportation routes. It is significant to understand ocean dynamics, the study of which includes water temperature, bathymetry, mesoscale feature identification, ocean productivity, as well as wind and wave retrieval. Moreover, prediction of global circulation and forecasting and monitoring storms can help to reduce the impacts of extreme phenomena, such as El Niño, as well as the impact of disasters on marine navigation, ship routing, offshore exploration, fish stock assessment and coastal settlements.

The coastal zone is a region subject to increasing stress from human activity, since it can be highly urbanized. Coastal ecosystems interface between the ocean and land and are affected by economic development leading to changes in land-use patterns. Indeed, it is estimated that over 60% of the world's population lives close to the ocean. Moreover, coastlines constitute biologically diverse intertidal zones. New data sources and approaches, such as remote sensing data and methods, are required in order to monitor such diverse changes as coastal erosion, loss of natural habitat, urbanization, effluents and offshore pollution.

Remote sensing data and methods can be used to map and monitor many of the dynamics of the open ocean and changes in the coastal region. Indeed, remote sensing applications for ocean and coastal ecosystems include ocean features, such as currents, regional circulation patterns, shears, internal waves, gravity waves, upwelling zones, shallow water and bathymetry. They may also include storm monitoring, which involves frontal zones, storm early warning systems, cyclones, marine fog, wind and wave retrieval. Another remote sensing application is ocean color analysis, which covers water temperature monitoring, sea surface temperature (SST), salinity, density, water quality, such as sediments and dissolved oxygen (DO), ocean productivity, phytoplankton, concentration and drift, aquaculture and fish stocks. Moreover, oil spills may be considered, which involve oil spills mapping, predicting oil spill extent and drift, emergency response decisions, or identification of natural oil seepage areas for exploration. Sea ice is another remote sensing application, which covers ice type, ice age, ice motion, ice condition, i.e. state of decay, navigation for safe shipping routes or rescue, ice and iceberg conditions and tracking, as well as pollution monitoring. Similarly, shipping is another remote sensing application, which involves navigation routing, traffic density, operational fisheries surveillance and near-shore bathymetry mapping. Coastal and intertidal zones constitute another remote sensing application, which involve land/water interface, coastal zoning monitoring, coastal vegetation, tidal and storm impacts, mapping shoreline features/beach dynamics, tsunami and the impact of human activity.

5.2 OCEAN FEATURES

Ocean features refer to currents, shears, internal waves, gravity waves, regional circulation patterns, upwelling zones, bathymetry and shallow waters. Ocean features also cover sea-floor modeling, which may involve waste disposal and resource extraction planning activities.

5.2.1 Currents

The main concern is the direction and strength of currents. The assessment of the general sea state, as expressed by currents, waves and winds, requires time sensitive information, which is valuable when the conditions exist.

The interaction of the upper ocean with the atmosphere at the marine boundary layer may result in surface roughness patterns, which can be detected by Synthetic aperture radar (SAR). There is a dynamic relationship between ocean and atmosphere. Specifically, Bragg scattering constitutes the principal scattering mechanism for ocean surface imaging. The surface waves, which result in Bragg scattering, are roughly equivalent to the wavelength used by C-band RADARSAT (5.3 cm). Indeed, the backscatter intensity is a function of the incidence angle and radar wavelength, as well as the sea state conditions at the time of imaging. Moreover, the short waves on the ocean surface create spatially varying surface patterns. Indeed, the wind stress at the upper ocean layer can cause the formation of these short waves. Furthermore, variable wind speed, long gravity waves and surface currents associated with upper ocean processes, such as eddies, fronts and internal waves can cause modulation in the short waves. As a result, there are spatially variable surface roughness patterns, which can be detected by SAR imagery.

Case studies. Three case studies are briefly presented as follows:

Case study 1. Figure 5.1 shows SAR images, which delineate atmospheric gravity waves. (Note: The description of Figure 5.1 is taken from Fan et al., 2019a). Specifically: (a) RADARSAT-2 (RS-2) SAR HH-polarized images (dB units) of an area of the West Coast of USA acquired at 13:53 UTC on May 24, 2012 (Fan et al., 2019a). The key parameters associated with the enlarged region of the yellow fragment in (a) are estimated as: (b) polarization ratio (PR) (linear units); (c) non-polarized (NP) contrast; (d) polarization difference (PD) contrast; and (e) cross-polarized wave breaking (CPwb) contrast. The incidence angle is about 44°, the averaged wind speed is 11 m/s and the angle between radar look direction and wind direction is 59°. The black and white arrows in (a) correspond to wind direction and radar look direction. Moreover, wind direction is from the NDBC buoy and wind speed from the C-band SAR model-2 geophysical model function (C_SARMOD2 GMF model).

Case study 2. Figure 5.2 shows SAR images, which delineate an oceanic internal solitary wave (IW) and a train of IWs. (Note: The description of Figure 5.2 is taken from Fan et al., 2019a). Specifically: (a) RS-2 SAR HH-polarized images (dB units) of an area of the West Coast of

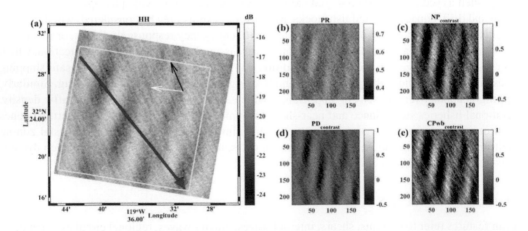

FIGURE 5.1 Synthetic aperture radar (SAR) images containing atmospheric gravity waves of an area of the West Coast of USA acquired at 13:53 UTC on May 24, 2012 (from Fan et al., 2019a).

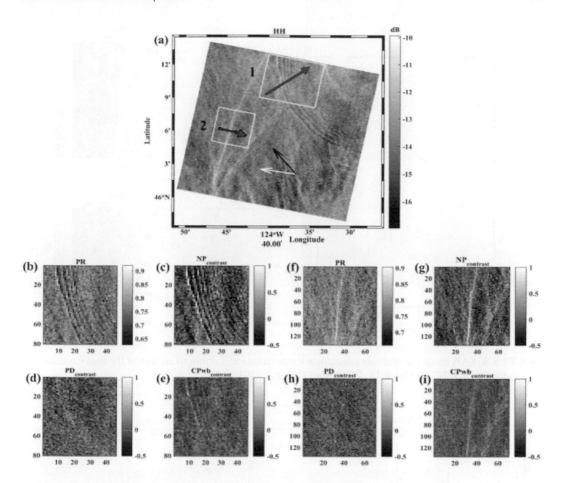

FIGURE 5.2 SAR images containing an oceanic internal solitary wave (IW) and a train of IWs of an area of the West Coast of USA acquired at 14:35 UTC on August 15, 2009 (from Fan et al., 2019a).

USA acquired at 14:35 UTC on August 15, 2009 (Fan et al., 2019a). The key parameters associated with the enlarged region of the yellow fragment in (a) are estimated as: (b,f) PR (linear units); (c,g) NP contrast; (d,h) PD contrast; (e,f) CPwb contrast. The incidence angle is about 29°, the averaged wind speed is 6 m/s and the angle between radar look direction and wind direction is 39°.

Case study 3. Figure 5.3 shows SAR images, which delineate sea surface temperature (SST) front. (Note: The description of Figure 5.2 is taken from Fan et al., 2019a). Specifically: (a) RS-2 SAR HH-polarized images (dB units) of an area of the East Coast of USA acquired at 22:07 UTC on March 18, 2009 (Fan et al., 2019a). The key parameters associated with the enlarged region of the yellow fragment in (a) are estimated as: (b) PR (linear units); (c) NP contrast; (d) PD contrast; and (e) CPwb contrast. The incidence angle is about 33°, the averaged wind speed is 4 m/s and the angle between radar look direction and wind direction is 112°.

The study of ocean dynamics, currents, tides, as well as ship safety are based on bathymetry maps. Indeed, only small fractions of the Arctic Ocean have ever been covered, although there have been several campaigns to map sea-floor bathymetry through ship soundings. An Arctic bathymetry map using marine gravity has been recently developed (Abulaitijiang et al., 2019). Specifically, the height of the ocean surface mimics the rise and fall of the ocean floor due to gravitational pulls. Indeed, areas of greater mass, such as underwater mountains, have a higher gravity and attract more

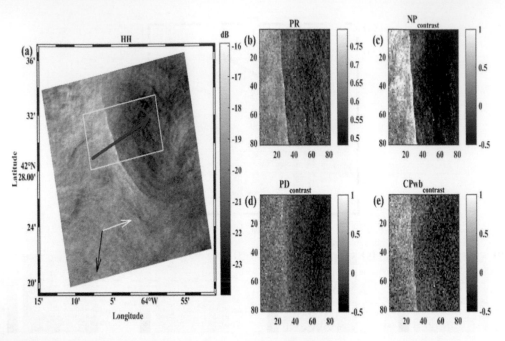

FIGURE 5.3 SAR images containing sea surface temperature (SST) front of an area of the East Coast of USA acquired at 22:07 UTC on March 18, 2009 (from Fan et al., 2019a).

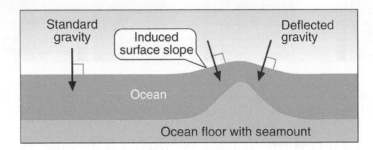

FIGURE 5.4 Interaction between gravity and sea level (from ESA, July 24, 2019a).

water creating a rise in the sea surface (Figure 5.4). Satellite gravity measurements have been used to calculate sea-floor bathymetry in the Arctic Ocean from the relationship between bathymetry and gravity. Specifically, an altimetric gravity model has recently been developed based on ERS-1 and ERS-2, Envisat and seven years of CryoSat data (Abulaitijiang et al., 2019). As a result, a new and improved hybrid bathymetry map of the Arctic Ocean has been produced by combining this model with the existing International Bathymetric Chart of the Arctic Ocean (IBCAO) bathymetry map. Indeed, Figure 5.5 over the Chukchi Cap in the Canadian Arctic is an example of how the hybrid bathymetry could improve the existing IBCAO model derived from sparse ship tracks in the region. This mapping shows the capability of satellites to provide us with new data, especially in more difficult areas, such as the unknown Arctic waters.

5.2.2 Winds

The emphasis is on intensity and direction of surface winds. Scatterometers collect information on wind speed and direction, whereas altimeters measure wave height and identify wind speed. SARs can receive signal information for the ocean surface under certain wind speed conditions.

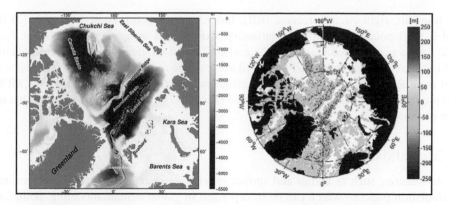

FIGURE 5.5 Bathymetry map Over the Chukchi Cap in the Canadian Arctic (from ESA, July 24, 2019a).

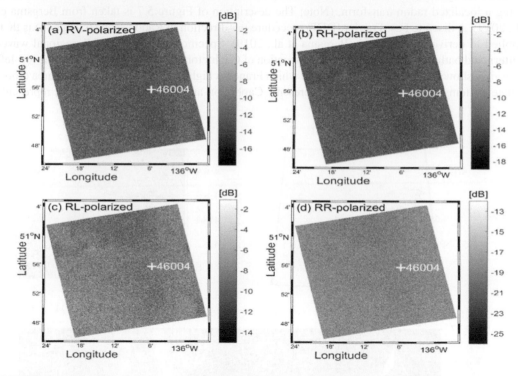

FIGURE 5.6 RADARSAT-2 for retrieval of ocean surface wind speeds by using four new channels. RADARSAT-2 Data and Products from MacDonald, Dettwiler, and Associates Ltd. All Rights Reserved (from Fang et al., 2019).

Specifically, at very high wind speeds (greater than 14 m/s) the SAR is masked by ocean clutter whatever surface features may be present and at very low wind speeds (2–3 m/s) the SAR is not sensitive enough to detect the ocean "clutter".

Case study. Figure 5.6 shows the use of C-band RADARSAT Constellation Mission (RCM) SAR for retrieval of ocean surface wind speeds. (Note: The description is taken from Fang et al., 2019). The image was taken on November 8, 2009, at 02:47:39 UTC, and the location of the Environment and Climate Change Canada (ECCC) buoy (#46004, 50°5504800 N 136°60W) collocated to the SAR images. Specifically, four new channels are used in compact polarimetry (CP) mode (Fang et al., 2019) as follows: (1) right circular transmit, vertical receive (RV); (2) right circular transmit, horizontal receive (RH); (3) right circular transmit, left circular transmit (RL); and (4) right circular transmit, right circular receive (RR). Indeed, Figure 5.6 presents: (1) RV-, (2) RH-, (3) RL- and (4) RR-polarized compact polarimetry SAR images using the CP simulator. Moreover, RCM CP data was simulated using a "CP simulator" based on 256 buoy measurements collocated with RADARSAT-2 fine beam quad-polarized scenes. The results indicate that wind speed can be retrieved from RV, RH and RL polarization channels using existing C-band model (CMOD) geophysical model function (GMF) and polarization ratio (PR) models, under the assumption the wind direction is known.

5.2.3 CIRCULATION PATTERNS

Mesoscale features, such as eddies and surface gravity waves, can facilitate the determination of ocean circulation patterns, which can be incorporated in global climate modeling, pollution monitoring, navigation and forecasting for offshore operations.

Case study. Figure 5.7 shows derived wave patterns from Sentinel-2 imagery, which are extracted using a localized radon transform. (Note: The description of Figure 5.7 is taken from Bergsma et al., 2019). A discrete Fast-Fourier (DFT) procedure per direction in Radon space (sinogram) is then applied to derive wave spectra (Bergsma et al., 2019). Specifically, in Figure 5.7, regional wave-pattern and bathymetry is presented at the location of Capbreton in South–West France. The top-left overview shows part of Western Europe in which France is highlighted as the darker gray area (coordinate system WGS84). The red dot represents Capbreton and the box around the dot represents

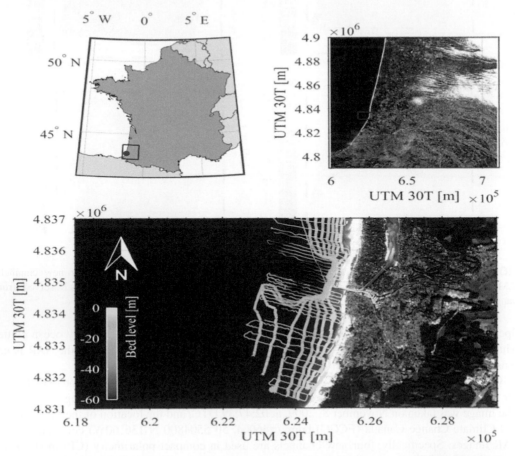

FIGURE 5.7 Regional Wave-Pattern and Bathymetry at the location of Capbreton in South–West France (from Bergsma et al., 2019).

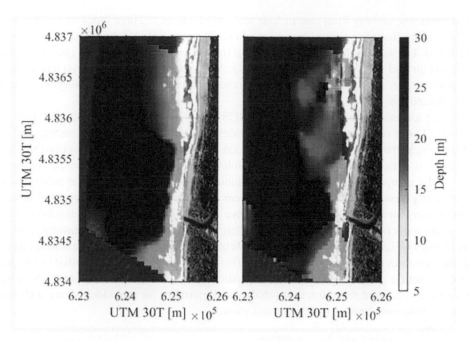

FIGURE 5.8 Measured (left) and Sentinel-2 estimated (right) water depths in the vicinity of the Capbreton harbour (from Bergsma et al., 2019).

Sentinel-2 tile 30TXP projected on the UTM-zone 30T (Universal Transverse Mercator). The tiled Sentinel-2 image is presented at the top right (image taken on November 20, 2017). The red box in the top-right figure highlights the zoomed-in area of the bottom plot. The small dots (mainly on cross-shore arrays) represent the echo-sounder measured depths. The thicker predominantly along-shore lines represent the depth iso-contours. The deep-water canyon (dark blue dots) is evident close to shore, West–Northwest of the harbor entrance. The echo-sounder was realistically capable of measuring from the shallowest as the boat would go until 60–70 m water depth. The RT-based wave-pattern extraction is tested on 10 × 10 m resolution Sentinel-2 imagery covering the region surrounding Capbreton—Sentinel-2 relative orbit 94, tile 30TXP—on two dates: (1) 2 days after the end field campaign on November 20, 2017 and (2) preferable wave conditions (March 30, 2018). The bathymetry is only derived for the latter to demonstrate the methods' performance (Figure 5.7). Moreover, Figure 5.8 shows the measured (left) and Sentinel-2 estimated (right) water depths in the vicinity of the Capbreton harbor. (Note: The description of Figure 5.8 is taken from Bergsma et al., 2019). Specifically, the measured bathymetry is interpolated on the depth estimation locations. This highly complex bathymetry, specifically the contours of the deep-water channel and near-shore zone south of the harbor entrance, show similarities.

5.3 STORM AND FOG MONITORING

5.3.1 Storm Forecasting and Monitoring

5.3.1.1 Storm Forecasting: Predictive Signals

The atmosphere appears to reflect more about the future state of the combined atmosphere-ocean system than does the ocean surface by itself. There are strong multi-season predictive signals for the Atlantic seasonal hurricane activity contained in prior-year West African rainfall, the strength of the Azores anticyclone, as well as the southern oscillation index (SOI). Combined atmospheric precur-sor signals allow an explanation as much as 40–60 percent of the year-to-year variance of Atlantic seasonal hurricane activity from one to four seasons in advance. Three types of low-level convergence boundaries that often initiate storms are: (1) cold fronts, (2) cold thunderstorm outflows, such as cold fronts, but with cold air mass originating from the storm, (3) drylines, i.e. change between warm, moist air and hot, dry air (Dalezios, 2017). A banded area of concentrated temperature contrast or gradient, i.e. a baroclinic zone, often coincides with strong air-flow convergence along these low-level boundaries.

Predicting a mesoscale convective system (MCS) may be achieved by anticipating three sequential events: (1) the outbreak of strong (often widespread) convection, (2) the subsequent upscale growth and organization, (3) the ultimate evolution and decay. In addition to successfully forecasting an MCS, it is necessary to anticipate the evolution of the following attributes: (1) the type or "mode" of the component convective cells (including modal evolution); (2) the nature and degree or mode of organization of the convective system as a whole; (3) the total area and dimensions, location, movement and duration of the convective system. There are new operational regional and smaller-scale numerical prediction models, which support severe storm forecasting. Specifically, there is higher resolution and more accurate fields of the critical atmospheric parameters that lead to severe thunderstorm and tornado development. The new technology seems to assist in moving from the era of detection of warnings to the era of prediction of warnings for tornadoes, which has contributed to the continued reduction in morbidity and mortality for tornadoes, along with improvements in warning coordination and communication.

5.3.1.2 Storm Tracking and Now-casting

Experimental forecasters are able to produce successful daily forecasts of deep convection. A largely unsolved problem in MCS forecasting involves anticipating the timing and location of convective initiation. Since storms often form within a range of 10–20 km of mesoscale boundaries, which may themselves be meso-γ-scale or narrower in width, variability in the boundary layer is probably critical for the storm initiation process along with formation, structure and movement (Dalezios, 2017). These methods include extrapolation, knowledge-based now-casting, numerical models, neural network models and further approaches, such as probability forecasts and modified turning band (MTB) models. Specifically, extrapolation methods involve steady-state assumption methods, including cross-correlation and feature-tracking methods, and echo size and intensity trending methods. In Chapter 2, these methods are briefly presented.

Case studies. Two case studies are briefly presented as follows:

Case study 1. Tropical cyclones (TCs) with maximum wind speeds higher than 32.7 m/s in the north-western Pacific are referred to as typhoons. Specifically, typhoons Sarika and Haima successively passed the observation array in the northern South China Sea in 2016 (Figure 5.9). (Note: The description of Figure 5.9 was taken from Zhang et al., 2019). In Figure 5.9, the winds (clouds and rainfall) are biased to the right (left) sides of the typhoon tracks, based on the satellite data. Indeed, Sarika and Haima cooled the sea surface ~4 and ~2°C and increased the salinity ~1.2 and ~0.6 psu, respectively, with the maximum sea surface cooling to occur almost one day after the passage of the two typhoons (Zhang et al., 2019). Figure 5.9 shows the tracks of typhoons Sarika and Haima, obtained from the Joint Typhoon Warning Center (JTWC) data (orange), Japan Meteorological Agency (JMA) data (brown), and China Meteorological Administration (CMA) data (black), with 6-hour tracking steps (dots) (Zhang et al., 2019). Specifically, the text boxes show the dates, pointing to the UTC 00.00 of the positions every day, and sustained maximum wind speed and central pressure obtained from the CMA data. Moreover, the red numbers indicate the positions of the observation stations, whereas the background shade indicates the topography.

Case study 2. Figure 5.10 presents typhoons BANYAN and SAOLA. (Note: The description of Figure 5.10 is taken from Hasan and Takewaka, 2015). Indeed, the shallow waters of a near-shore region are dynamic and often hostile. Moreover, prediction in this region is usually

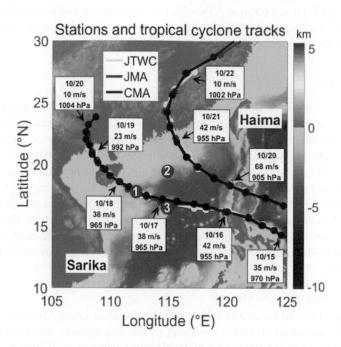

FIGURE 5.9 Tracks of Typhoons Sarika and Haima in the northern South China Sea in 2016 (from Zhang et al., 2019).

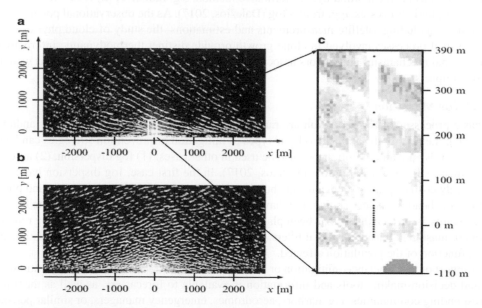

FIGURE 5.10 Typical radar echo image during the period of (a) typhoon BANYAN and (b) typhoon SAOLA (from Hasan and Takewaka, 2015)

difficult probably due to our limited understanding of the physics or lack of accurate field data. In this case, an X-band nautical radar system was employed for this study to examine alongshore propagation of low frequency run-up motion around the research pier HORS (Hasaki Oceanographical Research Station) in Hasaki beach, Japan (Hasan and Takewaka, 2015). Figure 5.10 shows typical radar echo image during the period of (a) typhoon BANYAN (typhoon # 7, T7, Hs = 3.69 m and Ts = 12.1 s); (b) typhoon SAOLA (typhoon # 17, T17, Hs = 4.87 m and Ts = 9.6 s) in 2005; (c) close-up view of an echo image showing wave gauge locations (black dots) along the pier.

5.3.2 Fog Hazard

Fog is defined as a cloud at or near the surface. Fog consists of water droplets suspended above or near the surface and/or ice particles suspended above or near the surface (Croft, 2013). Fog is a hazard to land, sea and air transportation. The production, maintenance and dissipation of fog constitutes significant risks and hazards for multiple modes of transportation, with impacts on populations (Dalezios, 2017). Moreover, supercooled fog can be modified, however, it refers only to a small proportion of fogs. Warm fog can also be modified using thermal energy, however, with high costs of installation and operation of dispersal systems and the pollution problem (Kunkel, 1980). Visibility is one of the most difficult meteorological phenomena to forecast. Several objective fog forecasting techniques have been attempted, but as yet no entirely reliable method exists (Dalezios, 2017). Publicity of the hazards and driver education remain the surest ways of reducing the dangers of fog on highways.

5.3.2.1 Fog Modeling and Assessment

The observed properties and behaviors of fog based on weather observation data and fog's associated features is a function of dynamic processes in the atmosphere. Weather data could include satellite observations of cloud types and meteorological parameters (Cermak, 2018; Egli et al., 2018). Fog prediction is developed in order to consider avoidance, mitigation or prevention of its impacts. Fog has been classified by: (1) formative method, e.g. radiative; (2) location, e.g. sea fog; or (3) atmospheric processes, e.g. frontal fog (Dalezios, 2017). As the observational potential of fog has matured, including satellite measurements and estimations, the study of cloud physics and the microphysical processes involved in cloud growth provide distinct thermodynamic information as to the internal mechanisms and factors associated with fog formation and dissipation, as well as its radiative impacts.

5.3.2.2 Fog Mitigation

The basic processes of fog formation are radiative, i.e. cooling, advective, i.e. cooling and/or lift, and mixing, i.e. thermodynamic. Two basic types of mitigation or prevention strategies can be used for fog hazards, and a third option can also be used in other cases: (1) fog dispersion, (2) air quality management, and (3) special cases (Dalezios, 2017). In the first case, fog dispersion has focused on relatively costly mechanical means, i.e. bulk mixing, as opposed to thermodynamic and physical methods, i.e. heating and/or seeding, that have been applied predominantly at aerodrome locations. For air quality issues, emphasis has been placed on reduced exposure, e.g. avoidance by remaining indoors or masking by using personal filters, reduction of contributing sources, or ventilation and is often a function of the population affected.

In support of the avoidance, mitigation or prevention of fog hazards, a variety of operational support and decision-making tools and information is available to forecasters, as well as the affected and responding communities, e.g. harbors, aerodromes, emergency managers, or similar personnel. These are particularly effective when tied to GIS and remote sensing databases and decision-support software or artificial intelligence and are more commonly used in a military or disaster-related type of response, e.g. applications in catastrophe modeling (Dalezios, 2017). The additional deployment of meso- and microscale surface-based and remotely sensed observing networks is expected to increase the spatial and temporal variability of data as related to fog occurrence and its evolution, so that it may be more accurately detected, assessed and compared with forecasts (Cermak, 2018; Egli et al., 2018). In some instances, highway systems have been deployed to provide "instant" warning signs to alert drivers to rapidly changing visibility in fog-prone regions.

Satellite and similar remote sensing platforms offer a variety of products, namely the website of "Nighttime Fog and Low Cloud Images" as produced by NOAA-NESDIS, USA: http://www.orbit.nesdis.noaa.gov/smcd/opdb/aviation/fog.html, and provide a gross estimate of fog occurrence, severity and coverage by channel differencing, as well as through examination of sounder data to construct vertical and near-surface profiles of temperature and moisture in the atmosphere. This information covers fog after its formation and thus allows the tracking of its movements and evolution. Although microwave sensors and ground-based radar, lidar, and profiler platforms may offer additional information and operational support, none are presently suited to fog detection or prediction, although there are operational automated highway alert systems (Cermak, 2018). Additional information on atmospheric chemistry and structure is also available through several remote sensing platforms (Egli et al., 2018), and they provide a real-time observation profile of the physics-chemistry of the atmosphere as related to fog, which also involves a diagnosis of the pre-fog environment.

5.4 OCEAN COLOR ANALYSIS

Remote sensing provides a near-surface view of the ocean. Ocean color analysis covers water temperature monitoring, SST, salinity, density, water quality (sediments, DO), ocean productivity, phytoplankton, concentration and drift, aquaculture, fish stocks (Lee et al., 2019). Indeed, ocean color analysis attempts to assess the "health" of the ocean by measuring oceanic biological activity using optical sensors. Optical data can detect dissolved organic matter, suspended sediments and distinguish between algal blooms and oil slicks. In addition, SAR data can detect current, wave and mesoscale features, as well as temporal trends when optical data are not available due to cloud cover. There are significant results and inferences about fundamental properties and processes in the marine biosphere from mapping the spatial and temporal of ocean color over regional and global scales. Moreover, mapping the changes in ocean color can assist in the management of fish stocks and other aquatic life and monitor the water quality, such as oil or algal blooms. In general, ocean productivity appears highest in coastal areas due to the proximity to nutrient upwelling and circulation conditions.

Phytoplankton is a significant component of the world's food chain being a function of sunlight and the pigment chlorophyll. Specifically, chlorophyll absorbs red light, leading to the ocean's blue-green color, and is considered a good indicator of the ocean's health and productivity. Multispectral data are employed for ocean color measurements, and wide spatial coverage provides the best synoptic view of the spatial distribution and variability of phytoplankton, water temperature and concentration of suspended matter. Moreover, hyperspectral data are obtained in several narrow ranges of the visible and infrared wavelengths and allow for greater precision in target spectral signatures. For fish harvesting and fish farming, information is required on a daily or weekly basis, whereas for modeling monthly and seasonal imaging data are required. There are currently advanced sensors, such as MODIS (NASA), collecting data on chlorophyll variability, primary productivity and SST using advanced algorithms. Specifically, scanning radiometers and microwave sounders collect SST data. These sensors are mainly used for regional coverage with relatively coarse (500–1200 m) resolution and wide fields of view. The design of spectral channels of these sensors is focused on optimizing target reflectance and supporting quantitative measurements of specific biophysical properties.

Remote sensing can be used to improve our understanding of global climate patterns by detecting the arrival of the El Niño current off the coast of Peru (Figure 4.4). El Niño is a warm water current that appears off the coast of South America approximately every seven years. Indeed, the arrival of a warm water current, such as El Niño, displaces the cold current further offshore, causing changes in the migration of the fish population, since nutrients in the ocean are associated with cold water upwelling. El Niño is a good example of remote sensing application in ocean productivity. A typical example is the El Niño of 1988, which caused a loss in anchovy stocks near Peru and then moved north, altering the regional climatic patterns and creating an unstable weather system. As a result,

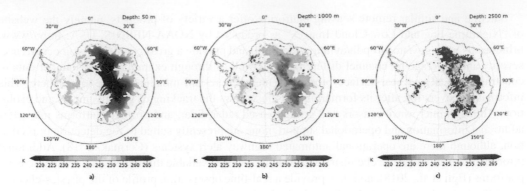

FIGURE 5.11 Antarctica's internal temperature in different depths using SMOS L-band passive microwave observations (from ESA, November 4, 2019e).

storms forced the jet stream further north blocking the southward flow of continental precipitation from Canada over Central America. Central and eastern USA suffered drought, a reduction in crop production, an increase in crop prices, and a rise in commodity prices on the international markets (CCRS, 1998).

Case studies. Several case studies on ocean color are briefly presented as follows:

Case study 1. Figure 5.11 presents Antarctica's internal temperature in different depths using SMOS L-band passive microwave observations. (Note: The description of Figure 5.11 is taken from ESA, November 4, 2019b). ESA's SMOS can measure how temperature varies according to the depth of the ice. Specifically, SMOS' L-band passive microwave observations over Antarctica are combined with glaciological and emission models to infer information on glaciological properties of the ice sheet at various depths, including temperature (Macelloni et al., 2019). The Antarctic ice sheet is, on average, about 2 km thick, although in some places the bedrock is almost 5 km below the surface of this huge polar ice cap. The surface of the ice sheet is cold, and the temperature increases with depth due to the basal geothermal heating from beneath the Earth's crust. Satellite data are used to measure changes in the height of ice sheets and consequently their "mass balance". Temperature information is fundamental for understanding the presence of aquifers inside or at the bottom part of ice sheets.

Case study 2. Figure 5.12a presents an example of the daily SST field from a Copernicus Imaging Microwave Radiometer (CIMR) level 3 collated (L3C) file, and Figure 5.12b shows the added error field for this CIMR L3C field (from Pearson et al., 2019). (Note: The description of Figure 5.12 is taken from Pearson et al., 2019). Specifically, estimation of SSTs derived from passive microwave (PMW) observations, which is presented in Chapter 4, benefit global ocean and SST analyses due to their near-all-weather availability. The Copernicus Imaging Microwave Radiometer (CIMR) is an instrument that would be capable of real aperture resolution < 15 km with low total uncertainties in the range 0.4–0.8 K for channels between 1.4 and 36.5 GHz, and a dual-view arrangement that further reduces noise. A comparative study of SST uncertainty and feature resolution with and without the availability of CIMR has been conducted (Pearson et al., 2019).

Case study 3. Figure 5.13a shows the Southeastern Pacific along continental Chile and Figure 5.13b shows the study area (red box) off central-southern Chile, which includes sea surface temperature (SST, color, °C) from October 21, 2013, which represents coastal upwelling in austral spring. (Note: The description of Figure 5.13 is taken from Pinochet et al., 2019). Specifically, along the Chilean coast there are several upwelling focal points. Moreover, the seasonal variability of coastal upwelling off central-southern Chile has been evaluated based

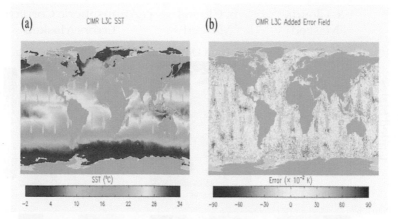

FIGURE 5.12 (a) Daily sea surface temperature (SST) field and (b) the added error field (from Pearson et al., 2019).

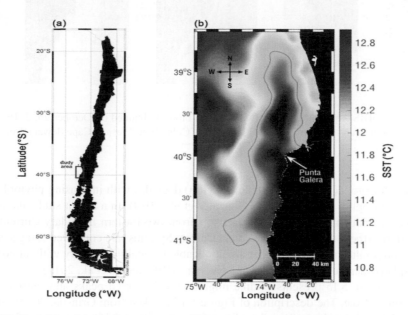

FIGURE 5.13 Upwelling focal points along the Chilean coast: (a) Southeastern Pacific along continental Chile, and (b) off central-southern Chile, from October 21, 2013 (from Pinochet et al., 2019).

on SST and sea surface wind (SSW), and 8-day composite chlorophyll-a concentration between 2003 and 2017 (Pinochet et al., 2019).

Case study 4. Figure 5.14 shows MODIS Aqua imagery of phytoplankton blooms obtained on October 26, 2016 in Monterey Bay, California (upper) and September 27, 2011 near Cape Columbine, Western Cape (lower). (Note: The description of Figure 5.14 is taken from Houskeeper and Kudela, 2019; Kudela et al., 2019). Specifically: (a) Pseudo-true color images with clouds masked in gray; (b) Red band difference (RBD) algorithm (a proxy for phytoplankton biomass) with clouds masked in gray; (c) RBD algorithm with clouds and suspect atmospheric correction (defined as maximum aerosol iterations reached and low water-leaving radiances) masked in gray. Indeed, satellite estimation of oceanic chlorophyll-a content has enabled characterization of global phytoplankton stocks, but the quality of retrieval for many

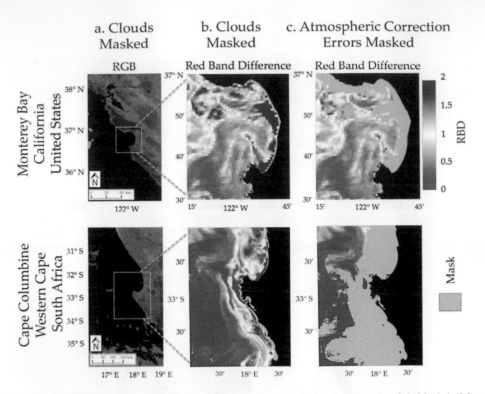

FIGURE 5.14 MODIS Aqua imagery of phytoplankton blooms obtained on October 26, 2016, in Monterey Bay, California (upper) and September 27, 2011 near Cape Columbine, Western Cape (lower) (from Houskeeper and Kudela, 2019).

ocean color products (including chlorophyll-a) degrades with increasing phytoplankton biomass in eutrophic waters (Soriano-González et al., 2019). In a recent study, the masked and unmasked fractions of ocean color data sets from two Eastern Boundary Current upwelling ecosystems (the California and Benguela Current Systems) are compared using satellite proxies for phytoplankton biomass that are applicable to satellite imagery without correction for atmospheric aerosols (Houskeeper and Kudela, 2019; Kudela et al., 2019).

Case study 5. Figure 5.15 presents sardine migration, chlorophyll peaks and sardine interannual variability. (Note: The description of Figure 5.15 is taken from Grinson, 2015). Specifically: Figure 5.15a shows the Indian oil sardine migration, Figure 5.15b presents the SeaWiFS satellite chlorophyll peaks utilized for explaining the trophic link between the upwelling bloom biology and Figure 5.15c shows sardine interannual variability along the southwest coastal waters of India (Grinson et al., 2012). Indeed, the use of modeled and satellite remote sensing (SRS) data is considered in supporting the research, technology-development and management of marine fishery resources (Grinson, 2015). Remote sensing (RS) data are used to locate fish stocks, locate areas of reef stress and delineate areas of high productivity in the wake of cyclone paths. Coupling SRS with models helps to manage fishery resources on an ecosystem scale, generate potential fishing zones (PFZ), forecast ocean state (OSF), detect meso-scale features such as eddies and track cyclones threatening coastal resources.

Case study 6. Figure 5.16 presents maps of global annual primary production (PP) and associated parameters for the period 1998–2018. (Note: The description of Figure 5.16 is taken from Kulk et al., 2020). Specifically: (A) Global annual primary production based on mean photosynthesis versus irradiance (P-I) parameters, (B) Linear trends in global annual primary production between 1998 and 2018 given as percentage change per year (dark gray color

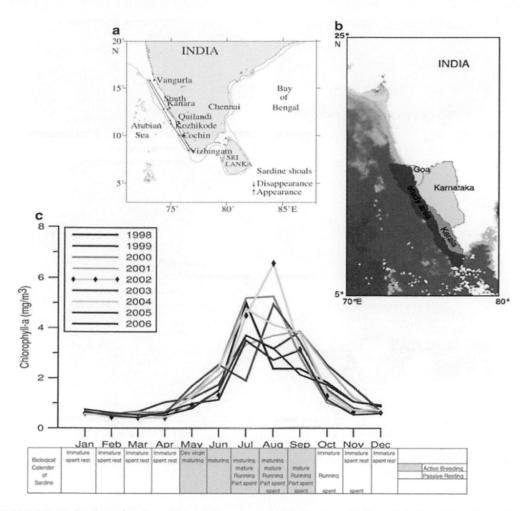

FIGURE 5.15 Sardine migration, chlorophyll peaks and interannual variability: (a) Indian oil sardine migration, (b) SeaWiFS satellite chlorophyll peaks and (c) sardine interannual variability along the south west coastal waters of India (from Grinson, 2015).

represents non-significant trends), (C) Remote sensing derived mean surface chlorophyll-a (Chl-a), (D) Difference in primary production between mean P-I parameters and -1 standard deviation (-1 SD) based estimations, (E) Remote sensing derived Photosynthetic Active Radiation (PAR, 400–700 nm), and (F) Difference in primary production between mean P-I parameters and +1 standard deviation (+1 SD) based estimations (from Kulk et al., 2020).

Primary production by marine phytoplankton is one of the largest fluxes of carbon on our planet. In the past few decades, considerable progress has been made in estimating globalprimary production at high spatial and temporal scales by combining in situ measurements of primary production with remote sensing observations of phytoplankton biomass. In this case study, a global database of in situ measurements of photosynthesis versus irradiance (P-I) parameters and a 20-year record of climate quality satellite observations were used to assess global primary production and its variability with seasons and locations as well as between years (Kulk et al., 2020). Surface chlorophyll-a concentrations at 9 km spatial resolution and monthly temporal resolution for the period 1998–2018 were obtained from the European Space Agency (ESA) Ocean Color Climate Change Initiative project (OC-CCIv4.1, https://esa-oceancolour-cci.org/). The data set contains merged

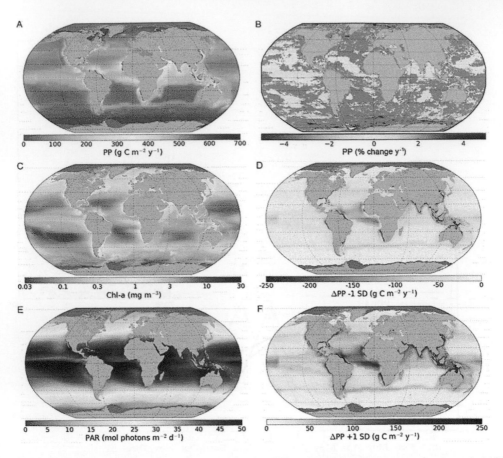

FIGURE 5.16 Maps of global annual primary production (PP) and associated parameters for the period 1998–2018 (from Kulk et al., 2020).

products of observations from the Sea-viewing Wide Field-of-View Sensor (SeaWiFS, 1997–2010), the Medium Resolution Imaging Spectrometer (MERIS, 2002–2012), the Moderate Resolution Imaging Spectroradiometer (MODIS, 2002–present) and the Visible Infrared Imaging Radiometer Suite (VIIRS, 2012–present) that are climate-quality controlled, bias-corrected and error-characterized (Kulk et al., 2020).

5.5 OIL SPILLS

The study of oil spills covers mapping oil spills, predicting oil spill extent and drift, emergency response decisions and identification of natural oil seepage areas for exploration. Large-area spills are usually spectacular in the extent of their environmental damage, such as the Exxon Valdez (Dalezios, 2017) or the recent BP oil spill in the Gulf of Mexico. Spills often occur in stormy weather conditions. Remote sensing can be used for the detection and monitoring of spills. For ocean spills, remote sensing data can detect and monitor the rate and direction of oil movement through frequent multitemporal imaging, and input to drift prediction modeling and may contribute to targeting clean-up and control efforts.

High-resolution sensors are generally required for spill identification, although wide area coverage is very important for initial detection and monitoring. Wind speed can affect the detection of an oil spill. Specifically, at wind speeds greater than 10 m/s, the slick will be broken up and dispersed, making it difficult to detect. Remote sensing devices used include the infrared video

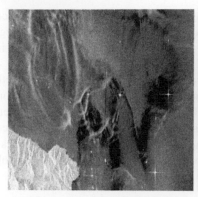 

(a) Synthetic aperture radar (SAR) image.   (b) Ground truth mask.

FIGURE 5.17 Extracted SAR image accompanied with the ground truth mask. (a) sample of a SAR image and (b) corresponding annotated image (from Krestenitis et al., 2019).

and photography from airborne platforms, thermal infrared imaging, airborne laser fluourosensors, airborne and spaceborne optical sensors, as well as airborne and spaceborne SAR. Airborne sensors have the advantage of frequent site-specific coverage on demand, however, at a high cost. It is stated that laser fluourosensors are the best sensors for oil spill detection (CCRS, 1998), with the capability of identifying oil on shores, ice and snow, and determining what type of oil has been spilled. However, they require relatively cloud-free conditions to detect the oil spill (Ilčev, 2019). On the other hand, SAR sensors have the advantage of all-weather penetration and during darkness over optical sensors.

Case study. Figure 5.17 shows a sample of a SAR image and the corresponding ground truth mask. (Note: The description of Figure 5.17 is taken from Krestenitis et al., 2019). Specifically: (a) Sample of a SAR image and (b) ground truth mask. Cyan color corresponds to oil spills, red to look-alikes, brown to ships, green to land and black is for sea surface. Indeed, efficient monitoring and early identification of oil slicks are vital to confine the environmental pollution and avoid further damage. SAR sensors are commonly used for this objective due to their capability for operating efficiently regardless of the weather and illumination conditions. Black spots probably related to oil spills can be clearly captured by SAR sensors, yet their discrimination from look-alikes poses a challenging objective. To overcome these limitations, semantic segmentation with deep convolutional neural networks (DCNNs) is proposed as an efficient approach (Krestenitis et al., 2019). An available SAR image data set is employed to review the performance of well-known DCNN segmentation models in the specific task.

5.6 SEA ICE

5.6.1 SEA ICE FEATURES

Ice covers a substantial part of the Earth's surface. Polar sea ice seasonally covers an even larger area, roughly equal in size to the North American continent, 25 million km². Sea ice contributes to the albedo of the earth through its extensive distribution. Albedo refers to the reflectivity of the Earth's surface. The changes in the distribution of ice and snow contribute to global climate change due to sea ice interactions with atmosphere and ocean.

These distributions affect the absorbed solar energy by the earth, since they both are highly reflective. During ice melting less incoming energy is reflected which leads to an increase in the warming trend. On the other hand, during cooler conditions ice reflects even more of the incoming solar energy which leads to even colder conditions. During winter months, ice creates a substantial barrier to vessels navigating through lakes and ocean. Specifically, icebergs, pack ice and ice floes may create hazards to navigation. To monitor these conditions, regular support for observation is provided, ice analysis charts and daily ice hazard bulletins are issued, along with seasonal forecasts.

Different ice types can be detected and mapped using remote sensing data and methods, as well as the identifying of large navigable cracks in the ice, and ice movement can be monitored using current technological advancements, such as web platforms. Typical examples of remote sensing contribution to sea ice information include surface topography, ice condition and concentration, ice type and age, ice motion, iceberg detection and tracking, ice and iceberg dynamics for planning purposes, routine identification of leads, safe shipping routes and rescue, as well as pollution monitoring, wildlife habitat and global change.

5.6.2 ICE DETECTION AND FORECASTING

5.6.2.1 Ice Detection

In winter when the ice is at a maximum, high latitude areas experience low solar illumination conditions. Ice is usually defined by its age. Surface texture is the main contributor to the radar backscatter, which is also used to infer ice age and thickness. Indeed, backscatter is influenced by dielectric properties of the ice (which depends on salinity and temperature), surface factors, such as roughness and snow cover, as well as internal microstructure and geometry. Moreover, the ice type, the areal extent of ice, as well as the concentration and distribution of each type are significant ice features. Specifically, new ice is usually thin (5–30 cm) and smooth, which tends to have a low return and therefore dark appearance on the imagery due to the specular reflection of incident energy off the smooth surface. First-year ice is older and thicker than new ice (30–200 cm) and is considered a significant hazard, which can cause damage to all vessels, including icebreakers. Coarse resolution optical sensors, such as NOAA's AVHRR, provide an excellent overview of pack ice extent if atmospheric conditions are optimal. Ice that survives into a second, and later, year, generally becomes thicker (>2m) and declines in salinity, increasing the internal strength. This ice is a dangerous hazard to ships and offshore structures.

Remote sensing technology can provide significant information for the identification of ice type and, as a result, for ice thickness. From this data, ice charts can be created, which are maps of different ice types and ice concentration. The areas of ice can be easily mapped from an image, and when georeferenced, provide a useful information source. Specifically, high-resolution data covering 1–50 km is useful for immediate ship navigation, whereas coarse resolution (1–50 km) images, which cover large area (100–2000 km²), are more useful for regional strategic route planning. Indeed, for navigation purposes, the value of this information is considered time limited. However, the data has long-term value, since it can increase our knowledge of climate dynamics and ice as an indicator of global climate change.

The all-weather/day-night capabilities of SAR systems make radar remote sensing the most useful for ice type and concentration mapping. Specifically, **passive** microwave sensing has a role in sea ice applications. Sea ice and water emit substantially different amounts of radiation. As a result, it is relatively easy to delineate the interface between the two, e.g. using the SSM/I. The main drawback of passive microwave sensors is their poor spatial resolution (approximately 25 km). On the other hand, **active** radar is an excellent sensor for the observation of ice conditions, because the microwave energy and imaging geometry are combined to provide measures of both surface and internal features. For example, RADARSAT has orbital parameters and a radar sensor designed to cover the ice requirements. Specifically, RADARSAT provides an image of the Arctic area once per day. This is efficiently downloaded and the data are transmitted from the ground processing station right to the vessel requiring the information in a time frame of four hours. Airborne radar sensors are also useful for targeting specific areas and providing high-resolution imagery.

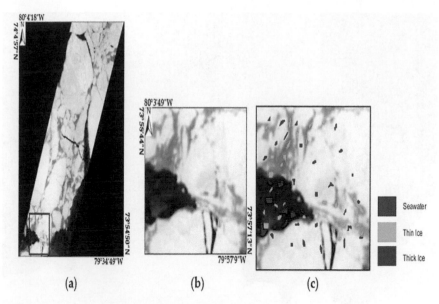

FIGURE 5.18 Sample database: (a) False-color composite, with a delineation of the experimental area (R: 29, G: 23, and B: 16), (b) experimental area, and (c) ground truth (from Han et al., 2019).

Case studies. A few case studies are presented as follows:

Case study 1. Figure 5.18 shows sample database: (a) False-color composite, with a delineation of the experimental area (R: 29, G: 23, and B: 16), (b) experimental area and (c) ground truth (from Han et al., 2019). (Note: The description of Figure 5.18 is taken from Han et al., 2019). Specifically, it is difficult to obtain the labeled samples in hyperspectral sea ice image classification. In addition, most of the current sea ice classification methods mainly use spectral features for shallow learning, which limits improvement of the sea ice classification accuracy. As a result, a hyperspectral sea ice image classification method is presented based on the spectral-spatial-joint feature with deep learning (Han et al., 2019). By manually labeling a certain number of labeled samples as the sample database, which is taken as ground truth, the sample library was randomly divided into training samples and test samples, as shown in Figure 5.18.

Case study 2. Figure 5.19 shows the D28 iceberg spotted breaking off from the Amery Ice Shelf in Antarctica (from ESA, October 28, 2019c, contains modified Copernicus Sentinel data (2019), processed by ESA, CC BY-SA 3.0 IGO). (Note: The description of Figure 5.19 is taken from ESA, October 28, 2019c). Specifically, the iceberg, which is around 1600 km² – about the size of Greater London – has taken a 90° turn. Captured by the Copernicus Sentinel-1 mission, this multitemporal false-color image shows the before and after location of the iceberg produced by this calving event. Blue shows the iceberg before separation, taken on September 20, 2019, while the red shows the location of the iceberg on October 19, 2019 after calving. Small red fragments of the iceberg can be seen floating in the vicinity of D28. Approximately 30 km wide and 60 km long, and with a thickness exceeding 200 m, the iceberg is estimated to contain over 300 billion ice tones (ESA, October 28, 2019c).

Case study 3. The Greenland ice sheet is losing mass seven times faster than in the 1990s, according to new research (IMBIE Team, 2019). (Note: The description of Case Study 3 is taken from IMBIE Team, 2019). In this research, it has been estimated that Greenland lost 3.8 trillion tonnes of ice between 1992 and 2018 – enough to push the global sea level up by 10.6 mm. Over the study period, the rate of ice loss was found to have increased seven-fold from

FIGURE 5.19 D28 iceberg spotted breaking off from the Amery Ice Shelf in Antarctica (from ESA, October 28, 2019c, contains modified Copernicus Sentinel data (2019), processed by ESA, CC BY-SA 3.0 IGO).

33 billion tonnes per year in the 1990s to 254 billion tonnes per year in the last decade. The Ice Sheet Mass Balance Intercomparison Exercise (IMBIE) compared and combined data from 11 satellites – including ESA's ERS-1, ERS-2, Envisat and CryoSat missions, as well as the EU's Copernicus Sentinel-1 and Sentinel-2 missions – to monitor changes in the ice sheet's volume, flow and gravity. This study condenses the available data and provides a consensus regarding Greenland's ice loss, enabling more accurate projections of future sea-level rise to be made. The findings will help densely populated communities in coastal areas to prepare, but also illustrate the urgent need for greenhouse-gas emissions to be curtailed worldwide.

The Intergovernmental Panel on Climate Change's (IPCC) central climate warming scenario predicted a 60-cm rise in global sea level by 2100, putting 360 million people at risk of coastal flooding every year. The faster-than-expected rate reported by the IMBIE Team shows that ice loss is following the IPCC's worst-case climate warming scenario, which predicts that sea level will rise by an additional 7 cm. As a rule of thumb, for every centimeter rise in global sea level, another six million people are exposed to coastal flooding around the planet. With this current trend, Greenland ice melting will cause 100 million people to be flooded each year by the end of the century – so 400 million in total due to sea-level rise. Using satellite observations in combination with regional climate models, the team shows that just over half of the ice loss was due to increased surface meltwater runoff, driven by warming air. The remaining losses were the result of increased glacier flow triggered by rising ocean temperatures. Ice losses reached a peak of 335 billion tonnes per year in 2011, a trend that since dropped to an average of 238 billion tonnes per year through to 2018 – but, nevertheless, remained seven times higher than observed during the 1990s (ESA, December 10, 2019d).

5.6.2.2 Ice Forecasting

Real-time information is required for operations. Low to moderate resolution is considered sufficient for ice forecasting over a large area to support navigation. Local high-resolution information is required in support of ice-breading or drilling operations by determining ice characteristics and features, such as ice cover, percent of multi-year, thickness, ridges, ice-islands, leads, bergs. This is a very complex detection and classification problem, since the ice surface changes in response to climate, season and location, i.e. surface meltwater, surface snow-cover, surface roughness, surface salinity. All change and greatly affect response to sensing. Indeed, there is no remote sensing system that can satisfy all the requirements, rather a combination of sensors is needed. Real-time ice monitoring significantly depends on all-water capability and daily coverage, leading to microwave sensors for operational ice monitoring.

Visible, near-infrared and thermal infrared wavelengths are useful for mapping sea ice extent, observing surface melting, and following the motion or deformation of ice cover. Clouds and darkness affect monitoring at these wavelengths, except infrared for darkness. Landsat cannot identify ice features, such as ridges, but is useful for large-area coverage.

Passive microwave sensors, namely radiometers, record the brightness temperature of a surface. Brightness temperature, T_B, is proportional to the temperature multiplied by the emissivity of the surface, $T_B = T * \varepsilon$, where emissivity can be easily deduced from radiometer data. There are characteristics, such as: (1) an all-weather system; (2) a simple classification described below can be used to estimate the concentration of ice cover on the ocean. The key is the great difference in emissivity between ice and water; (3) longer wavelengths (≈ 21cm) can penetrate an ice sheet to give an indicator of its thickness; (4) multichannel radiometers allow for more complex classification schemes, including the separation of multi-year ice from first-year ice, and the multi-year ice fraction; (5) radiometer resolution is generally poor compared to active microwave sensing techniques.

Active microwave (radar) systems have the following characteristics: (1) they are useful in determining the features of sea ice; (2) they can sense surface dielectric properties and roughness; (3) smooth ice and rough water can look similar; (4) scatterometer is a calibrated, downward-looking airborne radar, which can be useful to measure the backscatter coefficient, Γ, as a function of aspect angle along a strip of terrain; (5) SLAR (Side Looking Airborne Radars) is a reliable source of information independent of light or water, which provides information on the location and size of floes, and other ice features, even under heavy snow over a wide swath of area on each side of an airplane; (6) SAR (Synthetic Aperture Radar): high resolution, ice concentrations and floes sizes are easy, surface roughness can be inferred, can detect icebergs, real-time SAR and SLAR imaging has proven useful for tactical support of drilling operations and iceberg detection.

Combining active and passive sensors is promising, since it can be generally applicable in all regions in all seasons; however, due to the variability of ice surface characteristics, this is expected to be difficult.

Operational ice monitoring. (1) Satellite-borne sensors: SAR and Radiometers supplement by occasional Landsat information are useful for general ice monitoring and forecasting over the entire Arctic; airborne flightlines to groundtruth. (2) Airborne sensors, such as SLAR, SAR, Laser profiliometer, radiometer have good resolution, which allows for detailed analysis of ice features in a local area of special interest. Research trend is towards better information through improved sensors and improved classifiers (Table 7.2).

In summary, ice monitoring is essential for safe and economic Arctic marine operations; most remote sensors can detect ice; microwave sensors are best for applied sea ice monitoring, because of their all-weather capabilities; difference between multi-year and first-year ice can be detected by satellite; ice thickness, ridge delineation, etc. generally require airborne reconnaissance; operational ice monitoring utilizes both satellite and airborne sensing; and classification of sea ice is a complex variable task in space and time.

5.6.3 Ice Motion and Shipping

5.6.3.1 Ice Motion

Ice moves quickly in response to ocean currents and wind. Remote sensing provides a tangible measure of direction and rate of ice movement through mapping and change detection methods. Monitoring of ice movement requires frequent and reliable imaging. The revisit interval must be frequent enough to follow identifiable features before tracking becomes difficult due to excessive movement or change in appearance. Active microwave sensing (radar) provides a reliable source of imaging under all-weather and illumination conditions. RADARSAT provides this type of sensor and is a spaceborne platform, which is advantageous for routine imaging operations. The orbital path ensures that Arctic areas are covered daily, which meets the requirement for frequent imaging.

Ice floes move like tectonic plates, sometimes breaking apart or colliding in a style, such as the Asian plates, creating a series of ridges and blocky ice rubble. Vessels can be trapped or damaged by the pressure resulting from these moving ice floes. Even offshore structures can be damaged by the strength and momentum of moving ice. It is, thus, important to understand the ice dynamics in areas of construction or in the vicinity of a shipping or fishing route. The floes can be mapped and their movement monitored to facilitate the planning of optimum shipping routes, to predict the effect of ice movement on standing structures (bridges, platforms). The resolution and imaging frequency requirements for ice motion tracking vary with the size of floes and the ice dynamics in a region. For example, in areas of large slow-moving floes, such as the Beaufort Sea, 1 km resolution data over 10-day intervals are adequate, whereas in dynamic marginal ice zones, such as the Gulf of St. Lawrence, Canada, 100 m resolution data over 12 to 24-hour intervals are required.

5.6.4 Shipping

Ships navigating through high latitude seas (both northern and southern) are often faced with obstacles of pack ice and moving ice floes. Ice breakers are designed to facilitate travel in these areas, but they require knowledge about the most efficient and effective route through the ice. For ship routing it is useful to obtain information on currents, wind speed, tides, storm surges and surface wave height. For forecasting and ship routing, real-time data handling/turnaround facilities are necessary, requiring two-way data links for efficient dissemination between the forecast center and data user. Buoy-collected information can be combined with remote sensing data to produce image maps displaying information, such as hurricane structure with annotated wind direction and strength, and wave height. This information can be useful for offshore engineering activities, operational fisheries surveillance and storm forecast operations.

Case studies. A few case studies are briefly presented as follows:

Case study 1. Figure 5.20 shows an example of a ship imaged by TerraSAR-X satellite over the Gulf of Naples (August 3, 2012) (from Graziano et al., 2019). (Note: The description of Figure 5.20 is taken from Graziano et al., 2019). Specifically, the ship shows a negative azimuth offset with respect to its wake vertex (true ship position). Indeed, the integration of SAR images with data reported by the automatic identification system (AIS) is of high interest. Accurate matching of ships detected in SAR images with AIS data requires compensation of the azimuth offset, which depends on the ship's velocity. The existing procedures interpolate the route information gathered by AIS to estimate the ship's velocity at the epoch of the SAR data, to remove the offset. This case study proposes the use of SAR-based ship velocity estimations to improve the integration of AIS and SAR data. A case study has been analyzed, in which the method has been tested on TerraSAR-X images collected over the Gulf of Naples, Italy (Graziano et al., 2019).

Case study 2. Figure 5.21 shows illustrations of a compact polarimetric synthetic aperture radar (CP SAR) image (left), a sub-image (top right), corresponding automatic identification system

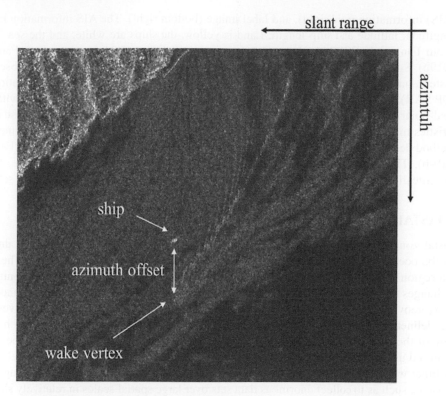

FIGURE 5.20 Example of a ship imaged by TerraSAR-X satellite over the Gulf of Naples, Italy (August 3, 2012) (from Graziano et al., 2019).

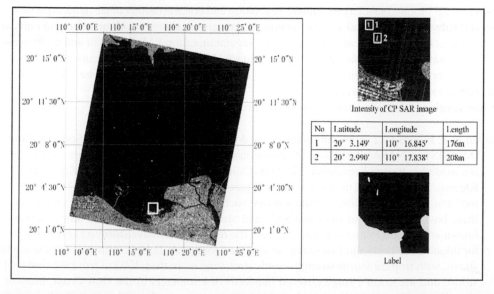

FIGURE 5.21 Illustrations of a compact polarimetric synthetic aperture radar (CP SAR) image (left), a sub-image (top right), corresponding automatic identification system (AIS) information (mid right), and label image (bottom right). Land is yellow, the ships are white and the sea is black (from Fan et al., 2019b).

(AIS) information (mid right), and label image (bottom right). The AIS information includes longitude, latitude and ship length. Land is yellow, the ships are white, and the sea is black (from Fan et al., 2019b). (Note: The description of Figure 5.21 is taken from Fan et al., 2019b). Specifically, compact polarimetric SAR (CP SAR), as a new method or observation system, has attracted much attention in recent years. Compared with quad-polarization SAR (QP SAR), CP SAR provides an observation with a wider swath, while, compared with linear dual-polarization SAR, retains more polarization information in observations. These characteristics make CP SAR a useful tool in marine environmental applications. A segmentation method designed specifically for ship detection from CP SAR images is proposed (Fan et al., 2019b). The pixel-wise detection is based on a fully convolutional network (i.e. U-Net). Specifically, three classes (ship, land and sea) were considered in the classification scheme.

5.7 COASTAL AND INTERTIDAL ZONE

The coastal zone is a complex system, since coastlines are environmentally sensitive interfaces between the ocean and land. Coastlines are also biologically diverse intertidal zones. The coastal zone is a region subject to increasing stress from human activity. There is a need to monitor such diverse changes as urbanization, effluents and offshore pollution, loss of natural habitat and coastal erosion. Moreover, the topics of intertidal zone include tidal and storm effects, coastal vegetation mapping, delineation of the land and water interface mapping, shoreline features and beach dynamics. Many of the dynamics of the open ocean and changes in the coastal systems can be mapped and monitored using remote sensing data and methods (La Barbera et al., 1994).Nevertheless, even under optimal weather conditions the use of remote sensing data and methods is advantageous for many reasons, such as to collect enormous data sets over large spatial scales in relatively short time spans.

Coastal ecosystems have high spatial complexity and temporal variability, requiring, thus, high spatial, spectral and temporal resolutions. Recent advances in remote sensing systems and data analysis methods can significantly contribute to effective monitoring of natural and man-made changes in coastal ecosystems. Specifically, high-resolution multispectral and hyperspectral imagers, radar systems and LiDAR are available for monitoring changes in beach profiles, coral reefs, coastal marshes, algal blooms, submerged aquatic vegetation and concentrations of suspended particles and dissolved substances in coastal waters. Imaging radars are sensitive to soil moisture and inundation and can detect hydrological features beneath the vegetation canopy. Multi-sensor and multi-seasonal data fusion methods are significantly improving coastal land cover mapping accuracy and efficiency. Using time-series of images enables the study of coastal ecosystems and the determination of short-term changes and long-term trends.

Case studies. A few case studies are briefly presented as follows:

Case study 1. Figure 5.22 shows an image of the Moderate Resolution Imaging Spectroradiometer (MODIS) on NASA's Terra satellite captured on September 26, 2008, 13 days after Hurricane Ike made landfall on September 13, 2008 (Color figure online) (Credits: NASA/GSFC) (from Klemas, 2015). (Note: The description of Figure 5.22 is taken from Klemas, 2015). Specifically, medium spatial resolution satellite sensors, such as MODIS on NASA's Terra and Aqua satellites, have been used to map wetlands and study their interaction with storm surges. The brown areas in the image of Figure 5.22 are the result of a massive storm surge that Ike pushed far inland over Texas and Louisiana causing a major marsh dieback. Hurricane Ike was a large storm, with tropical-storm-strength winds stretching more than 400 km from the center of the storm, and the storm's surge covered hundreds of kilometers of the Gulf Coast. Most of the shoreline in this region is coastal wetland. The salty water burned the plants, leaving them wilted and brown. The brown line corresponds to the location and extent of the wetlands. North of the brown line, the vegetation gradually transitions to pale green farmland and dark

FIGURE 5.22 Image of the Moderate Resolution Imaging Spectroradiometer (MODIS) on NASA's Terra satellite captured on September 26, 2008, 13 days after Hurricane Ike made landfall on September 13, 2008 (Color figure online) (Credits: NASA/GSFC) (from Klemas, 2015).

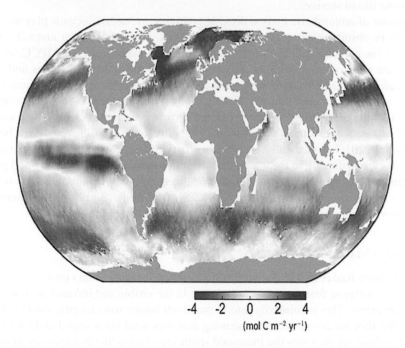

FIGURE 5.23 Global Atlas of Ocean CO_2 (from ESA, October 8, 2019c)

green natural vegetation untouched by the storm's surge. Some of the brown seen in the wetlands may be deposited sediment. Plumes of brown water are visible as sediment-laden water drains from rivers and the coast in general. The muddy water slowly diffuses, turning pale green, green and finally blue as it blends with clearer Gulf water (NASA/GSFC, 2010).

Case study 2. Figure 5.23 shows the Global Atlas of Ocean CO2 (from ESA, October 8, 2019c). (Note: The description of Figure 5.23 is taken from ESA, October 8, 2019c). Specifically, the recent IPCC special report on the ocean and cryosphere (IPCC, 2019) highlights our dependence on the oceans and ice, and how they are intrinsic to the health of our planet, although stresses the many ways in which they are being altered by climate change. It states, for example, that through the 21st century, the global ocean is projected to transition to unprecedented conditions, where seawater temperatures rise as they remove more heat from the air and undergo further acidification as they take in more atmospheric carbon dioxide. Estimating the size of the oceanic carbon sink depends on the calculation of upward and downward flows of carbon dioxide at the sea surface and, in turn, this flow is governed largely by turbulence.

It was previously estimated that around a quarter of the carbon dioxide released by anthropogenic activities into the atmosphere ends up in the ocean. To gain a more accurate figure on this downward flow, researchers used new knowledge of the transfer processes at the sea surface along with data from the Surface Ocean Carbon Dioxide Atlas, which is an ongoing large international collaborative effort to collect and compile measurements of carbon dioxide in the upper ocean (Figure 5.23). Based on ESA's SMOS satellite, the MetOp series and Copernicus Sentinel-3 that provide measurements of salinity, surface wind speeds and sea surface temperature, it was estimated that carbon per year is within 0.6 Giga-tones, and the earlier figure of around a quarter underestimated the role of the ocean in its ability to sequester carbon. This new finding may sound significant for climate change, but warming ocean waters are leading to issues, such as sea-level rise through thermal expansion and continental ice melt, and the more carbon dioxide that dissolves into the oceans, the more it leads to ocean acidification, which is a serious environmental problem that makes it difficult for some marine life to survive.

As the amount of atmospheric carbon dioxide continues to rise, our oceans play an increasingly important role in absorbing some of this excess. In fact, the global ocean annually draws down about a third of the carbon released into the atmosphere by human activities (IPCC, 2019). Recent advances in data capture have included state-of-the-art pH instruments on ships and floats, but a global view can be gained by taking measurements from space. However, at present there are no spaceborne sensors that can measure pH directly. The use of satellites has not yet been thoroughly explored as an option for routinely observing ocean surface chemistry, however, new ways of merging different data sets to estimate and ultimately monitor ocean acidification have been recently tested (Land et al., 2019). Specifically, marine chemistry can be studied using four parameters: partial pressure of carbon dioxide in the water, dissolved inorganic carbon, alkalinity and pH. Any two of these parameters, along with measurements of salinity and temperature, allow us to understand the complete carbon chemistry of the ocean. ESA's SMOS mission and NASA's Aquarius mission, which both provide information on ocean salinity, have been key to the research.

5.7.1 Wetland Mapping

Synthetic Aperture Radar (SAR) for wetland mapping. Imaging radars provide information that is fundamentally different from sensors that operate in the visible and infrared portions of the electromagnetic spectrum. This is primarily due to the much longer wavelengths used by SAR sensors and the fact that they are active sensors, meaning that they send out a signal and receive their own energy. SAR technology provides the increased spatial resolution that is necessary in regional wetland mapping and SAR data have been used extensively for this purpose (Novo et al., 2002). The sensitivity of microwave energy to water and its ability to penetrate vegetative canopies make SAR ideal for the detection of hydrological features below the vegetation (Wilson and Rashid, 2005). The presence of standing water interacts with the radar signal differently depending on the dominant vegetation type and structure, as well as the biomass and condition of vegetation.

Wetland Change Detection. Several coastal wetlands, such as tidal salt marshes, are generally within fractions of a meter of sea level and will be lost, especially if the impact of sea-level rise is amplified by coastal storms. Man-made interventions are expected to further limit the ability of wetlands to survive sea-level rise. To identify long-term trends and short-term variations, such as the impact of rising sea levels and storm surges on wetlands, time-series of remotely sensed imagery should be analyzed. High temporal resolution, precise spectral bandwidths and accurate georeferencing procedures are factors that contribute to the frequent use of satellite image data for change detection analysis (Baker et al., 2007). One way to approach this problem is to reduce the spectral information to a single index, reducing the multispectral imagery into one single field of the index for each time step. The most common index used is the Normalized Difference Vegetation Index (NDVI), which is expressed as the difference between the red and near-infrared (NIR) reflectance divided by their sum (Domenikiotis et al., 2003). These two spectral bands represent the most detectable spectral characteristic of green plants. Detecting the actual changes between two registered and radiometrically corrected images from different dates can be accomplished by employing one of several techniques, including post-classification comparison (PCC), spectral image differencing (SID) and change vector analysis (CVA).

Submerged Aquatic Vegetation (SAV). Coral reef ecosystems usually exist in clear water and can be classified to show different forms of coral reef, coral rubble, dead coral, lagoons, sand, different densities of seagrasses, or algal cover. SAV often grows in somewhat turbid waters and thus is more difficult to map. Aerial hyperspectral scanners and high-resolution multispectral satellite imagers, such as IKONOS and QuickBird, have been used to map SAV with accuracies of about 75% for classes including high-density seagrass, low-density seagrass, and unvegetated bottom (Mishra et al., 2006). The mapping of submerged aquatic vegetation (SAV), coral reefs and general bottom characteristics from satellites has become more accurate, since high-resolution (0.4–4 m) multispectral imagery became available (Purkis, 2005).

Case study. Figure 5.24 presents study areas in tropical and subtropical southern Florida: LADS (Laser Airborne Depth Sounding) and IKONOS surveys of coral reefs (from Finkl et al., 2015)

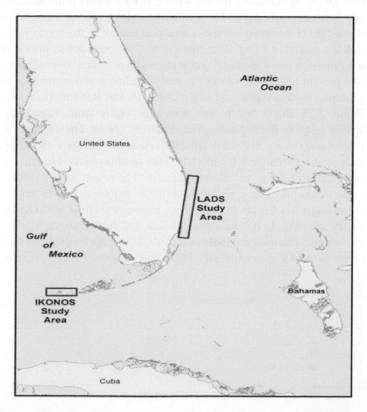

FIGURE 5.24 Study areas in tropical and subtropical southern Florida: LADS (Laser Airborne Depth Sounding) and IKONOS surveys of coral reefs (from Finkl et al., 2015) (in Finkl and Makowski, 2015).

(in Finkl and Makowski, 2015). (Note: The description of Figure 5.24 is taken from Finkl et al., 2015). Specifically, modern advancements like high-density airborne laser bathymetry and multispectral satellite imagery can now detect sea-floor reflectance at depths ranging to 50–60 m. Carbonate banks in the south Florida region provide nearly ideal conditions for mapping submarine topography and interpreting geomorphological and biophysical environments. A hierarchical open-ended classification system was developed for both open ocean and key (low carbonate islands) environments. These classification systems, which are based on cognitive recognition of sea-floor features interpreted from Laser Airborne Depth Sounding (LADS) and IKONOS imagery, are directly applied in GIS cartography programs to create comprehensive, informative and interactive products. This new technology and its associated classification systems permit major advancements in the detailed mapping of seafloors that have been achieved for margins of regional seas. The LADS survey covers an area of about 600 km² on the southeast open ocean coast, where there are shelf-edge coral reefs, carbonate rock reefs and sediment flats on the narrow continental shelf. The IKONOS survey covers an area of about 422 km² on platform coral reefs and carbonate sediment banks in the Florida Keys.

5.7.2 BEACH PROFILING AND SHORELINE CHANGE DETECTION

Beach profiles and shoreline positions show slow changes due to littoral drift and sea-level rise, although they can change rapidly with seasons and after storms (Stockdon et al., 2009).

Case studies. A few case studies are briefly presented as follows:

Case study 1. Figure 5.25 presents changes in beach profiles between summer and winter due to changes in wave climate (from Klemas, 2015). (Note: The description of Figure 5.25 is taken from Klemas, 2015). Specifically, during winter, storms waves remove sand from the beach and deposit it offshore, typically in bar formation. Indeed, during winter storms the beach is eroded and seaward cross-shore sediment transport results in the formation of offshore bars (Purkis and Klemas, 2011). Long-term changes of shorelines due to littoral drift or sea-level rise can be aggravated by man-made structures, such as jetties, seawalls and groins (Wang, 2010). At the present time, beach profiling and shoreline position analysis are based on the Global Positioning System (GPS) and Light Detection and Ranging (LiDAR) systems.

Case study 2. Figure 5.26 shows beach and dune topography from Stereoscopic Pleiades-1A Satellite Optical Imagery (from Salameh et al., 2019). (Note: The description of Figure 5.26 is taken from Salameh et al., 2019). Specifically, reliable topography and bathymetry information are fundamental parameters for modeling the morpho-hydrodynamics of coastal areas, for flood forecasting and for coastal management. The recent advancements of spaceborne remote sensing techniques, along with their ability to acquire data over large spatial areas and to provide high frequency temporal monitoring, has made them very attractive for topography and bathymetry mapping. In this case study, beach and dune topography was retrieved using Pleiades-1A stereoscopic imagery (Salameh et al., 2019). An evaluation of the precision and accuracy of the derived 2-meter Pleiades DEM was performed through comparison between

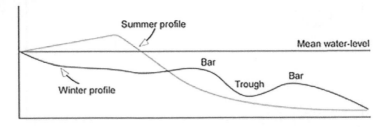

FIGURE 5.25 Changes in beach profiles between summer and winter due to changes in wave climate (from Klemas, 2015).

FIGURE 5.26 Beach and Dune Topography from Stereoscopic Pleiades-1A Satellite Optical Imagery (from Salameh et al., 2019).

the satellite derived topography and traditional survey methods, such as airborne light detection and ranging (LiDAR) and RTK-GPS surveys. The pilot site selected for this comparison was a 40 km stretch of coastline in southwest France (Figure 5.26). Specifically, Figure 5.26 presents: (A) Satellite image of Western France (source: Google Earth Pro, 2018) showing the location where the Pleiades stereo pair was obtained (orange rectangle located in southwest France) on November 14, 2017; (B) Close-up of the Pleiades mosaic showing the area where the light detection and ranging (LiDAR) airborne topographic survey was carried out (green dotted line polygon) and the area where real-time kinematic global positioning system (RTK-GPS) topographic measurements were taken; (C) The RTK-GPS survey lines (in red) and a photograph of the surveyor with the mobile GPS (from Salameh et al., 2019).

Case study 3. Figure 5.27 presents four applications of Landsat-8 of September 23, 2013: (a) Long-term evolution of the Bugio ephemeral ebb delta island; (b) Seasonal evolution of the Prainha transient beach; (c) Shoreline detection and extraction; and (d) Bathymetric data retrieval on Sado ebb delta (from Lira and Taborda, 2015). (Note: The description of Figure 5.27 is taken from Lira and Taborda, 2015). Specifically, remote sensing emerges as a very effective technique to capture the dynamics of the coastal system, as it provides a holist view of the system at a wide range of spatial and temporal scales. In mesoscale coastal environment applications and considering the universe of the solutions, the Landsat program arises as a good compromise between spectral, radiometric, spatial and temporal resolutions, combined with free data access, supported by an efficient data sharing platform. The capabilities of the Landsat program have been recently extended with the launch of Landsat-8, with improved radiometric and spectral resolution, opening the door to new studies.

5.7.3 BATHYMETRY

The application of geospatial data to coastal environments has been limited until recently by the lack of accurate near-shore bathymetric data (Malthus and Mumby, 2003).

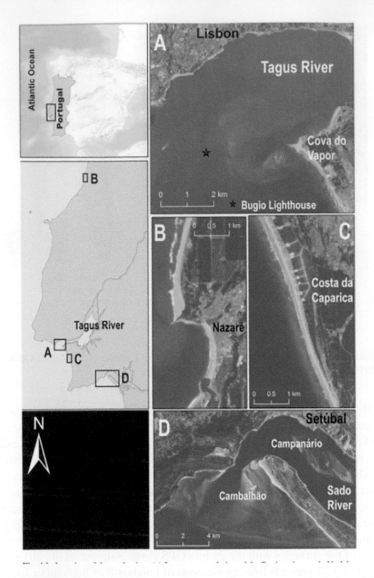

FIGURE 5.27 Applications of Landsat-8 of September 23, 2013: (a) Long-term evolution of the Bugio ephemeral ebb delta island; (b) Seasonal evolution of the Prainha transient beach; (c) Shoreline detection and extraction; and (d) Bathymetric data retrieval on Sado ebb delta (from Lira and Taborda, 2015).

Case study. Figure 5.28 presents the coastal topography for a section of the Assateague Island National Seashore acquired using the Experimental Advanced Airborne Research LiDAR (EAARL) (from Klemas, 2015). (Note: The description of Figure 5.27 is taken from Klemas, 2015). Specifically, remote sensing techniques that have been successfully used to map coastal water depth include LiDAR and acoustic depth sounding (Bonisteel et al., 2009). In LiDAR bathymetry, a laser transmitter/receiver mounted on an aircraft transmits a pulse that reflects off the air-water interface and the sea bottom. Since the velocity of the light pulse is known, the water depth can be calculated from the time lapse between the surface return and the bottom return. Because laser energy is lost due to refraction, scattering and absorption at the water surface, the sea bottom and inside the water column, these effects limit the strength of the bottom return and limit the maximum detectable depth.

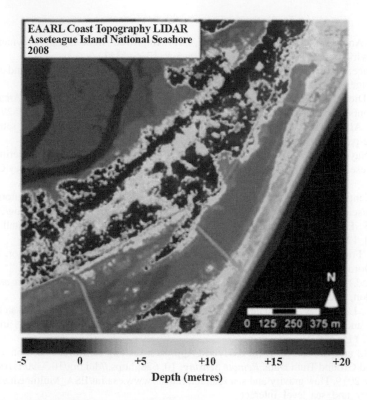

FIGURE 5.28 Coastal topography for a section of the Assateague Island National Seashore acquired using the Experimental Advanced Airborne Research LiDAR (EAARL) (Credit: USGS) (Klemas, 2015).

5.8 SUMMARY

Remote sensing applications for ocean and coastal ecosystems have been presented, which included ocean features, such as currents, winds and circulation patterns. They have also included storm and fog monitoring, which involves frontal zones, storm early warning systems, cyclones, marine fog, wind and wave retrieval. Ocean color analysis was also presented, which covers water temperature monitoring, sea surface temperature (SST), salinity, density, water quality, ocean productivity, phytoplankton. Moreover, oil spills have been considered, which involve oil spills mapping, predicting oil spill extent and drift. Sea ice was also presented, involving sea ice features, ice detection and forecasting, as well as ice motion and shipping.

Remote sensing can also be applied to coastal and intertidal zones, which also constitute habitats for a wide variety of plants, fish, shellfish, and other wildlife. Moreover, several natural and man-induced alterations affect coastal wetlands and estuaries. Among these are mentioned eutrophication, pollutant runoff, hydrological modifications, impoundments and fragmentation by roads and ditches, where remote sensing data and methods can be applied. In addition, remote sensing can contribute to beach erosion, wetland destruction, algal blooms, hypoxia and anoxia, fish kills, the release of pollutants, and spread of pathogens, which can be caused by coastal storms or sea-level rise.

Remote sensing high spatial, temporal and spectral resolutions are required due to high spatial and temporal variability of coastal ecosystems. There are recent technological and computational advances in remote sensing systems for monitoring and mapping changes in the coastal zone. Specifically, high-resolution multispectral and hyperspectral imagers, LiDAR and radar systems are available for monitoring changes in coastal marshes, submerged aquatic vegetation, coral reefs and beach profiles. Moreover, imaging radars are sensitive to soil moisture and inundation and can detect hydrological features beneath the vegetation canopy.

REFERENCES

Abulaitijiang, A., Andersen, O.E., and Sandwell, D., 2019. Improved Arctic Ocean bathymetry derived from DTU17 gravity model. *Earth and Space Science*, 6, 8, 1336–1347.

Baker, C., Lawrence, R.L., Montagne, C., and Patten, D., 2007. Change detection of wetland ecosystems using Landsat imagery and change vector analysis. *Wetlands*, 27, 610–619.

Baumhoer, C.A., Dietz, A.J., Kneisel, C., and Kuenzer, C., 2019. Automated extraction of Antarctic glacier and ice shelf fronts from Sentinel-1 imagery using deep learning. *Remote Sensing*, 11, 2529.

Bergsma, E.W.J., Almar, R., and Maisongrande, P., 2019. Radon-augmented Sentinel-2 satellite imagery to derive wave-patterns and regional bathymetry. *Remote Sensing*, 11, 1918.

Bonisteel, J.M., Nayegandhi, A., Wright, C.W., Brock, J.C., and Nagle, D.B., 2009. Experimental Advanced Airborne Research LiDAR (EAARL) data processing manual. US Geological Survey OpenFile Report, 2009–1078.

CCRS (Canada Centre for Remote Sensing), 1998. "Fundamentals of Remote Sensing," tutorial, https://www.nrcan.gc.ca/science-and-data/research-centres-and-labs/canada-centre-remote-sensing/21749

Cermak, J., 2018. Fog and low cloud frequency and properties from active-sensor satellite data. *Remote Sensing*, 10, 1209. doi:10.3390/rs10081209

Croft, P.J., 2013. Fog Hazards, pp. 342–346, in *Encyclopedia of Natural Hazards*. Editor: P.T. Bobrowsky, Springer, Dordrecht.

Dalezios, N.R. (ed.), 2017. *Environmental Hazards Methodologies for Risk Assessment and Management*. IWA, London, UK, 534.

Domenikiotis, C., Loukas, A., and Dalezios, N.R., 2003. The use of NOAA/AVHRR satellite data for the monitoring and assessment of forest fires and floods. EGS Journal Natural Hazards and Earth Systems *Sciences*, 3, 115–128.

Egli, S., Thies, B., and Bendix, J., 2018. A Hybrid Approach for Fog Retrieval Based on a Combination of Satellite and Ground Truth Data. *Remote Sensing*, 10, 628, https://doi.org/10.3390/rs10040628

ESA, 2019a. July 2019. How gravity and sea interact, https://www.esa.int/ESA_Multimedia/Images/2019/07/How_gravity_and_sea_level_interact

ESA, 2019b. July 2019. Bathymetry of Chukchi Cap, https://www.esa.int/ESA_Multimedia/Images/2019/07/Bathymetry_of_Chukchi_Cap

ESA, 2019c. October 2019. D28 iceberg takes a turn, https://www.esa.int/ESA_Multimedia/Images/2019/10/D28_iceberg_takes_a_turn

ESA, 2019d. October 2019. Characterising how carbon flows between ocean and atmosphere, https://www.esa.int/ESA_Multimedia/Videos/2020/09/Characterising_how_carbon_flows_between_ocean_and_atmosphere

ESA, 2019e. November 2019. Antarctica's internal temperature, https://www.esa.int/ESA_Multimedia/Images/2019/10/Antarctica_s_internal_temperature

ESA, 2019f. December 2019. Greenland ice loss much faster than expected, https://www.esa.int/Applications/Observing_the_Earth/Space_for_our_climate/Greenland_ice_loss_

Fan, S., Kudryavtsev, V., Zhang, B., Perrie, W., Chapron, B., and Mouche, A., 2019a. On C-Band quad-polarized synthetic aperture radar properties of ocean surface currents. *Remote Sensing*, 11, 2321.

Fan, Q., Chen, F., Cheng, M., Lou, S., Xiao, R., Zhang, B., Wang, C., and Li, J., 2019b. Ship detection using a fully convolutional network with compact polarimetric SAR images. *Remote Sensing*, 11, 2171.

Fang, H., Perrie, W., Zhang, G., Xie, T., Khurshid, S., Warner, K., Yang, J., and He, Y., 2019. Ocean surface wind speed retrieval using simulated RADARSAT constellation mission compact polarimetry SAR data. *Remote Sensing*, 11, 1876.

Finkl, C.W., and Makowski, C. (eds.), 2015. *Remote Sensing and Modeling: Advances in Coastal and Marine Resources*. Springer, Boca Raton, FL, USA, 502.

Finkl, C.W., Makowski, C., and Vollmer, H., 2015. Advanced Techniques for Mapping Biophysical Environments on Carbonate Banks Using Laser Airborne Depth Sounding (LADS) and IKONOS Satellite Imagery, pp. 31–63, in *Remote Sensing and Modeling: Advances in Coastal and Marine Resources*. Editors: C.W. Finkl and C. Makowski. Google Earth Pro

Graziano, M.D., Renga, A., and Moccia, A., 2019. Integration of Automatic Identification System (AIS) data and single-channel Synthetic Aperture Radar (SAR) images by SAR-based ship velocity estimation for maritime situational awareness. *Remote Sensing*, 11, 2196.

Grinson, G., 2015. Numerical Modelling and Satellite Remote Sensing as Tools for Research and Management of Marine Fishery Resources, pp. 431–452, in *Remote Sensing and Modeling: Advances in Coastal and Marine Resources*. Editors: C.W. Finkl and C. Makowski.

Grinson, G., Bharathiamma, M., Mini, R., Srinivasa, K., Vethamony, P., Babu, M.T., and Xivanand, V., 2012. Remotely sensed chlorophyll: A putative trophic link for explaining variability in Indian oil sardine stocks. *Journal of Coast Research*, 28, 1A, 105–113.

Han, Y., Gao, Y., Zhang, Y., Wang, J., and Yang, S., 2019. Hyperspectral sea ice image classification based on the spectral-spatial-joint feature with deep learning. *Remote Sensing*, 11, 2170.

Hasan, G.M.J., and Takewaka, S., 2015. Foreshore Applications of X-band Radar, pp. 161–190, in *Remote Sensing and Modeling: Advances in Coastal and Marine Resources*. Springer, Boca Raton, FL, USA.

Houskeeper, H.F., and Kudela, R.M., 2019. Ocean color quality control masks contain the high phytoplankton fraction of coastal ocean observations. *Remote Sensing*, 11, 2167.

Ilčev, S.D., 2019. *Global Satellite Meteorological Observation (GSMO) Applications: Volume 2*. Springer, 401.

IMBIE Team, 2019. Mass balance of the Greenland ice sheet from 1992 to 2018. *Nature*

IPCC Special Report on the Ocean and Cryosphere in a Changing Climate, 2019. [Pörtner, H.-O., Roberts, D.C., Masson-Delmotte, V., Zhai, P., Tignor, M., Poloczanska, E., Mintenbeck, K., Nicolai, M., Okem, A., Petzold, J., Rama, B., and Weyer, N. (eds.)], September, 1203p.

Jahid Hasan, G.M., and Takewaka, S., 2015. Foreshore Applications of X-band Radar, pp. 161–191, in *Remote Sensing and Modeling: Advances in Coastal and Marine Resources*. Springer, Boca Raton, FL, USA.

Klemas, V.V., 2015. Remote Sensing of Coastal Ecosystems and Environments, pp. 3–30, in *Remote Sensing and Modeling: Advances in Coastal and Marine Resources*. Editors: C.W. Finkl and C. Makowski.

Krestenitis, M., Orfanidis, G., Ioannidis, K., Avgerinakis, K., Vrochidis, S., and Kompatsiaris, I., 2019. Oil spill identification from satellite images using deep neural networks. *Remote Sensing*, 11, 1762.

Kudela, R.M., Hooker, S.B., Houskeeper, H.F., and McPherson, M., 2019. The influence of signal to noise ratio of legacy airborne and satellite sensors for simulating next-generation coastal and inland water products. *Remote Sensing*, 11, 2071.

Kulk, G., Platt, T., Dingle, J., Jackson, T., Bror, F., Jönsson, H.A., Bouman, M.B., Brewin, R.J.W., Doblin, M., Estrada, M., Figueiras, F.G., Furuya, K., González-Benítez, N., Gudfinnsson, H.G., Gudmundsson, K., Huang, B., Isada, T., Kovac, Ž., Lutz, V.A., Marañón, E., Raman, M., Richardson, K., Rozema, P.D., van de Poll, W.H., Segura, V., Tilstone, G.H., Uitz, J., van Dongen-Vogels, V., Yoshikawa T., and Sathyendranath, S., 2020. Primary production, an index of climate change in the ocean: Satellite-based estimates over two decades. *Remote Sensing*, 12, 826

Kunkel, B.A., 1980. Controlling fog. *Weatherwise*, 33, 3, 117–123.

La Barbera, P., Lanza, L., and Siccardi, F., 1994. Application of geographical information systems in the analysis of hydrological inputs to a shallow Mediterranean coastal aquifer. *IAHS Publications*, 220, 337–346.

Land, P.E., Findlay, H.S., Shutler, J.D., Ashton, I.G.C., Holding, T., Grouazel, A., Girard-Ardhuin, F., Reul, N., Piolle, J-F., Chapron, B., Quilfen, Y., Bellerby, R.G.J., Bhadury, P., Salisbury, J., Vandemark, D., and Sabia, R., 2019. Optimum satellite remote sensing of the marine carbonate system using empirical algorithms in the global ocean, the Greater Caribbean, the Amazon Plume and the Bay of Bengal. *Remote Sensing of Environment*, 235, 11469.

Lee, S., Meister, G., and Franz, B., 2019. MODIS Aqua reflective solar band calibration for NASA's R2018 ocean color products. *Remote Sensing*, 11, 2187.

Lira, C., and Taborda, R., 2015. Advances in Applied Remote Sensing to Coastal Environments Using Free Satellite Imagery, pp. 77–102, in *Remote Sensing and Modeling: Advances in Coastal and Marine Resources*. Editors: C.W. Finkl and C. Makowski.

Macelloni, G., Leduc-Leballeur, M., Montomoli, F., Brogioni, M., and Picard, G., 2019. On the retrieval of internal temperature of Antarctica ice sheet by using SMOS observations. *Remote Sensing of Environment*, 233, 111405

Malthus, T.J., and Mumby, P.J., 2003. Remote sensing of the coastal zone: an overview and priorities for future research. *International Journal of Remote Sensing*, 24, 2805–2815.

Mishra, D., Narumalani, S., Rundquist, D., and Lawson, M., 2006. Benthic habitat mapping in tropical marine environments using QuickBird multispectral data. *Photogrammetic Engineering and Remote Sensing*, 72, 1037–1048.

NASA/GSFC, 2010. Hurricane Ike: storm surge flooding image of the Gulf Coast. NASA image courtesy Jeff Schmaltz, MODIS Rapid Response team at NASA GSFC.

Novo, E.M.L.M., Costa, M.P.F., Mantovani, J.E., and Lima, I.B.T., 2002. Relationship between macrophyte stand variables and radar backscatter at L and C band, Tucurui reservoir, Brazil. *International Journal of Remote Sensing*, 23, 1241–1260.

Pearson, K., Good, S., Merchant, C.J., Prigent, C., Embury, O., and Donlon, C., 2019. Sea surface temperature in global analyses: Gains from the Copernicus Imaging Microwave Radiometer. *Remote Sensing*, 11, 2362

Pinochet, A., Garcés-Vargas, J., Lara, C., and Olguín, F., 2019. Seasonal variability of upwelling off Central-Southern Chile. *Remote Sensing*, 11, 1737

Purkis, S.J., 2005. A "reef-up" approach to classifying coral habitats from IKONOS imagery. *IEEE Transactions on Geoscience and Remote Sensing*, 43, 1375–1390.

Purkis, S.J., and Klemas, V., 2011. *Remote sensing and Global Environmental Change*. Wiley-Blackwell, Oxford.

Salameh, E., Frappart, F., Almar, R., Baptista, P., Heygster, G., Lubac, B., Raucoules, D., Almeida, L.P., Bergsma, E.W.J., Capo, S., De Michele, M., Idier, D., Zhen Li, Marieu, V., Poupardin, A., Silva, P.A., Turki, I., and Laignel, B., 2019. Monitoring beach topography and nearshore bathymetry using space-borne remote sensing: A review. *Remote Sensing*, 11, 2212

Soriano-González, J., Angelats, E., Fernández-Tejedor, M., Diogene, J., and Alcaraz, C., 2019. First results of phytoplankton spatial dynamics in two NW-Mediterranean bays from chlorophyll-a estimates using Sentinel 2: Potential implications for aquaculture. *Remote Sensing*, 11, 1756

Stockdon, H.F., Doran, K.S., and Sallenger, A.H., 2009. Extraction of LiDAR-based dune-crest elevations for use in examining the vulnerability of beaches to inundation during hurricanes. *Journal of Coastal Research*, 35, 59–65.

Wang, Y., 2010. Remote Sensing of Coastal Environments: An Overview, in *Remote Sensing of Coastal Environments*. Editor: J. Wang, CRC Press, Boca Raton.

Wilson, B.A., and Rashid, H., 2005. Monitoring the 1997 flood in the Red River Valley using hydrologic regimes and RADARSAT imagery. *The Canadian Geographer*, 49, 100–109.

Zhang, H., Liu, X., Wu, R., Liu, F., Yu, L., Shang, X., Qi, Y., Wang, Y., Song, X., Xie, X., Yang, C., Tian, D., and Zhang, W., 2019. Ocean response to successive typhoons Sarika and Haima (2016) based on data acquired via multiple satellites and moored array. *Remote Sensing*, 11, 2360

6 Environmental Hazards

6.1 INTRODUCTION: REMOTE SENSING CONCEPTS OF ENVIRONMENTAL HAZARDS

The study of environmental hazards and risks has a very important spatial component. At the present time, Earth observation (EO) provides one of the most promising avenues for providing information at global, regional and even basin scales related to environmental hazards. The general circumstances that make EO technology attractive for this purpose in comparison to traditional techniques consists of their ability to provide repetitive, inexpensive and synoptic views of large areas in a spatially contiguous fashion without disturbing the area to be surveyed and without site accessibility issues (Petropoulos et al., 2009). Additional research is required to incorporate geographic information systems (GIS), remote sensing, simulation models and other computational techniques into an integrated multi-hazard risk management framework for sustainable environment, which includes early warnings of natural disasters (Sivakumar et al., 2005; Dalezios, 2017). There should also be more research into the impacts of potentially increasing severity and frequency of extreme events related to global change and mitigation measures.

The use of EO products and GIS has become an integrated approach in disaster-risk management. GIS and remote sensing and, in general, geoinformatics are increasingly employed due to the complex nature of databases to facilitate strategic and tactical applications at farm and policy levels. Hazard and risk assessments are carried out at multiple scales with specific objectives. Moreover, there are spatial data requirements for environmental data, triggering or causal factors, hazard inventories and elements-at-risk (Dalezios, 2017, Dalezios et al., 2020). Remote sensing is nowadays an essential tool in monitoring changes in the earth's surface, atmosphere and oceans, with steadily increasing use in early warning systems for hazardous events. Moreover, remote sensing provides the input for thematic information used in hazard modeling, such as topography, lithology and land cover (Van Westen, 2013). Satellite remote sensing, combined with ground measurement networks and various types of modeling, has now emerged as a powerful operational and readily applicable tool to provide information on environmental hazards and its dynamics.

In this chapter, the remote sensing potential is considered in environmental hazards affecting several sectors of the economy under increasing climate variability. Specifically, the environmental hazards examined include hydrometeorological hazards, such floods and excess rain, droughts and land desertification; biophysical hazards, such as frost and heatwaves, biological and health hazards and wildland fires; and geophysical hazards, such as geological hazards (landslides, snow avalanches), earthquakes and volcanoes. The emphasis is on remotely sensed methodologies in the three stages of hazard development, namely early warning systems (before), monitoring (during) and assessment (after).

6.2 HYDROMETEOROLOGICAL HAZARDS

In this section, hydrometeorological hazards are considered, namely floods and excess rain, droughts and desertification. Storms and hail are covered in Chapters 2, 3, 4 and 5. For each hazard some concepts are presented, along with methodologies on the three stages of hazard development, namely forecasting-now-casting (before), monitoring (during) and assessment (after). A brief description follows.

6.2.1 Floods and Excess Rain

In this section, flood typology and the remote sensing potential in flood forecasting and monitoring and mapping of floods and excess rain is considered. Flood forecasting and monitoring is based on hydrological simulation and rainfall-runoff modeling at watershed level, involving the use of weather radar and/or satellite data and methods, which is covered in Chapter 7. In mapping of floods and excess rain, recent remote sensing scientific contributions have increased considerably, and science has presented innovative research and methods for retrieving information content from multiscale coverages of disastrous flood events all over the world (Schumann, 2018). Moreover, the information obtained from remote sensing of floods is becoming mature enough not only to be integrated with computer simulations of flooding to allow better prediction, but also to support operational assistance to flood response agencies.

6.2.1.1 Flood Typology and Forecasting

Floods can be generated by a range of processes related to catchment state and the properties of heavy precipitation. The flood regime is the combined effect of rainfall variability at different time scales and the interactions with landscape dynamics, namely soil moisture and snow processes. Flood processes often show seasonal occurrence. On this basis, five flood process types are identified and briefly described (Merz and Blöschl, 2003; Destro et al., 2017), namely long-rain floods, short-rain floods, flash floods, rain-on-snow floods and snowmelt floods.

Flood forecasting and monitoring can be achieved through hydrological simulation. The prediction or prognosis of flood events covers not only the estimation of the frequency of a hydrological episode of a certain size, but also the forecast of the size and time of a flood peak. Flood frequency analysis is used to assess design floods. There are several methods to estimate the frequency of occurrence of floods of different magnitude, which depend on the availability of hydrometric data (Dalezios and Eslamian, 2016).

6.2.1.2 Mapping of Floods and Excess Rain

The area and extent of floods can be obtained through most satellite imaging platforms leading to mapping of permanent water bodies (Figure 6.1). Nevertheless, the area and extent of surface water may be measured with a variety of visible band sensors, such as Landsat, Moderate Resolution Imaging Spectroradiometer (MODIS) and Sentinel-2, with different repeat frequencies, as well as by SAR, such as COSMO-SkyMed, TerraSAR-X, Radarsat-2, and Sentinel-1A and -1B, with varying degrees of success and very limited operational applications (Schumann et al., 2016; Zinno et al., 2017; Schumann, 2018). There is an exception of the near real-time (NRT) MODIS flood mapping effort by the National Aeronautics and Space Administration (NASA) and the Dartmouth Flood Observatory (DFO) (http://floodobservatory.colorado.edu). There is a strong inverse relationship between spatial resolution and revisit time for satellites, which currently limits the monitoring of floods from space in NRT or operationally only through either low-resolution imagery or satellite constellations. As a result, this type of spaceborne data can be used for monitoring major floods on medium to large rivers. For finer-resolution systems, the same issue occurs, although revisit times can be up to 35 days. In most river basins, imaging a flood with a high-resolution system is possible for constellation or through the International Charter (https://www.disasterscharter.org).

The **Visible and infrared** ranges are very widely used for imaging floods. Indeed, low spatial resolution for sensors with high revisit time, cloud cover and restriction to daylight operations constitute common problems with optical imagery (Schumann, 2018). These sensors are successfully applied to global monitoring of flooded areas, as demonstrated by the DFO, using MODIS images as much as twice daily. This database currently represents the only global observed record of flood events (Schumann and Domeneghetti, 2016). Moreover, global multi-satellite historical observations cover open water surfaces over a large spectrum, ranging from visible (advanced very high-resolution radiometer: AVHRR) to microwave wavelengths, both passive (special sensor

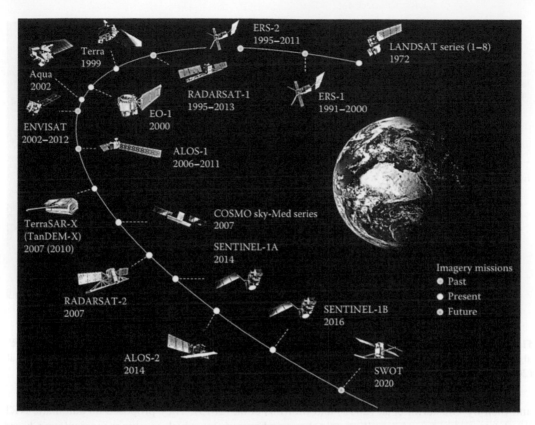

FIGURE 6.1 Past, current and future satellite missions that carry instruments for imaging floods (from Schumann, 2018).

microwave/imager: SSMI) and active (ERS scatterometer). These systems have been used to map temporal wetland dynamics, albeit at spatial resolution too coarse to be of significance to local decision-making, floodplain management, or disaster assistance, unless downscaled to the appropriate spatial resolution.

The **SAR** is considered suitable for flood mapping, due to its relatively high (1–3 m is now possible) spatial resolution and its near-all-weather, as well as day and night operating capabilities. However, radar sensors are invaluable for monitoring floods at microwave range, since, although there are advantages during flood events, SAR image geometry and processing restrict widespread civil applications as compared to optical sensors. Indeed, the magnitude of the deteriorating effects in a SAR (flood) image depends on wavelength, polarization and incidence angle, which impact the ability to discriminate features or conditions of the Earth's surface. Specifically, polarization refers to the direction at which materials reflect signals and SAR sensor receives these signals, whereas incidence angle refers to the angular deviation of the incident signal from nadir (Ulaby et al., 1982; Schumann, 2018). There are many image-processing algorithms to map flooding on a SAR image (Matgen et al., 2011). However, there are interpretation difficulties in a SAR image, which may arise from different sources, such as complex signal backscatter (e.g. diffuse and volume backscatter), inadequate wavelength and/or polarizations, remaining geometric image distortions, and multiplicative noise.

Case study. Probably the best-known SAR flood image research studies that propose an automated classification algorithm are based on the TerraSAR-X image of the 2007 summer flood in and around the town of Tewkesbury, England (Schumann, 2018). The first study using this case study to illustrate a fully automated flood mapping procedure was performed by Mason et al. (2012)

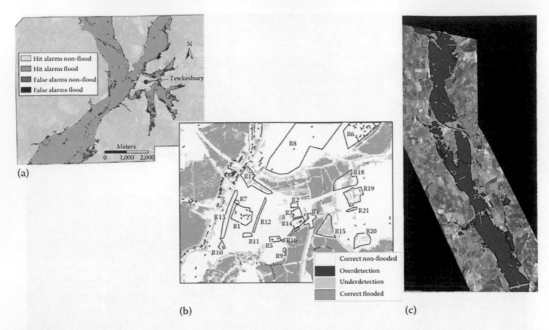

FIGURE 6.2 Flood maps produced by three different automated SAR flood mapping algorithms: (a) DLR's algorithm by Martinis et al. (2013); (b) Mason et al.'s (2012) algorithm; and (c) the algorithm by Matgen et al. (2011), as applied by Giustarini et al. (2013) (from Schumann, 2018).

and then by Giustarini et al. (2013) and Martinis et al. (2013). They proposed an automatic NRT flood detection approach (resulting map shown in Figure 6.2a), which combines segmentation-based classification and histogram thresholding, specifically oriented to the analysis of single-polarized, very-high-resolution SAR satellite images. Indeed, the use of a LiDAR digital surface model (DSM) within the urban area, including buildings, is considered to account for areas of misclassification due to layover and shadow effects in the SAR scene. Specifically, a fully automated approach has been applied (Giustarini et al., 2013), based on a developed operational procedure (Matgen et al., 2011), which relies on the comparison of the actual flood image backscatter distribution function with a theoretical gamma distribution, thus, objectively selecting the threshold backscatter value that distinguishes flooded from non-flooded surfaces (Schumann, 2018).

6.2.1.3 Operational Flood Mapping and Monitoring Systems

There are several current operational flood mapping and monitoring systems, such as the global flood monitoring system, the DFO, NOAA's VIIRS sensor, NASA's JPL ARIA, ESA's G-POD and DLR's ZKI (Schumann, 2018). A brief description follows.

The Global Flood Monitoring System (GFMS). The University of Maryland established the GFMS (Wu et al., 2012) (http://flood.umd.edu), funded by NASA. This is an experimental system that uses the iMERG product from the Global Precipitation Measurement (GPM) mission and real-time TRMM Multi-satellite Precipitation Analysis (TMPA) data. The system performs real-time quasi-global hydrological calculations at 1/8th degree and 1 km resolution inundation simulations. Moreover, the system issues flood forecasts with 4- to 5-day lead time, based on numerical weather prediction (NWP) precipitation, and mapping of inundation at a 3-hour time step.

The Dartmouth Flood Observatory (DFO) (http://floodobservatory.colorado.edu). The DFO conducts and archives global remote sensing flood mapping and measurements in NRT. The DFO performs rapid flood mapping with MODIS (Brakenridge and Anderson, 2006). In addition, during high-impact flood disasters, it also maps flooding from other satellites, such as EO-1, the Landsat series, and SAR satellite missions. The DFO offers two map series accessible from the global index:

Current Flood Conditions, providing daily, satellite-based updates of surface water extent, and the **Global Atlas of Floodplains**, a remote sensing record of floods, from 1993 to 2015.

NOAA's VIIRS Sensor. The Visible Infrared Imaging Radiometer Suite (VIIRS) instrument is a major (EO) instrument onboard the National Oceanic and Atmospheric Administration's (NOAA's) S-NPP and JPSS satellites. With a very large swath width of 3060 km, it provides full daily coverage, both in the day and night sides of the Earth. The VIIRS has 22 spectral bands, including 16 moderate spatial resolution bands at 750 m pixel spacing at nadir, 5 imaging resolution bands at 375 m at nadir, and 1 panchromatic band with 750 m spatial resolution (Wang et al., 2016).

NASA JPL's ARIA. The ARIA Center (https://aria.jpl.nasa.gov) is a joint venture co-sponsored by the California Institute of Technology (Caltech) and by NASA through the Jet Propulsion Laboratory (JPL). The ARIA Center plans to provide automated imaging and analysis capabilities necessary to keep up with the imminent increase in raw data from geodetic imaging missions. Data set analyses are currently handcrafted following each event. Moreover, the ARIA Center plans to provide the infrastructure to generate imaging products in NRT that can improve situational awareness for disaster response.

ESA's G-POD. The European Space Agency (ESA) hosts an automated SAR-based flood mapping tool on its Grid Processing on Demand (G-POD) system (http://gpod.eo.esa.int), which is currently operated in testing mode. The G-POD is expected to be freely available to end users, who can search the ESA SAR database for a flood image and retrieve an automatically generated flood map.

DLR's ZKI. The DLR's ZKI monitors flood disasters by tasking its TerraSAR-X satellite, and, as part of the International Charter, it has access to other tasked satellites that may provide relevant disaster data. In addition, the data archive is searched for matching pre-disaster satellite scenes. During the first 6 hours after the activation of ZKI, reference maps based on archive satellite data can be made available to relief organizations and provide a first overview of the affected area and estimate of damages (Schumann, 2018).

6.2.2 DROUGHTS

Drought is a natural phenomenon recurring at a regional scale throughout history. Essentially, the main cause of droughts is the deficiency or lack of precipitation in a region over an extended period of time, which is referred to as an extreme climatic event associated with water resources deficit (Dalezios et al., 2017a; Dalezios et al., 2017b; Dalezios et al., 2017c; Dalezios, 2018; Dalezios et al., 2018a). Satellite-based data are consistently available and can be used to develop drought indices and detect several drought features, such as severity, periodicity, duration, onset and end time, as well as areal extent. Indeed, the growing number and improving reliability of relevant earth observation satellite systems present a wide range of new capabilities, which can be used to assess and monitor drought hazard and its effects (Dalezios et al., 2017c, Dalezios et al., 2017d). Remote sensing data and methods can delineate the quantitative spatial and temporal variability of several drought features (Dalezios et al., 2014). This section focuses on the remote sensing potential and capabilities in drought analysis. Drought definitions and concepts, including types, factors and features, are initially presented. Then, remote sensing capabilities, in terms of data and methods, are explored in drought analysis and assessment. Representative remotely sensed drought indices are presented, namely the Reconnaissance Drought Index (RDI) (Tsakiris and Vangelis, 2005; Dalezios et al., 2012) and the Vegetation Health Index (VHI) (Kogan, 1995; Dalezios et al., 2014). Applications of both indices, RDI and VHI, are considered in drought early warning systems (DEWS) and monitoring.

6.2.2.1 Remotely Sensed Drought Indices

Drought quantification is accomplished through indicators and indices (Mishra and Singh, 2010; Zargar et al., 2011). Specifically, the approaches to drought quantification and assessment are essentially three (Svoboda et al., 2002): (1) single indicator or index (parameter); (2) multiple indicators or indices; and (3) composite or hybrid indicators, which integrate several indicators or indices and converge an evidence approach. There are numerous remote sensing drought indices. Moreover, new sensors have improved information due to higher spatial resolution, which constitutes a current shortcoming in drought indices products (Niemeyer, 2008). The thematic accuracy of remote sensing data sets has been steadily improved by novel noise reduction algorithms and other atmosphere correction algorithms. Remotely sensed drought indices rely on information from remote sensing sensors to map the condition of the land and to detect several drought features based on parameters, such as precipitation or temperature.

Any meteorological drought index can be converted to a remotely sensed index provided that meteorological variables, such as precipitation and/or temperature, are computed by remotely sensed algorithms or methods. Moreover, the use of remotely sensed drought indices is expected to increase drastically in the forthcoming years due to the availability of precipitation and temperature data platforms at a global scale, besides the increasing number of remote sensing systems, as well as increasing technological and computational advances. Meteorological drought indices can be used in the context of DEWS in order to provide timely information on drought for decision-making. It should be stated that a meteorological drought index value is essentially considered far more useful than raw data, especially in the case of drought monitoring for NRT decision-making. Moreover, a meteorological drought index can also be used as ground-truth information for modeling efforts or remotely sensed detection of several drought features.

Table 6.1 presents an indicative list of available and commonly used meteorological drought indices in different classes (Dalezios, 2018). References for individual indices can be found in the aforementioned publication. There are precipitation-only drought indices, however, additional meteorological variables and indices have been considered in order to account for temporal temperature trends. Specifically, there are modifications to Standardized Precipitation Index (SPI)

TABLE 6.1

Indicative List of Meteorological Drought Classes and Indices (from Dalezios, 2018)

Classification of Drought Indices

1. *Atmospheric Drought Indices*	4. *Recursive Indices*
1.1 Saturation Deficit	4.1 Fooley Anomaly Index (FAI)
2. *Precipitation Anomaly Indices*	4.2 Bhalme-Mooley Drought Index (BMDI)
2.1 Precipitation Index	4.3 Palmer Drought Severity Index (PDSI)
2.2 Relative Precipitation Sum	4.4 Standardized Precipitation Index
2.3 Relative Anomaly	4.5 Surface Water Supply Index (SWSI)
2.4 Standardized Anomaly Index (SAI)	4.6 Reclamation Drought Index (RDI)
2.5 Average Standard Anomaly	4.7 Palmer Drought Index (PDI)
3. *Aridity Indices*	4.8 Palmer Crop Moisture Index (CMI)
3.1 Lang's Rainfall Index	4.9 Keetch-Byram Drought Index (KBDI)
3.2 De Martone Aridity Index	4.10 Effective Drought Index
3.3 Ped's Drought Index (PDI1)	4.11 Reconnaissance Drought Index (RDI)
3.4 Selyaninov's Hydrothermal Coefficient	5. *Remotely Sensed Information*
3.5 Thornthwaite Index	5.1 Crop Water Stress Index (CWSI)
3.6 Potential Water Deficit	5.2 Vegetation Index
3.7 Potential Evaporation Ratio	5.3 Normalized Difference Vegetation Index
3.8 Aridity Index: Moisture Available Index	5.4 Stress Degree Days
3.9 Relative Evaporation	
3.10 Surface Energy Balance	
3.11 Bowen Ratio	

(McKee et al., 1993) to develop the more comprehensive RDI (Tsakiris and Vangelis, 2005), which incorporates evapotranspiration leading to better association with impacts from agricultural and hydrological droughts. There is a similar index, namely Standardized Precipitation Evapotranspiration Index (SPEI) (Vicente-Serrano et al., 2010), which also account for long-term trends in temperature change. If such trends are absent, SPEI performs similarly to SPI. A brief description follows of the classes of meteorological drought indices (Table 6.1). *Atmospheric drought indices*: Low humidity is considered the standard signal of dry spell. Atmospheric drought is usually described by the water vapor saturation deficit. *Indices of precipitation anomaly*: Several existing precipitation anomaly indices are listed in Table 6.1. *Aridity indices*: The aridity index uses the evapotranspiration/precipitation ratio. There are several types of aridity indices, some of which are listed in Table 6.1. *Recursive drought indices*: There are several recursive drought indices consisting of the family of PDSI, which are listed in Table 6.1. *Remotely sensed information:* In this class there are indices, which are also considered indices of agricultural drought.

Most of the existing and widely used remotely sensed drought indices are based on spectral reflectance of vegetation and, thus, are mainly used as agricultural drought indices, also known as vegetation indices. There are several indices based on remotely sensed information, some of which are listed in Table 6.1. Agricultural drought is essentially based on monitoring soil-water balance and the resulting deficit in case of drought. The traditionally used bands are near-infrared (NIR), red and short-wavelength infrared (SWIR). Moreover, the land surface temperature (LST) has been used as an additional source along with NDVI to improve drought quantification accuracy. Table 6.2 presents an indicative list of internationally used conventional and remotely sensed agricultural drought indices (Dalezios, 2018), including soil-water balance models. References for individual indices can be found in the aforementioned publication. Remote sensing indices are diverse, and new indices are frequently proposed. It is worth mentioning the Vegetation Condition Index (VCI) and the Temperature Condition Index (TCI), which jointly lead to VHI (Kogan, 1995; Dalezios et al., 2014), being operationally used (NOAA, 2011) for monitoring agricultural drought.

Remotely sensed Composite Drought Indices (CDIs). Combined drought indices, which are also termed *hybrid* or *aggregate* or *composite* (Waseem et al., 2015), are derived by incorporating

---

TABLE 6.2
Conventional and Satellite-based Agricultural Drought Indices
(from Dalezios, 2018)

Conventional Drought Indices	Satellite-based Drought Indices
1. Agricultural Drought Index (DTx)	1. Normalized Difference Vegetation Index
2. Bhalme-Mooley Drought Index (BMDI)	2. Deviation NDVI index
3. Corn Drought Index	3. Enhanced Vegetation Index (EVI)
4. Crop Moisture Index (CMI)	4. Vegetation Condition Index (VCI)
5. Crop Specific Drought Index.	5. Monthly Vegetation Condition Index
6. Evapotranspiration Deficit Index (ETDI)	6. Temperature Condition Index (TCI)
7. Global Vegetation Water Moisture Index	7. Vegetation Health Index (VHI)
8. Leaf Water Content Index (LWCI)	8. Normalized Difference Temperature Index
9. Moisture Availability Index (MAI)	9. Crop Water Stress Index (CWSI)
10. Reclamation Drought Index (RDI)	10. Drought Severity Index (DSI)
11. Soil Moisture Anomaly Index (SMAI)	11. Temperature- Vegetation Dryness Index
12. Soil Moisture Deficit Index (SMDI)	12. Normalized Difference Water Index
13. Soil Moisture Drought Index (SMDI)	13. Remote Sensing Drought Risk Index
14. Standardized Vegetation Index (SVI)	14. Vegetation Drought Response Index
15. Computed Soil Moisture.	
16. Agro-Hydro Potential	

existing drought indicators and indices into a single measure. In general, CDIs can provide a stronger correlation with actual impacts sustained in the ground. The predicted climatic non-stationarity (IPCC, 2012) has instigated research for including future temporal patterns in drought characterization. Indeed, combining drought indices has been increasingly considered to incorporate and more effectively exploit information that is readily available and proven to be useful in field-specific drought indices (Niemeyer, 2008). It is worth mentioning that more than 20 years ago an international effort was initiated to develop a web-based environment for the computation and integration of continental and regional drought monitors leading to spatially consistent systems (Svoboda et al., 2002; Dalezios et al., 2018a). Indeed, this is an ongoing applied research effort, which has indicated, among others, the research needs for CDIs towards a global drought risk modeling system (Zargar et al., 2011; Dalezios et al., 2018a).

There are currently several regional/continental drought monitor systems and models. The objective is to develop a Global Drought Monitor (GDM), which coordinates and exchanges information towards a Global Drought Information System (GDIS) (Svoboda et al., 2002; Dalezios, 2017; Dalezios et al., 2018a). Specifically, at the present time, there are four major regional/continental models, namely the North American Drought Monitor (NADM), which consists of US Drought Monitor (USDM), Canada and Mexico; the European Drought Observatory (EDO) model; the African Drought Monitor (ADM); and the Australian Drought Monitor model (Dalezios et al., 2017a; Dalezios et al., 2017b; Dalezios et al., 2017c; Dalezios, 2018; Dalezios et al., 2018a). For example, the USDM system uses a composite of multiple drought indicators covering various short- and long-term time frames within a region or climatic zone, to develop a percent ranking methodology for drought analysis leading to a single CDI (Svoboda et al., 2002; Dalezios et al., 2018a).

6.2.2.2 Drought Early Warning Systems (DEWS)

Two remotely sensed drought indices are considered in DEWS and drought monitoring, namely the meteorological RDI (Tsakiris and Vangelis, 2005) and the agricultural VHI (Kogan, 1995), respectively. RDI is a physically based general meteorological index, which is based on precipitation and potential evapotranspiration, and provides information for the water deficit in a region. RDI can be used in a variety of climatic conditions. The computation of RDI relies on the Blaney-Criddle method (Blaney and Criddle, 1950) for potential evapotranspiration of semi-arid climates, such as the Mediterranean region, instead of the Thornthwaite method (Thornthwaite, 1948). The steps for the estimation of RDI include pre-possessing of satellite data to estimate Brightness Temperature (BT) and LST, calculation of air temperature, estimation of potential evapotranspiration, rain map extraction and remotely sensed estimation of RDI (Dalezios et al., 2012).

Similarly, agricultural drought is computed through VHI, which combines VCI and TCI, and is one of the most reliable and widely used indices (Kogan, 1995; Dalezios et al., 2014; Dalezios et al., 2018a). For the computation of VHI, remotely sensed monthly VHI images are produced. Specifically, pre-processing of the initial satellite images is conducted, involving geometric and atmospheric correction of all images. Moreover, certain filters are used for smoothing the data, leading to improved VHI values (Dalezios et al., 2014). Then, monthly VHI images are produced on a pixel basis.

For illustrative purposes, two case studies using empirical models and leading to DEWS, one based on RDI (Dalezios et al., 2012) and the other based on VHI (Dalezios et al., 2014), respectively, are briefly presented (Dalezios et al., 2018a).

DEWS: RDI. Considering all the drought episodes over 20 years in Thessaly, Greece, the cumulative monthly areal extent values (pixels) of the extreme RDI drought class are selected and plotted (Dalezios et al., 2012; Dalezios et al., 2018a). As a result, two figures are produced, namely Figure 6.3 for droughts of large areal extent and Figure 6.4 for droughts of small areal extent, respectively. Furthermore, curve fitting is conducted for each of these figures resulting in the

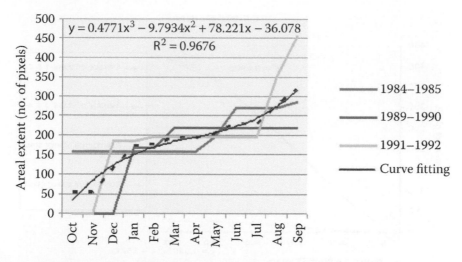

FIGURE 6.3 Cumulative large areal extent (No of pixels 8X8 km²) of extreme drought (>2.0) during drought years based on remotely sensed RDI (from Dalezios et al., 2012).

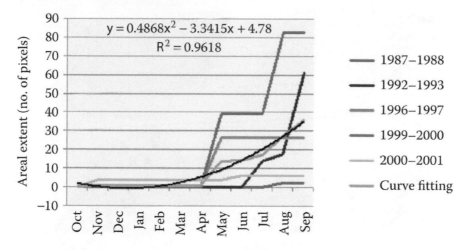

FIGURE 6.4 Cumulative small areal extent (No of pixels 8X8 km²) of extreme drought (>2.0) during drought years based on remotely sensed RDI (from Dalezios et al., 2012).

following polynomials, namely Equation (6.1) for droughts of large areal extent and Equation (6.2) for droughts of small areal extent, respectively, both with high coefficient of determination.

$$y = 0.4771x^3 - 9.7934x^2 + 78.221x - 36.078 \left(R^2 = 0.9676\right) \tag{6.1}$$

$$y = 0.4868x^2 - 3.3415x + 4.78 \left(R^2 = 0.9618\right) \tag{6.2}$$

It should be noted that for droughts of large areal extent (Figure 6.4), drought starts during the first three months of the hydrological year (October to December), whereas for droughts of small areal extent (Figure 6.5), drought starts in spring (April). This finding justifies the use of the fitted curves of Figures 6.3 and 6.4, along with the corresponding Equations (6.1) and (6.2), respectively, for drought prognostic assessment or DEWS (Dalezios et al., 2018a).

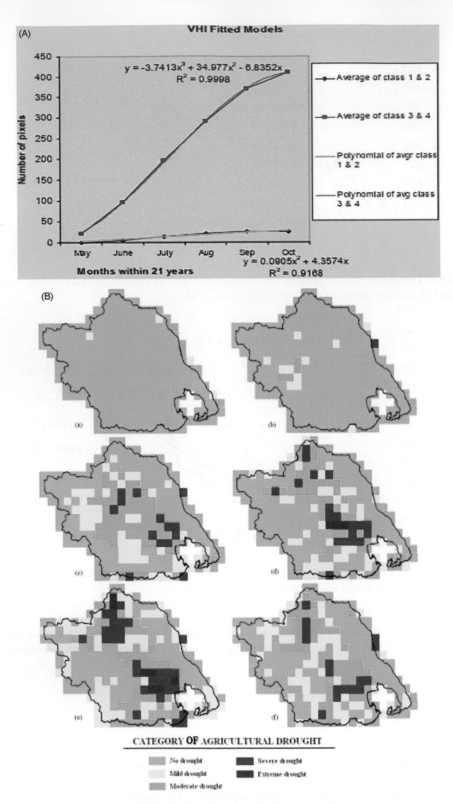

FIGURE 6.5 (A). Fitted models of cumulative areal extent (number of pixels) of average monthly drought VHI values for the two sums of severity classes. (B). VHI map of Thessaly for 6 months (April–September 1985). (a) April 1985, (b) May 1985, (c) June 1985, (d) July 1985, (e) August 1985, (f) September 1985. (from Dalezios et al., 2014).

DEWS: VHI. For agricultural drought, two cumulative monthly areal extent curves, which correspond to merged high and low severity classes of VHI, respectively, is shown in Figure 6.5 (Dalezios et al., 2014). Moreover, curve fitting is conducted for each of these curves resulting in the following polynomials, namely Equation (6.3) for high severity areal extent drought and Equation (6.4) for low severity areal extent drought, respectively, both with high coefficient of determination (Dalezios et al., 2014).

$$y = 0.0905\,x^2 + 4.3574\,x\left(R^2 = 0.9168\right) \tag{6.3}$$

$$y = -3.7413\,x^3 + 34.977\,x^2 - 6.8352\,x\left(R^2 = 0.9998\right) \tag{6.4}$$

The range of values for agricultural drought during the warm season every year is depicted in the two curves of Figure 6.5. These findings justify the use of the fitted curves of Figure 6.5, along with the corresponding Equations (6.3) and (6.4), respectively, for first-guess drought prognostic and monitoring assessment leading to DEWS (Dalezios et al., 2014; Dalezios et al., 2018a).

6.2.3 Land Desertification

Desertification is defined as "land degradation in arid, semi-arid, and dry sub-humid areas, collectively known as drylands, resulting from many factors, including climatic variations and human activities" (UNEP, 1994). There is a diachronic increase of the range and intensity (severity) of desertification in some dryland regions over the past several decades. Figure 6.6 shows a recent global dryland classification map based on the Aridity Index (AI). Although desertification processes are frequently grouped into physical, chemical, biological and anthropogenic (Perez-Trejo, 1992), the mechanism of desertification is characterized by the reduction of available soil water to the growing plants resulting in critical low plant productivity.

6.2.3.1 Factors and Processes Leading to Desertification

Desertification is the consequence of a series of significant degradation processes in semi-arid and arid regions, where water is the main limiting factor of land use performance on ecosystems (IPCC, 2012). The identification of the **factors** contributing to desertification is a complex process. There are many factors that trigger desertification, which are grouped into categories, such as climate, geology and soils, hydrology and water resources, biology and human activities. Indicatively,

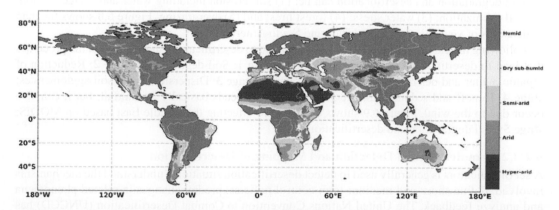

FIGURE 6.6 Geographical distribution of drylands, delimited based on the Aridity Index (AI). The classification of AI is: Humid AI > 0.65, Dry sub-humid 0.50 < AI ≤ 0.65, Semi-arid 0.20 < AI ≤ 0.50, Arid 0.05 < AI ≤ 0.20, Hyper-arid AI < 0.05. Data: TerraClimate precipitation and potential evapotranspiration (1980–2015) (from Abatzoglou et al., 2018).

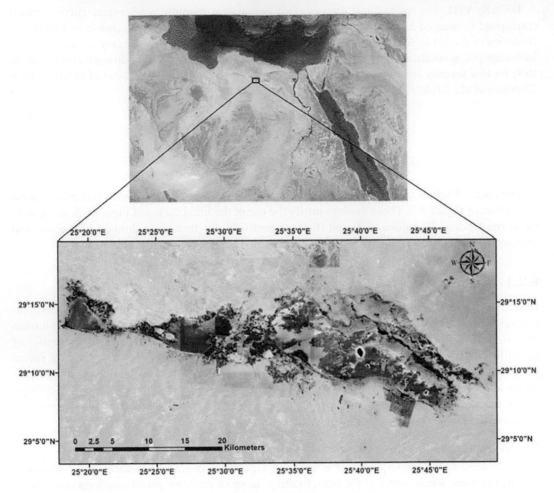

FIGURE 6.7 The Satellite Image of the Siwa Oasis, Egypt. (Source: Google Maps).

Figure 6.7 shows the satellite image of the Siwa Oasis in Egypt, where the main processes or causes of land degradation and desertification can be: (1) soil erosion including water and tillage erosion; (2) soil salinization; (3) water stress; (4) forest fires; and (5) overgrazing. Figure 6.8 shows the map of densities of vegetation cover in Africa and South America.

Following the causes and factors, which contribute to desertification, the stages and the processes that lead to desertification are briefly presented. **Stage 1**: Soil degradation; **stage 2**: Reduction of organic matter and deterioration of the soil structure; **stage 3**: Dispersion of the soil agglomerates; **stage 4**: Runoff and sediment transfer. This process depends basically on rainfall, but it can also occur due to the wind. Figure 6.9 shows a sandstorm affecting the Middle East based on MODIS; **stage 5**: Soil degradation and desertification.

6.2.3.2 Remotely Sensed Detection and Mapping of Desertification

A set of indicators is generally used to detect desertification situations, understand the mechanisms involved, define short- and medium-term control policies, predict changes in related phenomena and analyze feedback. The United Nations Convention to Combat Desertification (UNCCD) has identified the most representative and relevant indicators at the national level, i.e. the percentage of degraded land in relation to total land, the extent of vegetation covers and the soil organic carbon content. Remote sensing is a major source of information for calculating environmental

FIGURE 6.8 Map of densities of vegetation cover in Africa and South America.

indicators, such as albedo, vegetation indices, surface roughness, surface temperature or soil moisture (Sieza et al., 2019).

Albedo. Albedo is the ratio of the amount of light reflected by an object to the total amount of light it receives. It is expressed by a number between 0 (no reflected light) and 1 (total reflected light). Albedo is a major factor in energy and radiation balances as it controls the amount of solar energy absorbed by a surface. The interpretation of this value and its spatiotemporal variations is linked to desertification processes. The albedo of bare ground decreases as its water content or roughness increases. Similarly, the albedo of land with vegetation depends on the extent of vegetation cover and its chlorophyll activity.

Vegetation index. Green vegetation, due to its chlorophyll activity, thus has low reflectance in the red spectral wavelength domain (600–700 nm, denoted R), as well as high reflectance in the near-infrared domain (0.8–1.1 μm, denoted IR) due to the structure of green plant tissues (incomplete parenchyma). When chlorophyll activity decreases, R increases while IR decreases. This remarkable property of green plants has facilitated the development of several vegetation indices using these bands, with the Normalized Difference Vegetation Index (NDVI) being the most commonly used. Arid ecosystems are characterized by the relatively sparse and inactive vegetation cover. NDVI is a desertification monitoring indicator recommended by UNCCD. Similarly, the extent of vegetation cover is an indicator for monitoring the state of desertification in terrestrial areas, as recommended by UNCCD. Caution is therefore required when using NDVI, which has been shown to be

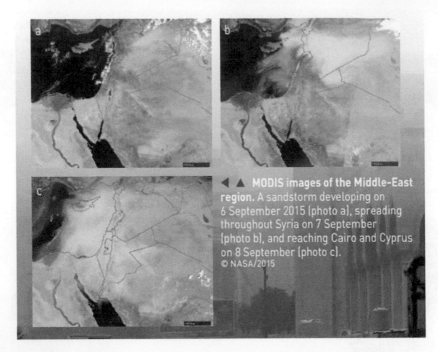

FIGURE 6.9 Wind effect: MODIS in sandstorm MODIS images of the Middle East region: a sandstorm developing on September 6, 2015.

more suitable for areas with greener medium-to-dense vegetation cover than for semi-arid regions. Figure 6.10 shows the trend in the Annual Maximum NDVI 1982–2015 (Global Inventory Modeling and Mapping Studies NDVI3g v1) calculated using the Theil-Sen estimator, which is a median based estimator and is robust to outliers, where non-dryland regions (Aridity Index > 0.65) are masked in gray (Escadafal and Bégni, 2019).

Surface roughness. The roughness parameter serves to quantify the irregularity of a surface, which can be due to the ground surface or to the vegetation cover. The extent of surface roughness depends on the extent of its irregularity, which is generally assessed by radar remote sensing. Radar emits microwaves and measures the power at which an object reflects them (so-called "backscatter"). The backscatter increases with the surface irregularity. Roughness is an important parameter in calculating sensitivity to wind and water erosion and surface water flow.

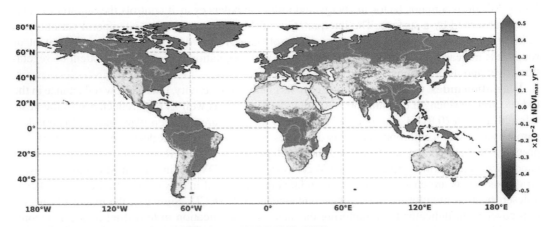

FIGURE 6.10 Trend in the Annual Maximum NDVI 1982–2015.

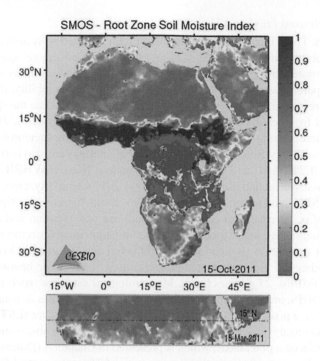

FIGURE 6.11 Daily ET: SMOS. SMOS: Map of soil moisture indices in Africa in October 2011 (1 = maximum). The resolution is around 40 km. The thumbnail image at the bottom is an excerpt for conditions in the Sahel in March 2011.

Surface temperature. The surface temperature is the result of energy exchanges at the Earth's surface and is estimated by thermal infrared radiation measurement. Its value depends directly on many factors, such as the albedo, the soil surface moisture, the amount of active vegetation present and its water stress level, which determines evapotranspiration. With geostationary meteorological satellites, surface temperature patterns may be monitored over large areas but at low resolution. Higher resolution observations (60 to 30 m) are possible with Landsat spacecraft.

Soil moisture. Soil surface moisture can be estimated by radar (ERS, Envisat, Sentinel-1 satellites) provided that the roughness impact can be isolated, or more recently by passive remote sensing in the microwave domain, but at very low resolution (ESA SMOS satellite) (Figure 6.11). Soil moisture governs energy exchanges with the atmosphere, particularly through its role in evapotranspiration. In arid and semi-arid regions this parameter is generally low and it is a desertification early warning sign.

6.3 BIOPHYSICAL HAZARDS

6.3.1 FROST AND HEATWAVES

Frost is a natural environmental risk, which occurs when the air temperature at the surface becomes equal to °C or below, with impacts on various human activities (Webb and Snyder, 2013). On the other hand, heat is a feeling of discomfort, which is related to exceptionally high temperatures. If heat lasts for several days or longer it is called a heatwave. Heatwave is characterized by higher-than-normal temperatures, low winds and generally good weather. In the tropics, heatwaves are endemic, whereas in the mid and high latitudes, they are embedded in the usual course of summer weather.

6.3.1.1 Remotely Sensed Frost and Heatwaves Quantification

Frost quantification. There is an increasing number of relevant satellite systems, which, along with their improving reliability and capabilities, lead to a fast growth of remote sensing applications to frost assessment. Remote sensing methodologies and techniques can be employed in several aspects of frost, such as preparedness and disaster prevention, warning, vulnerability, damage assessment, as well as relief. Moreover, remote sensing data and methods can delineate the spatial and temporal variability of several frost features quantitatively (Dalezios and Petropoulos, 2018). Indeed, there are two types of remote sensing systems for frost assessment, namely meteorological and environmental (or resource) satellites. Specifically, meteorological satellites are also two types, namely geostationary, such as METEOSAT, and geosynchronous, such as NOAA/AVHRR, and can contribute to operational frost assessment and monitoring (Dalezios, 2017). Similarly, environmental satellites, such as LANDSAT, SPOT and recently Quickbird or WV-2 with high to very high spatial resolution but low frequency of occurrence, can contribute to land-use classification and qualitative features of frost and less to quantitative assessments. Frost quantification involves remotely sensed temperature measurement or estimation (Becker and Li, 1990; van de Griend and Owe, 1993). Table 6.3 presents a list of indices of extreme air temperature, used for frost or heatwaves (http://cccma. seos.uvic.ca/ETCCDMI/list_27_indices.html). Meteorological satellites, such as NOAA/AVHRR, METEOSAT, or MODIS, are used. Specifically, brightness temperature is usually observed from thermal IR channels on a pixel basis, from which land surface temperature (LST) can be computed, and through regression analysis with conventional surface temperature observations air temperature can be computed, again on a pixel basis, which is presented in Chapter 4 (Dalezios et al., 2012). For illustrative purposes, Figure 6.12 shows the classification of Thessaly, central Greece, according to the minimum air temperature under conditions, which favor radiation frost in cloudless nights during the spring season in frost-prone valleys, where temperature is recorded by six meteorological stations. Brightness temperatures are recorded around midnight time (UTC) from NOAA/AVHRR satellite (Domenikiotis et al., 2006).

Heatwaves quantification. This section covers two groups of heatwaves indices, namely the thermal indices and the joint distribution indices (Dalezios, 2017).

1. **Thermal indices** (Epstein and Moran, 2006)**:** (1) The effective temperature (ET) index is defined as the temperature of a still, saturated atmosphere and it indicates the combined effects of relative humidity, air velocity, air temperature and clothing; (2) The WetBulb Globe Temperature (WBGT) is a measure of the heat stress in direct sunlight, which takes into account: temperature, humidity, wind speed, sun angle and cloud cover (solar radiation); (3) Physiologically Equivalent Temperature (PET) is defined as the physiological equivalent temperature at a given place (outdoors or indoors); (4) Universal Thermal Climate Index (UTCI) is defined as the equivalent ambient temperature (°C) of a reference environment providing the same physiological response of a reference person as the actual environment (Weihs et al., 2012; Dalezios, 2017).

2. **Joint distribution indices** (Beniston, 2009; Matzarakis and Nastos, 2011)**:** The four joint distribution climate indices, based on air temperature and precipitation, concern Cool/Dry days (CD), Cool/Wet days (CW), Warm/Dry days (WD), Warm/Wet days (WW): (1) The CD index is defined as the number of days with the daily mean air temperature (T) below the 25th percentile of the daily mean temperature (T25) and simultaneously the daily precipitation (P) below the 25th percentile of the daily precipitation (P25), thus (T < T25 and P < P25); (2) The CW index is defined as the number of days with the daily mean air temperature (T) below the 25th percentile of the daily mean temperature (T25) and simultaneously the daily precipitation (P) above the 75th percentile of the daily precipitation (P75), thus T < T25 and P > P75; (3) The WD index is defined as the number of days with the daily mean air temperature (T) above the 75th percentile of the daily mean temperature (T75) and simultaneously the daily

TABLE 6.3

Extreme air temperature indices recommended by the ETCCDMI. (See also https://www. researchgate.net/figure/Excerpt-of-the-extreme-indices-recommended-by-the-ETCCDMI-http-cccmaseosuvicca_tbl1_225107102 (from Dalezios, 2017)

ID	Indicator name	Indicator definitions	UNITS
TXx	Max Tmax	Let Tx_{kj} be the daily maximum temperatures in month k, period j. The maximum daily maximum temperature each month is then $TXx_{kj} = \max(Tx_{kj})$	°C
TNx	Max Tmin	Let Tn_{kj} be the daily minimum temperatures in month k, period j. The maximum daily minimum temperature each month is then $TNx_{kj} = \max(Tn_{kj})$	°C
TXn	Min Tmax	Let Tx_{kj} be the daily maximum temperatures in month k, period j. The minimum daily maximum temperature each month is then $TXn_{kj} = \min(Tx_{kj})$	°C
TNn	Min Tmin	Let Tn_{kj} be the daily minimum temperatures in month k, period j. The minimum daily minimum temperature each month is then $TNn_{kj} = \min(Tn_{kj})$	°C
TN10p	Cold nights	Let Tn_{ij} be the daily minimum temperature on day i in period j and let $Tn_{in}10$ be the calendar day 10th percentile centred on a 5-day window (Zhang et al., 2005b). The percentage of time is determined where $Tn_{ij} < Tn_{in}10$	Days
TX10p	Cold days	Let Tx_{ij} be the daily maximum temperature on day i in period j and let $Tx_{in}10$ be the calendar day 10th percentile centred on a 5-day window (Zhang et al., 2005b). The percentage of time is determined where $Tx_{ij} < Tn_{in}10$	Days
TN90p	Warm nights	Let Tn_{ij} be the daily minimum temperature on day i in period j and let $Tn_{in}90$ be the calender day 90th percentile centred on a 5-day window (Zhang et al., 2005b). The percentage of time is determined where $Tn_{ij} > Tn_{in}90$	Days
TX90p	Warm days	Let Tx_{ij} be the daily maximum temperature on day i in period j and let $Tx_{in}90$ be the calender day 90th percentile centred on a 5-day window (Zhang et al., 2005b). The percentage of time is determined where $Tx_{ij} > Tx_{in}90$	Days
DTR	Diurnal temperature range	Let Tx_{ij} and Tn_{ij} be the daily maximum and minimum temperature respectively on day i in period j. If I represents the number of days in j, then $$DTR_j = \frac{\sum_{i=1}^{I}(Tx_{ij} - Tn_{ij})}{I}$$	°C
FD0	Frost days	Let Tn_{ij} be the daily minimum temperature on day i in period j. Count the number of days where $Tn_{ij} < 0°C$	Days
SU25	Summer days	Let Tx_{ij} be the daily maximum temperature on day i in period j. Count the number of days where $Tx_{ij} > 25°C$	Days
ID0	Ice days	Let Tx_{ij} be the daily maximum temperature on day i in period j. Count the number of days where $Tx_{ij} < 0°C$	Days
TR20	Tropical nights	Let Tn_{ij} be the daily maximum temperature on day i in period j. Count the number of days where $Tn_{ij} > 20°C$	Days
GSL	Growing season Length	Let T_{ij} be the mean temperature on day i in period j. Count the number of days between the first occurrence of at least 6 consecutive days with $T_{ij} > 5°$ C and the first occurrence after 1st July (1st January in SH) of at least 6 consecutive days with $T_{ij} < 5°$ C	Days
WSDI*	Warm spell duration indicator	Let Tx_{ij} be the daily maximum temperature on day i in period j and let $Tx_{in}90$ be the calendar day 90th percentile centred on a 5-day window (Zhang et al., 2005b). Then the number of days per period is summed where, in intervals of at least 6 consecutive days:- $Tx_{ij} > Tx_{in}90$	Days

precipitation (P) below the 25th percentile of the daily precipitation (P25), thus T > T75 and P < P25; (4) The WW index is defined as the number of days with the daily mean air temperature (T) above the 75th percentile of the daily mean temperature (T75) and simultaneously the daily precipitation (P) above the 75th percentile of the daily precipitation (P75), thus T > T75 and P > P75. Figure 6.13 shows the bioclimate diagram for Athens for the period 1955–2001 (Matzarakis and Nastos, 2011; Dalezios, 2017).

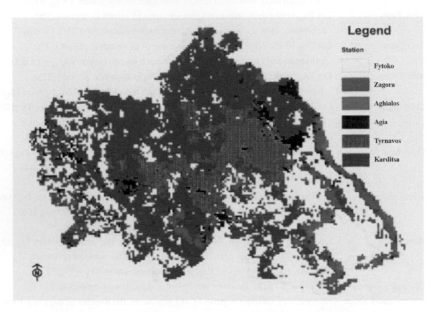

FIGURE 6.12 The classification of Thessaly, central Greece, according to the minimum air temperature under conditions which favor radiation frost in cloudless nights during the spring (from Domenikiotis et al., 2006).

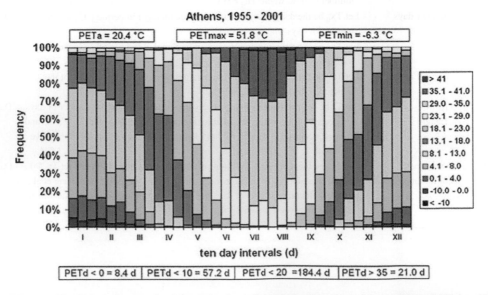

FIGURE 6.13 Bioclimate diagram for Athens for the period 1955–2001 (from Matzarakis and Nastos, 2011).

6.3.1.2 Remotely Sensed FEWS and HEWS

Frost early warning systems (FEWS). Frost monitoring is based on the development of FEWS. Indeed, frost occurrence and warning are based on temperature and its spatiotemporal variability (Pouteau et al., 2011). Indicative examples of FEWS and frost monitoring are presented based on remote sensing data and methods.

1. **Kalman filtering approach.** For monitoring and forecasting or now-casting frost a so-called phenomenological approach is used based on Kalman filtering, which belongs to estimation and control theory (Dalezios, 1987). Specifically, a one step-ahead forecasting on a pixel basis using 2D satellite temperature series images records (e.g. METEOSAT) is considered. The adopted approach comes from the optimal estimation theory (Dalezios, 1987). Specifically, as already mentioned, the system model is the so-called phenomenological temperature model, which assumes that the daily temporal variability of temperature follows a sinusoidal function.

2. **Frost risk mapping model.** A deterministic model is developed to predict frost hazard in agricultural land based on remotely sensed imagery and GIS (Louka et al., 2012). The model is based on the main factors that govern frost risk including environmental parameters, such as land surface temperature and geomorphology. Its implementation is based primarily on Earth Observation (EO) data from MODIS and ASTER polar-orbiting sensors, supported also by ancillary ground observation data. Topographical parameters required in the model include the altitude, slope, steepness, aspect, topographic curvature, and extent of the area influenced by water bodies. Additional data required include land use and vegetation classification (i.e. type and density).

Heatwaves early warning systems (HEWS). Heatwaves monitoring is based on the development of HEWS. Heatwaves hazard is based on temperature variability. Quantification of heatwaves hazard is based on the maximum temperature consideration. The database consists of series of satellite records (e.g. LANDSAT, METEOSAT, NOAA/AVHRR) from which temperature is extracted on a pixel basis. For heatwaves monitoring, a similar approach to FEWS is followed, namely the so-called phenomenological approach based on Kalman filtering (Dalezios, 1987), which assumes that the daily variability of temperature follows a sinusoidal function. There are also several other HEWS methods based on either statistical regression analysis or on physically based energy (heat) balance methods.

Case study. Once again an extreme heatwave hit Europe for a week, following the hot spell of June 2019 (Figure 6.14). Temperatures reached as high as 39–40°C, with Netherlands, Belgium and Germany recording their highest ever temperatures. Paris reached a sweltering 41°C, breaking its previous record in 1947. This animation of two images shows the land surface temperature from July 25, 2019, compared to data recorded during the previous heatwave on June 26, 2019. The map has been generated using the Copernicus Sentinel-3's Sea and Land Surface Temperature Radiometer. Weather forecasts use predicted air temperatures, whereas the satellite measures the real amount of energy radiating from Earth. As a result, this map better represents the real temperature of the land surface. Clouds are visible in white in the image, while the light blue represent snow-covered areas.

6.3.2 BIOLOGICAL AND HEALTH HAZARDS

Climate has an effect on several environmental components, such as hydrological cycle, ecosystems, food species, as well as disease agents and vectors, thus, influencing human health indirectly (Dalezios, 2017). Large short-term fluctuations in weather can cause acute adverse effects, usually resulting into increased death rates (WHO/WMO/UNEP, 1996). A global climate change is expected to affect the structure and functioning of many ecosystems, as well as the biological health of many plants and living organisms. Typical case studies are briefly presented.

FIGURE 6.14 Heatwave of July 25, 2019-Europe (from ESA, July 26, 2019) (contains modified Copernicus Sentinel data (2019), processed by ESA, CC BY-SA 3.0 IGO).

Satellite Earth Observation (EO) Data in Epidemiological Modeling. Environmental variables that affect the transmission cycle of the pathogens leading to mosquito-borne diseases (MBDs) can be estimated through EO data (Parselia et al., 2019). The state-of-the-art is examined in order to identify knowledge gaps on the latest methods, which are based on satellite EO data, for epidemiological models with emphasis on malaria, dengue and West Nile Virus (WNV). Indeed, the need has emerged to leverage furthermore new powerful modeling approaches, such as artificial intelligence and ensemble modeling. The objective is to explore new and enhanced EO sensors for the analysis of large volumes of satellite data, in order to develop accurate epidemiological models and contribute to the reduction of the burden of MBDs.

EO and indicators pertaining to determinants of health. Environmental determinants (EVDs) have been identified as a key determinant of health (DoH) for the emergence and re-emergence of several vector-borne diseases (Kotchi et al., 2016). Maintaining the ongoing acquisition of data related to EVDs at local scale and for large regions constitutes a significant challenge. Earth observation (EO) satellites offer a framework to overcome this challenge. However, EO image analysis methods commonly used to estimate EVDs are time and resource consuming. Moreover, the effectiveness of climatic variables derived from EO is limited due to variations of microclimatic conditions combined with high landscape heterogeneity. A description of DoH and EVDs is required, along with the impacts of EVDs on vector-borne diseases in the context of global environmental change. Specifically, there is a need to characterize EVDs of vector-borne diseases at local scale and its challenges, and the proposed approach is based on EO images to estimate at local scale indicators pertaining to EVDs of vector-borne diseases.

6.3.3 WILDLAND FIRES

Remote sensing is a useful tool for providing information before, during and after a wildfire through the visible, infrared and microwave portion of the electromagnetic spectrum (ESA, 2004, Dalezios et al., 2018b). Specifically, in the visible and infrared, meteorological satellites, such as the sun-synchronous NOAA-N series, or the geostationary Meteosat and geostationary operational environmental satellite (GOES), and environmental satellites, such as Landsat or satellite pour l'observation de la terre (SPOT), which are polar or near-polar low-orbit satellites, are mainly used. Moreover, in the microwave portion, there are active sensors, such as synthetic aperture radar (SAR) and light detection and ranging (LiDAR), which can provide significant data in wildfire analysis, because they can map the vertical structure of vegetation. Similarly, satellite sensors European remote sensing (ERS-1), Japanese earth resources satellite (JERS-1), and RADARSAT, as well as airborne sensors, have been widely used in wildfires (Zhang et al., 2016). New types of remote sensing systems can provide online open information for web platforms. Moreover, massive cloud computing resources and analytical tools dealing with big data sets enable the extraction of new information from environmental satellites with varying spatial resolution, such as Landsat-8 (15 m), QuickBird, Ikonos, RapidEye (5 m), Pleiades (0.5 m), or Worldview-3 (0.31 m). Specifically, remote sensing data and method can be employed in several aspects of wildfire, such as vulnerability and damage assessment, as well as relief, which involves assistance and/or intervention during or after a wildfire event. Similarly, a potential contribution of remote sensing could be focused on wildfire preparedness and disaster prevention or early warning.

6.3.3.1 Remotely Sensed Fuel Modeling

Fuel types and models. Satellite remote sensing has been proved to effectively assist in fuel type mapping of large areas with low costs. Both passive and active sensors can be used based on various algorithms with high accuracy, but each method presents both advantages and limitations. The main approach that is widely used is the extraction of the vegetation types and their reclassification into surface fuel models, based on fuel characteristics. Supervised classification, unsupervised classification, principal component analysis and tasseled cap transformation of medium resolution imagery along with ancillary data have been widely used (Palaiologou et al., 2013) at low or no cost nowadays. Multitemporal images were used to identify the different fuel types based on their phenology (Chuvieco et al., 2003). Sensors with very high resolution, such as QuickBird and Ikonos, have also been used (Mallinis et al., 2008) based mainly on object-oriented classification algorithms. However, the major drawback of passive sensors is the fact that they cannot see underneath the canopy and under cloudy conditions. Thus, fuel structural characteristics cannot be quantified for all fuel layers.

Sensors with high temporal and low spatial resolution have also been used in fuel models retrieval. For example, MODIS was used based on a pixel-explicit approach to generate maps of postfire biomass recovery and fuel development (Uyeda et al., 2015). Moreover, Fernández-Manso et al. (2014) used fraction images from linear spectral mixture analysis to estimate above-ground biomass (AGB) based on ASTER data. Spectral mixture analysis methods were also applied in hyperspectral data. Jia et al. (2006) used airborne visible infrared imaging spectrometer (AVIRIS) data to estimate forest canopy cover, discriminating among the dominant canopy types (Douglas-fir and ponderosa pine). Active sensors, such as SAR and LiDAR, can provide significant data for mapping stand biomass, since they can discriminate the vertical vegetation structure. Specifically, SAR and LiDAR can quantify the fuel potential of the stands, and they can estimate surface fuel models, crown bulk densities, and canopy dimensions (Keane, 2015).

The most promising results of fuel properties mapping are provided by airborne LiDAR. LiDAR data consist of a discrete point measurement of ranges, that is, distance between the sensor and the target. From these measurements, one can calculate elevations coupled with the strength of the return signal; and the fuel strata can be described in three dimensions. During the last decade, LiDAR data have been extensively used in conjunction with multispectral data. Erdody and Moskal (2010) used LiDAR and near-infrared imagery as explanatory variables in regression models to correlate them with field canopy fuel metrics. A small increase of the accuracy was also found by Ruiz et al. (2016), who compared data fusions of low-density LiDAR, WorldView-2, and the new and cost-free Sentinel-2. It should be mentioned that the usage of only a multispectral imagery presented an overall accuracy level below 70%. With the exception of the cost-free Sentinel-2, another low-cost approach was proposed based on LiDAR and Landsat (Hudak et al., 2016).

Fuel moisture content (FMC). Ignition and behavior of wildfires have great sensitivity in FMC (Vejmelka et al., 2016). In wildfires, it is critical to know both live and dead FMCs, which is a key parameter in risk assessment. Live fuel moisture (LFM) is based on the process of transpiration and soil water dynamics, whereas dead fuel moistures are influenced by the process of evaporation. The most common technique for the estimation of fuel moisture is the gravimetric sampling, that is, the ratio of weight of the water in the sample or material to the dry weight of the sample. Through remote sensing, vegetation conditions can be estimated directly, since water stress affects vegetation electromagnetic behavior. Both optical and thermal parts of the electromagnetic spectrum are used based on the reflectance properties of vegetation (Sow et al., 2013). Indeed, short-wave infrared (SWIR) is the most sensitive channel to water absorption (Chuvieco, 2009).

However, for vegetation analysis, the best approach is to use vegetation indices, with NDVI being the most widely used, which has been analyzed at multitemporal series (Sow et al., 2013). For example, there is a potential relationship between NDVI and FMC, due to the strong correlation between leaf chlorophyll content and leaf moisture content (Pettorelli, 2013). Usually, the estimation of live FMC leads to FMC retrieval from NDVI. Moreover, plants have high reflectance in NIR channels, and indices using NIR and SWIR estimate water content directly, since these wavelengths are related more to water absorption channels. Moreover, the combination of NDVI and LST justifies the significance of temperature on vegetation water stress detection. Opposed to optical sensors (infrared and thermal), radars have been rarely used in fuel moisture estimation.

6.3.3.2 Remotely Sensed Fire Early Warning Systems (FEWS)

FEWS. The type of vegetation is very significant in fire risk mapping and forecasting. Parameters, such as fuel type, fuel moisture, wind and topography, constitute inputs, among others, to fire danger predicting systems that have been developed for fire prevention and suppression. Two of the most effective operational and widely used systems are the National Fire Danger Rating System (NFDRS) in the United States (US Forest Service, 2012) and the Canadian Forest Fire Danger Rating System (CFFDRS) in Canada (Canadian Forest Service, 1992); both rely on remote sensing data through the estimation of indices used for prefire risk detection models. Fire danger condition systems could be broadly classified into two major groups: fire danger forecasting and monitoring systems. Indeed, most of the monitoring systems determine the danger during and/or after the period of image acquisition. A few examples and case studies are briefly presented.

A simple FEWS for wildfire detection in Malaysia has been developed (Chowdhury and Hassan, 2015), where thermal bands of MODIS are used to extract hot spot information and to generate a fire risk map, which is also based on MODIS NDVI values.

The Canadian Fire Weather Index (FWI) is one of the most widely known indices, which provides numerical rating of relative wildland fire potential in a standard fuel type on level terrains. Essentially, FWI includes six components that individually and/or collectively account for the effects of fuel moisture and wind on fire behavior. In this example, a procedure is followed that monitors fire danger over the Mediterranean daily (DaCamara et al., 2014). The developed products are based on integrated use of weather data, vegetation cover and fire activity as detected by satellite remote sensing. Moreover, statistical models based on two-parameter generalized Pareto (GP) distributions adequately fit the observed samples of fire duration and are significantly improved when the FWI is integrated as a covariate of scale parameters of GP distributions. Furthermore, FWI is calibrated based on probabilities of fire duration exceeding specified thresholds leading to five classes of fire danger. Specifically, fire duration is estimated on the basis of 15 min data provided by METEOSAT Second Generation (MSG) satellites and corresponds to the total number of hours in which fire activity is detected in a MSG pixel in a day.

The AEGIS platform is based on remote sensing (Kalabokidis et al., 2016) and incorporates early fire warning, fire planning, fire control and coordination of firefighting forces by providing online information, which is essential for wildfire management (Figure 6.15). Moreover, land use/land cover maps are produced based on field inventory data combined with high-resolution multi-spectral satellite images, namely RapidEye. Wildfire simulation tools are then developed based on these data for potential fire behavior and hazard with the minimum travel time fire spread algorithm. In addition, the system uses information, such as weather data, fire duration and ignition point, to conduct three types of fire simulations, namely single-fire propagation, point-scale calculation of potential fire behavior, and burn probability analysis, such as the FlamMap fire behavior modeling software (Kalabokidis et al., 2016). Furthermore, artificial neural networks (ANNs) are utilized for

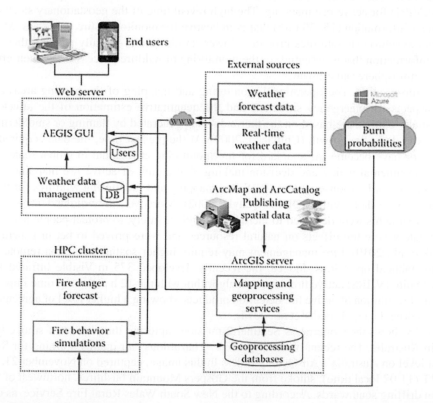

FIGURE 6.15 Architectural components of the AEGIS platform showing the linkages among data and computing resources (from Kalabokidis et al., 2016).

wildfire ignition risk assessment based on several parameters, training methods, activation functions, pre-processing methods and network structures. In addition, the combination of ANNs and expected burned area maps are used to generate integrated output map of fire hazard prediction. The system and associated computation algorithms utilize parallel processing methods, such as high-performance computing and cloud computing, which ensure the required computational power for real-time applications. In addition, all AEGIS functionalities are accessible to authorized end users through a web-based graphical user interface. Finally, an innovative smartphone application, AEGIS App, also provides mobile access to the web-based version of the system (Kalabokidis et al., 2016).

Wildfire detection. There are several automatic and semiautomatic detection and monitoring systems used globally (Alkhatib, 2014). Satellite-mounted sensors, such as Envisat's Advanced Along Track Scanning Radiometer, can measure infrared radiation emitted by fires and identify hot spots with temperatures higher than 39°C. Similarly, the NOAA's Hazard Mapping System combines remote sensing data from satellite sources, such as GOES, MODIS, and advanced very high-resolution radiometer (AVHRR), for detection of fire and smoke plume locations (NOAA, 1998). However, satellite detection is prone to offset errors, anywhere from 2 to 3 km for MODIS and AVHRR data and up to 12 km for GOES data (Ramachandran et al., 2008). Remote sensing techniques can be considered fully operational for wildfire detection (Leblon et al., 2016). Specifically, at local scale, they are mainly based on the use of visible and infrared cameras for the detection of active fires or smoke plumes, where fire detection at this scale is focused on support to wildfire fighting operations. Moreover, active sensors, such as the ERS-1, SAR and RADARSAT, can be used for monitoring fires under all-weather conditions.

On the other hand, at large scale, information is provided by geostationary satellite sensors (GOES), spinning enhanced visible and infrared imager (SEVIRI) or sun-synchronous sensors (AVHRR, advanced along track scanning radiometer (AATSR), MODIS) with operational capabilities (e.g. SEVIRI) for active fire mapping. The high revisit time of the geostationary satellites provides frequent information (15–30 min) that is indicative for monitoring fire processes. Moreover, although sun-synchronous satellites provide a lower revisit time (1–2 daily passes), they provide global fire information that is essential for the monitoring of wildfire processes and their effects on ecosystems, atmosphere and climate.

Wildfire monitoring. The operational monitoring and mapping of the burning areas are very important aspects for emergency situations and the quantitative estimation of the affected area. Methods usually applied are based on the thermal signal generated by flaming or smoldering combustion and the daily fire growth (Chuvieco, 2009). At the present time, the data acquired by the MODIS sensor have become the standard for fire monitoring at regional to global scales and are used for environmental policy and decision-making. A program of active fire mapping is operational, namely a rapid response system (RRS) for mapping and monitoring wildfires for the entire United States twice daily with MODIS (Quayle, 2002). Moreover, the wildland–urban interface (WUI) fuel treatments were designed to increase fire fighter safety, to protect people and property, and to mitigate severe fire effects on natural resources and were proved to be, in general, effective (Hudak et al., 2010). Fire management may require higher spatial resolution remote sensing products for tactical operations. Indeed, the recently developed 375 m Visible Infrared Imaging Radiometer Suite (VIIRS) active fire detection algorithm with the 12 h revisiting time improves the current spatial resolution of active fire detection products, showing a higher level of agreement with available airborne data (Schroeder et al., 2014).

Figure 6.16 shows the Copernicus Sentinel-2 mission capturing the plumes of smoke from the bushfires in Australia. The recent blazes triggered a "hazardous" air quality warning for Sydney – the highest level on Australia's Air Quality Index. In this image, captured on November 21, 2019 at 00:02 GMT (11:02 local time), smoke from the Gospers Mountain bushfires, northwest of Sydney, can be seen drifting southwards. According to the New South Wales Rural Fire Service, as of 21:00 local time, there were over 60 bush and grass fires burning in New South Wales, of which over 20 still needed to be contained. In Victoria, another 60 blazes were burning.

FIGURE 6.16 Bush fires in Australia (from ESA, November 21, 2019c).

6.3.3.3 Remote Sensing of Postfire Assessment

Available tools to create spatially explicit information on past wildland fire events are restricted to the availability of satellite data, such as the Landsat satellites where available multispectral scanner system (MSS) images exist from 1972 and Thematic Mapper images exist from 1984. Remote sensing, especially nowadays when satellites of improved spectral and spatial resolution are in orbit, is an ideal alternative for collecting and processing the required information in a relatively inexpensive and timely fashion. This technology, especially after the free release of Landsat archives from USGS and European space agency (ESA) and the availability also of Sentinel-2 by ESA, can be used to provide data of higher spatial resolutions at global scales, along with periodic spectral data in the visible and infrared part of the electromagnetic spectrum. Although satellite remote sensing appears to be a sui approach to monitor and map burned areas, this method is not free of errors, due to several existing limitations.

An example is presented for mapping the affected burned forested area of about 63 km² (Domenikiotis et al., 2002). The applied method attempted to assess the agreement of the NDVI,

(a)   *(b)*

FIGURE 6.17 Landsat TM color composites images (7,4,1) before (a) and after (b) the 1995 wildfire in Penteli, Greece. In image (a) the Penteli area appears as green brown, and in (b) as light to dark red (from Domenikiotis et al., 2002).

extracted by NOAA/AVHRR, and evaluate that by comparing it with the NDVI produced with Landsat TM data, delineating the burned areas. Figure 6.17a and 6.17b show color composites of the Landsat images before and after the forest fire, respectively. As expected, Landsat TM presented more details of the burned area. NDVI abrupt changes before and after the fire were the basis for mapping the extent of burned area and estimating the damage in near real-time (Domenikiotis et al., 2002). The magnitude of such changes depends on the amount of burned area per pixel, the vegetation density, and the dominating species. It should be emphasized that although the overall agreement of both data sets was similar, the Landsat TM was much more accurate when it came to the estimation of the burned areas only (Domenikiotis et al., 2002).

In a recent review paper on "methods for mapping forest disturbance and degradation from optical earth observation data" (Hirschmugl et al., 2017), two main change detection approaches were reported: (1) the classical image-to-image approach, and (2) the time series analysis approach. The time series approach can be very useful to create temporal profiles extracted from the spectral signal of time series satellite images, which can be used to characterize vegetation phenology, and thus support monitoring vegetation recovery in fire-affected areas. Vegetation phenology is an important element of vegetation characteristics that can be useful in vegetation monitoring, especially when satellite remote sensing observations are used.

6.4 GEOPHYSICAL HAZARDS

6.4.1 GEOLOGICAL HAZARDS

6.4.1.1 Remote Sensing of Landslides

Landslides are defined as the movement of a mass of rock, debris or earth down a slope, and can result in enormous casualties and huge economic losses (Dalezios, 2017). As a result, significant research effort has been undertaken to study landslide processes, including the type, location, stability, trigger factors, susceptibility and risk (Zhao and Lu, 2018). Different methods have been employed to generate landslide inventory maps, monitor landslide deformation and model landslide susceptibility. In addition, diverse remote sensing data have been well utilized, including optical remote sensing and spaceborne SAR, airborne light detection and ranging (LiDAR), ground-based SAR and terrestrial LiDAR, incorporating in-situ surveying measurements and observations of environmental factors.

Remote sensing data. SAR: SAR data have been widely used in landslide research, due to their broad coverage and high spatial (and to some extent, temporal) resolution, and the ability to operate under all-weather conditions. Satellite SAR data may include archived ERS and Envisat ASAR (Bozzano et al., 2017), ALOS/PALSAR (Kang et al., 2017), COSMO-SkyMed constellation, TerraSAR-X, TerraSAR-X/TanDEM-X, and Sentinel-1. **Optical Remote Sensing**: Optical remote sensing images are mainly applied to generate landslide inventory. Optical sensors may include long time-series of Landsat TM/ETM, SPOT 1-5, ASTER, IRS-1C LISS III, RapidEye, ZY-3 satellite high spatial resolution satellite images (Chen et al., 2017), and GF-1 (Bru et al., 2017). **LiDAR**: Multitemporal LiDAR images and ortho-photos can be compared to quantify landscape changes caused by an active landslide. The ground-based terrestrial laser scanner (TLS) LiDAR can produce highly detailed 3-dimensional (3D) images within minutes, allowing the study of 3D surface changes of landslides. **DEM**: Digital elevation models (DEMs) can be generated from the Indian Remote Sensing Satellite (IRS) P5 images by stereo-photogrammetry (Ren et al., 2017), and TerraSAR-X/TanDEM-X images by InSAR, respectively. Then, the DEM difference can be used to quantify the erosion and cliff recession and evaluate the landslide volume (Ren et al., 2017).

Landslides Detection. Landslide mapping (LM) has recently become an important research topic in remote sensing and geohazards. A case study is considered in the area near the Three Gorges Reservoir (TGR) along the Yangtze River in China, which is one of the most landslide-prone regions in the world, and the area has suffered widespread and significant landslide events in recent years (Figure 6.18). In a recent study, an object-oriented landslide mapping (OOLM) framework was proposed for reliable and accurate LM from "ZY-3" high spatial resolution (HSR) satellite images. The framework was based on random forests (RF) and mathematical morphology (MM) (Chen et al., 2017). RF was first applied as an object-feature information reduction tool to identify the significant features for describing landslides, and it was then combined with MM to map the landslides. Three object-feature domains were extracted from the "ZY-3" HSR data: layer information, texture and geometric features. The results showed that the feature selection (FS) method had a positive influence on effective landslide mapping (Chen et al., 2017). Figure 6.18 shows the location of the study area: (A) Site map of the Three Gorges area of the Yangtze River, China; (B) Digital elevation model (DEM) overlaid with landslides; the red hatched areas represent landslides areas. Moreover, Figure 6.18C shows the true color composite (R: 3, G: 2, B: 1) image overlaid by the segments for the Fanjiaping and Tanjiahe Landslides (Chen et al., 2017).

Moreover, rainfall-induced landslides are a major threat in the hilly and gully regions. A case study is considered in the Loess Plateau of China, where an automated detection algorithm has been developed to recognize loess landslides by combining spectral, textural and morphometric information with auxiliary topographic parameters based on high-resolution multispectral satellite data and high-precision DEM (Sun et al., 2017). In this study, loess landslides have been recognized through the development of an object-oriented method (OOA), which combines spectral, textural and morphometric information with auxiliary topographic parameters based on high-resolution multispectral satellite data (GF-1, 2 m) and a high-precision DEM (5 m) (Sun et al., 2017). The NDVI was used to discriminate landslides from vegetation cover and water areas. This approach shows great potential for quickly producing accurate results for loess landslides that are induced by extreme rainfall events in hilly and gully regions, such as the Loess Plateau. Figure 6.19 shows the location of the study area on the Loess Plateau in China: (a) areas with high concentrations of rainstorms in the Loess Plateau in July 2013; (b) variation in the elevation of the Yangou watershed; and (c) loess landslides in the GF-1 image of the study area (Sun et al., 2017).

Landslide monitoring. Time series surface deformation based on spaceborne satellite data are computed with persistent scatterer interferometry (PSI). Moreover, the integration of PS and Quasi-PS (QPS) takes advantage of two different image connection graphs in time-series InSAR

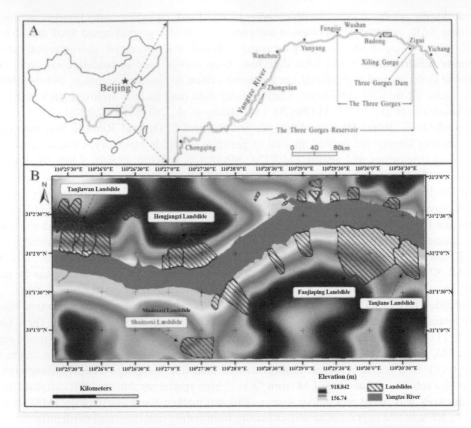

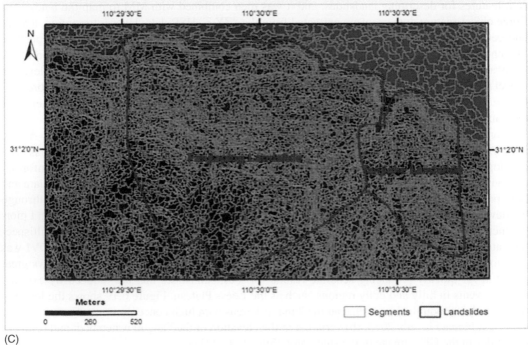

(C)

FIGURE 6.18 The location of the study area: (a) Site map of the Three Gorges area of the Yangtze River, China; (b) Digital elevation model (DEM) overlaid with landslides; the red hatched areas represent landslides areas; (c) True color composite (R: 3, G: 2, B: 1) image overlaid by the segments for the Fanjiaping and Tanjiahe Landslides (from Chen et al., 2017).

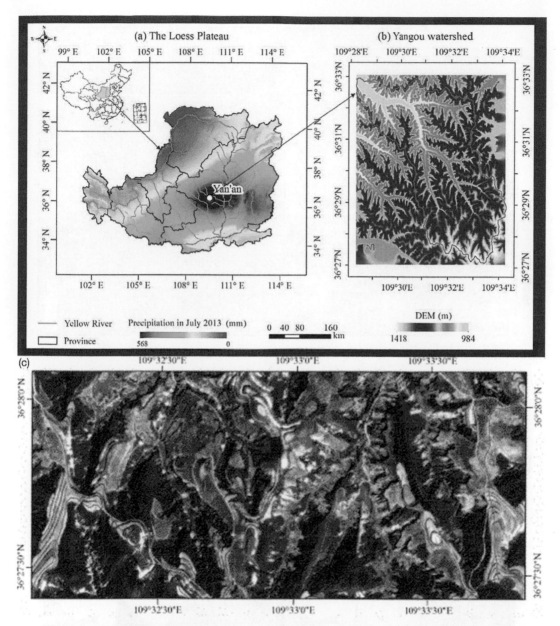

FIGURE 6.19 The location of the study area on the Loess Plateau in China: (a) areas with high concentrations of rainstorms in the Loess Plateau in July 2013; (b) variation in the elevation of the Yangou watershed; and (c) loess landslides in the GF-1 image of the study area (from Sun et al., 2017).

processing (Bozzano et al., 2017). Indeed, the vertical and the horizontal (east-west) components of the movement are derived through the combination of the dual-geometry InSAR data sets. Finally, the mapping of the temporal evolution of a landslide system was possible through the integrated use of satellite, aerial and drone images and ground-based GPS surveys. A case study is considered, which includes an extensive investigation of more than 90 landslides affecting a small river basin in Central Italy (Bozzano et al., 2017). Specifically, satellite SAR archive data (acquired by ERS and Envisat from 1992 to 2010) have been analyzed by means of A-DInSAR (Advanced Differential Interferometric Synthetic Aperture Radar) methods to evaluate landslides past displacements patterns. Moreover, multitemporal assessment of landslides' state of activity has been performed based on geomorphological evidence criteria and past ground displacement measurements obtained by A-DInSAR (Bozzano et al., 2017). Figure 6.20a shows the location of the study area. Specifically,

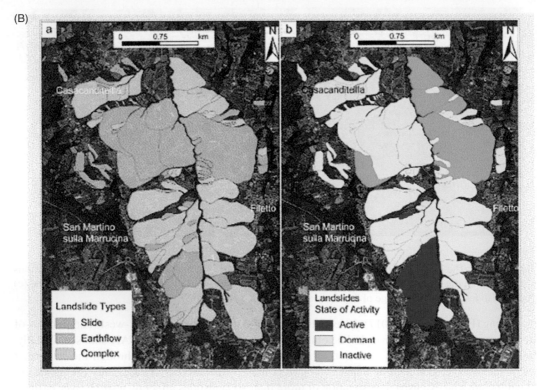

FIGURE 6.20 (A) The location of the study area. The red box identifies the study area: (a) A 3D image shows the watershed (blue line) bounding the basin under investigation. The images are extracted from Google Earth®. (B) (a) Landslides mapped in the study area; and (b) landslide state of activity inferred by geomorphological elements observed by aerial photos interpretation and field surveys. Images in background are extracted from Google Earth®(from Bozzano et al., 2017).

the red box identifies the study area: (a) A 3D image shows the watershed (blue line) bounding the basin under investigation. Figure 6.20b shows: (a) Landslides mapped in the study area; and (b) landslide state of activity inferred by geomorphological elements observed by aerial photos interpretation and field surveys (Bozzano et al., 2017).

The deformation retrieved from pre-installed corner reflectors imaged by X-band COSMO-SkyMed constellation SAR data can obtain high-precision deformation results by overcoming the temporal decorrelation (Ren et al., 2017). Moreover, potential landslides over large areas can be detected based on surface deformation maps from InSAR and small baseline subset (SBAS) InSAR (Bru et al., 2017; Kang et al., 2017). A case study of the Guanling landslide in China has been considered, where archived ALOS/PALSAR data was used to acquire the deformation prior to the landslide occurrence through stacking and time-series InSAR methods. First, the deformation structure from InSAR was compared to the potential creep bodies identified using the optical remote sensing data. Then, through observation and analysis, the deformation pattern of one creep body located within the source area can be segmented into three sections: a creeping section in the front, a locking section in the middle and a cracking section in the rear (Kang et al., 2017). The conclusion of this study was that a strong rainstorm caused a sudden sheer failure in the locking segment of one creeping body located within the source area, which triggered the Guanling landslide (Kang et al., 2017). Figure 6.21 shows the study region and SAR data coverage. (a) Location and topography of the Guanling landslide; (b) Aerial photogrammetry image after the landslide occurrence; (c) Geological

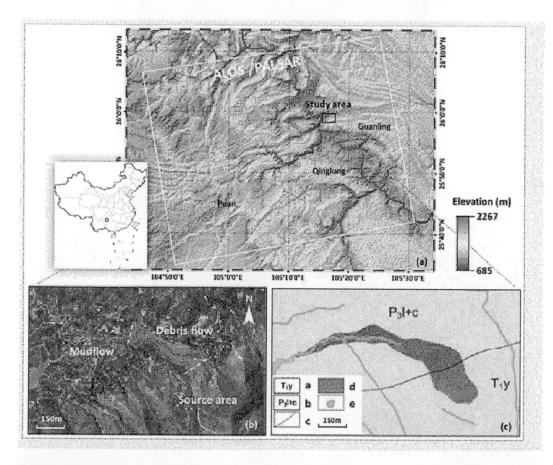

FIGURE 6.21 The study region and SAR data coverage. (a) Location and topography of the Guanling landslide; (b) Aerial photogrammetry image after the landslide occurrence; (c) Geological map of the study area, where a: Early Triassic Yelang sandstone; b: Late Permian Longtan sandy shale; c: stratigraphic boundary; d: landslide area; e: hydrographic net (Kang et al., 2017).

map of the study area, where a: Early Triassic Yelang sandstone; b: Late Permian Longtan sandy shale; c: stratigraphic boundary; d: landslide area; e: hydrographic net (Kang et al., 2017).

Landslide Susceptibility Modeling. Landslide activity can be assessed by the means of an activity matrix derived from InSAR measurements from descending and ascending passes (Bozzano et al., 2017). Moreover, a physically based model has been applied in the rainfall-induced shallow landslide susceptibility analysis, due to its ability to reproduce the physical processes governing landslide occurrence (Park et al., 2017). In this case study, fuzzy set theory, coupled with the vertex method and the point estimate method, was adopted for regional landslide susceptibility assessment, and the analysis results were compared with landslide inventory to evaluate the performance of the

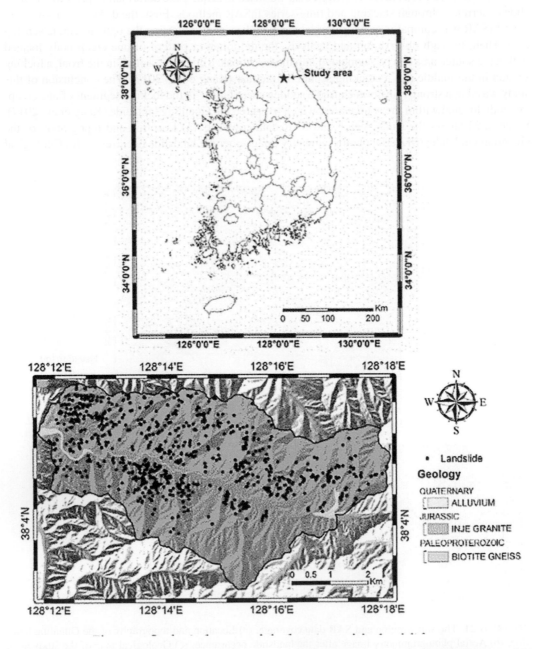

FIGURE 6.22 (A) The geological map and the locations of the landslides, and (B) The distributions of (a) slope angle and (b) elevation (from Park et al., 2017).

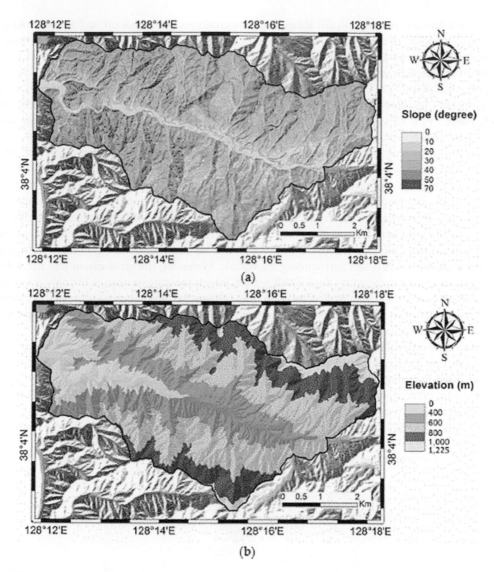

FIGURE 6.22 (*Continued*)

proposed approach (Park et al., 2017). The results indicate that the fuzzy approach shows similar performance with the probabilistic analysis, although it is more robust against variation of input parameters. As a result, at catchment scale, the fuzzy approach can respond appropriately to the uncertainties inherent in physically based landslide susceptibility analysis and is especially advantageous when the amount of quality data is very limited (Park et al., 2017). Figure 6.22a shows the geological map and the locations of the landslides, and Figure 6.22b shows the distributions of (a) slope angle and (b) elevation.

6.4.1.2 Remote Sensing of Snow Avalanches

Snow avalanches, which are commonly called avalanches, are an inherent consequence of the dynamic and variable snow cover in steep mountainous terrain (Eckerstorfer et al., 2016). Avalanches are rare natural hazards, however, they are the cause of most winter fatalities, as well as significant infrastructure loss worldwide. Thus, avalanche research includes risk reduction by trying to understand avalanche formation in space and time, relative to meteorological and snowpack triggering factors (Schweizer, 2008). Moreover, ground-based, air-, and spaceborne remote sensing can be used, which has the potential to fill these large uncertainties and temporal and spatial data gaps concerning mapping of avalanche activity, providing an improved quantitative measure of avalanche activity and dynamics (Eckerstorfer et al., 2016). This section covers spaceborne optical remote sensing of avalanches, ground radar and spaceborne radar remote sensing of avalanches.

Optical, LiDAR (Light Detection and Ranging) and Radar (Radio Detection and Ranging) sensors all use specific wavelength parts of the electromagnetic spectrum. Dry snow appears bright in the visible range (VIS: 350–750 nm wavelength) due to its high and constant reflectance (Eckerstorfer et al., 2016). Moreover, in the NIR wavelengths, ice is moderately absorptive, thus snow reflectance is most sensitive to grain size, which means that penetration is less than a few cm. In the SWIR absorption by ice is very high, thus, snow reflectance is comparably small (Eckerstorfer et al., 2016). In the microwave region of the electromagnetic spectrum 0.3 GHz–300 GHz in frequency (1 mm–1 m in wavelength), used by LiDAR and Radar sensors, the dielectric constant of ice is the driving force for permeability, where increasing liquid water content leads to decrease in backscatter and permeability (Eckerstorfer et al., 2016). In optical remote sensing, there is limiting avalanche detection potential, thus, the extent of avalanches can be detected by using contrast differences between avalanche debris and surrounding snowpack. Moreover, the areal extent and the volume of avalanches can be detected by the LiDAR sensor, which measures a change in snow cover mass balance. Furthermore, in radar, avalanche debris can be detected, since there is increased backscatter signal that stems from increased surface roughness in comparison to the undisturbed surrounding snowpack (Eckerstorfer et al., 2016).

Spaceborne optical remote sensing of avalanches. Spaceborne optical remote sensing products range from low to very high resolution, and are used in landslide mapping, earthquake deformation and flood mapping. However, spaceborne optical data in avalanche research is sparsely used. An avalanche detection algorithm was applied on spaceborne imagery from the QuickBird satellite over an area in Western Norway (Lato et al., 2012). Moreover, the potential to use Landsat-8 panchromatic images (15 m spatial resolution) has been examined to detect medium to large avalanche debris manually from the county of Troms in Northern Norway (Eckerstorfer et al., 2016). In Figure 6.23 an example is presented of a Landsat-8 image from the valley Lavangsdalen, Northern

FIGURE 6.23 Extract of a Landsat-8 image from March 18, 2014, showing the valley Lavangsdalen in Northern Norway, where the road E8 leads through avalanche-prone terrain. Six avalanches are detectable in the Landsat-8 image, of which five were validated in the field. Landsat-8 data was downloaded from http://earthexplorer.usgs.gov Copyright © NASA Earth Observatory/Landsat-8 (from Eckerstorfer et al., 2016).

Norway. Due to the easy accessibility of the valley via the main road E8, the remotely sensed avalanches could be validated.

Ground-based radar remote sensing of avalanches. Initial remote sensing studies of avalanches have been carried using a LISA (Linear SAR) ground-based SAR with 5.8 GHz frequency (C-band). Recently, Wiesmann et al. (2014) and Caduff et al. (2015) deployed a GAMMA Portable Radar Interferometer (GPRI) for two short, continuous campaigns, acquiring images with 2 and 3 min repeat intervals, respectively (Figure 6.24). It was found that coherence was strongly affected due to increasing liquid water content induced by solar radiation already after 15 min. Unwrapping the interferograms, they were able to assign displacement in the line-of-sight (LOS) to snow creep (Caduff et al., 2015).

Spaceborne radar remote sensing of avalanches. The E-SAR, which has been developed by the German Aerospace Center (DLR), seems to be a potential airborne SAR system, capable of detecting avalanches. E-SAR operates in four frequency bands (X-, C-, L- and P-band), covering a wavelength range from 3–85 cm (Eckerstorfer et al., 2016). The potential of spaceborne SAR to detect avalanche debris is considered by using ERS 1/2, C-band SAR data. Bühler et al. (2014) used two TerraSAR-X, X-band SAR scenes to detect avalanches based on backscatter change detection between the two acquisition dates. In the resulting RGB composite, five areas with avalanche debris

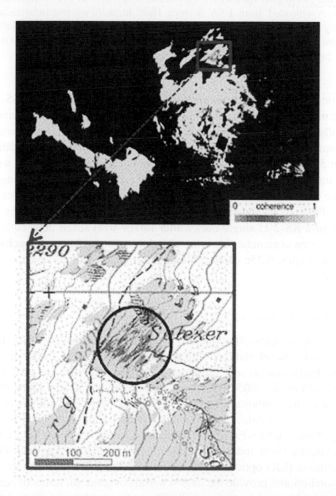

FIGURE 6.24 Radar coherence map showing sudden spatially localized coherence loss between 11:00 and 11:05 am on February 23, 2014 in Davos, Switzerland, due to an avalanche release (from Bühler et al., 2014) Copyright Montana State University Library (http://arc.lib.montana.edu/snow-science/item.php?id=2036). (from Eckerstorfer et al., 2016).

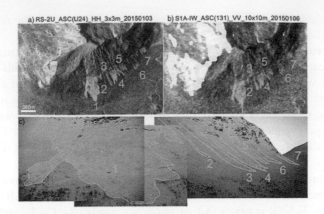

FIGURE 6.25 Avalanche debris detection using satellite borne SAR data. a) Seven avalanche debris are detectable in the single backscatter Radarsat-2 Ultrafine Mode image; b) Seven avalanche debris are also detectable in the single backscatter Sentinel-1A IW Mode image, with an acquisition date three days later than the RS-2 U image; c) Oblique photograph with all seven avalanches debris visualized with green outlines. Radarsat-2 data was acquired under the Norwegian Rardarsat-2 agreement. Copyright ©: MDA/NSC/KSAT, 2014 and 2015. Sentinel-1 data was downloaded from ESA Science Hub: https://www.researchgate.net/figure/Avalanche-debris-detection-using-satellite-borne-SAR-data-a-Seven-avalanche-debris-are_fig9_284095768. Copyright © European Space Agency, 2014 and 2015. (For interpretation of the references to color in this figure legend, the reader is referred to the web version of this article.) (Eckerstorfer et al., 2016).

could be visually identified. Moreover, Eckerstorfer et al. (2016) collected 12 RS-2 U images during an avalanche cycle in March 2014 in the county of Troms, Northern Norway. Small-sized avalanche debris was manually detected using the high backscatter contrast between avalanche debris and surrounding, undisturbed snowpack. In the single backscatter images, 546 features were initially detected, of which 57 were counted multiple times in overlapping images, 7 were radar shadows or layover effects and 15 were eliminated by a topographic GIS model, distinguishing avalanche from non-avalanche terrain (Eckerstorfer et al., 2016). Of the remaining 467 features, classified as avalanche debris, 37% were validated by fieldwork, or in optical remote sensing data. Figure 6.25a shows an example of a RS-2 U single backscatter image with seven detectable avalanche debris, Figure 6.25b shows seven avalanche debris are also detectable in the single backscatter Sentinel-1A IW Mode image, and Figure 6.25c shows oblique photograph with all seven avalanches debris, all validated in the field.

6.4.2 TECTONIC HAZARDS: EARTHQUAKES

Satellite remote sensing systems offer spatially continuous information of the tectonic landscape. Combined with ground network data, remote sensing can contribute to the understanding of specific fault systems, displacements and validation of slip models that are cast in a regional setting of tectonic strain (Tralli et al., 2005). Moreover, remote sensing observations provide information on the energy released by earthquakes, transfer of stress between fault systems or fault failure. In addition, satellite data can detect global gravity perturbations produced by earthquakes, such as the Challenging Mini-satellite Payload (CHAMP), Gravity Recovery and Climate Experiment (GRACE) and the Gravity Field and Steady-State Ocean Circulation Explorer (GOCE).

The lowest resolution (LR) optical (e.g. LANDSAT) and SAR (e.g. Envisat) satellites have 15 to 30 m spatial resolution and provide the largest aerial coverage (in terms of scene size), such that fewer scenes are required to evaluate the overall effects of an earthquake over a large area. Similarly, the moderate resolution (MR) optical satellites have higher resolution 2.5 to 15 m, but smaller scene sizes. Indeed, the high-resolution (HR) optical satellites provide the most detailed data, but, due to the smaller aerial coverage (100 km² per scene), it is difficult and expensive to obtain high-resolution

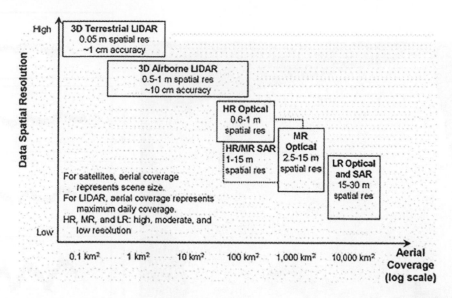

FIGURE 6.26 Comparison of resolution and aerial coverage of various types of remote sensing data (Rathje and Adams, 2008).

imagery over the entire area affected by an earthquake. A comparison of the different optical and SAR satellite sensors and LiDAR instruments is presented with respect to aerial coverage and data resolution/quality (Figure 6.26) (Rathje and Adams, 2008). Moderate and high-resolution SAR is available with TerraSAR-X (https://directory.eoportal.org/web/eoportal/satellite-missions/t/terrasar-x). This sensor allows for more detailed SAR imaging and the ability to obtain detailed imagery when clouds may hinder acquisition of moderate to high-resolution optical imagery. Moreover, airborne LiDAR can produce three-dimensional (3D) digital terrain models (DTM) over large to moderate areas (1–100 km² per day), but at significant cost. Finally, terrestrial LiDAR can provide the most detailed, three-dimensional models of damage, however, time is required for a large area.

Moreover, interferometric synthetic aperture radar (InSAR), can be combined with GPS and modern digital seismic data to provide spatially continuous deformation with sub-centimeter accuracy (Tralli et al., 2005). Specifically, Figure 6.27 presents a spatiotemporal comparison of an InSAR-derived deformation time series with GPS network time series data, which allows a better understanding of the dynamics of the area. It is recognized that InSAR methods constitute an integral component of an Earth observation capability for seismic risk mitigation and response (Tralli et al., 2005).

Building damage methods can be classified in two categories: multitemporal techniques that evaluate the changes between the pre- and post-event data and mono-temporal techniques that interpret only the post-event data (Dong and Shan, 2013). In both categories, remote sensing data can be used, including optical, LiDAR and SAR data. Moreover, high-resolution satellite remote sensing with InSAR and airborne LiDAR is subject of research for imaging and classifying the built environment through land cover and DTMs (Tralli et al., 2005). This contributes to vulnerability assessments and to rapid post-disaster damage assessment. Specifically, rapid damage assessment is critical for effective allocation of disaster response and relief resources. Available satellite remote sensing systems, such as Ikonos, OrbView or QuickBird, are used frequently in operational disaster management and research.

A case study is considered in Southeast France, which was hit by a magnitude 5 earthquake with tremors felt between Lyon and Montélimar (Figure 6.28). The Copernicus Sentinel-1 radar mission was used to map the shift of the ground as a result of the quake. Although earthquakes are unusual in this part of France, part of the Auvergne-Rhône-Alpes region was rocked by a quake on November 11, 2019 at noon (local time), resulting in evacuation and buildings damaged. A Copernicus Sentinel-1 was obtained on November 12, 2019, one day after the event, and was ready to process on ESA's Geohazards Exploitation Platform (GEP), which is a cloud-based processing environment

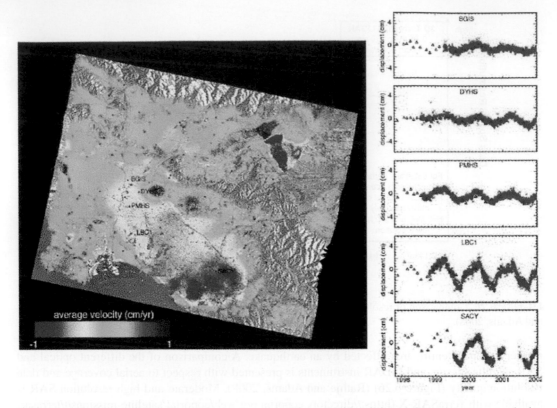

FIGURE 6.27 (Left) InSAR time series inversion map of the radar line-of-sight deformation average velocity, overlying the multi-look SAR amplitude image (gray scale). Small black squares mark Southern California Integrated GPS Network (SCIGN) GPS site locations. (Right) InSAR time series (black triangles) at selected points. Plots compare the InSAR time series to SCIGN GPS (red*) time series for indicated sites (produced from European Space Agency remote sensing data, ERS-1 and ERS-2. Figure courtesy of P. Lundgren, NASA/JPL. (Tralli et al., 2005).

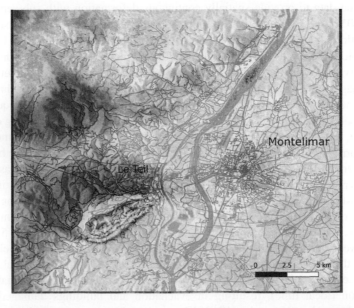

FIGURE 6.28 Southeast France: a magnitude 5 earthquake with tremors felt between Lyon and Montélimar (ESA, November 17, 2019c).

with on-demand terrain motion mapping services (Figure 6.28, ESA November 17, 2019c). The intensity of the ground motion was felt by the inhabitants and measured from space, where the epicenter was shallow and put at between 1 km and 3.5 km below the surface by seismic data.

6.4.3 TECTONIC HAZARDS: VOLCANOES

Several ground and space-based remote sensing instruments observe volcanic activity worldwide. Indeed, a multi-sensor approach is required, since there is no single system that can provide a comprehensive description of eruptive activity. Case studies are considered for remote sensing volcanic ash retrievals and volcaniclastic deposits.

Case study 1. This case study integrates infrared and microwave volcanic ash retrievals obtained from the geostationary METEOSAT Second Generation (MSG)-Spinning Enhanced Visible and Infrared Imager (SEVIRI), the polar-orbiting Aqua-MODIS and ground-based weather radar (Corradini et al., 2016). The expected outcomes are improvements in satellite volcanic ash as follows: (1) cloud retrieval in terms of altitude, mass, aerosol optical depth and effective radius; (2) the generation of new satellite products, such as ash concentration and particle number density in the thermal infrared; and (3) better characterization of volcanic eruptions, such as plume altitude, total ash mass erupted and particle number density from thermal infrared to microwave (Corradini et al., 2016). The Mt. Etna (Sicily, Italy) volcano lava fountaining event of November 23, 2013 was considered a test case. The results of the integration show the presence of two volcanic cloud layers at different altitudes. Measurements are obtained from the geostationary Spinning Enhanced Visible and Infrared Imager (SEVIRI) aboard the EUMETSAT-METEOSAT Second Generation (MSG) (Figure 6.29), the polar Moderate Resolution Imaging

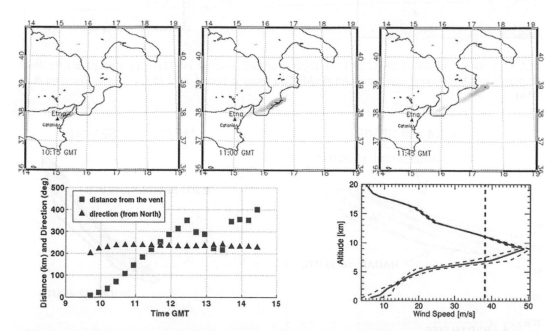

FIGURE 6.29 Top panels: ash cloud center of mass at different times. Bottom left plot: distance (red squares) and direction (blue triangles, defined as the direction from which the wind was blowing) of the volcanic ash cloud center of mass considering the different SEVIRI images. Bottom right plot: comparison between the wind speed retrieved (vertical dashed line) with the National Centers for Environmental Prediction (NCEP)/ National Center for Atmospheric Research (NCAR) mean wind speed profile extracted in the area occupied by the volcanic cloud (red solid line). The dashed blue lines represent the standard deviation of the mean NCEP/ NCAR wind speed profile (from Corradini et al., 2016).

Spectroradiometer (MODIS) aboard the NASA-Aqua (Figure 6.30) and the ground-based X-band weather radar (DPX4) systems, and they are merged to fully characterize volcanic ash clouds from source to atmosphere. Measurements collected from the Infrared Atmospheric Sounding Interferometer (IASI) aboard EUMETSAT-Metop, a ground visible (VIS) camera, the HYSPLIT model and ground tephra deposits are used to support the analysis outcomes. A presentation of the merging systems is provided in Figure 6.31.

Case study 2. Geological surfaces on Earth can be characterized by Orbital thermal infrared (TIR) remote sensing. However, the accuracy of TIR compositional analysis is reduced if the deposition of material from volcanic activity results in bedrock surfaces becoming significantly mantled over time. Indeed, Apparent Thermal Inertia (ATI) is the measure of the resistance to temperature change and has been used to determine parameters, such as grain/block size, density/mantling, and the presence of subsurface soil moisture/ice. The objective of this case study is to document the quantitative relationship between ATI derived from orbital visible/near-infrared (VNIR) and thermal infrared (TIR) data and tephra fall mantling of the Mono Craters and Domes (MCD) in California (Price et al., 2016). The ATI

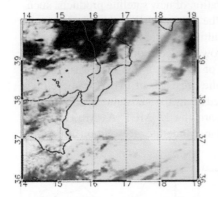
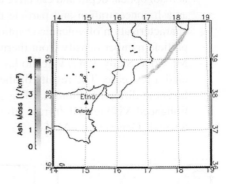

FIGURE 6.30 RGB image (R:28, G:29, B:31, upper left panel), volcanic ash Ma (upper right panel), Re (lower left panel) and AOD (lower right panel) for the MODIS image collected at 12:45 UTC (from Corradini et al., 2016).

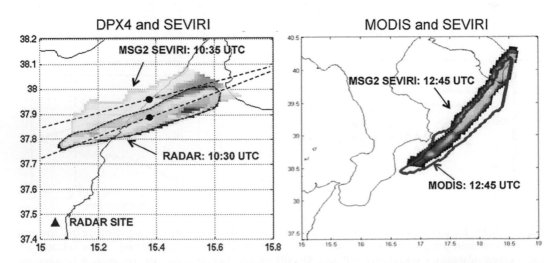

FIGURE 6.31 Example of parallax displacement when retrievals from SEVIRI and DPX4 radar (left), and MODIS and SEVIRI (right) are compared to each other at nearly simultaneous acquisition times (from Corradini et al., 2016).

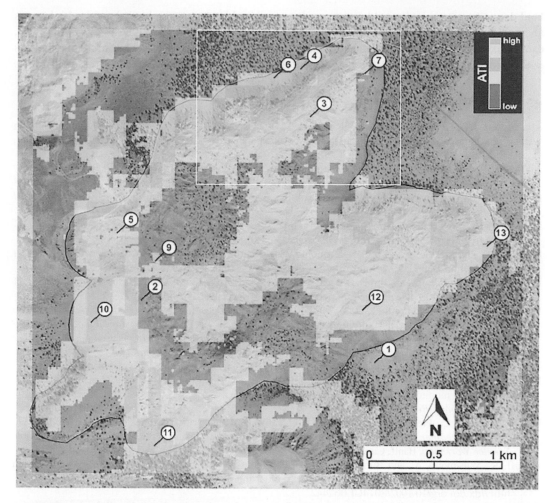

FIGURE 6.32 Relative apparent thermal inertia image of North Coulee (outlined in black) derived from July 10, 2011 ASTER data. The colorized scale image is overlain on a Digital Globe (within Google Earth) high-resolution color image for detail. This image was used as a guide for later fieldwork and the numbered pins refer to visited field sites. The white box denotes the area of ATI analysis. Data courtesy of NASA/GSFC/METI/Japan Space Systems. US/Japan ASTER Science Team (from Price et al., 2016).

data were created from two images collected about 12 h apart by the Advanced Spaceborne Thermal Emission and Reflection Radiometer (ASTER) instrument. This is presented in Figure 6.32.

6.5 SUMMARY

Earth observation (EO) provides one of the most promising avenues for providing information at global, regional and even basin scales related to environmental hazards. Remote sensing is considered an essential tool for monitoring changes in the earth's surface, atmosphere and oceans, and is increasingly used in early warning systems for hazardous events. Indeed, satellite remote sensing, combined with ground measurement networks and various types of modeling, has now emerged as a powerful operational and applicable tool to provide information on environmental hazards and its dynamics.

In this chapter, the remote sensing potential has been considered in environmental hazards affecting several sectors of the economy under increasing climate variability. Specifically, the environmental hazards examined include hydrometeorological hazards, such as floods and excess rain, droughts and land desertification; biophysical hazards, such as frost and heatwaves, biological and health hazards and wildland fires; and geophysical hazards, such as geological hazards (landslides, snow avalanches), earthquakes and volcanoes. The emphasis has been placed on remotely sensed hazard methodologies in the three stages, namely early warning systems (before), monitoring (during) and assessment (after).

REFERENCES

Abatzoglou, J.T., Dobrowski, S.Z., Parks, S.A., and Hegewisch, K.C., 2018. TerraClimate, a high-resolution global dataset of monthly climate and climatic water balance from 1958–2015. *Scientific Data*, 5, 170191, https://doi.org/10.1038/sdata.2017.191

Alkhatib, A.A.A., 2014. A review on forest fire detection techniques. Hindawi Publishing Corporation. *International Journal of Distributed Sensor Networks* 597368, 12. http://dx.doi.org/10.1155/2014/597368

Becker, F., and Li, Z.L., 1990. Towards a local "split window" method over land surface. *International Journal of Remote Sensing*, 3, 369–393.

Beniston, M., 2009. Trends in joint quantiles of temperature and precipitation in Europe since 1901 and projected for 2100. *Geophysical Research Letters*, 36, L07707.

Blaney, H.F., and Criddle, W.D., 1950. Determining Water Requirements in Irrigated Areas from Climatological and Irrigation Data. USDA Soil Conservation Service, Technical Paper, No. 96, 48p.

Bozzano, F., Mazzanti, P., Perissin, D., Rocca, A., Pari, P., and Discenza, M., 2017. Basin scale assessment of landslides geomorphological setting by advanced InSAR analysis. *Remote Sensing*, 9, 267.

Brakenridge, R., and Anderson, E., 2006. MODIS-based Flood Detection, Mapping and Measurement: The Potential for Operational Hydrological Applications, pp. 1–12, in *Transboundary Floods: Reducing Risks through Flood Management, NATO Science Series IV Earth and Environmental Sciences, NATO Advanced Research Workshop on Transboundary Floods— Reducing Risks through Flood Management*, Volume 72. Editors: J. Marsalek, G. Stancalie and G. Balint. NATO: Baile Felix, Romania.

Bru, G., González, P.J., Mateos, R.M., Roldán, F., Herrera, G., Béjar-Pizarro, M., and Fernández, J., 2017. A-DInSAR monitoring of landslide and subsidence activity: A case of urban damage in Arcos de la Frontera, *Spain. Remote Sensing*, 9, 787.

Bühler, Y., Bieler, C., Pielmeier, C., Wiesmann, A., Caduff, R., Frauenfelder, R., Jaedicke, C., and Bippus, G., 2014. *All-weather avalanche activity monitoring from space? Proceedings, 2014 International Snow Science Workshop*, Banff, Canada, 795–802.

Caduff, R., Wiesmann, A., and Bühler, Y., 2015. Continuous monitoring of snowpack displacement at high spatial and temporal resolution with terrestrial radar interferometry. *Geophysical Research Letters*, 01. http://dx.doi.org/10.1002/2014GL062442

Canadian Forest Service, (1992). *Development and Structure of the Canadian Forest Fire Behaviour Prediction System. Canadian Forest Service, Information Report ST-X-3*. ONT, Ottawa, 63

Chen, T., Trinder, J.C., and Niu, R., 2017. Object-oriented landslide mapping using ZY-3 satellite imagery, random forest and mathematical morphology, for the Three-Gorges Reservoir, *China. Remote Sensing*, 9, 333.

Chowdhury, E.H., and Hassan, Q.K., 2015. Operational perspective of remote sensing-based forest fire danger forecasting systems. *ISPRS Journal of Photogrammetry and Remote Sensing*, 104, 224–236.

Chuvieco, E., 2009. *Earth Observation of Wildland Fires in Mediterranean Ecosystems*. Springer, Dordrecht, the Netherlands, 87.

Chuvieco, E., Riano, D., van Wagtendonk, J., and Morsdof, F., 2003. Fuel Loads and Fuel Type Mapping, pp. 119–142, in *Wildland Fire Danger Estimation and Mapping: The Role of Remote Sensing Data*. E. Chuvieco. World Scientific, Singapore.

Corradini S., Montopoli, M., Guerrieri, L., Ricci, M., Scollo, S., Merucci, L., Marzano, F.S., Pugnaghi, S., Prestifilippo, M., Ventress, L.J., Grainger, R.G., Carboni, E., Vulpiani, G., and Coltelli, M., 2016. A multi-sensor approach for volcanic ash cloud retrieval and eruption characterization: The 23 November 2013 Etna Lava Fountain. *Remote Sensing*, 40, 58, doi:10.3390/rs8010058

DaCamara, C.C., Calado, T.J., ErmidaA, S.L., Trigo, I.F., Amraoui, M., and Turkman, K.F., 2014. Calibration of the fire weather index over Mediterranean Europe based on fire activity retrieved from MSG satellite imagery. *International Journal of Wildland Fire*, 23, 7, 945–958.

Dalezios, N.R., 1987. *Development of a Watershed System Using Estimation Theory. Proceedings, 3rd Greek Hydrotechnical Conference, Greek Hydrot Union*, Thessaloniki, 621–630.

Dalezios, N.R. (ed.), 2017. *Environmental Hazards Methodologies for Risk Assessment and Management*. IWA, London, UK, 534.

Dalezios, N.R., 2018. Drought and Remote Sensing: An Overview, pp. 3-32, in *Remote Sensing of Hydrometeorological Hazards*. Editors: G.P. Petropoulos and T. Islam. Taylor and Francis.

Dalezios, N.R., and Eslamian, S., 2016. Regional design storm for Greece within the flood risk management framework. *International Journal of Hydrology Science and Technology (I.J.H.S.T)*, 6, 1, 82–102.

Dalezios, N.R., and Petropoulos, G.P., 2018. Frost and Remote Sensing: An Overview of Capabilities, pp. 105–128 in *Remote Sensing of Hydrometeorological Hazards*. Editors: G.P. Petropoulos and T. Islam. Taylor and Francis.

Dalezios, N.R., Blanta, A., and Spyropoulos, N.V., 2012. Assessment of remotely sensed drought features in vulnerable agriculture. *Natural Hazards and Earth System Sciences*, 12, 3139–3150.

Dalezios, N.R., Dunkel, Z., and Eslamian, S., 2017a. *Meteorological Drought Indices: Definitions, in Handbook of Drought and Water Scarcity (HDWS)*, Volume 1. Editor: S. Eslamian. Taylor & Francis Group, Abingdon, UK.

Dalezios, N.R., Tarquis, A.M., and Eslamian, S., 2017c. Drought Assessment and Risk Analysis, in *Handbook of Drought and Water Scarcity (HDWS)*. Editor: S. Eslamian. Abingdon, UK: Taylor & Francis Group.

Dalezios, N.R., Spyropoulos, N.V., and Eslamian, S., 2017d. Remote Sensing in Drought Quantification and Assessment, in *Handbook of Drought and Water Scarcity (HDWS)*, Volume 1. Editor: S. Eslamian. Abingdon, UK: Taylor & Francis Group.

Dalezios, N.R., Dercas, N., and Eslamian, S., 2018a. Water scarcity management: Part 2: Satellite-based composite drought analysis. *International Journal of Global Environmental*, 172/3, 267–295.

Dalezios, N.R., Petropoulos, G.P., and Faraslis, I., 2020. Concepts and Methodologies of Environmental Hazards Affecting Agriculture and Agroecosystems, in Techniques for Disaster Risk Management and Mitigation. *AGU-Wiley*, 3-22.

Dalezios, N.R., Blanta, A., Spyropoulos N.V., and Tarquis, A.M., 2014. Risk identification of agricultural drought for sustainable agroecosystems. *Natural Hazards and Earth System Sciences*, 14, 2435–2448.

Dalezios, N.R., Gobin, A., Tarquis, A.M., and Eslamian, S., 2017b. Agricultural Drought Indices: Combining Crop, Climate and Soil Factors, in *Handbook of Drought and Water Scarcity (HDWS)*, Volume 1. Editor: S. Eslamian. Abingdon, UK: Taylor & Francis Group.

Dalezios, N.R., Kalabokidis, K., Koutsias, N., and Vasilakos, C., 2018b. Wildfires and Remote Sensing: An Overview, pp. 211–236, in *Remote Sensing of Hydrometeorological Hazards*. Editors: Prof. G.P. Petropoulos and T. Islam. Taylor and Francis.

Destro, E., Nikolopoulos, E.I., Creutin, J.-D. and Borga, M., 2017. Floods, pp. 137–176, in *Environmental Hazards Methodologies for Risk Assessment and Management*. Editor: N.R. Dalezios. IWA, London, UK.

Domenikiotis, C., Dalezios, N.R., Loukas, A., and Karteris, M., 2002. Agreement assessment of NOAA/AVHRR NDVI with Landsat TM NDVI for mapping burned forested areas. *International Journal of Remote Sensing*, 23, 4235–4246.

Domenikiotis C., Spiliotopoulos, M., Kanellou, E., and Dalezios, N.R., 2006. *Classification of NOAA/AVHRR Images for Mapping of Frost Affected Areas in Thessaly, Central Greece. International Symposium "GIS and Remote Sensing: Environmental Applications", University of Thessaly (UTH)*, Volos, November 7–9, 2003, 25–32.

Dong L., and Shan, J., 2013. A comprehensive review of earthquake-induced building damage detection with remote sensing techniques. *ISPRS Journal of Photogrammetry and Remote Sensing*, 84, 85–99.

Eckerstorfer M., Bühler, Y., Frauenfelder, R., and Malnes, E., 2016. Remote sensing of snow avalanches: Recent advances, potential, and limitations. *Cold Regions Science and Technology*, 121, 126–140.

Epstein, Y., and Moran, D.S., 2006. Thermal comfort and the heat stress indices. *Industrial Health*, 44, 3, 388–398.

Erdody, T.L., and Moskal, L.M., 2010. Fusion of LiDAR and imagery for estimating forest canopy fuels. *Remote Sensing of Environment*, 114, 4, 725–737.

ESA, 2004. Satellites are tracing Europe's forest fire scars, http://www.esa.int/Our_Activities/Observing_the_Earth/Satellites_are_tracing_Europe_s_forest_fire_scars

ESA, 2019a. July 2019. Extreme heatwave, https://www.esa.int/ESA_Multimedia/Images/2019/07/Extreme_heatwave

ESA, 2019b. November 2019. Bushfires rage in Australia, https://www.esa.int/ESA_Multimedia/Images/2019/11/Bushfires_rage_in_Australia

ESA, 2019c. November 2019. French earthquake interferogram, https://www.esa.int/ESA_Multimedia/Images/2019/11/French_earthquake_interferogram

Escadafal, R., and Bégni, G., 2019. Desertification monitoring by remote sensing. Les dossiers thématiques du CSFD. N°12, March 2019. CSFD/Agropolis International, Montpellier, France. 44pp.

Fernández-Manso, O., Fernández-Manso, A., and Quintano, C., 2014. Estimation of aboveground biomass in Mediterranean forests by statistical modelling of ASTER fraction images. *International Journal of Applied Earth Observation and Geoinformation*, 31, 1, 45–56.

Giustarini, L., Hostache, R., Matgen, P., Schumann, G.J.P., Bates, P.D., and Mason, D.C., 2013. A change detection approach to flood mapping in urban areas using TerraSAR-X. *IEEE Transactions on Geoscience and Remote Sensing*, 51, 4, 2417–2430, doi:10.1109/TGRS.2012.2210901

Hirschmugl, M., Gallaun, H., Dees, M., Datta, P., Deutscher, J., Koutsias, N., and Schardt, M., 2017. Methods for mapping forest disturbance and degradation from optical earth observation data: A review. *Current Forestry Reports*, 3, 1–14.

Hudak, A.T., Jain, T.B., Morgan, P., and Clark J.T., 2010. *Remote sensing of WUI fuel treatment effectiveness following the 2007 wildfires in central Idaho. Proceedings of 3rd Fire Behavior and Fuels Conference, October 25–29, Spokane, Washington, USA.* International Association of Wildland Fire, Birmingham, AL, 1–11.

Hudak, A.T., Bright, B.C., Pokswinski, S.M., Loudermilk, E.L., O'Brien, J.J., Hornsby, B.S., Klauberg, C., and Silva, C.A., 2016. Mapping forest structure and composition from low-density LiDAR for informed forest, fuel, and fire management at Eglin Air Force Base, Florida, USA. *Canadian Journal of Remote Sensing*, 42, 5, 411–427.

IPCC, 2012. Managing the Risks of Extreme Events and Disasters to Advance Climate Change Adaptation. Special Report of IPCC, 582p.

Jia, G.J., Burke, I.C., Kaufmann, M.R., Goetz, A.F.H., Kindel, B.C., and Pu, Y., 2006. Estimates of forest canopy fuel attributes using hyperspectral data. *Forest Ecology and Management*, 229, 27–38.

Kalabokidis, K., Ager, A., Finney, M., Athanasis, N., Palaiologou, P., and Vasilakos, C., 2016. AEGIS: A wildfire prevention and management information system. *Natural Hazards Earth System Sciences*, 16, 643–661.

Kang, Y. Zhao, C., Zhang, Q., Lu, Z., and Li, B., 2017. Application of InSAR techniques to an analysis of the Guanling landslide. *Remote Sensing*, 9, 1046.

Keane, R.E., 2015. *Wildland Fuel Fundamentals and Applications*. Switzerland: Springer International Publishing.

Kogan, F.N. 1995. Application of vegetation index and brightness temperature for drought detection, *Advances in Space Research*, 15, 91–100.

Kotchi, S.O., Brazeau, S., Ludwing, A., Aube, G., and Berthiaume, P., 2016. *Earth Observation and Indicators pertaining to determinants of health – An approach to support local scale characterization of environmental determinants of vector-borne diseases. Proc. "Living Planet Symposium 2016"*, Prague, Czech Republic, May 9–13, 2016, 8.

Lato, M.J., Frauenfelder, R., and Buhler, Y., 2012. Automated detection of snow avalanche deposits: Segmentation and classification of optical remote sensing imagery. *Natural Hazards and Earth System Sciences*, 12, 2893–2906.

Leblon, B., San-Miguel-Ayanz, J., Bourgeau-Chavez, L., and Kong, M., 2016. Remote Sensing of Wildfires, pp. 55–95, in *Land Surface Remote Sensing: Environment and Risks*. Editors: N. Baghdadi and M. Zhibi. Elsevier.

Louka, P., Papanikolaou, I., Petropoulos, G.P., and Stathopoulos, N., 2012. A Deterministic Model to Predict Frost Hazard in Agricultural Land Utilizing Remotely Sensed Imagery and GIS. Chapter 13.

Mallinis, G., Mitsopoulos, I.D., Dimitrakopoulos, A.P., Gitas, I.Z., and Karteris, M., 2008. Local-scale fuel-type mapping and fire behavior prediction by employing high-resolution satellite imagery. *IEEE Journal of Selected Topics in Applied Earth Observations and Remote Sensing*, 1, 4, 230–239.

Martinis, S., Twele, A., Strobl, C., Kersten, J., and Stein, E., 2013. A multi-scale flood monitoring system based on fully automatic MODIS and TerraSAR-X processing chains. *Remote Sensing*, 5, 5598–5619.

Mason, D.C., Davenport, I.J., Neal, J.C., Schumann, G.J.P., and Bates, P.D., 2012. Near real-time flood detection in urban and rural areas using high-resolution synthetic aperture radar images. *IEEE Transactions on Geoscience and Remote Sensing*, 50, 8, 3041–3052, 10.1109/TGRS.2011.2178030

Matgen, P., Hostache, R., Schumann, G., Pfister, L., Hoffmann, L., and Savenije, H. 2011. Towards an automated SAR-based flood monitoring system: Lessons learned from two case studies. *Physics and Chemistry of the Earth*, 36, 241–252.

Matzarakis, A., and Nastos, P.T., 2011. Human-biometeorological assessment of heat waves in Athens. *Theoretical and Applied Climatology*, 105, 99–106, 10.1007/s00704-010-0379-3

McKee, T.B., Doesken, N.J., and Kleist, J., 1993. *The relationship of drought frequency and duration to timescales. 8th Conference on Applied Climatology*, Anaheim, CA, 179–184.

Merz, R., and Blöschl, G., 2003. A process typology of regional floods. *Water Resources Research*, 39, 12, 1340, doi:10.1029/2002WR001952

Mishra, A.K. and Singh, V.P., 2010. A review of drought concepts. *Journal of Hydrology*, 39, 1–2, 202–216.

Niemeyer, S., 2008. New drought indices. Options Méditerranéennes. *Série A: Séminaires Méditerranéens*, 80, 267–274.

NOAA, 1998. Hazard mapping system fire and smoke product. Satellite and information service, http://www.ospo.noaa.gov/Products/land/hms.html

NOAA, 2011. STAR – Global Vegetation Health Products. NOAA, http://www.star.nesdis.noaa.gov/smcd/emb/vci/VH/vh_browse.php

Palaiologou, P., Kalabokidis, K., and Kyriakidis, P., 2013. Forest mapping by geoinformatics for landscape fire behavior modelling in coastal forests, Greece. *International Journal of Remote Sensing*, 34, 12, 4466–4490.

Park, H.J., Jang, J.Y., and Lee, J.H., 2017. Physically based susceptibility assessment of rainfall-induced shallow landslides using a fuzzy point estimate method. *Remote Sensing*, 9, 487.

Parselia, E., Kontoes, C., Tsouni, A., Hadjichristodoulou, C., Kioutsioukis, I., Magiorkinis, G., and Stilianakis, N.I., 2019. Satellite Earth observation data in epidemiological modeling of malaria, dengue and West Nile Virus: A scoping review. *Remote Sensing*, 11, 1862

Perez-Trejo, F., 1992. Desertification and Land Degradation in the European Mediterranean. *European Commission Report, EUR* 14850, p. 63.

Petropoulos, G., Carlson, T.N., Wooster, M.J., and Islam, S., 2009. A review of Ts/VI remote sensing based methods for the retrieval of land surface energy fluxes and soil surface moisture. *Progress in Physical Geography: Earth and Environment*, 33, 224–250

Pettorelli, N., 2013. *The Normalized Difference Vegetation Index*. Oxford University Press, New York. pp. 1–224.

Pouteau R., Rambal, S., Ratte, J-P., Gogé, F., Joffre, R., and Winkel, T., 2011. Downscaling MODIS-derived maps using GIS and boosted regression trees: The case of frost occurrence over the arid Andean highlands of Bolivia. *Remote Sensing of Environment*, 115, 117–129

Price, M.A., Ramsey, M.S., and Crown, D.A. 2016. Satellite-based thermophysical analysis of volcaniclastic deposits: A terrestrial analog for mantled lava flows on Mars. *Remote Sensing*, 8, 152, doi:10.3390/rs8020152

Quayle, B., 2002. Rapid mapping of active wildland fires: Integrating satellite remote sensing, GIS, and internet technologies. Joint interim report, College Park, MD: USDA-NASA-University of Maryland, 7p.

Ramachandran, C., Misra, S., and Obaidat, M.S., 2008. A probabilistic zonal approach for swarm-inspired wildfire detection using sensor networks. *International Journal of Communication Systems*, 21, 10, 1047–1073, doi:10.1002/dac.937

Rathje, E.M., and Adams, B.J., 2008. The role of remote sensing in earthquake science and engineering: *Opportunities and challenges. Earthquake Spectra*, 24, 2, 471–492.

Ren, Z., Zhang, Z., and Yin, J., 2017. Erosion associated with seismically-induced landslides in the Middle Longmen Shan Region, Eastern Tibetan Plateau, China. *Remote Sensing*, 9, 864.

Ruiz, L.Á., Recio, J.A., Crespo-Peremarch, P., and Sapena, M., 2016. An object-based approach for mapping forest structural types based on low-density LiDAR and multispectral imagery. *Geocarto International*, 31, 1–15.

Schroeder, W., Oliva, P., Giglio, L., and Csiszar, I., 2014. The new VIIRS 375 m active fire detection data product: Algorithm description and initial assessment. *Remote Sensing of Environment*, 143, 85–96.

Schumann, G.J.-P. 2018. Satellite Remote Sensing of Floods for Disaster Response Assistance, pp. 317–335 in *Remote Sensing of Hydrometeorological Hazards*. Editors: G.P. Petropoulos and T. Islam.. Taylor and Francis.

Schumann, G.J.-P., and Domeneghetti, A., 2016. Exploiting the proliferation of current and future satellite observations of rivers. *Hydrological Processes*, 30, 2891–2896, doi:10.1002/hyp.10825.

Schumann, G.J.-P., Frye, S., Wells, G., Adler, R., Brakenridge, R., Bolten, J., Murray, J. et al., 2016. Unlocking the full potential of Earth Observation during the 2015 Texas flood disaster. *Water Resources Research*, 52, 5, 3288–3293.

Sieza, Y., Gomgnimbou, P.K.A., Belem, A., and Serme, I., 2019. Use of satellite imagery for pastoral resources monitoring in Kossi Province (Burkina Faso) *Journal of Agricultural Studies*, 7, 1–10.

Sivakumar, M.V.K., Motha, R.P., and Das, H.P. (eds.), 2005. *Natural Disaster and Extreme Events in Agriculture*. Springer, 367.

Sow, M., Mbow, C., Hély, C., Fensholt, R., and Sambo, B., 2013. Estimation of herbaceous fuel moisture content using vegetation indices and land surface temperature from MODIS data. *Remote Sensing*, 5, 2617–2638.

Sun, W., Tian, Y., Mu, X., Zhai, J., Gao, P., and Zhao, G., 2017. Loess Landslide inventory map based on GF-1 satellite imagery. *Remote Sensing*, 9, 314.

Svoboda, M., LeComte, D., Hayes, M., Heim, R., Gleason, K., Angel, J., Rippey, B., Tinker, R., Palecki, M., Stooksbury, D., Miskus, D., and Stephens, S., 2002. The drought monitor. *Bulletin of the American Meteorological Society*, 83, 8, 1181–1190.

Thornthwaite, C.W., 1948. An approach toward a rational classification of climate. *Geographical Review*, 38, 1, 55–94.

Tralli, D.M., Blom, R.G., Zlotnicki, V., Donnellan, A., Evans, D.L., 2005. Satellite remote sensing of earthquake, volcano, flood, landslide and coastal inundation hazards. *ISPRS Journal of Photogrammetry & Remote Sensing*, 59, 185–198.

Tsakiris, G., and Vangelis, H., 2005. Establishing a drought index incorporating evapotranspiration. *European Water*, 9, 10, 3–11.

Ulaby, F.T., Moore, R.K., and Fung, A.K., 1982. *Microwave Remote Sensing: Active and Passive, Vol. II – Radar Remote Sensing and Surface Scattering and Emission Theory*. Addison-Wesley, Advanced Book Program, Reading, MA, 609.

UNEP – United Nations Environment Programme, 1994, Use of terms – cf. United Nations Convention to Combat Desertification. Article 1, p. 4, available: http://www.unccd.int

US Forest Service, 2012. The Wildland Fire Assessment System, http://www.wfas.net/

Uyeda, K.A., Stow, D.A., and Riggan, P.J., 2015. Tracking MODIS NDVI time series to estimate fuel accumulation. *Remote Sensing Letters*, 6, 8, 587–596.

Van de Griend, A.A., and Owe, M., 1993. On the relationship between thermal emissivity and the normalized difference vegetation index for natural surfaces. *International Journal of Remote Sensing*, 14, 1119–1137.

Van Westen, C.J., 2013. Remote sensing and GIS for natural hazards assessment and disaster risk management, pp. 259–298, in *Treatise on Geomorphology*, Volume 3. Editors: J. Shroder and M.P. Bishop. Academic Press, San Diego, CA.

Vejmelka, M., Kochanski, A.K., and Mandel, J., 2016. Data assimilation of dead fuel moisture observations from remote automated weather stations. *International Journal of Wildland Fire*, 25, 5, 558–568.

Vicente-Serrano, S.M., Begueria, S., and Lopez-Moreno, J.I., 2010. A multiscalar drought index sensitive to global warming: The standardized precipitation evapotranspiration index. *Journal of Climate*, 23,1696–1718.

Wang, L., Han, Y., and Chen, Y., 2016. *Combination of VIIRS measurements and products with CrIS toward extending data ulilization. IEEE International Geoscience and Remote Sensing Symposium (IGARSS)*, Beijing, 2016, 3960–3962, 10.1109/IGARSS.2016.7730029.

Waseem, M., Ajimal, M., and Kim, T-W., 2015. Development of a new composite drought index for multivariate drought assessment. *Journal of Hydrology*, 527, 30–37.

Webb, L., and Snyder, R.L., 2013. Frost hazards, pp. 363–366, in *Encyclopedia of Natural Hazards*. Editor: P.T. Bobrowsky. Springer, Dordrecht.

Weihs, P., Staiger, H., Tinz, B., Batchvarova, E., Rieder, H., Vuilleumier, L., Maturilli, G., and Jendritzky, G., 2012. The uncertainty of UTCI due to uncertainties in the determination of radiation fluxes derived from measured and observed meteorological data. *International Journal of Biometeorology*, 56, 537–555.

WHO/WMO/UNEP, 1996. *Climate Change and Human Health*. McMichael, A.J., Haines, A., Slooff, R., and Kovats, S. (eds.), WHO, Geneva.

Wiesmann, A., Caduff, R., Strozzi, T., Papke, J., and Mätzler, C., 2014. *Monitoring of dynamic changes in alpine snow with terrestrial radar imagery. IGARSS 2014. IEEE*, 3662–3665.

Wu, H., Adler, R.F., Hong, Y., Tian, Y., and Policelli, F., 2012. Evaluation of global flood detection using satellite-based rainfall and a hydrologic model. *Journal of Hydrometeorology*, 13, 1268–1284.

Zhang, Y., He, C., Xu, X., and Chen, D., 2016. Forest vertical parameter estimation using PolInSAR imagery based on radiometric correction. *ISPRS International Journal of Geo-Information*, 5, 10, 186, doi:10.3390/ijgi5100186.

Zhao, C., and Lu, Z., 2018. Remote sensing of landslides – A review. *Remote Sensing*, 10, 279 doi:10.3390/rs10020279.

Zargar, A., Sadiq, R., Naser, B., and Khan, F.I., 2011. A review of drought indices. *Environmental Reviews*, 19, 333–349.

Zinno, I., Casu, F., De Luca, C., Elefante, S., Lanari, R., and Manunta, M., 2017. A cloud computing solution for the efficient implementation of the P-SBAS DInSAR approach. *IEEE Journal of Selected Topics in Applied Earth Observations and Remote Sensing*, 10, 3, 802–817.

Part 3

Applications in Earth System Sciences

7 Agriculture

7.1 INTRODUCTION: REMOTE SENSING CONCEPTS OF AGRICULTURE

Remote sensing applications to agriculture fall broadly into three categories: (1) soil/land classification and crop mapping; (2) identification of stress in crops and general vegetation; (3) monitoring and forecasting of crop production (Steven and Jaggard 1995; Dalezios, 2015). Moreover, remote sensing data can be used to complement weather data in agricultural hazards assessment and agroclimatic analysis (Quarmby et al., 1993; Domenikiotis et al., 2004a; Dalezios et al., 2018; Dalezios et al., 2020). Remote sensing has contributed to agricultural studies since the early seventies using satellite optical data for the classification of agricultural crops and monitoring crop growth and crop development. Specifically, the early seventies is characterized by simple qualitative observations, differences in color and estimates by interpolation or extension. The second period, the eighties, presents quantitative land use analyses, geometric shapes, geographical locations, extent, length of characteristics, and quantification of the area covered by different land cover categories. Finally, the third period, from the nineties to the present day, is characterized by direct quantitative assessments of environmental, agrometeorological and hydrological parameters, initially by spatial correlation methods.

Moreover, remote sensing is a useful tool to analyze the vegetation dynamics on local, regional or global scales (Kogan, 2001; Dalezios et al., 2014) and to determine the impact of climate on vegetation. Satellite data can be used to monitor the vegetation status and provide phenological assessments during the growing season. In addition, satellite systems provide continuous data over the globe in time and space (Kogan, 2001) and, thus, they constitute potentially better tools for regional applications, such as monitoring vegetation, mapping of agricultural land, assessing agroenvironmental conditions, or contributing to climate analysis than conventional weather data (Domenikiotis et al., 2004a). Studies have been carried out in several parts of the world (Domenikiotis et al., 2004b; Basso et al., 2013; Wu et al., 2015) showing the potential of satellite-based data in agricultural monitoring and climate impact assessment. Satellite and airborne images are used as mapping tools to classify crops, assess their health and viability, and monitor farming practices.

Remote sensing applications to agriculture are covered in this chapter and include, among others, crop condition assessment, crop type classification, mapping of soil characteristics, mapping of soil management practices, crop yield estimation, compliance monitoring (farming practices). In addition, remote sensing methods for estimating land surface temperature (LST) and sea surface temperature (SST), vegetation indices, energy and water balance, soil moisture, water deficit, land use and crop classification, crop and biomass monitoring and estimation of risks, as well as precision farming.

7.2 CROP MAPPING AND CLASSIFICATION

7.2.1 CROP IDENTIFICATION AND MAPPING

Remote sensing offers an efficient and reliable means of collecting the information required for mapping crop type and acreage. This information can be used for yield prediction and forecasting, collecting crop production statistics, mapping water and soil productivity, facilitating crop rotation records, assessing crop damage due to extreme events, identifying factors causing crop stress, as well as monitoring farming activity. Remote sensing can also provide information about the vegetation health, since the spectral reflection of a field can be affected by variations in phenology, stage type and crop health, which leads to multispectral sensors for measuring and monitoring crop parameters.

Specifically, radar is sensitive to the crop structure, alignment and moisture content, and thus can provide complementary information to optical data. As a result, the combination of active and passive sensors increases the information available in order to distinguish each target class and its respective signature, leading to a more accurate classification. Moreover, results from image processing and analysis can be inputted into a geographic information system (GIS) and crop rotation systems, which can be combined with ancillary data to provide information of ownership, management practices and similar aspects.

As already mentioned, key activities include the identification of crop types and the delineation of their areal extent, where traditional methods of obtaining this information are census and ground surveying. Remote sensing can provide common strategies for data collection and information extraction, which may facilitate the standardization of measurements, specifically for international agencies. Again, multitemporal imagery can significantly contribute to crop identification and mapping and can facilitate classification by considering changes in reflectance as a function of plant phenology. This, in turn, requires a sequence of frequently reoccurring images throughout the growing season properly calibrated.

It is worth mentioning that some crops, such as cotton, may be easier to identify when the plant is fully covered by leaves, due to spectral reflectance change and the timing of full plant coverage. Moreover, multisensor data are very useful for increasing classification accuracies, since they provide more information than a single sensor.

Specifically, VIR (Visible InfraRed) sensing provides information, which is related to plants' chlorophyll content and the canopy structure, whereas radar provides information, which is related to plant moisture and structure. In addition, under cloud coverage or haze, radar can monitor and distinguish crop type, since it is an active remote sensor with long wavelengths and all-weather capabilities, penetrating through atmospheric water vapor.

7.2.2 LAND-USE AND CROP CLASSIFICATION

One of the first applications of satellite image processing, especially environmental satellites, is land-use classification. Generally, the term land-use refers to the way people use the land, with the emphasis on the role of land in economic activities. Accordingly, land-cover refers to the categorization of the various natural and anthropogenic soil elements based mainly on visible land-use indications. Land-use and land-cover classification systems have been developed internationally, leading to corresponding planetary cover mappings (Pirnazar et al., 2018).

In agriculture, land-use and land-cover classification contribute significantly to the development of simulation models and methodologies for each individual area or catchment, such as urban, rural, forest, snow-covered or even mountainous (Domenikiotis et al., 2004a). An important application of land-use classification in agriculture is also considered the crop classifications, which in order to be successful must consider the type of crop and the phenological stage of development. It goes without saying that crop classification varies by crop, and different satellites may be used on a case-by-case basis. Crop classification is also considered a very important component of agricultural research and agricultural practice, such as agro-meteorological simulation, estimation of agricultural production, estimation of pests and diseases, as well as crop restructuring and agro-climate zones.

Special reference is made to SAR satellite data. SAR images help to identify the different types of land-use (urban, agricultural, forests, water bodies), mainly due to the sensitivity of land-use mapping systems, but fieldwork is also required, which combines data of optical systems, leading to reliable results. A key advantage of SAR data over others (e.g. optical data) for crop recognition is the ability to obtain data at any given time to distinguish crops during the phenological cycle, regardless of the prevailing weather conditions. The reflected energy from an agricultural area can be divided into sub-segments, such as the part reflected from the vegetation, the part of multiple

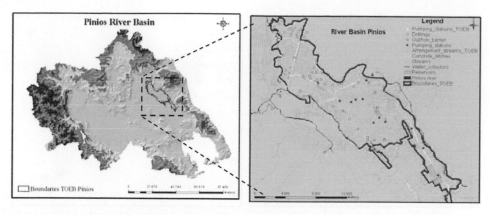

FIGURE 7.1 Study area (from Dalezios et al., 2019a).

reflections from the ground and vegetation, and the part that reflects only from the ground. Color Composite SAR (time series) images are important for monitoring and identifying crops.

Case study on crop classification and crop fractional cover (Dalezios et al., 2019a). The region of Thessaly in central Greece is characterized by a highly variable landscape, where high mountains surround the plain, which is the largest in the country. Specifically, the region of Thessaly has a total area of 14,036 km², which roughly represents 10.6% of the whole country (Figure 7.1).

One of the factors related to crop classification are the crop growth stages. As the crop develops, changes occur in ground cover, height and leaf area. Indeed, the growing period can be divided into four major and distinct growth phenological stages: initial, development, mid-season and late season (FAO, 1998). Specifically, phenological stages and fractional cover for cotton crop for the study period (2004–2010) are recorded (Dalezios et al., 2019a). For the determination of initial and development stages of cotton for the study area one experimental station is set up. In the field, two more complementary cotton fields are selected. In each experimental sampling site, two polygons are created 93 cm × 93 cm for cotton (Figure 7.2). Sampling is conducted twice a week for the period from May to September. In all pilot areas, the following are recorded and measured: photographs, localization of the fields, crop height, irrigation data, pesticide use, crop calendar and meteorological data. The percentage of ground cover is estimated based on GIS analysis. The ground cover methodology for cotton crop is delineated (Figure 7.2). The mid-season and the late season stages are determined based on the phenology of cotton. Fifty ground control points (GCPs) have been used for crop classification for the main and most cultivated crops in the region as signatures of the

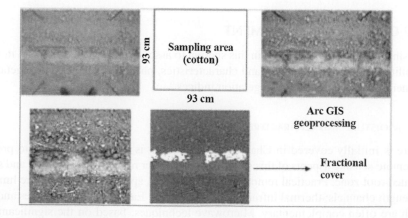

FIGURE 7.2 The process applied to field based on canopy characteristics.

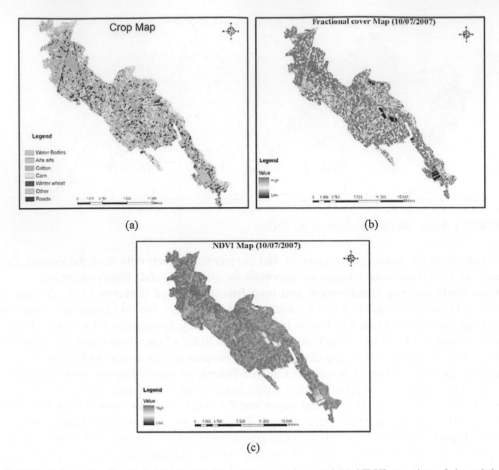

FIGURE 7.3 (a) crop classification; (b) fraction cover mapping; and (c) NDVI mapping of the sub-basin (TOEB Pinios) (from Dalezios et al., 2019a).

conducted supervised classification. The GCPs have been uniformly distributed within the area of interest. The most common crops in Pinios river basin are cotton, alfalfa, corn and winter wheat. The crop classification for the sub-basin is shown in Figure 7.3. For illustrative purposes, Figure 7.2 presents the process applied to the field based on canopy characteristics.

7.3 CROP CONDITION ASSESSMENT

Remote sensing aspects are considered in this section to assess the crop condition and its spatiotemporal variability, such as soil moisture and characteristics, phenology, water and vegetation stress, water and energy balance, as well as vegetation indices.

7.3.1 SOIL MOISTURE AND CHARACTERISTICS

Soil moisture is initially covered in Chapter 6. However, it is deemed necessary to present some additional remote sensing aspects of this significant parameter related to agriculture and specifically the unsaturated root zone. Practical remote sensing methods, applicable to crops, are limited to two main wavelength channels: thermal infrared and microwave. Each has its advantages and disadvantages, which are often complementary. Microwave techniques, based on the significant impact of water on the dielectric constant of the soil, have the advantage of the large potential range in signals between wet and dry soils. They also allow measurements through non-rain producing clouds. On the other hand, the advantage of soil moisture in thermal infrared is that no demanding sensor systems are required.

Significant steps have been taken to improve soil moisture and surface energy balance (SEB) estimation procedures by satellite measurements (Wetzel et al., 1984). Specifically, it has been found that morning temperature change is the best predictor of soil moisture, where a one-dimensional atmospheric boundary layer has been used along with a surface model. Exploiting temperature changes rather than absolute values is a very useful approach to alleviate some of the problems of using absolute contact (skin) satellite-based surface temperature due to lack of knowledge of the surface emission capability. The ground heat flux and evapotranspiration have been formulated as a function of part of the total moisture in the root zone. Cultivation is treated in a general way, which can be used in large heterogeneous surfaces. Sensitivity tests have shown that the variation of the meridian surface temperature with respect to the absorbed solar radiation is particularly sensitive to soil moisture. The model results have also shown that the soil moisture can be accurately calculated. Linear interpolation is used to associate soil moisture with surface temperature and other variables, such as wind speed, vegetation covers and convergence in the lower layers. A useful parameter for describing the soil water deficit is the available moisture (M), which defines the fraction of evapotranspiration rate to potential evapotranspiration. The available humidity has been estimated (Wetzel et al., 1984) as the necessary parameter, which determines the separation between sensible and latent heat flux on the surface. Correlations between precipitation and available humidity are high (~ 0.8). The spatial variation of humidity is much greater than the existing uncertainty of the method.

7.3.2 PHENOLOGY CONCEPTS AND MODELING

Remote sensing data and methods can contribute to the identification of the phenological stages of different crops. Phenology information can be used for crop and pasture monitoring. Indeed, the start of the growing season can be derived early in the season from satellite data and can be forecasted using climatic indices. Moreover, phenological models generally aim to assess, monitor and predict the development of the life cycle of living organisms (Dalezios, 2015). The main drivers of phenological development are light (photoperiod) and temperature, while the impact of water shortage on the growth of the mass of the plants is significant. Temperature is undoubtedly the main life-controlling factor and the main driver of the phenological development of all living organisms. Available solar radiation of a specific wavelength practically identical with the visible light (photosynthetically active radiation) is also important. Thus, the length of the day plays a significant role in controlling the phenological development of plants.

Remote sensing data and methods can also contribute to the assessment of these parameters in phenological models. A distinction can be made between modeling of the plant development, which represents modeling of the development of annual and perennial plants, and modeling of the animal life cycle. A basic principle applied in different modifications to almost all phenological models assumes a certain amount of accumulated heat to start the specific phenological phase. In practice, Growing Degree Days (GDDs) are calculated to describe the stages of the phenological development. The main inputs into the phenological models consist of climate data coming from the in situ meteorological measurement. Nevertheless, the availability of space data enables an aerial view, and an increasing number of models is based on satellite data covering much bigger areas. This brings a new approach (Fu et al., 2014) and enables the monitoring and modeling of the phenological development at regional and continental scale.

Case study of the Sahel region, Africa (Meroni et al., 2014). At the present time, there is limited knowledge about the relationship between phenology and gross primary production (GPP) in the Sahel. Moreover, it is not clear what the link is between the timing of the start of the season, as derived from remote sensing indicators, and vegetation production at the end of the growing season. The retrieval of phenology in the Sahel is challenging because satellite observations may be

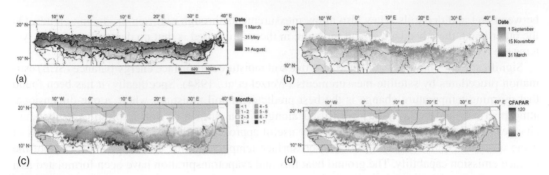

FIGURE 7.4 Average start (SOS) (a), end (EOS) (b) and length (c) of the growing season and the cumulative Fraction of Absorbed Photosynthetically Active Radiation (CFAPAR) value (d) during the growing season. Average values are computed over the set of phenological variables extracted from the FAPAR time series (n = 15, years 1998–2012). Phenological variables and CFAPAR are shown for the herbaceous and cropland land covers (other classes in white) and the five main eco-regions of the Sahel (other regions in gray) (from Meroni et al. 2014).

hampered by the presence of scattered clouds, especially during the wet season, and this can interfere with the detection of specific events, such as the SOS. Additionally, the rain-fed arid ecosystem can exhibit false starts, or complete season failures.

In this case study, the model-fit approach has been used to retrieve phenological variables from SPOT-VEGETATION time series of the Fraction of Absorbed Photosynthetically Active Radiation (FAPAR). Briefly, the analysis of the autocorrelation of the FAPAR time series helps to detect the seasonality (being uni- or bi-modal). Then, the seasonal FAPAR trajectory is fitted, based on a double hyperbolic tangent model. Finally, the start of season (SOS) is deemed to occur whenever the value of the modeled time series exceeds the initial base value by 20% of the growing amplitude. Similarly, the end of season (EOS) is deemed to occur whenever the modeled value drops below the final base value plus 20% of the decay amplitude. The fitted model can be used to compute the length of the growing season (GSL), the maximum FAPAR value (Peak), as well as the cumulative FAPAR value during the growing season (CFAPAR). When required, SOS is expressed in terms of anomaly (ΔSOS, the actual SOS minus its mean value). In this case study, the objective was to visualize the relative importance of such phenological features in determining the seasonal CFAPAR over different geographical settings in the Sahel region. Thus, the correlations of the seasonal biomass production were analyzed, represented here by CFAPAR, with the retrieved key phenology parameters (GSL and ΔSOS) and the maximum productivity indicator (Peak). The plausibility of using CFAPAR as a proxy of biomass production was checked using field measurements. Figure 7.4 shows the average start (SOS) (a), end (EOS) (b) and length (c) of the growing season and the cumulative Fraction of Absorbed Photosynthetically Active Radiation (CFAPAR)value (d) during the growing season.

7.3.3 Vegetation and Water Stress

7.3.3.1 Vegetation Stress: Damage Assessment

High agricultural productivity can be assured when crops are healthy. Early detection of crop infestations, such as moisture deficiencies, weed infestations, insects and fungal, are necessary in order to mitigate crop stress. Moreover, crops may show significant spatial variability within one field leading to uneven distribution of crop yield mainly due to soil moisture or nutrient deficiencies or other forms of stress. In such cases, remote sensing imagery can be useful, if it is provided to farmers at least on a weekly basis, since remote sensing can be used to monitor the health of crops. Optical sensing, namely visible and infrared (VIR), and especially the infrared (IR) wavelengths are highly sensitive to crop vigor, as well as crop stress and crop damage. Specifically, remote sensing can support the identification of crops affected by extremely dry or wet conditions, or crops affected by insect, weed or fungal infestations or weather-related damage. Remote sensing imagery can be obtained throughout the growing season to detect problems and to monitor the success of the treatment. Moreover, remote sensing imagery provides the necessary spatial overview of the land, which allows field observations and timely decisions about crop management due to advances in communication and technology.

Healthy vegetation contains large quantities of chlorophyll, which is responsible for the vegetation green color. In healthy crops, chlorophyll absorbs the energy in the blue and red parts of the spectrum leading to low reflectance, whereas in the green and near-infrared regions of the spectrum reflectance is high. On the other hand, stressed or damaged crops show a decrease in chlorophyll content, which results in a decrease in reflectance in the green region. Moreover, internal leaf damage results in a decrease in near-infrared reflectance, which allows early detection of crop stress. Indeed, the ratio of reflected infrared to red wavelengths is a measure of vegetation health, which leads to vegetation indices, such as the NDVI, where healthy plants have a high NDVI value due to their high reflectance of infrared light, and relatively low reflectance of red light. A typical example is an irrigated crop, where the image appears bright green, whereas in non-irrigated land the image appears darker. It should be stated that high-resolution and multispectral imagery is required for damage detection and crop monitoring. Remote sensing provides a supplementary tool for monitoring field work, where images are required at specific times during the growing season, and on a frequent basis.

Remote sensing can support precision farming, which allows spatial variability within a field and the identification of areas with difficulties, leading, for example, to the application of the correct type and amount of irrigation water, fertilizer, pesticide or herbicide (Dalezios et al., 2019b). Specifically, variations in crop growth within a field can be delineated, with areas of consistently healthy and vigorous crop appearing uniformly bright, whereas stressed vegetation would appear dark among the brighter, healthier crop areas. Moreover, by georeferencing remote sensing data and using a GPS (global position system) unit, it is possible to directly identify the exact area of the problem, by matching the coordinates of the location to that on the image. This approach leads to improvement in agricultural productivity from this land, as well as reduction of the farm input costs and minimization of environmental impacts. Besides, there is the sector of trading, pricing and selling of crops, which requires information regarding crop health worldwide to set prices and to negotiate trade agreements. In such cases, the practice is to rely on products, such as a crop assessment index, to compare growth rates and productivity temporally and to assess each country's agricultural production industry. Through remote sensing it is possible to identify regions affected by disasters damaging crops, such as the extended drought periods in Africa in the eighties, which facilitates in planning and directing humanitarian aid and relief efforts.

7.3.3.2 Water Stress and Deficit

Remote sensing data and methods can be used to detect water stress and deficit, such as the detection of plant water deficit caused by drought (Jackson et al., 1986). Indeed, the interaction of vegetation with radiation, namely near-infrared (NIR, 0.7–1.3 μm) and medium infrared (MIR, 1.3–2.5 μm), depends partly on the volume of water in leaf cells. Remote sensing methods for detecting water stress based on plant physiology are desirable, since they can be readily available with minor modifications depending on the type of vegetation. Methods using NIR and MIR data can distinguish the stress areas satisfactorily. For example, the ratio of LANDSAT/TM bands 5 and 4 has been used to extract the Moisture Stress Index (MSI) for the detection of coniferous forest destruction (Rock et al., 1986). Vegetation indices based on red (R 0.65–0.70 μm) and near-infrared have also been used successfully in plant productivity. These indicators are related to the leaf mass per unit area. The ratio (NIR/R) is related to the amount of chlorophyll in plants, which is absorbed for photosynthesis and decreases with plant growth, and not by the low relative water content (RWC) or the leaf water potential, and the ratio is detected by these indices (Jackson et al., 1986). As stomatal conductivity decreases, there is a loss of latent heat due to transpiration resulting in an increase in leaf temperature, which is detected by thermal infrared sensors. Based on this temperature change, the magnitude of the water deficit can be quantified (Jackson et al., 1986).

The change in temperature due to sudden water deficit is faster than the change in reflection in the NIR. Methods, combining thermal data with NIR and R, are advantageous for identifying areas of vegetation water deficit. There is strong absorption of radiation in the MIR range of the spectrum and it is a major factor affecting the spectral properties of the leaves. Reflection increases as the water content of the leaves decreases (Carlson et al., 1981). Thus, the MIR wavelengths (mainly 1.55–1.75 µm) are preferable for the detection of leaf water content by remote sensing methods. In order to adjust the radiometric differences in a region, indices, such as MSI, have been developed to compare two regions using one image. Research has also shown that NDVI is related to the vegetation water content and the ratio of channels 5–7 is related to soil and vegetation water content. The main objective remains to examine the ability of the LWCI (Leaf Water Content Index) to correlate with the RWC of the leaves for different species and different leaf morphology. Differences in vegetation reflectance between normal and stress vegetation conditions, can only be detected if they are large and can be verified by ground-level measurements. Therefore, it is not possible to detect plant water deficit from NIR and MIR. Only coniferous forests at almost lethargic levels due to water stress seem to show significant differences detectable by remote sensing methods.

Water deficit over an extended period, coupled with intense plant transpiration, results in the breakdown of chlorophyll in the leaves and, thus, at low values of the vegetation index. However, this effect may be temporary and may have no impact on the final production (Steven and Jaggard, 1995). The loss of the leaves' green color due to lack of chlorophyll may not be caused by water deficit but could be due to a disease that leads to symptoms of chlorination or a chlorophyll-destroying virus.

In the framework of ESA Fluorescence campaigns 2010, the objective of the EDOCROS (Early Detection Of CROp Stress) campaign was to detect early crop stress due to water and nitrogen (N) deficit, by means of advanced hyperspectral remote sensing techniques (Panigada et al., 2010). Moreover, high-spatial resolution remote sensing thermal imagery was used with the Airborne Hyperspectral Scanner (AHS) acquiring imagery in 38 spectral bands in the 0.43–12.5 mm spectral range at 2.5 m spatial resolution to conduct an investigation of the detection of water stress in non-homogeneous crop canopies, such as orchards (Sepulcre-Canto et al., 2006). Specifically, the spatial and diurnal variability of temperature as a function of water stress was investigated by flying the AHS sensor at 7:30, 9:30 and 12:30 GMT July 25, 2004 over an olive orchard with three different water-deficit irrigation treatments. The split-window algorithm, separating pure crowns from shadows and sunlit soil pixels using the reflectance bands was employed to assess a total of 10 AHS bands located within the thermal-infrared region for the retrieval of LST.

In another study, non-water-stressed and non-transpiring baselines for irrigated maize in a semi-arid region of Colorado in the USA were developed based on infrared thermometry in conjunction with a few weather parameters. A remote sensing-based Crop Water Stress Index (CWSI) was then estimated for four hourly periods each day between August 5 and September 2, 2011 (29 days) (Taghvaeian et al., 2012).

Similarly, a methodology is presented for water stress detection in crop canopies based on the Photochemical Reflectance Index (PRI) and a radiative transfer modeling approach (Suárez et al., 2009). Specifically, airborne imagery was used with a 6-band multispectral camera yielding 15 cm spatial resolution and 10 nm FWHM over three crops comprising two tree-structured orchards and a corn field. The methodology was based on water stress using the PRI indicator, and a radiative transfer modeling approach to simulate PRI baselines for non-stress conditions as a function of canopy leaf area index (LAI), leaf structure, and chlorophyll concentration (Cab).

Furthermore, quantitative assessment of vegetation vulnerability under drought stress is essential for the implementation of drought preparedness and mitigation. In a recent study, a bivariate probabilistic framework is developed to assess vegetation vulnerability and map drought-prone ecosystems more informatively (Fang et al., 2019). Specifically, a correlation is initially considered between NDVI and the Standardized Precipitation Index (SPI) at contrasting timescales in order to evaluate the degree of vegetation dependence on water availability and assess the vegetation response time. Then, the monthly NDVI series is connected to the most correlated SPI to derive joint distributions using a copula approach.

7.3.4 Water and Energy Balance: Thermal Inertia

The energy balance of a surface can be represented by the equation:

$$R_n - ETo - J - G - P = 0 \left[Watts\,m^{-2} \right] \tag{7.1}$$

where R_n is the net radiation emission density, ETo is the reference evapotranspiration, J is the soil heat flux and P is the radiation flow density used in photosynthesis. The variable P is very small and is considered negligible. The contribution of remote sensing to the estimation of the energy balance on the earth's surface consists mainly of the use of skin temperature. For a vegetated surface, the complex energy balance equation can be written (Dalezios, 2015):

$$R_n - LE - H - G - L_pF_p + A_h = \frac{\theta Q}{\theta t} \tag{7.2}$$

where H is the sensible heat flux, LE the latent heat flux, G the soil heat flux, L_p is the CO_2 thermal conversion factor, F_p is the specific CO_2 flux, A_h is the horizontal transfer energy and Q is the stored energy at t time. Energy absorption by L_pF_p photosynthesis, usually less than 1% of total radiation, is within the limits of experimental error and, as a result, it is not considered (Dalezios, 2015). The energy storage rate is important at sunrise and sunset when the net radiation is very low, but it can be omitted during daytime. When the satellite's orbit is early afternoon, due to the absence of horizontal energy transfer, the equilibrium of the surface energy balance is mainly governed by net radiation, ground flux, latent heat flux and sensible heat flux. Thus, the energy balance equation becomes:

$$R_n = G + H + LE \tag{7.3}$$

where R_n, G and H are calculated from empirical formulae using satellite and terrestrial observations and LE is estimated as the residual term of Equation 7.3. Thermal infrared temperature is commonly used in the calculation of H, which usually leads to an overestimation of H under unstable conditions. Alternatively, aerodynamic temperature can be used instead of surface temperature, however, this change results in systematic errors in H that require adjustments.

The most critical disadvantage of using LST satellite measurements for quantification of Surface Energy Balance (SEB) or soil moisture calculation is related to the fact that for a given surface balance, there are many possible surface temperatures that can be measured by remote sensing. One of the major challenges in using remote sensing to determine LST for SEB and soil moisture is the relationship of the measured temperature to the aerodynamic temperature and these in relation to heat transfer from a bare ground composite surface and vegetation. Additional data is therefore sought. For example, a simultaneous estimation of LST and vegetation index (NDVI) has been developed. A method has also been developed for calculating SEB, which uses the daily variation of the atmospheric boundary layer depth, in combination with surface temperatures, measured by geostatistical satellites. Also, using the relationship between the Leaf Area Index (LAI) and the Meridian ground heat flow fraction to the net radiation (G/R_n) and two vegetation indices (VI), equations have been developed that describe the relationships between G/R_n and VI (Kustas et al., 1993) and show that nonlinear approaches are more appropriate, except in the case of NDVI, where linear relationships appear to fit the data.

Thermal inertia. Thermal inertia is a natural variable that describes the trend in temperature fluctuations. High thermal inertia values result in small temperature changes for a given heat transfer, while low thermal inertia values result in large temperature changes for the same amount of heat transfer. Thermal inertia cannot be measured directly, but it is deduced from measurements of temperature fluctuations during the daily cycle, combined with knowledge of the heating processes occurring during this cycle and the visible and near-infrared reflection processes, during the day. It is noted that a method has been developed for estimating thermal inertia from the Daily Temperature Range (DTR) from infrared (IR) satellite images. This model has been applied with data from Nimbus 3 and Nimbus 4 (Phon et al., 1974), where it has been found that soil moisture has a large influence on the thermal inertia of the porous soil. An elementary atmospheric correction model has been used to calculate soil moisture in bare soil directly from the DTR and a much more complex model has also been used (Carlson et al., 1981). However, none of the aforementioned models considered the effect of cultivation on the soil moisture-surface relationship. It is noted that useful qualitative information on soil moisture in the root zone can be obtained from temperature measurements under the shade of the crop, where the soil is completely covered. To this end, the aforementioned model (Carlson et al., 1981) with GOES data has been applied and a good correlation has been found between these results and the previous rainfall index.

The satellite of the Heat Capacity Mapping Mission (HCMM – NASA, 1987) has been designed to obtain 500 m thermal infrared data and methods have been developed based on this data (Price, 1980). The HCMM satellite has collected data on the feasibility of using infrared temperature data to calculate thermal inertia on the earth's surface. The so-called "apparent thermal inertia" has been derived from the satellite's dynamic values with surface data. The NOAA satellite series and geostatistical meteorological satellites, such as METEOSAT or GOES, provide global images with a high repetition rate in visible and thermal infrared. Complex models have been developed for estimating thermal inertia by DTR, and the results have shown that there is a significant reduction in evaporation and available humidity with a corresponding increase in sensible heat flux in urban and low vegetation areas (Carlson et al., 1981). A theoretical relationship has also been developed for estimating daily heat capacity with respect to thermal inertia and soil moisture in bare soil (Price, 1980).

7.3.5 VEGETATION INDICES

There are differences in the reflectance of green vegetation in parts of the electromagnetic spectrum (visible and near-infrared), which lead to an innovative approach for monitoring healthy vegetation from space (Dalezios, 2013, Martín-Sotoca et al., 2019). Specifically, the spectral behavior of vegetation cover in the visible (0.4–0.7 µm) and near-infrared(0.74–1.1 µm, 1.3–2.5 µm) offers the possibility to monitor the changes from space in various stages of cultivated and non-cultivated plants, while considering the corresponding behavior of the surrounding microenvironment, e.g. ground. It should be mentioned that the visible (VIS) part of the electromagnetic spectrum could be used to draw conclusions about the rate of photosynthesis, whereas the near-infrared (NIR) is suitable to draw conclusions about the density of chlorophyll, the amount of foliage in the vegetable mass and the water contained in the leaves, which is directly related to the rate of transpiration and therefore its effect on the normal process of photosynthesis (Roujean and Lacaze, 2002, García-Haro et al., 2005). According to the above, there are four spectral regions that are of most interest, because they are dominated by very important physiological phenomena, as shown in Table 7.1.

TABLE 7.1

Spectral areas of normal plant cell processes

1	Radiation absorption by carotenoids and chlorophyll	< 0.5 mm
2	Powerful absorption of chlorophyll radiation	0.62–0.7 mm
3	Reflection of radiation from the cell walls of the mesophyll	0.74–1.1 mm
4	Absorption of radiation from water	1.3–2.5 mm

In this section, a few vegetation indices are briefly presented, which are widely used in agriculture, namely Normalized Difference Vegetation Index (NDVI), fractional vegetation cover (FVC), fraction of Absorbed Photosynthetically Active Radiation (FAPAR), Leaf Area Index (LAI) and the Normalized Difference Water Index (NDWI).

7.3.5.1 Normalized Difference Vegetation Index (NDVI)

Traditionally two types of satellite data are used to monitor vegetation, namely high-resolution images with low repeatability, such as SPOT and Landsat, as well as low resolution images with high repeatability, such as NOAA/AVHRR. The number of new satellite systems has been increasing in recent years with improving resolution and repeatability, such as the Copernicus system. Using SPOT and Landsat images, mainly qualitative information is obtained, which may be complementary to other sources with more accurate information. Also, one of the major developments of the Landsat Multispectral Scanner (MSS) is the use of visible and near-infrared channels 5 and 7 to monitor the health and development of vegetation and crops. NOAA/AVHRR polar orbit meteorological data provide low resolution images (1 x 1 sq. km) with high repeatability (at least twice daily). NOAA has an Advanced Very High-Resolution Radiometer (AVHRR) and five visible and infrared spectral bands. The AVHRR's visible and near-infrared channels, channels 1 and 2, have a spectral response, such as the MSS channels 5 and 7. The advantage of NOAA satellites for monitoring green vegetation is that they provide daily observation, while Landsat has a repeating time of 16 days. Various mathematical combinations of channels 1 and 2 have been found to be sensitive indicators of green vegetation. Table 7.2 presents a summary of remote sensing vegetation indices used in the seasonal yield forecast.

NDVI is extensively used for total vegetation monitoring, because it partially offsets variations in lighting conditions, terrain slope and viewing orientation. Clouds, water and snow are also more reflective in the visible than in the near-infrared and therefore have a negative NDVI. Still, the bare and rocky soil gives vegetation index values near zero. In addition, NDVI is a measure of the degree of absorption by chlorophyll at wavelengths of red in the electromagnetic spectrum. Therefore, NDVI is a reliable indicator of chlorophyll density in the leaf, as well as the percentage of leaf area density in the soil, and it is therefore a reliable measure of the evaluation of dry plant matter (biomass) in various plant species.

In summary, NDVI is an indispensable index associated with plant growth and development and practically applicable for monitoring of healthy vegetation from space. The NDVI increase over time during the vegetative period reflects vegetative and reproductive growth (flowering, fruiting) due to intense photosynthetic activity, as well as showing a good correlation with the final biomass production at the end of the vegetative period. In contrast, a gradual decrease in NDVI indicates water stress or excessively high temperatures for the season or for plants, leading to a decrease in photosynthesis rate and ultimately to a qualitative and quantitative degradation of the plants. The NDVI values range from −1 to +1 in theory, as deduced from the mathematical equation that defines it. Values above zero indicate green vegetation (chlorophyll) or bare soil (values near zero), while values below zero indicate water, snow, ice and clouds. Indicative values are presented in Table 7.3.

7.3.5.2 Fractional Vegetation Cover (FVC)

FVC is defined as the fraction of green vegetation seen from nadir, which can characterize the growth conditions and horizontal density of land surface live vegetation (Zhang et al., 2013). As a significant biophysical parameter involved in surface processes, the FVC is widely used for studies of the atmosphere, pedosphere, hydrosphere, ecology, and their interactions. Thus, accurate and stable FVC products with regional and global scales are critical for related studies, such as those on climate change, numerical weather predictions and land-surface processes. Remote sensing technology is a feasible and reliable way for large-scale and long-term FVC generation due to its excellent ability to provide land surface observations.

TABLE 7.2

Summary of remote sensing vegetation indices used in the seasonal yield forecast (from Basso and Liu, 2019)

Remote sensing index	Equations/descriptions[a]	Sources[b]
Greenness index		
Enhanced vegetation index (EVI)	$G * (NIR - red)/(NIR + CI * Red - C2 * Blue + L)$	Figueiredo et al. (2016) and Johnson et al. (2016)
Two-band enhanced vegetation index (EVI2)	$G * (NIR - Red)/(NIR + C1 * Red + 1)$	Bolton and Friedl (2013) and Dempewolf et al. (2014)
Excess green index	$2 * Green-red-blue$	Geipel et al. (2014)
Greenness index	$a * MSS4 + b * MSSS + c * MSS6 + d * MSS7$	Das et al. (1993)
Greenness normalized difference vegetation index (GNDVI)	$(NIR - Green)/(NIR + Green)$	Li et al. (2011)
IN-season estimated yield (INSEY)	(Vegetation index)/(growing degree days), where vegetation index represents NDVI or NDRE	Bu et al. (2017) and Sharma et al. (2015)
Normalized difference vegetation index (NDVI)	$(NIR-red)/(NIR+red)$	Johnson (2014)
Normalized difference red edge index (NDRE)	$(NIR-RE)/(NIR+RE)$	Peralta et al. (2016) and Sharma et al. (2017)
Soil-adjusted vegetation index (SAVI)	$(1 + L) * (NIR-red)/(NIR+red+L)$	Al-Gaadi et al. (2016)
Transformed soil-adjusted vegetation index (TSAVI)	$a * (NIR - a * red - b)/(red+a * NIR - a * b)$	Shanahan et al. (2001)
Transformed vegetation index "six" (TVI 6)	$(0.5+(MSS6-MSS5)/(MSS6+MSS5))^{1/2}$	Idso et al. (1980)
Vegetation health index (VHI)	$a * VCI+b * TCI$	Kussul et al. (2014)
Wide dynamic range vegetation index (WDRVI)	$(a * NIR-red)/(a * NIR+red)$	Maresma et al. (2016)
Saturation adjusted normalized difference vegetation index (SANDVI)	$SANDVI=NDIV$, if $NDVI \leq 0.78$, else $0.03 * RVI+0.5363$	Dempewolf et al. (2014)
Chlorophyll index		
Chlorophyll index red-edge	$(NIR/RE)-1$	Torino et al. (2014)
Fraction of absorbed photosynthetically active radiance (FAPAR)	Satellite imagery product	Kolotii et al. (2015) and Kussul et al. (2014)
Photochemical index		
Photochemical reflectance index (PRI)	$(Green-Blue)/(Green+Blue)$	Geipel et al. (2014)
Structural independent pigment index (SIPI)	$(R_{445}-R_{800})/(R_{680}-R_{800})$	Royo et al. (2003)
Fluorescence retrieval index		
Fluorescence (SIF760)	$(E_{out}*L_{in}-L_{out})/(E_{out}-E_{in})$	Quemada et al. (2014)
Dryness index		
Crop water stress index (CWSI)	$((T_c-T_a)-(T_c-T_a)_{ll})/((T_c-T_a)_{ul}-(T_c-T_a)_{ll})$, (the subscript ll is the nonwater stressed lower baseline, and ul is the non-transpiring upper baseline)	Orta et al. (2004)
Normalized difference water index (NDWI)	$(NIR-SWIR)/(NIR+SWIR)$	Bolton and Friedl (2013) and Wang et al. (2014)
Normalized water index (NWI)	$(R_{970}-R_{900})/(R_{970}+R_{900})$, $(R_{970}-R_{850})/(R_{970}+R_{850})$, $(R_{970}-R_{920})/(R_{970}+R_{920})$, $(R_{970}-R_{880})/(R_{970}+R_{880})$	Bandyopadhyay et al. (2014)
Temperature condition index (TCI)	$100 * (BT_{max}-BT)/(BT_{max}-BT_{min})$	Arshad et al. (2013)

(Continued)

TABLE 7.2 (*Continued*)

Summary of remote sensing vegetation indices used in the seasonal yield forecast (from Basso and Liu, 2019)

Remote sensing index	Equations/descriptions[a]	Sources[b]
Temperature vegetation dryness index (TVDI)	$(LST-LST_{min})/(LST_{max}-LST_{min})$	Holzman and Rivas (2016)
Standardized crop-specific drought index (S-CSDI)	$\prod_{i}^{n}\left(\Sigma\, AET\, /\, \Sigma\, PET\right)_{i}^{\lambda i}$, where AET is estimated by the SEBAL model	Arshad et al. (2013)
Water index (WI)	R_{970}/R_{900}	Bandyopadhyay et al. (2014)
Temperature index		
Land surface temperature (LST)	MODIS product	Johnson (2014)
Vegetation condition index (VCI)	$100 * (NDVI-NDVI_{min})/(NDVI_{max}-NDVI_{min})$	Kowalik et al. (2014) and Kuri et al. (2014)
Ratio index		
Area weighted radiance ratio	—	Navalgund et al. (2000)
Green ratio vegetation index (GRVI)	NIR/green	Maresma et al. (2016)
Inverse simple ratio red-edge (ISRre)	RE/NIR	Torino et al. (2014)
Simple ratio (SR) as known as ratio vegetation index (RVI)	NIR/red	Kancheva et al. (2007)
Simple ratio red-edge (SRre)	NIR/RE	Torino et al. (2014)
Difference index		
Near-infrared and red difference	NIR–red	Das et al. (1993)
Reflectance/backscatter		
Optical sensors: reflectance from green NIR and Red, etc.	N/A	Fieuzal et al. (2017) and Lin and Kuo (2013)
Synthetic aperture radar sensors: backscatters from C, X, and L band	N/A	Fieuzal et al. (2017)

[a] Abbreviations in the equations: NIR, near-infrared; RE, red-edge; MSS4, Landsat Multispectral Scanner band 4 (0.5–0.6 μm); MSS5, Landsat Multispectral Scanner band 5 (0.6–0.7 μm); MSS6, Landsat Multispectral Scanner band 5 (0.7–0.8 μm); MSS7, Landsat Multispectral Scanner band 7 (0.8–1.1 μm); T_c, canopy temperature; T_a, air temperature; BT, brightness temperature; AET, actual ET; and PET, potential ET.

[b] The column listed up to five sources where the index was used to forecast crop yield.

TABLE 7.3

Indicative values of NDVI in different types of land cover

	Land cover type	NDVI scale −1 to +1	NDVI scale 0–255
1	Dense vegetation	$0,500 =< NDVI =< 1$	$210 =< NDVI =<255$
2	Green vegetation	$0,140 =< NDVI < 0,500$	$118 =< NDVI < 210$
3	Spare vegetation	$0,090 =< NDVI < 0,140$	$105 =< NDVI <118$
4	Bare soil	$0,025 =< NDVI < 0,090$	$88 =< NDVI<105$
5	Clouds	$0,002 =< NDVI < 0,025$	$83 =< NDVI < 88$

Currently, several FVC estimation algorithms have been developed based on remote sensing data, which can be divided into three major types: empirical methods, pixel unmixing models and machine learning methods (Liu et al., 2019). Empirical methods build the statistical relationships between the FVC and vegetation indices or specific bands' reflectance through sufficient and reliable sample data. Empirical methods can achieve satisfactory accuracy at the regional scale with specific vegetation types. However, empirical methods become invalid over large-scale regions, in which the various vegetation types and land conditions increase uncertainties in the established relationships. Pixel unmixing models assume that each pixel is composed of several components, and the fraction of vegetation composition is the corresponding FVC value of the pixel (Jia et al., 2015; Liu et al., 2019). The dimidiate pixel model, as a widely used method of pixel unmixing models, assumes that each pixel can be divided into two parts: vegetation and non-vegetation. However, the main limitation of pixel unmixing models is the determination of representative endmembers because of the complex land surface conditions and various spectral characteristics over a large scale.

Recently, machine learning methods have also been widely used to retrieve FVC values because of their computational efficiency and stable performances in nonlinear fitting. In general, machine learning methods estimate the FVC through training on a representative sample database containing pre-processed reflectance and corresponding simulated land surface parameters data. Several algorithms of machine learning methods are proposed for FVC product generation over regional and global scales with satisfying results (Jia et al., 2015; Liu et al., 2019). With these developed FVC estimation algorithms, some FVC products over regional and global scales are generated from remote sensing data, which are summarized in Table 7.4. Figures 7.1 to 7.3 present an example of FVC calculation in a rural basin.

7.3.5.3 Leaf Area Index (LAI)

Empirical relationships between LAI and nadir-viewing measurements in the red and infrared bands based on simple and feasible approaches have been recognized. In these methods, it is assumed that all factors, which affect the spectral response of canopy, are fixed, except LAI. The following model has the advantage of being expressed by a semi-empirical relationship between LAI and WDVI (Weighted Difference Vegetation Index), defined as follows:

$$WDVI = \rho_i - \rho_r \left(\rho_{si} / \rho_{sr} \right) \tag{7.4}$$

where ρ_r and ρ_i refer to the reflectance of observed canopy in the red and infrared bands, respectively, ρ_{sr} and ρ_{si} are the corresponding values for bare soil conditions. The ratio ρ_{si}/ρ_{sr} can be considered as a constant, in analogy with the "soil line concept". Indeed, the WDVI index has the advantage of significantly reducing the impact of soil background on the surface reflectance values. The LAI is then related to WDVI of the observed surface through the expression:

$$LAI = -\left(1/\alpha\right)\ln\left[1-\left(WDVI/WDVI\infty\right)\right] \tag{7.5}$$

where α is an extinction coefficient to be determined from simultaneous measurements of LAI and WDVI; WDVI∞ is the asymptotical value of WDVI for LAI→∞, determined from the WDVI image histogram (value between 0.55 and 0.70). The value of ρsi/ρsr has resulted in the range [1.1–1.3], with α in the range [0.21–0.41], with lower values occurring in tree crops, higher values in herbaceous. This approach allows for image-based adjustments, which partially compensate for the uncertainty in the atmospheric correction of satellite data that has been validated by field measurements and numerical models, which simulate the reflectance of leaf and canopy in a wide range of field conditions. There are also vegetation indices useful for LAI retrieval (Song et al., 2016).

7.3.5.4 Soil-Adjusted Vegetation Index (SAVI)

SAVI measures the density and vigor of green vegetation by eliminating the reflectivity of the ground beneath the canopy using the red and near-infrared bands (Huete, 1988). The data should be atmospherically corrected before employing SAVI. SAVI is calculated from the following equation:

$$SAVI = a_0 - a_1 * \exp\left(-a_2 * LAI\right) \tag{7.6}$$

TABLE 7.4

The major fractional vegetation cover (FVC) products and their characteristics (from Liu et al., 2019)

Products	Sensor	Methods	Spatial Resolution	Temporal Resolution	Spatial Coverage	Temporal Coverage
CNES/POLDER	POLDER	Empirical model	6 km	10 days	Global	1996–1997, 2003
EUMETSAT/LSA SAF	SEVIRI	The dimidiate pixel model	3 km	Daily	Europe, Africa, South American	2005–present
EPS/CYCLOPES	SPOT VGT	Machine learning methods	1/112*	10 days	Global	1998–2007
ESA/MERIS	MERIS	Machine learning methods	300 m	Month/10days	Global	2002–2012
GEOV2 FVC	SPOT VGT, PROBA-V	Machine learning methods	1/112*	10 days	Global	1999–present
GEOV3 FVC	PROBA-V	Machine learning methods	300 m	10 days	Global	2014–present
GLASS FVC	MODIS	Machine learning methods	500 m	8 days	Global	2000–present

where a_0, a_1 and a_2 are three parameters which can be specified by the user. The default values that were used are: $a_0 = 0.75$, $a_1 = 0.65$, and $a_2 = 0.6$, respectively.

7.3.5.5 Fraction of Absorbed Photosynthetically Active Radiation (FAPAR)

FAPAR is a biophysical variable, which is correlated with the primary productivity of vegetation. Indeed, FAPAR represents the fraction of the solar energy, which is absorbed by vegetation, since the intercepted PAR is the energy (carried by photons) underlying the biochemical productivity processes of plants. FAPAR is one of the Essential Climate Variables recognized by the UN Global Climate Observing System (GCOS) and by the FAO Global Terrestrial Observing System (GTOS) to characterize the climates of the Earth. Moreover, FAPAR has been considered as a drought indi-cator, since it is sensitive to vegetation stress (Gobron et al., 2008). Therefore, it can provide stake-holders and farmers with potentially useful information for water and agricultural management.

FAPAR is estimated from models which describe the transfer of solar radiation in plant cano-pies based on Earth Observation (EO) information as input data, since direct Vergermeasurement is very difficult (Baret et al., 2006; Baret et al., 2007; Verger et al., 2013; Baret et al., 2019; Verger et al., 2020). Specifically, numerically inverting physically based models are employed based on EO information to retrieve FAPAR estimates. Indeed, the FAPAR estimates can be produced from multispectral images acquired by the Medium Resolution Imaging Spectrometer (MERIS) onboard ENVISAT by means of the MERIS Global Vegetation Index (MGVI) algorithm, developed at the EU-JRC (European Union-Joint Research Center) (Gobron et al., 2008).

Operationally, 10-day FAPAR estimates were regularly produced by the European Space Agency (ESA) as MGVI Level-3 Aggregated Products, until 2011. From 2012, the produced FAPAR images are the result of applying the same algorithm to the images acquired by the VEGETATION sen-sor onboard SPOT. These images are created by the Flemish Institute for technological research (VITO). The objective of the MGVI algorithm is to reach the maximum sensitivity to the presence and changes in healthy live green vegetation, while at the same time minimizing the sensitivity to atmospheric scattering and absorption effects, to soil color and brightness effects, and to temporal and spatial variations in the geometry of illumination and observation. Specifically, MGVI is a physically based index, which transforms the calibrated multispectral directional reflectance into a single numerical value, while minimizing possible disturbing factors. It is constrained by means of an optimization procedure to provide an estimate of the FAPAR of a plant canopy. The MGVI level-3 aggregation processor has been developed and is maintained by the EU-JRC (Gobron et al., 2008). FAPAR-anomalies are produced every 10-days as follows:

$$\text{FAPAR anomaly}_t = \left(X_t - X_{\text{mean}}\right)/\sigma \tag{7.7}$$

where X_t is the FAPAR of the 10-day period t of the current year, X_{mean} is the long-term average FAPAR and σ is the standard deviation, both calculated for the same 10-day period t using the available time series. From 2012, the archive of SPOT-VEGETATION FAPAR covers the period from October 1998 to the present day. FAPAR and FAPAR-anomalies can be presented in the form of maps and graphs, which provide information both on the temporal evolution over longer time periods and the spatial distribution of the vegetation activity. Specifically, the FAPAR product is dimensionless and ranges from 0 to 1 in terms of real values, and from yellow to green on the map, with 1 corresponding to a maximum of vegetation activity. Moreover, the FAPAR anomaly product is given in standard deviation units and commonly ranges from -4 to +4 and from red to green, red showing negative anomalies. This variation can be mainly the outcome of a rainfall/soil moisture deficit and indicates a variation in the vegetation health and/or cover. Figure 7.5 shows an example of the FAPAR (left) and FAPAR-anomaly (right) images produced by the processing chain within the European Drought Observatory (EDO) of the EU-JRC for July 1–11, 2011.

Other method. The FAPAR can also be estimated from empirical relationships between field measured FAPAR and LAI (Liu et al., 2019). Beer-Lambert law has been widely used to describe the

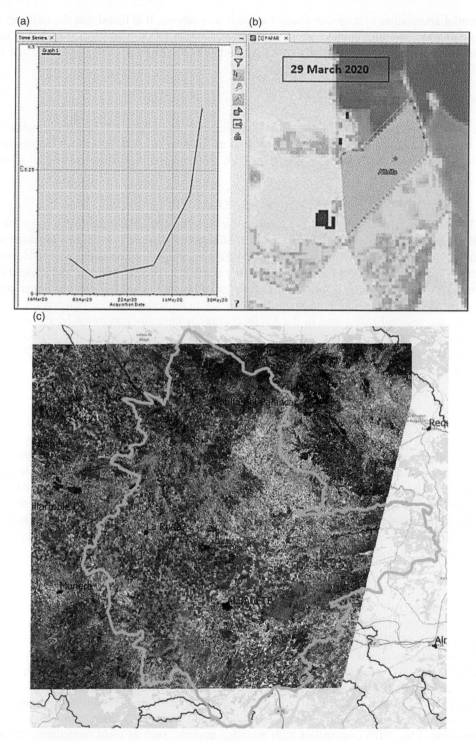

FIGURE 7.5 (a) Change detection of FAPAR from March to May 2020, for AlfaAlfa crop in central Spain; (b) Sentinel-2 image of March 29, of 2020, for the plot of AlfaAlfa crop. (c) Map of plot area in central Spain - eastern Mancha covered by two sentinel- 2 scenes.

exponential attenuation of monochromatic radiation in a canopy. It is found that the ratio between transmitted and incoming photosynthetically active radiation (PAR; 0.480.7μm) under a canopy decreases approximately exponentially with LAI, and, thus, the FAPAR can be expressed as a function of LAI as follows (Liu et al., 2019):

$$FAPAR = c \times \left[1 - a \times \exp\left(-b \times LAI\right) \right] \tag{7.8}$$

where FAPAR is fraction of PAR intercepted by the canopy, and LAI is leaf area index, where a, b, and c are the FAPAR parameters and default values as follows: a: 1.0; b: 0.4; c: 1.0. There are also empirical models for the Fraction of Absorbed PAR with spectral vegetation indices, as shown in Table 7.5.

7.3.5.6 Normalized Difference Water Index (NDWI)

NDWI (Gao, 1996) is a satellite-based index from the Near-Infrared (NIR) and Short-Wave Infrared (SWIR) channels. Indeed, the SWIR reflectance is affected by changes in both the vegetation water content and the spongy mesophyll structure in vegetation canopies, whereas the NIR reflectance is affected by leaf internal structure and leaf dry matter content, but not by water content. Specifically, the combination of the NIR with the SWIR removes variations caused by leaf dry matter content and leaf internal structure, which improves the accuracy in retrieving the vegetation water content. Moreover, the spectral reflectance in the SWIR channel is basically controlled by the amount of water available in the internal leaf structure. There is, thus, a negative relationship between SWIR reflectance and leaf water content. NDWI is computed using the near-infrared (NIR – MODIS band 2) and the short-wave infrared (SWIR – MODIS band 6) reflectances.

$$NDWIt = \frac{NIRt \quad SWIRt}{NIRt + SWIRt} \tag{7.9}$$

For each tile, all TERRA or AQUA passes of each day are combined into the daily MODIS composite and then produce a maximum value time composite every 10 days. Each month is split into three periods: from the 1st to the 10th, from the 11th to the 20th, and from the 21st to the end of the month. The 10-day composites are then resampled from 250 m to 1 km and the ratio between band 2 and band 6 is then calculated following the above equation. Then, a unique map of Europe is produced by mosaicking the different tiles. Finally, a mask derived from GLC 2000 is applied to recode water bodies, cities, permanent snow and desert as no data. NDWI anomalies are produced for every 10-day period as follows:

$$NDWI \; anomaly_t = \left(X_t - X_{mean} \right) / \sigma \tag{7.10}$$

where X_t is the NDWI of the 10-day period t of the current year and X_{mean} is the long-term average NDWI and σ is the standard deviation, both calculated for the same 10-day period t using the available time series. NDWI anomalies are produced only for pixels that have at least five years of data for the given 10-day period.

NDWI and NDWI-anomalies can be presented in the form of maps and graphs, which provides information both on the spatial distribution of the vegetation water stress and its temporal evolution over longer time periods. Gridded data can easily be aggregated over administrative or natural entities, such as watersheds. This allows for the qualitative and quantitative comparison of the intensity and duration of the NDWI anomalies with recorded impacts, such as yield reductions, low flows, lowering of groundwater levels. The NDWI product is dimensionless and varies between -1 to +1, depending on the leaf water content, but also on the vegetation type and cover. Indeed, high values of NDWI correspond to high vegetation water content and to high vegetation fraction cover. On the other hand, low NDWI values correspond to low vegetation water content and low vegetation fraction cover. During periods of water stress, NDWI is expected to decrease. Figure 7.6 shows the NDWI of the third 10-day period of May 2011.

TABLE 7.5
Empirical Models for the Fraction of Absorbed PAR with Spectral Vegetation Indices (from Song et al., 2016)

Index	FPAR Model	Notes	References
SR	0.34SR–0.63	Various crops	Kumar and Monteith (1981)
SR	0.369ln(SR)–0.0353	Field spectrophotometer/sugar beet	Steven et al. (1983)
SR	0.0026SR^2+0.102SR–0.006	Landsat/corn	Gallo et al. (1985)
SR	0.0294SR+0.3669	SPOT/wheat	Steinmetz et al. (1990)
SR	Not specified	Modeling study	Sellers (1987)
NDVI	1.253NDVI–0.109	Landsat/spring wheat	Asrar et al. (1984)
NDVI	1.200NDVI–0.184 (growing) 0.257NDVI+0.684 (senescence)	Landsat/wheat	Hatfield et al. (1984)
NDVI	2.9NDVI^2–2.2NDVI+0.6 (growing)	Landsat/corn	Gallo et al. (1985)
NDVI	1.00NDVI–0.2	Landsat/coniferous	Peterson et al. (1987)
NDVI	1.23NDVI–0.06	Wheat	Baret and Olioso (1989)
NDVI	1.33NDVI–0.31	Modeling study	Baret and Olioso (1989)
NDVI	1.240NDVI–0.228	Modeling study	Baret et al. (1989)
NDVI	0.229exp(1.95NDVI)–0.344 (growing) 1.653NDVI–0.450 (senescence)	SPOT/cotton and corn	Wiegand et al. (1991)
NDVI	1.222NDVI–0.191[*]	Modeling study	Asrar et al. (1992)
NDVI	1.254NDVI–0.205	Field spectroradiometer/corn and soybean	Daughtry et al. (1992)
NDVI	1.075NDVI–0.08	Modeling study	Goward and Huemmrich (1992)
NDVI	1.386NDVI–0.125	Cereal crop/modeling study	Bégué (1993)
NDVI	1.164NDVI–0.143	Modeling study	Myneni and Williams (1994)
NDVI	1.21NDVI–0.04	AVHRR/mixed forests	Goward et al. (1994)
NDVI	0.95NDVI–0.02	Landsat/modeling study	Friedl et al. (1995)
Greenness	Not specified	Landsat/corn	Daughtry et al. (1983)
PVI	0.036PVI–0.015 (growing) 0.037PVI+0.114 (senescence)	SPOT and videography/cotton and corn	Wiegand et al. (1991)

SR, simple ratio; NDVI, normalized difference vegetation index, greenness; PVI, perpendicular vegetation index.

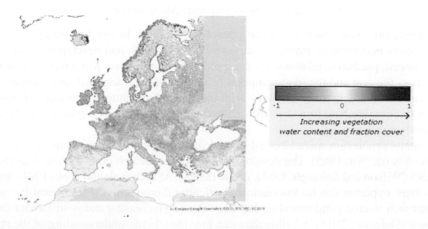

FIGURE 7.6 NDWI of the 3rd 10-day period of May 2011 (EDO model).

7.4 CROP YIELD ASSESSMENT AND MONITORING

7.4.1 CONCEPTS OF CROP YIELD ASSESSMENT AND MONITORING

Crop yield assessment and monitoring is an important component in the field of contemporary agriculture and agrometeorology internationally. Crop yield assessment consists of the simulation of the crop photosynthesis cycle and the energy balance with mathematical equations and technological systems, such as remote sensing. In fact, the accuracy of these methods is constantly improving. The simulation may concern only individual phenological stages, or the whole growing season, as well as the estimation of pests and diseases, but also of the corresponding effects on crops (Linkosalo et al., 2008). Initially, crop yield modeling allows for a better understanding of the behavior of crops on climate variables and weather. It also contributes to improving agricultural production and adds a great added value to agriculture, while allowing for better production management, which includes storage and disposal of production, export rates and overall significant benefit to national or regional economy. Open-source model frameworks, such as BIOMA and DSSAT (http://dssat.net/), should be favored, since they provide solutions whose robustness is guaranteed by the large community of users. Extreme weather events are expected to become more frequent based on climate change estimates (IPCC, 2012), whereas historical data poorly represented the emerging conditions in recent years. As a result, sensitivity to extreme events becomes an issue of high priority for reliable forecasting, since "normal" years are less in need.

At the present time, crop yield/production forecasts are widely used at global, national, regional and field levels, however, there are different objectives, methodologies, data needs, timeliness, costs and reliability (Lin and Kuo, 2013; Rembold et al., 2013). The general objective of crop yield and production forecast should be the reduction of the risks associated with local or national food systems. The selection of the spatiotemporal scale affects the food system. At the farmer level, "prescriptive farming" is applied in the USA with the objective to perform in-field yield modeling to improve management techniques and boost actual yield. Moreover, the GEOGLAM project (https://cropmonitor.org/) monitors current year conditions and contributes national crop production forecasts (for wheat, soybeans, corn and rice) computed by 30 national partners on a monthly basis. Similarly, the Agricultural Model Inter-comparison and Improvement Project (AGMIP) (Rosenzweig et al., 2013) attempts to improve agricultural models for the medium- and long-term effects of climate change on crop yields. Moreover, the development of Crop Data Layers (CDL) is an efficient solution in regions with large field sizes. ESA has recently initiated the operational production of 20m monthly land cover maps of Africa from Sentinel-2, which is a free-of-charge solution for the African continent (54 countries).

7.4.2 REMOTELY SENSED CROP YIELD ASSESSMENT AND MONITORING

Accurate production assessment at critical phenological stages before harvesting on regional to national scales is becoming increasingly important in developing and developed countries (Wu et al., 2015). Indeed, predictive relationships are quite difficult to derive. Moreover, several logistical factors, such as the cost and the spatiotemporal resolution of the associated satellite data, drive the incorporation of the vegetation physiology with the spectral signatures. The methods for estimating crop production are based on objective methodologies, such as crop growth and development modeling and remote sensing.

Crop biomass production was addressed several years ago, with the development of detailed biophysical models (de Wit, 1965). The development of space research programs, such as the USDA-AgRISTARS (Wilson and Sebaugh, 1981), or the ISRO-CAPE (Navalgund et al., 1991), has offered new assessment opportunities for international agricultural production. The interest for the remote sensing approach in crop yield modeling has been steadily increasing and is still under continuous improvement (Dalezios, 2015). Satellite data can provide a better understanding of the spatial and temporal evolution of the parameters incorporated into models. Remote sensing methods can be classified as: statistical, deterministic or process-based and combined.

7.4.2.1 Statistical Models

The statistical approach is based on indices derived from satellite sensors averaged over a region or country and entered in a regression analysis based on the general formula:

$$Yield = a + b\{mean\ areal\ NDVI\} \tag{7.11}$$

Statistical crop models are simple and entail low costs. Indeed, statistical models are usually regression models (simple or multiple, linear or nonlinear, static or dynamic) that link the variables of interest (i.e. the yields) to the predictors known for the current season. The predictors are selected from the domains of meteorology (rainfall, temperature, solar radiation) and/or the remote sensing (vegetation indices, LAI, soil moisture). The main drawback of statistical models is the smallest prediction interval, which is around the average of the reference data set, whereas the interest is mostly on abnormal years. Moreover, these models cannot be extrapolated in time or space.

Crop yield forecast based on statistical models using remote sensing data contains at least one variable that is derived from remote sensors (Basso and Liu, 2019). Active and passive sensors can collect within-season crop, soil and weather information. Specifically, active sensors have light sources and can emit radiations at various wavelengths to the earth's surface and detect radiance reflected to the sensors. On the other hand, passive sensors receive and measure radiance at specific wavelengths returned by objects, and they rely on external light sources, often sunlight (Basso and Liu, 2019). Satellite imageries, including SPOT-Vegetation, AVHRR, Landsat, MODIS and recently Copernicus, are the most popular venues to obtain within-season information to forecast final yield. Indeed, both multispectral and hyperspectral imageries have been applied to crop yield forecasting. Recently, the use of radar satellites and the information in the microwave band has also been considered in yield forecasting (Basso and Liu, 2019). The spectral information at various bands, such as red and near-infrared (NIR) band, can be directly used to quantify vegetation conditions. Moreover, vegetation indices, which are a product of combining reflectance from multiple bands, are more relevant to crop biomass, compared to reflectance from the individual band (Basso and Liu, 2019).

There are many remote sensing applications in agricultural production estimation over time with an emphasis on the use of NDVI. However, other indices, such as VCI and TCI, are also important to assess crop yield. At a mixed farmland, Emilia-Romagna, Italy, 10-day NDVI (MVC) images from NOAA/AVHRR were used to evaluate wheat yield, as well as a crop development model to estimate dry weight for four main crops, namely soybean, maize, beets and wheat. Accumulated NDVI on cultivated land in Greece was used to evaluate the yield of significant crops (Quarmby et al., 1993). Ten-day MODIS-NDVI images were also used to estimate winter wheat production in China. In addition, a NOAA/AVHRR (4 x 4 km) NDVI weekly images time-series 1985–1992 was used to extract the VCI for wheat and maize (Hayes and Decker, 1996). Moreover, VCI and TCI were used to develop a statistical model of predicting grain yield in Poland (Dabrowska-Zielinska et al., 2002).

Case study 1. Biomass estimation. Biomass or phytomass is defined as the weight of plant dry matter per unit area. There is a relationship from which the dry biomass can be calculated from satellite data. One such relationship is (Dalezios, 2015):

$$Dry\ Biomass = -321.7 + 429.02(ch10 / ch8) \tag{7.12}$$

where ch8 and ch10 are the values of the channels of electromagnetic spectrum with ranges of 0.66–0.72 μm and 0.8–1.0 μm, respectively. Dry biomass can lead to production forecasts, as there is a high correlation between dry biomass and final production. There is also a direct relationship of NDVI with crop phytomass, which is dependent on the plant species, the leaf area, that is, the greater leaf area the higher the NDVI values are estimated, as well as the amount of chlorophyll on leaves. A model has been developed showing that dry biomass (dW) production per unit area is proportional to photosynthetically active radiation (S), the percentage (f) absorbed by the crop and the yield (n) conversion of energy absorbed into biomass (Steven and Jaggard, 1995), as follows:

$$dW / dt = n.f.S \tag{7.13}$$

where S can be derived from meteorological data. The yield n is not easily calculated, but it has been found to be slightly different if taken as an average throughout the growing season, except in the case of stress in the crop. The f can be calculated from empirical formulas. Several linear relationships have been developed between In (P_{mass}) or ln (GP_{mass}), where GP_{mass} is the above ground biomass of various vegetation types and NDVI, calculated from NOAA-AVHRR channel data. By using $ln(GP_{mass})$ as a variable in place of $ln(P_{mass})$, the correlation coefficients were higher, confirming the need to measure the ratio of green to non-green parts of the plant to improve the relationship of plant growth and vegetation indices. A relationship is used (Dalezios, 2015 – Chapter 11), linking NDVI vegetation index (from LANDSAT data) and P_{mass} phytomass (in kg per square meter).

$$P_{mass} = -0.2088 + 2.0212 * (NDVI)^2 \tag{7.14}$$

The biomass estimation model (Equation 7.23) was applied to cotton at Karditsa area, central Greece. The data used in this work came from images of NOAA satellites (Dalezios, 2015). Each day, two NOAA images were taken, one during the day and one at night, over three years (1994, 1995 and 1996) to produce 10-day maximum value composites (MVC) per pixel. The next step involved calculating the "maximum" images for each year, and each region. The model of Equation 7.23 (11.12) was then applied to each pixel of the image. The yield data for the Karditsa region are presented in Figure 7.7, showing the model's average values and the corresponding yields.

Case study 2: Remotely sensed assessment of agricultural production. Assessment of cotton cultivation in Central and Northern Greece (Domenikiotis et al., 2004a) is examined. The Vegetation Condition Index (VCI) is used, which is an extension of NDVI. The VCI is associated with the long-term minimum and maximum of NDVI and is expressed by the following equation:

$$VCI = 100 x (NDVI - NDVI_{min}) / (NDVI_{max} - NDVI_{min}) \tag{7.15}$$

where NDVI, $NDVI_{max}$, $NDVI_{min}$ are the 10-day normalized values of NDVI, the maximum–minimum values of NDVI over time, for each pixel, respectively. VCI values range from zero for

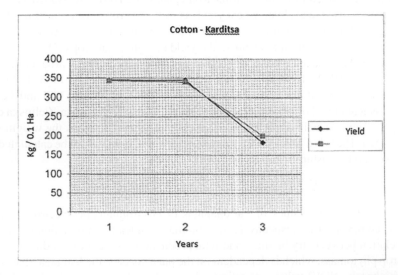

FIGURE 7.7 Fluctuations of yields and model values for cotton cultivation in the Karditsa region (from Dalezios, 2015).

extremely adverse conditions, up to 100 for optimal conditions. Therefore, higher VCI values represent healthy stress-free vegetation.

The database used consists of NOAA daily image time series with a resolution of 8 x 8 km for Central and Northern Greece and for the period 1982–1999. The NDVI daily values are extracted for each pixel, the 10-day maximum value composite (MVC) of the NDVI, from which the region's NDVI images are generated, which are used in the calculation of the VCI. The 1982–1997 data are used to develop statistical regression models for estimating cotton production, while the 1998 and 1999 data are used for certification. A "4352" moving median filter is used to smooth the NDVI time series. Also, a correlation analysis between the 10-day VCI values during the growing season and the official cotton production prices is applied. This analysis identifies the critical periods with a significant impact on final production, which coincide with the period of maximum cotton cover, i.e. the 10-day period at the end of July and the beginning of August. The statistical models that have emerged are the following:

$$\text{For Northern Greece} \quad y = 16530(\text{VCI}) - 1\text{E} + 06 \quad (R2 = 0,7423) \tag{7.16}$$

$$\text{For Central Greece} \quad y = 21581(\text{VCI}) - 888809 \quad (R2 = 0,8451) \tag{7.17}$$

$$\text{For all of Greece} \quad y = 40877(\text{VCI}) - 2\text{E} + 06 \quad (R2 = 0,8707) \tag{7.18}$$

where y is the estimated cotton production in tonnes. At certification, the results of the models showed an absolute percentage difference from the official cotton production of more than 20% in about 84% of the cases. Subsequently, cluster analysis was used to group areas with similar phenological characteristics based on VCI values. The groups eventually showed a mean absolute difference between estimated and official cotton production price, ranging from 5% to 19% at regional level and 7.6% at national level, respectively, as shown in Table 7.6. These results are considered extremely satisfactory internationally.

7.4.2.2 Deterministic or Process-based Models

Most of the existing models are deterministic, although some recent developments refer to stochastic approaches (Chipanshi et al., 2015). Deterministic models are based on mechanistic models and replace the theoretical relations with empirical functions. The deterministic approach consolidates the computation of NDVI, LAI, FAPAR and biomass before yield assessment. Most deterministic models require information on crop management, such as variety, planting dates, phenology, nutrient availability (soil parameters or fertilization), water availability (soil moisture, or evapotranspiration) and energy received (radiation). At the present time, most models make use of remote sensing information, such as crop phenology. Reference validated data sets are becoming available for meteorological data (NCAR, 2014), soil information (FAO, 2009), satellite imagery (USGS, 2015) and land cover mapping. Indeed, most yield models perform crop yield assessments, however, very

TABLE 7.6
Certification of Estimated Cotton Production (tonnes) (from Dalezios, 2015)

Territory	Production 1998	Production 1999	Estimation 1998	Estimation 1999	Deviation 1998(%)	Deviation 1999 (%)
Central Greece	723370	733862	689512	693145	4,5	3,5
Northern Greece	310738	458875	375549	370158	20,0	19,0
Greece	1033108	1192737	1195560	1192335	15,0	0,03

TABLE 7.7

Descriptions of the different approaches to synthesizing the input weather data in crop simulation models to forecast crop yield. (from Basso and Liu, 2019)

Approaches	Selected references
Historical weather scenarios: the assembled weather input was composed by real-time weather data till the forecasting date and historical weather scenarios from the forecasting date to maturity	Li et al. (2015) and Machakaire et al. (2016)
Mean historical weather: the assembled weather input was composed by real-time weather data till the forecasting date and the averaged historical weather from the forecasting date to maturity	Dumont et al. (2014)
Weather generators: the assembled weather input was composed by real-time weather data till the forecasting date and weather from weather generators (e.g. LARS, SIMMETEO, and WGEN) for the remainder	Dumont et al. (2015)
Climate forecast models: the assembled weather input was composed by real-time weather data till the forecasting date and weather output from climate forecast models	Singh et al. (2017) and Togliatti et al. (2017)
Satellite-derived climatic variables: the unknown weather input was derived from satellites, such as the METEOSAT	Manatsa et al. (2011)

few of them include biomass production of pastures land, although grasslands occupy 70 percent of agricultural land. Although most of the solutions proposed refer to the regression of pasture biomass with remote sensing vegetation indices, some process-based models have also been developed (Taubert et al., 2012).

In crop simulation models for crop yield forecasting, different approaches have been used. Specifically, averaged historical weather, historical weather scenarios, weather data from weather generators and climate model output, as well as weather derived from satellites, have been included to prepare the required weather input to run crop simulation models and to obtain end-of-season crop yield (Basso and Liu, 2019). Table 7.7 presents descriptions of the different approaches to synthesizing the input weather data in crop simulation models to forecast crop yield. Finally, the combined methods integrate remotely sensed information into crop growth models. The latter two methods require a large amount of data, such as plant physiology, soil and site-specific characteristics, or daily weather conditions, whereas the statistical methods require minimum data.

7.4.3 EO-ASSISTED METHODOLOGY IN PRECISION AGRICULTURE

Precision agriculture or precision farming or operational agriculture can be defined as a management strategy which uses technologies that provide information from multiple sources, to produce data and make decisions related to production. The most important consideration of precision farming seems to be the way management decisions address spatial and temporal variability in production systems. A key difference between conventional management and precision farming is the application of modern information technologies, such as IoT (Internet of Things), ICT, GIS or remote sensing, which increase the flow of information in agriculture. Applications of precision agriculture with the aim of managing both the crop water and fertilizer (nitrogen) needs (cotton, fruit and vegetable, olive trees, etc.) is presented (González-Piqueras et al., 2017; Dalezios et al., 2019a; Dalezios et al., 2019b; Spyropoulos et al., 2019).

7.4.3.1 Remotely Sensed Crop Water Requirements

The present case study aims to calculate crop water requirements in vulnerable ecosystems of the Mediterranean region based on remote sensing data and methods, and to use precision agriculture approaches for optimal crop water monitoring and irrigation management (Dalezios et al., 2019b). A drought-prone area is selected, namely Thessaly in central Greece. Crop water requirements are considered within the precision agriculture context by assessing crop evapotranspiration (ET) through the synergistic use of WV-2 satellite images with ground-truth data sets (Dalezios et al., 2019b).

Estimation of ETo. Reference evapotranspiration (ETo) is computed through the Penman-Monteith method based on FAO-56 using conventional meteorological data:

$$ET_0 = \frac{0.408(R_n - G)\Delta + \gamma\left(\dfrac{900}{T_{mean} + 273}\right)u_2(e_s - e_a)}{\Delta + \gamma(1 + 0.34u_2)} \tag{7.19}$$

where T_{mean} is mean temperature (°C) and u^2 is 2 m height wind speed.

Estimation of Kc. The crop coefficient (Kc) is developed from NDVI index using red, NIR1 and NIR2 channels of WV-2. The Kc integrates the effect of features that distinguish a typical crop cultivation from the reference type of grass reference, having a uniform and a completely ground cover appearance (Dalezios et al., 2019b). The Kc values depend on crop type and growth stages, climate and soil evaporation. In the study area, the following relationship was developed as follows (number of observations = 1,000, standard error = 0.12):

$$Kc = 1.33 * (redNDVI) + 0.21 \left(R^2 = 0.92\right) \tag{7.20}$$

It should be mentioned that the behavior of RedEdge NDVI index (NIR–RedEdge)/(NIR + RedEdge) differs from that of NDVI. Indeed, the WV-2 red edge channel is more sensitive to the change of vegetation reflectance, so it was also generated to monitor cotton phenological cycle (Spyropoulos et al., 2019).

Estimation of Crop Evapotranspiration ETc. Crop Evapotranspiration ETc is the evapotranspiration from a crop with good health, optimum conditions of irrigation, fertilization, growing in large fields, and achieving high production yields under specified climatic conditions. In FAO method, ETc is calculated as follows:

$$ETc = Kc * ETo \tag{7.21}$$

where ETo is calculated by ground-based meteorological observations. The satellite-based ETcsat is calculated by utilizing the reference ETo from FAO-56 Penman-Monteith formula derived from meteorological data and Kc extracted using redNDVI of WV-2, respectively. Figure 7.8 presents

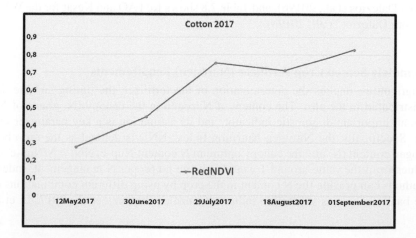

FIGURE 7.8 Cotton 2017 RedNDVI variation along the cultivation period (from Dalezios et al., 2019b).

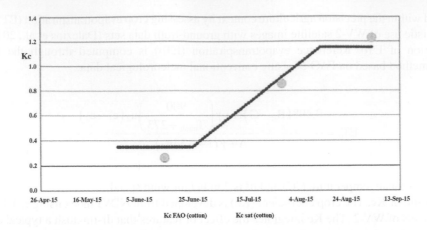

FIGURE 7.9 Comparison of Kcsat from WV-2 and Kc produced by FAO for the cotton field in 2015 (from Dalezios et al., 2019b).

TABLE 7.8
Kc FAO and Kc sat for the WV-2 acquisition days (from Dalezios et al. 2019b)

Cotton 2015			Cotton 2016			Cotton 2017		
Acquisition date	ETc FAO (mm)	ETc sat (mm)s	Acquisition date	ETc FAO (mm)	ETc sat (mm)	Acquisition date	ETc FAO (mm)	ETc sat (mm)
						May 12	1,96	3,19
June 13	2,42	2,42				June 30	9,46	7,07
July 29	6,65	6,11	July 8	6,24	5,46	July 29	7,19	7,56
			August 12	5,66	5,17	August 18	6,06	6,11
Sept 2	6,64	7,39				Sept. 1	5,35	6,31

cotton 2017 redNDVI variation along the cultivation period (from Dalezios et al., 2019b). Moreover, Figure 7.9 shows the comparison of Kcsat from WV-2 and Kc produced by FAO for the cotton field in 2015 (from Dalezios et al., 2019b), and Table 7.8 shows Kc FAO and Kcsat for the WV-2 acquisition days (from Dalezios et al., 2019b).

7.4.3.2 Remotely Sensed Crop Fertilizer (Nitrogen) Requirements

Precision agriculture implies the determination of the fertilizer, the timing and the doses to be properly distributed in the plot. The content of Nitrogen in the canopy (N, grams of N/grams of biomass) is an important diagnostic indicator, and its estimation is a key parameter in precision agriculture. Specifically, the Nitrogen Nutrition Index (NNI) is defined as the ratio between the actual nitrogen content (N) and the canopy optimum N content Nup-c, NNI = N/Nup-c. Deviations of NNI values from the value around 1 would indicate no proper N management. Indeed, remote sensing methods can provide the N content in the crop by using different combinations of spectral reflectance bands known as vegetation indices (González-Piqueras et al., 2017; Li et al., 2011).

TABLE 7.9

Definitions of the Red-Edge vegetation indices applied. The table includes some indications about their use (from González-Piqueras et al. 2017)

Vegetation Index	Formula	Use	Reference
Red Edge Position	$R_{700} + 40 \times \dfrac{\left[\dfrac{R_{670} + R_{780}}{2} - R_{700} \right]}{R_{740} - R_{700}}$	Sensitive in variations of Clorophyll and N.	González-Piqueras et al. (2017)
Normalized difference red edge index (NDRE)	$\left(R_{790} - R_{720} \right) \Big/ \left(R_{790} + R_{720} \right)$	Sensitive in variations of Clorophyll and N.	Fitzgerald et al. (2010)
Red edge chlorophyll index	$\dfrac{R_{790}}{R_{720}} - 1$	Estimation of N plant uptake at different bandwidths.	Gitelson et al. (2005) and Dash and Curran (2004)
MERIS terrestrial chlorophyll index (MTCI)	$\left(R_{750} - R_{710} \right) \Big/ \left(R_{710} - R_{680} \right)$	N plant concentration after heading. N uptake before heading.	Dash and Curran (2004)
Canopy chlorophyll content index (CCCI)	$\left(NDRE - NDRE_{MIN} \right) \Big/ \left(NDRE_{MAX} - NDRE_{MIN} \right)$	N plant concentration after heading. N uptake across growth stages.	Fitzgerald et al. (2010)
Angular Insensitivity Vegetation Index (AIVI)	$\dfrac{R_{445} \cdot \left(R_{720} + R_{735} \right) - R_{573} \cdot \left(R_{720} - R_{735} \right)}{R_{720} \cdot \left(R_{573} + R_{445} \right)}$	Stability estimating N at different view zenith angles.	He et al. (2016)

The red edge spectral band (700 ± 40 nm) is a transition region with a rapid change in leaf reflectance caused by the strong absorption in the red, and leaf scattering in the NIR spectrum. Due to this crop spectral behavior of the red, red-edge and near-infrared region, sensors are used capable of acquiring reflectance at different spatial, spectral and temporal resolutions in order to estimate the Chlorophyll and N content in crops, such as the recently launched multispectral medium resolution Sentinel-2. Table 7.9 presents a few Red-Edge vegetation indices used in N content estimation.

In summary, the process for assessing nitrogen requirements includes the following steps, which are presented as a flow chart in Figure 7.10.

1. Image pre-processing is conducted for atmospheric correction in order to obtain the reflectance in the base of the atmosphere (BOA). For Sentinel-2, the L2 product is usually provided with the atmospheric correction. For WV-2, the L2 product may also be provided, if not, the FLAASH module implemented in ENVI can be applied to obtain the BOA values.
2. Red-edge vegetation indices are obtained from Sentinel-2 or WV-2 (Table 7.8).
3. FAPAR is obtained based on empirical linear relationship. For example, for central Spain the following regression equation has been developed: FAPAR=1.25·NDVI - 0.11 (González-Piqueras et al., 2017). Similarly, transpiration can be obtained by using empirical relationship, such as the following regression equation for central Spain: $T = K_T \cdot Eto$, where $K_T = 1.51 \cdot NDVI - 0.23$ (González-Piqueras et al., 2017).
4. The biomass is obtained by applying a crop growth model, e.g. CROPWAT, or by integrating the FAPAR or transpiration, which give similar results (Campos et al., 2018).

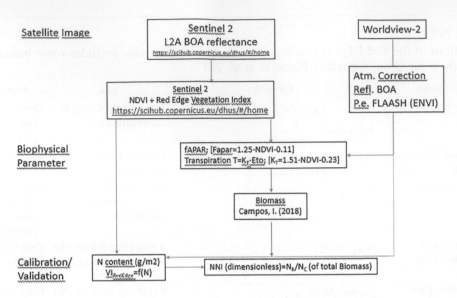

FIGURE 7.10 Flow chart of image processing methodology for the estimation of nitrogen requirements.

5. Once the red-edge vegetation indices are obtained from Sentinel-2 or WV-2, which are a function of N, then they are calibrated/validated with field observations of total nitrogen content N (g/m²) and nitrogen nutrition index NNI (dimensionless) (González-Piqueras et al., 2017). Specifically:

$$N_R\left(g\,/\,100\,g\,\text{total biomass}\right) = \left[N\left(g\,/\,m^2\right)\,/\,\text{biomass}\left(g\,/\,m\right)\right] \qquad (7.22)$$

$$NNI\left(\text{dimensionless}\right) = N_R\left(g\,/\,100\,g\,\text{total biomass}\right)\,/\,Nc\left(g\,/\,100\,g\,\text{total biomass}\right) \qquad (7.23)$$

where N_R is the computed nitrogen N from satellites (Equation 7.22) and N_C is the optimal N (reference), which for wheat receives the values: $N_C = 5.35$ (g/100g), if the biomass $W_R \leq 1$ t/ha, and $N_C = 5.35 \times W_R^{-0.44}$, if $W_R \geq 1$ t/ha. Then, the deficit of nitrogen N for the plant is $\Delta N = N_C - N_R$, based on the curve of Figure 7.11.

7.5 AGROCLIMATIC CLASSIFICATION AND ZONING

Agroclimatic classification is very useful in identifying sustainable production zones within a climatic region. Recently, a methodology has been developed following three steps based on remote sensing data and methods and applied in a semi-arid region (Tsiros et al., 2009; Dalezios et al., 2018). This approach results in optimal production, which can be computed for each sustainable production zone (e.g. high, medium, low productivity zones), at a multi-scale level, within a major climatic region (Figure 7.12).

7.5.1 HYDROCLIMATIC ZONES

The first step is the so-called hydroclimatic classification, which addresses the Water Limited Growth Environment (WLGE) and identifies zones adequate for sustainable farming (Dalezios et al., 2018).

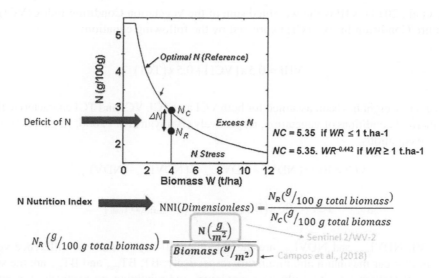

FIGURE 7.11 Curve showing the relationship between plant nitrogen demand and biomass.

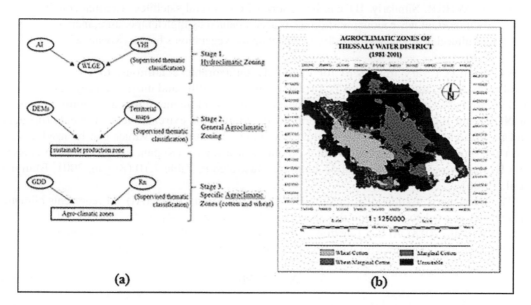

FIGURE 7.12 (a) Flow chart of the agroclimatic zoning methodology (from Dalezios et al., 2018). (b) Map of Thessaly showing specific agroclimatic zones (3rd stage) (from Dalezios et al., 2018).

In order to determine such zones, satellite-derived Vegetation Health Index (VHI) and Aridity Index (AI) are used (Tsiros et al., 2009). The WLGE zones are identified through superposition of the images of the two indices over an area and describe the hydroclimatic component of the agroclimatic zoning (Dalezios et al., 2018).

Water limited growth environment (WLGE). The first index used to identify WLGE is VHI. VHI represents overall vegetation health (moisture and thermal conditions) and is used for identification of vegetative stress and areas affected by agricultural drought (Kogan, 2001;

Dalezios et al., 2014). VHI is the weighted sum of the Vegetation Condition Index (VCI) and the Temperature Condition Index (TCI) expressed by the following equation:

$$VHI = 0.5 \times (VCI) + 0.5 \times (TCI) \tag{7.24}$$

where an equal weight has been assumed for both VCI and TCI. VCI and TCI characterize the moisture and thermal conditions of vegetation, respectively (Bhuiyan et al., 2006; Kogan, 2001), and are given by the equations:

$$VCI = 100 \times (NDVI - NDVI_{min}) / (NDVI_{max} - NDVI_{min}) \tag{7.25}$$

$$TCI = 100 \times (BT_{max} - BT) / (BT_{max} - BT_{min}) \tag{7.26}$$

where NDVI, $NDVI_{max}$ and $NDVI_{min}$ are the smoothed 10-day normalized difference vegetation index, its multi-year maximum and minimum, respectively; BT, BT_{max} and BT_{min} are the smoothed 10-day radiant temperature, its multi-year maximum and minimum, respectively, for each pixel, in an area (Dalezios et al., 2014). NDVI has been previously described and can be obtained from several satellites, e.g. by combining the channels 1 and 2, the visible and near-infrared, respectively, of NOAA/AVHRR. Similarly, BT can be obtained from several satellites. Thermal conditions are especially important when moisture shortage is accompanied by high temperature, increasing severity of agricultural drought, having a direct impact on the vegetation's health (Kogan, 2001; Dalezios et al., 2014).

VCI and TCI vary from zero for extremely unfavorable conditions, to 100 for optimal conditions. Thus, higher VCI and TCI values represent healthy and unstressed vegetation. Both indices are based on the same concept. Therefore, the absolute maximum and minimum values of NDVI and BT, calculated from several years, contain the extreme weather events (drought and no drought conditions). The resulted maximum and minimum values can be used as criteria for quantifying the environmental potential of a region (Kogan, 2001). The five classes of VHI that represent agricultural drought are illustrated in Table 7.10 (Kogan, 2001; Bhuiyan et al., 2006).

The other index used to identify WLGE zones is the Aridity Index (AI). AI represents climatic aridity and is used to determine the adequacy of rainfall in satisfying the water needs of crops (Dalezios et al., 2018). AI is a function of the ratio of the cumulative monthly precipitation (P) to monthly potential evapotranspiration (PET) based on the following equation:

$$AI = P / PET \tag{7.27}$$

TABLE 7.10
VHI drought classification scheme (from Dalezios et al., 2018)

VHI values	Vegetative drought classes
<10	Extreme drought
<20	Severe drought
<30	Moderate drought
<40	Mild drought
>40	No drought

TABLE 7.11

Dryland categories (UNESCO, 1979)

Aridity Index: P/PET	Rainfall (mm)	Classification
PET>P		Desert climate
<0.03	<200	Hyper-arid
0.03 to <0.20	<200 (winter) <400 (summer)	Arid
0.20 to <0.50	200–500 (winter) 400-600 (summer)	Semi-arid
0.50 to <0.65	500–700 (winter) 600-800 (summer)	Dry sub-humid
>0.65		No desertification

TABLE 7.12

WLGE generalized classification scheme (from Dalezios et al., 2018)

Agricultural drought classes	Aridity classes	WLGE classes
Extreme drought	Extremely dry	
Severe drought	Dry	Limited environment
Moderate drought	Semi-dry	
Mild drought	Semi-wet	Partially limited environment
No drought	Wet	No limitations

The index is calculated on multi-year basis, using monthly values. The calculation of PET is based on the Blaney-Criddle method (Tsiros et al., 2009) using monthly air temperature data, the ratio of daytime hours (month/year) and a weighted crop coefficient (Kc). Kc values are defined according to land use provided by the CORINE database. In PET calculations, LST is used instead of air temperature. The generation of LST maps is based on the "split-window" algorithm (Tsiros et al., 2009). In order to avoid over-estimating PET, LST is converted to air temperature using a linear empirical relationship by applying a regression analysis to the LST and air temperature data of the time series for one or more stations in the area (Dalezios et al., 2018). The categories as they are defined by the values of AI are illustrated in Table 7.11 (UNESCO, 1979).

The derived map is combined with the climatic aridity map and leads to the definition of WLGE zones. The generalized thematic classification scheme is shown in Table 7.12.

7.5.2 Agroclimatic Zones: Sustainable Production Zones

General Agroclimatic Zones. The second step is to identify sustainable production zones, which characterize the general agroclimatic zones in terms of water efficiency, fertility (appropriate or not for agricultural use), desertification vulnerability and altitude restrictions. Thus, WLGE zones are combined with Soil Maps and a Digital Elevation Model (DEM) (Tsiros et al., 2009). Soil maps and DEM overlap WLGE zones, which leads to the identification of sustainable crop production regions, where agriculture is the best suited agronomic use. Soil types are digitized according to fertility (appropriate or not for sustainable agricultural use) and desertification vulnerability. Three major crop growth zones are selected according to altitude limitations (Dalezios, 2015). The first, ranging from 0 to 600 m, is appropriate for most of the crops. The second, ranging from 600 m to 900 m is appropriate for non-tropic crops and fruit trees. The last one, having altitudes higher than 900 m, is not appropriate for crops (Dalezios et al., 2018).

Specific Agroclimatic Zoning: Sustainable Production zones. Two indices are used to identify areas suitable for cultivation, namely Growing Degree Days (GDD) and Net Radiation (Rn) (Tsiros et al., 2009). Rn is used to define areas, where crop growth is not restricted due to radiation limitations. This leads to production classification zones, i.e. high, medium and low productivity zones, where, for instance, in low productivity zones energy crops could be considered (Tsiros et al., 2009; Dalezios et al., 2018). Figure 7.13 summarizes the flow chart of the various steps of the agroclimatic zoning methodology.

7.5.3 MULTI-SCALING AGROCLIMATIC ZONES

Agroclimatic classification at meso-scale level based on additional seven bioclimatic indices has been achieved leading to zones with different suitability and their possible geographical shift under future climate change scenarios using regional climate model simulations (Kogan, 2001; Malheiro et al., 2010). These indices include length of the growing season (LGS), growing season precipitation (GSP), Huglin heliothermal index (HI), cool night index (CI), hydrothermal index (HyI), dryness index (DI) and composite index (CompI) (Malheiro et al., 2010). The combination of all eleven (11) indices leads to finer scale agroclimatic classification even in hilly terrain. Malheiro et al. (2010) also considered climate change scenarios applied to viticultural zoning in Europe.

7.6 SUMMARY

There is a gradually increasing trend for the use of remote sensing in the broad field of agriculture (Dalezios, 2015). Moreover, a major consideration for remote sensing use in agriculture is the extent to which operational users can rely on a continued supply of data. Indeed, satellite systems provide temporally and spatially continuous data over the globe and, thus, they are potentially better and relatively inexpensive tools for regional applications, such as quantification, monitoring and assessment of agricultural aspects, than conventional environmental and weather data. For these types of applications, appropriate remote sensing systems are weather radars and satellites that provide low spatial and high temporal resolution data, in cases that daily coverage and data acquisition is required. The series of geosynchronous, polar-orbiting meteorological satellites fulfill the above requirements and there are already long series of data sets.

In Chapter 8, remote sensing applications to agriculture are considered, which fall broadly into three major categories: (1) soil/land classification and crop mapping; (2) identification of stress in crops and generally vegetation; (3) monitoring and forecasting of crop production. Specifically, crop type classification, crop condition assessment, crop yield estimation, mapping of soil characteristics, mapping of soil management practices, compliance monitoring in terms of farming practices are covered. In addition, remote sensing methods are considered and presented for estimating land surface temperature (LST) and sea surface temperature (SST), vegetation indices, energy and water balance, soil moisture, water stress and deficit, land use and crop classification, crop and biomass monitoring and estimation of risks, as well as precision farming.

In addition, in Chapter 8, it is also highlighted that remote sensing is a useful tool to analyze the vegetation dynamics in agriculture on local, regional or global scales and to determine the impact of climate on vegetation (Kogan, 2001). Specifically, it is presented that satellite data can be used to monitor the onset of the agrometeorological conditions to vegetation and provide details about its condition. Moreover, satellite systems provide temporally and spatially continuous data over the globe and, thus, they are potentially better and relatively inexpensive tools for regional applications, such as monitoring vegetation's condition, mapping of agricultural land and crop yield estimation than conventional weather data (Domenikiotis et al., 2004a). Studies have been carried out in several parts of the world (Domenikiotis et al., 2004b; Basso et al., 2013; Wu et al., 2015) showing the potential of satellite-based data in agricultural monitoring and climate impact assessment.

REFERENCES

Al-Gaadi, K.A., Hassaballa, A.A., Tola, E., Kayad, A.G., Madugundu, R., Alblewi, B., and Assiri, F., 2016. Prediction of potato crop yield using precision agriculture techniques. *PLOS ONE*, 11,9, e0162219, https://doi.org/10.1371/journal.pone.0162219

Arshad, S., Morid, S., Mobasheri, M.R., Alikhani, M.A., and Arshad, S., 2013. Monitoring and forecasting drought impact on dryland farming areas. *International Journal of Climatology*, 33, 2068–2081, doi:10.1002/joc.3577

Asrar, G., Fuchs, M., Kanemasu, E.T., and Hatfield, J.L., 1984. Estimating absorbed photosynthetic radiation and leafarea index from spectral reflectance in wheat. *Agronomy Journal*, 76, 300–306.

Asrar, G., Myneni, R.B., and Choudhury, B.J., 1992. Spatial heterogeneity in vegetation canopies and remote-sensing of absorbed photo photosynthetically active radiation – A modeling study. *Remote Sensing of Environment*, 41, 85–103.

Bandyopadhyay, K.K., Pradhan, S., Sahoo, R.N., Ravender, S., Gupta, V.K., Joshi, D.K., and Sutradhar, A.K., 2014. Characterization of water stress and prediction of yield of wheat using spectral indices under varied water and nitrogen management practices. *Agricultural Water Management*, 146, 115–123, 10.1016/j.agwat.2014.07.017

Baret, F., and Olioso, A., 1989. Estimation à partir de mesures de réflectance spectrale du rayonnement photosynthetiquement actif absorbé par une culture de blé. *Agronomie*, 9, 885–895.

Baret, F., Guyot, G., and Major, D., 1989. Crop biomass evaluation using radiometric measurements. *Photogrammetria*, 43, 241–256.

Baret, F., Pavageau, K., Béal, D., Weiss, M., Berthelot, B., and Regner, P., *Algorithm Theoretical Basis Document for MERIS Top of Atmosphere Land Products (TOA_VEG)*. INRA-CSE: Avignon, France, 2006.

Baret, F., Hagolle, O., Geiger, B., Bicheron, P., Miras, B., Huc, M., Berthelot, B., Niño, F., Weiss, M., and Samain, O., 2007. LAI, fAPAR and fCover CYCLOPES global products derived from VEGETATION: Part 1: Principles of the algorithm. *Remote Sensing of Environment*, 110, 275–286.

Baret, F., Weiss, M., Verger, A., and Smets, B., 2019. ATBD for LAI, FAPAR and FCOVER from PROBA-V products at 300m resolution (GEOV3). Available online: https://land.copernicus.eu/global/sites/cgls.vito.be/files/products/ImagineS_RP2.1_ATBD-FCOVER300m_I1.73.pdf [accessed: September 22, 2019].

Basso, B., and Liu L., 2019. Seasonal crop yield forecast: Methods, applications, and accuracies. *Advances in Agronomy*, 154, 201–255.

Basso, B., Cammarano, D., and Carfagna, E., 2013. *Review of crop yield forecasting methods and early warning systems. Report to the Global Strategy to Improve Agricultural and Rural Statistics*. FAO Publication: Rome.

Bégué, A., 1993. Leaf area index, intercepted photosynthetically active radiation, and spectral vegetation indices: A sensitivity analysis for regular-clumped canopies. *Remote Sensing of Environment*, 46, 45–59.

Bhuiyan, C., Singh, R.P., and Kogan, F.N., 2006. Monitoring drought dynamics in the Aravalli region (India) using different indices based on ground and remote sensing data. *International Journal of Applied Earth Observation and Geoinformation*, 8, 289–302.

Bolton, D.K., and Friedl, M.A., 2013. Forecasting crop yield using remotely sensed vegetation indices and crop phenology metrics. *Agricultural and Forest Meteorology*, 152, 223–232.

Bu, H., Sharma, L.K., Denton, A., and Franzen, D.W., 2017. Comparison of satellite imagery and ground-based active optical sensors as yield predictors in sugar beet, spring wheat, corn, and sunflower. *Agronomy Journal*, 109, 299–308.

Campos, I., González-Gómez, L., Villodre, J., González-Piqueras, J., Suyker, A.E., and Calera, A., 2018. Remote sensing-based crop biomass with water or light-driven crop growth models in wheat commercial fields. *Field Crops Research*, 216, 175–188.

Carlson, T.N., Dodd, J.K., Benjamin, S.G., and Cooper, J.N., 1981. Satellite estimation of the surface energy balance, moisture availability and thermal inertia. *Journal of Applied Meteorology*, 20, 67–81.

Chipanshi, A., Zhang, Y., Kouadio, L., Newlands, N., Davidson, A., Hill, H., Warren, R., Qian, B., Daneshfar, B., Bedard, F., and Reichert, G., 2015. Evaluation of the ICCYF model for in-season prediction of crop yield across the Canadian agricultural landscape. *Agricultural and Forest Meteorology*, 206, 137–150.

Dabrowska-Zielinska, K., Kogan, F.N., Ciolkosz A., Gruszczynska, M., and Kowalik, W., 2002. Modelling of crop growth conditions and crop yield in Poland using AVHRR-based indices. *International Journal of Remote Sensing*, 23, 1109–1123.

Dalezios, N.R., 2013. The Role of Remotely Sensed Vegetation Indices in Contemporary Agrometeorology. Invited paper in Honorary Special Volume in memory of late Prof. A. Flokas. Hellenic Meteorological Association, 33–44.

Dalezios, N.R., 2015. Agrometeorology: Analysis and Simulation (in Greek). KALLIPOS: Libraries of Hellenic Universities (also e-book), ISBN: 978-960-603-134-2, 481 pages, November 2015.

Dalezios, N.R., Blanta, A., Spyropoulos, N.V., and Tarquis, A.M., 2014. Risk identification of agricultural drought for sustainable agroecosystems. *Natural Hazards And Earth System Sciences*, 14, 2435–2448.

Dalezios, N.R., Mitrakopoulos, K., and Manos, B., 2018. Multi-scaling Agroclimatic Classification for Decision Support towards Sustainable Production, pp. 1–42, in *Multi-criteria Analysis in Agriculture*. Springer

Dalezios, N.R., Blanta, A., Loukas, A., Spiliotopoulos, M., Faraslis, I.N., and Dercas, N., 2019a. Satellite methodologies for rationalizing crop water requirements in vulnerable agroecosystems. *International Journal of Sustainable Agricultural Management and Informatics*, 5, 1, 37–58.

Dalezios, N.R., Dercas, N., Spyropoulos, N.V., and Psomiadis, M., 2019b. Remotely sensed methodologies for water availability and requirements in precision farming of vulnerable agriculture. *Water Resources Management*, 33, 1499–1519.

Dalezios, N.R., Petropoulos, G.P., and Faraslis, I.N., 2020. Concepts and Methodologies of Environmental Hazards and Disasters, pp. 3–22, in *Techniques for Disaster Risk Management and Mitigation*. Editors: P.K. Srivastava, S.K. Singh, U.C. Mohanty and T. Murty. AGU-Wiley, 352.

Das, D.K., Mishra, K.K., and Kalra, N., 1993. Assessing growth and yield of wheat using remotely sensed canopy temperature and spectral indices. *International Journal of Remote Sensing*, 14, 17, 3081–3092, doi: 10.1080/01431169308904421

Dash, J., and Curran, P.J., 2004. The MERIS terrestrial chlorophyll index. *International Journal of Remote Sensing*, 25, 5403–5413.

Daughtry, C.S.T., Gallo, K., and Bauer, M.E., 1983. Spectral estimates of solar radiation intercepted by corn canopies. *Agronomy Journal*, 75, 527–531.

Daughtry, C.S.T., Gallo, K., Goward, S., Prince, S., and Kustas, W., 1992. Spectral estimates of absorbed radiation and phytomass production in corn and soybean canopies. *Remote Sensing of Environment*, 39, 141–152.

de Wit, C.T., 1965. *Photosynthesis of Leaf Canopies. Agricultural Research Report No. 683*. Center for Agriculture Publications and Documentation: Wageningen, The Netherlands.

Dempewolf, J., Adusei, B., Becker-Reshef, I., Hansen, M., Potapov, P., Khan, A., and Barker, B., 2014. Wheat yield forecasting for Punjab Province from vegetation index time series and historic crop statistics. *Remote Sensing*, 6, 9653–9675.

Domenikiotis, C., Spiliotopoulos, M., Tsiros, E., and Dalezios, N.R., 2004a. Early cotton yield assessment by the use of the NOAA/AVHRR derived drought vegetation condition index in Greece. *International Journal of Remote Sensing*, 25, 14, 2807–2819.

Domenikiotis, C., Spiliotopoulos, M., Tsiros, E., and Dalezios, N.R., 2004b. Early cotton production assessment in Greece based on the combination of the drought vegetation condition index (VCI) and Bhalme and Mooley drought index (BMDI). *International Journal of Remote Sensing*, 25, 5373–5388.

Dumont, B., Leemans, V., Ferrandis, S., Bodson, B., Destain, J.-P., Destain, M.-F., 2014. Assessing the potential of an algorithm based on mean climatic data to predict wheat yield. *Precision Agriculture*, 15, 255–272.

Dumont, B., Basso, B., Leemans, V., Bodson, B., Destain, J.P., and Destain, M.F., 2015. A comparison of within-season yield prediction algorithms based on crop model behaviour analysis. *Agricultural and Forest Meteorology*, 204, 10–21.

Fang W., Huang, S., Huang, Q., Huang, G., Wang, H., Leng, G., Wang, L., and Guo, Y., 2019. Probabilistic assessment of remote sensing-based terrestrial vegetation vulnerability to drought stress of the Loess Plateau in China. *Remote Sensing of Environment*, 232, 111290.

FAO, 1998. Crop Evapotranspiration, in Allen, R., Pereira, L.A., Raes, D., and Smith, M., *FAO Irrigation and Drainage*, Paper No. 56, FAO, Rome.

FAO, 2009. *Harmonized World Soil Database*. FAO, Rome.

Fieuzal, R., Sicre, C.M., and Baup, F., 2017. Estimation of corn yield using multi-temporal optical and radar satellite data and artificial neural networks. *International Journal of Applied Earth Observation and Geoinformation*, 57, 14–23.

Figueiredo, G.K.D.A., Brunsell, N.A., Higa, B.H., Rocha, J.V., and Lamparelli, R.A.C., 2016. Correlation maps to assess soybean yield from EVI data in Paraná State, Brazil. *Scientia Agricola*, 73, 5, 462–470, https://dx.doi.org/10.1590/0103-9016-2015-0215

Fitzgerald, G.J., Rodriguez, D., and O'Leary, G., 2010. Measuring and predicting canopy nitrogen nutrition in wheat using a spectral index – the canopy chlorophyll content index (CCCI). *Field Crops Research*, 116, 318–324.

Friedl, M.A., Davis, F.W., Michaelsen, J., and Moritz, M.A., 1995. Scaling and uncertainty in the relationship between the NDVI and land surface biophysical variables: An analysis using a scene simulation model and data from FIFE. *Remote Sensing of Environment*, 54, 233–246.

Fu, Y., Zhang, H., Dong, W., Yuan, W., 2014. Comparison of phenology models for predicting the onset of growing season over the Northern Hemisphere. *PLoS ONE*, 9, 10, e109544, https://doi.org/10.1371/journal.pone.0109544

Gallo, K., Daughtry, C., and Bauer, M.E., 1985. Spectral estimation of absorbed photosynthetically active radiation in corn canopies. *Remote Sensing of Environment*, 17, 221–232.

Gao, B.-C., 1996. NDWI – A normalized difference water index for remote sensing of vegetation liquid water from space. *Remote Sensing of Environment*, 58, 257–266.

García-Haro, F., Camacho-de Coca, F., Meliá, J., and Martínez, B., 2005. *Operational derivation of vegetation products in the framework of the LSA SAF project*. In *Proceedings of the 2005 EUMETSAT Meteorological Satellite Conference*, Dubrovnik, Croatia, September 19–23, pp. 19–23.

Geipel, J., Link, J., and Claupein, W., 2014. Combined spectral and spatial modeling of corn yield based on aerial images and crop surface models acquired with an unmanned aircraft system. *Remote Sensing*, 6, 10335–10355.

Gitelson, A.A., Vina, A., Ciganda, V., Rundquist, D.C., and Arkebauer, T.J., 2005. Remote estimation of canopy chlorophyll content in crops. *Geophysical Research Letters*, 32, 8.

Gobron, N., Pinty, B., AussÈdat, O., Taberner, M., Faber, O., MÈlin, F., Lavergne, T., Robustelli, M., and Snoeij, P., 2008. Uncertainty estimates for the FAPAR operational products derived from MERIS – Impact of top-of-atmosphere radiance uncertainties and validation with field data. *Remote Sensing of Environment*, 112, 1871–1883.

González-Piqueras J., López, H., Sánchez, S., Villodre, J., Bodas, V., Campos, I., Osann, A., and Calera, A., 2017. Monitoring crop N status by using red edge-based indices. Advances in Animal Biosciences, The Animal Consortium 2017 doi:10.1017/S2040470017000243

Goward, S.N., and Huemmrich, K.F., 1992. Vegetation canopy PAR absorptance and the normalized difference vegetation index: An assessment using the SAIL model. *Remote Sensing of Environment*, 39, 119–140.

Goward, S.N., Waring, R.H., Dye, D.G., and Yang, J.L., 1994. Ecological remote-sensing at OTTER – Satellite macroscale observations. *Ecological Applications*, 4, 322–343.

Hatfield, J., Asrar, G., and Kanemasu, E., 1984. Intercepted photosynthetically active radiation estimated by spectral reflectance. *Remote Sensing of Environment*, 14, 65–75.

Hayes, M.J., and Decker, W.L., 1996. Using NOAA AVHRR data to estimate maize production in the United States corn belt. *International Journal of Remote Sensing*, 17, 3189–3200.

He, L., Song, X., Feng, W., Guo, B.B., Zhang, Y.S., Wang, Y.H., Wang, C.Y., and Guo, T.C., 2016. Improved remote sensing of leaf nitrogen concentration in winter wheat using multi-angular hyperspectral data. *Remote Sensing of Environment*, 174, 122–133.

Holzman, M.E., and Rivas, R.E., 2016. Early maize yield forecasting from remotely sensed temperature/vegetation index measurements. *IEEE Journal of Selected Topics in Applied Earth Observations and Remote Sensing*, 9, 507–519.

Huete, A., 1988. A soil-adjusted vegetation index (SAVI). *Remote Sensing of Environment*, 25, 3, 295–309.

Idso, S.B., Pinter, P.J., Jackson, R.D., and Reginato, R.J., 1980. Estimation of grain yields by remote sensing of crop senescence rates. *Remote Sensing of Environment*, 9, 1, 87–91

IPCC, 2012. Managing the Risks of Extreme Events and Disasters to Advance Climate Change Adaptation. Special Report of IPCC, 582p.

Jackson, R.D., Pinter, P.J., Reginato, R.J., and Idso, S.B., 1986. Detection and evaluation of plant stresses for crop management decision. *IEEE Transactions on Geoscience and Remote Sensing*, 24, 99–106.

Jia, K., Liang, S., Liu, S., Li, Y., Xiao, Z., Yao, Y., Jiang, B., Zhao, X., Wang, X., and Xu, S., 2015. Global land surface fractional vegetation cover estimation using general regression neural networks from MODIS surface reflectance. *IEEE Transactions on Geoscience and Remote Sensing*, 53, 4787–4796.

Johnson, D.M., 2014. An assessment of pre- and within-season remotely sensed variables for forecasting corn and soybean yields in the United States. *Remote Sensing of Environment*, 141, 116–128.

Johnson, M.D., Hsieh, W.E., Cannon, A.J., Davidson, A., and Bedard, F., 2016. Crop yield forecasting on the Canadian prairies by remotely sensed vegetation indices and machine learning methods. *Agricultural and Forest Meteorology*, 218, 74–84.

Kancheva, R., Borisova, D., and Georgiev, G., 2007. Spectral predictors of crop development and yield. IEEE.

Kogan, F.N., 2001. Operational space technology for global vegetation assessment. *Bulletin of the American Meteorological Society*, 82, 1949–1964.

Kolotii, A., Kussul, N., Shelestov, A., Skakun, S., Yailymov, B., Basarab, R., Lavreniuk, M., Oliinyk, T., and Ostapenko, V., 2015. Comparison of biophysical and satellite predictors for wheat yield forecasting in Ukraine International Archives of the Photogrammetry, Remote Sensing and Spatial Information Sciences. *ISPRS Archives*, 40, 7W3, 39–44.

Kowalik, W., Dabrowska-Zielinska, K., Meroni, M., Raczka, T.U., and de Wit, A., 2014. Yield estimation using Spot-vegetation products: A case study of wheat in European countries. *International Journal of Applied Earth Observation and Geoinformation*, 32, 228–239.

Kumar, M., and Monteith, J., 1981. *Remote sensing of crop growth*, pp. 133–144, in *Plants and the Daylight Spectrum: Proceedings of the First International Symposium of the British Photobiology Society*, Leicester, UK.

Kuri, F., Murwira, A., Murwira, K.S., and Masocha, M., 2014. Predicting maize yield in Zimbabwe using dry dekads derived from remotely sensed vegetation condition index. *International Journal of Applied Earth Observation and Geoinformation*, 33, 39–46.

Kussul, N., Kolotii, A., Skakun, S., Shelestov, A., Kussul, O., and Oliynuk, T., 2014. *Efficiency estimation of different satellite data usage for winter wheat yield forecasting in Ukraine*. IEEE Geoscience and Remote Sensing Symposium, Quebec City, QC, 5080–5082, doi: 10.1109/IGARSS.2014.6947639.

Kustas, W.P., Craig, S., Daughtry, T., and Van Oevelen, P.J., 1993. Analytical treatment of the relationship between soil heat flux/net radiation ratio and vegetation indices. *Remote Sensing of Environment*, 46, 319–330.

Li, G., Zhang, H., Wu, X., Shi, C., Huang, X., and Qin, P., 2011. Canopy reflectance in two castor bean varieties (Ricinus communis L.) for growth assessment and yield prediction on coastal saline land of Yancheng District, China. *Industrial Crops and Products*, 33, 2, 395–402.

Li, Z., Song, M., Feng, H., and Zhao, Y., 2015. Within-season yield prediction with different nitrogen inputs underrain-fed condition using CERES-wheat model in the northwest of China. *Journal of the Science of Food and Agriculture*, 96, 2906–2916.

Lin, W.-S., and Kuo, B.-J., 2013. Using the orthogonal projections methods for predicting rice (Oryza sativa L.) yield with canopy reflectance data. *International Journal of Remote Sensing*, 34, 1428–1448.

Linkosalo, T., Lappalainen, H.K., and Hari, P., 2008. A comparison of phenological models of leaf bud burst and flowering of boreal trees using independent observations. *Tree Physiology*, 28, 1873–1882.

Liu, D., Jia, K., Wei, X., Xia, M., Zhang, X., Yao, Y., Zhang, X., and Wang, B., 2019. Spatiotemporal comparison and validation of three global-scale fractional vegetation cover products. *Remote Sensing*, 11, 2524, https://doi.org/10.3390/rs11212524.

Machakaire, A.T.B., Steyn, J.M., Caldiz, D.O., and Haverkort, A.J., 2016. Forecasting yield and tuber size of processing potatoes in South Africa using the LINTUL-potato-DSS model. *Potato Research*, 59, 195–206.

Malheiro, A.C., Santos, J.A., Fraga, H., and Pinto, J.G., 2010. Climate change scenarios applied to viticultural zoning in Europe. *Climate Research*, 83, 163–177.

Manatsa, D., Nyakudya, I.W., Mukwada, G., and Matsikwa, H., 2011. Maize yield forecasting for Zimbabwe farming sectors using satellite rainfall estimates. *Natural Hazards*, 59, 447–463.

Maresma, Á., Ariza, M., Martínez, E., Lloveras, J. and Martínez-Casasnovas, J.A., 2016. Analysis of vegetation indices to determine nitrogen application and yield prediction in maize (Zea mays L.) from a standard UAV service. *Remote Sensing*, 8, 973.

Martín-Sotoca, J.J., Saa-Requejo, A., Moratiel, R., Dalezios, N., Faraslis, J.N., and Tarquis, A.M., 2019. Statistical analysis for satellite-index-based insurance (SIBI) to define damaged pasture thresholds. NHESS, *Natural Hazards and Earth System Sciences*, 19, 1685–1702.

Meroni, M., Rembold, F., Verstraete, M.M., Gommes, R., Schucknecht, A., and Beye, G., 2014. Investigating the relationship between the inter-annual variability of satellite-derived vegetation phenology and a proxy of biomass production in the Sahel. *Remote Sensing*, 6, 5868–5884, doi:10.3390/rs6065868.

Myneni, R.B., and Williams, D.L., 1994. On the relationship between FAPAR and NDVI. *Remote Sensing of Environment*, 49, 200–211.

NASA, 1987. *Heat Capacity Mapping Mission User's Guide. NASA*, Goddard Space Flight Center, Maryland, USA.

Navalgund, R.R., Parihar, J.S., and Nageshwara Rao, P.P., 1991. Crop inventory using remotely sensed data. *Current Science*, 61, 162–171.

Navalgund, R.R., Parihar, J.S., and Nageshwara Rao, P.P., 2000. Crop inventory using remotely sensed data. *Indian Journal of Agricultural Economy*, 55, 96.

NCAR, 2014. Precipitation Data Sets: Overview & Comparison Table. Available at: https://climatedataguide.ucar.edu/climate-data/precipitation-data-sets-overview comparison-table, [accessed: September 4, 2015].

Orta, A.H., Başer, I., Şehirali, S., Erdem, T., and Erdem, Y., 2004. Use of infrared thermometry for developing baseline equations and scheduling irrigation in wheat. *Cereal Research Communications*, 32, 363–370.

Panigada, C., Busetto, L., Meroni, M., Amaducci, S., Rossini, M., Cogliati, S., Boschetti, M., Picchi, V., Marchesi, A., Pinto, F., Rascher, U., and Colombo, R., 2010. *EDOCROS: Early detection of crop water and nutrition stress by remotely sensed indicators. 4th International Workshop on Remote Sensing of Vegetation Fluorescence*, Valencia, Spain, 7.

Peralta, N.R., Assefa, Y., Du, J., Barden, C.J., and Ciampitti, I.A., 2016. Mid-season high-resolution satellite imagery for forecasting site-specific corn yield. *Remote Sensing*, 8, 848.

Peterson, D.L., Spanner, S.W., Running, S.W., and Teuber, K.B., 1987. Relationship of thematic mapper simulator data to leaf area index of temperate coniferous forests. *Remote Sensing of Environment*, 22, 323–341.

Pirnazar, M., Haghighi, N., Azhand, D., Ostad-Ali-Askari, K., Eslamian, S., Dalezios, N.R., and Singh, V.P., 2018. Land use change detection and prediction using Markov-CA and publishing on the web with Platform Map Server, Case Study: Qom Metropolis, Iran. *Journal of Geography and Cartography*, 1, 1–16

Price, J.C., 1980. The potential of remotely sensed thermal infrared data to infer soil moisture and evaporation. *Water Resources Research*, 16, 787–795.

Quarmby, N.A., Milnes, M., Hindle, T.L., and Silleos, N., 1993. The use of multi-temporal NDVI measurements from AVHRR data for crop yield estimation and prediction. *International Journal of Remote Sensing*, 14, 199–210.

Quemada, M., Gabriel, J.L., and Zarco-Tejada, P., 2014. Airborne hyperspectral images and ground-level optical sensors as assessment tools for maize nitrogen fertilization. *Remote Sensing*, 6, 2940–2962.

Rembold, F., Atzberger, C., Savin, I., and Rojas, O., 2013. Using low resolution satellite imagery for yield prediction and yield anomaly detection. *Remote Sensing*, 5, 4, 1704–1733.

Rock, B.N., Vogelmann, J.E., Williams, D.L., Vogelmann, A.F., and Hoshizaki, T., 1986. Remote detection of forest damage. *Bioscience*, 36, 439–445.

Rosenzweig, C., Jones, J.W., Hatfield, J.L., et al., 2013. The agricultural model intercomparison and improvement project (AgMIP): Protocols and pilot studies. *Agricultural and Forest Meteorology*, 170, 166–182.

Roujean, J.L., and Lacaze, R., 2002. Global mapping of vegetation parameters from POLDER multiangular measurements for studies of surface-atmosphere interactions: A pragmatic method and its validation. *Journal of Geophysical Research: Atmospheres*, 107, 1–6.

Royo, C., Aparicio, N., Villegas, D., Casadesus, J., Monneveux, P., and Araus, J.L., 2003. Usefulness of spectral reflectance indices as durum wheat yield predictors under contrasting Mediterranean conditions. *International Journal of Remote Sensing*, 24, 4403–4419.

Sellers, P.J., 1987. Canopy reflectance, photosynthesis and transpiration. 2. The role of biophysics in the linearity of their interdependence. *Remote Sensing of Environment*, 21, 143–183.

Sepulcre-Canto, G., Zarco-Tejada, P.J., Jimenez-Munoz, J.C., Sobrino, J.A., de Miguel, E., Villalobos, F.J., 2006. Detection of water stress in an olive orchard with thermal remote sensing imagery. *Agricultural and Forest Meteorology*, 136, 31–44.

Shanahan, J.F., Schepers, J.S., Francis, D.D., Varvel, G.E., and Wilhelm, W.W., 2001. Use of remote sensing imagery to estimate corn grain yield. *Agronomy Journal*, 93, 583–589.

Sharma, L.K., Bu, H., Denton, A., and Franzen, D.W., 2015. Active-optical sensors using red NDVI compared to red edge NDVI for prediction of corn grain yield in North Dakota, U.S.A. *Sensors*, 15, 27832–27853.

Sharma, L.K., Bali, S.K., Dwyer, J.D., Plant, A.B., and Bhowmik, A., 2017. A case study of improving yield prediction and sulfur deficiency detection using optical sensors and relationship of historical potato yield with weather data in Maine. *Sensors*, 17, 1095.

Singh, P.K., Singh, K.K., Singh, P., Balasubramanian, R., Baxla, A.K., Kumar, B., Gupta, A., Rathore, L.S., and Kalra, N., 2017. Forecasting of wheat yield in various agro-climatic regions of Bihar by using CERES-wheat model. *Journal of Agrometeorology*, 19, 346–349.

Song, C., Chen, J.M., Hwang, T., Gonsamo, A., Croft, H., Zhang, Q., Dannenberg, M., Zhang, Y., Hakkenberg, C., and Li, J., 2016. Ecological Characterization of Vegetation Using Multisensor Remote Sensing in the Solar Reflective Spectrum, pp. 533–575, in *Land Resources Monitoring, Modeling, and Mapping with Remote Sensing*. Taylor & Francis Group LLC.

Spyropoulos, N.V., Dalezios, N.R., Kaltsis, Y., and Faraslis, I., 2019. Very high-resolution satellite-based monitoring of crop (olive trees) evapotranspiration in precision agriculture. *International Journal of Sustainable Agricultural Management and Informatics* (accepted, in press).

Steinmetz, S., Guerif, M., Delecolle, R., and Baret, F., 1990. Spectral estimates of the absorbed photosynthetically active radiation and light-use efficiency of a winter wheat crop subjected to nitrogen and water deficiencies. *Remote Sensing*, 11, 1797–1808.

Steven, M.D., and Jaggard, K.W., 1995. Advances in Crop Monitoring by Remote Sensing, pp. 143–156, in *Advances in Remote Sensing*. Editors: M. Danson and S.E. Plununer. John Wiley & Sons Ltd., Chichester, West Sussex, England.

Steven, M., Biscoe, P., and Jaggard, K., 1983. Estimation of sugar beet productivity from reflection in the red and infrared spectral bands. *International Journal of Remote Sensing*, 4, 325–334.

Suárez, L., Zarco-Tejada, P.J., Berni, J.A.J., González-Dugo, V., Fereres, E., 2009. Modelling PRI for water stress detection using radiative transfer models. *Remote Sensing of Environment*, 113 730–744.

Taghvaeian, S., Chávez, J.L., and Hansen, N.C., 2012. Infrared thermometry to estimate crop water stress index and water use of irrigated maize in Northeastern Colorado. *Remote Sensing*, 4, 3619–3637, doi:10.3390/rs4113619

Taubert, F., Frank, K., and Huth, A., 2012. A review of grassland models in the biofuel context. *Ecological Modelling*, 245, 84–93.

Togliatti, K., Archontoulis, S.V., Dietzel, R., Puntel, L., and VanLoocke, A., 2017. How does inclusion of weather forecasting impact in-season crop model predictions? *Field Crops Research*, 214, 261–272.

Torino, M.S., Ortiz, B.V., Fulton, J.P., Balkcom, K.S., and Wood, C., 2014. Evaluation of vegetation indices for early assessment of corn status and yield potential in the Southeastern United States. *Agronomy Journal*, 106, 1389–1401.

Tsiros, E., Domenikiotis, C., and Dalezios, N.R., 2009. Sustainable production zoning for agroclimatic classification using GIS and remote sensing. *IDŐJÁRÁS*, 113, 1–2, 55–68.

UNESCO, 1979. Map of the world distribution of Arid regions: Explanatory note. Man and the Biosphere. Technical Notes 7, Paris.

USGS, 2015. Landsat 8 (L8) Data Users Handbook. LSDS-1574 Version 1.0. USGS Publication.

Verger, A., Baret, F., and Weiss, M., *GEOV2/VGT: Near real time estimation of global biophysical variables from VEGETATION-P data*. In *Proceedings of the MultiTemp 2013: 7th International Workshop on the Analysis of Multi-temporal Remote Sensing Images*, Banff, AB, Canada, 2013, pp. 1–4.

Verger, A., Baret, F., and Weiss, M., 2020. Algorithm Theoretical Basis Document of GEOV2 FVC. CGLOPS1_ATBD_FCOVER1kmV2_I1.41.pdf [accessed: September 22, 2019]. Available at: https://land.copernicus.eu/global/sites/cgls.vito.be/files/products/

Wang, M., Tao, F., and Shi, W., 2014. Corn yield forecasting in Northeast China using remotely sensed spectral indices and crop phenology metrics. *Journal of Integrative Agriculture*, 13, 7, 1538–1545.

Wetzel, P.J., Atlas, D., and Woodward, R.H., 1984. Determining soil moisture from geosynchronous satellite infrared data: A feasibility study. *Journal of Applied Meteorology and Climatology*, 23, 375–391.

Wiegand, C., Richardson, A., Escobar, D., and Gerbermann, A., 1991. Vegetation indices in crop assessments. *Remote Sensing of Environment*, 35, 105–119.

Wilson, W., and Sebaugh, J.L., 1981. *Established Criteria and selected Methods for Evaluating Crop Yield Models in the AgRISTARS Program*. In *American Statistical Association 1981. Proceedings of the Section on Survey Research Methods*, 24–31.

Wu, B., Gommes, R., Zhang, M., Zeng, H., Yan, N., Zou, W., Zheng, Y., Zhang, N., Chang, S., Xing, Q., and van Heijden, A., 2015. Global crop monitoring: A satellite-based hierarchical approach. *Remote Sensing*, 7, 3907–3933.

Zhang, X., Liao, C., Li, J., and Sun, Q., 2013. Fractional vegetation cover estimation in arid and semi-arid environments using HJ-1 satellite hyperspectral data. *International Journal of Applied Earth Observation and Geoinformation*, 21, 506–512.

Phon, H.A., Offield, T.W., and Watson, K., 1974. Thermal inertia mapping from satellite discriminations of geotogic units in Oman. *Journal Research of United States Geological Survey*, 2, 141–158.

8 Forestry

8.1 INTRODUCTION: REMOTE SENSING CONCEPTS OF FORESTRY

Remote sensing data and methods constitute a fundamental technological tool to support forest management. It is recognized that major issues in forest management may include forest depletion due to natural causes, such as fires and infestations or human activity, such as land conversion, burning or clear-cutting, as well as monitoring of forest health and growth for effective conservation and exploitation, since forestry production is an important global industry. In general, forests constitute one of the most valuable resources globally. Indeed, forests are exploited by humans to provide fuel, food, shelter, wildlife habitat, as well as daily supplies, such as paper or medicinal ingredients, among others. Moreover, forests contribute to balancing the Earth's CO_2 exchange and supply, being a key component in one of the most significant current environmental issues, namely global warming and greenhouse effect. In addition, tropical rainforests cover viable agricultural land and host a very large diversity of species, which also provides habitat for numerous animal species, and is an important source of medicinal ingredients.

Applications and uses of remote sensing data and methods can include cadaster information, forest sustainability, biodiversity, deforestation monitoring, reforestation monitoring and management, biophysical monitoring, such as wildlife habitat assessment, protection of watersheds and coastal zones, commercial logging operations, as well as other environmental concerns. Moreover, forest cover information is valuable to developing countries with limited previous recorded information on their forestry resources. In such cases, remote sensing technology can be part of the information solutions internationally, which, again, may include general cover type mapping, protection of coastal zones and watersheds through monitoring and mapping, fire/burn mapping, as well as monitoring of cutting practices and regeneration.

8.2 RECONNAISSANCE MAPPING

The section of reconnaissance mapping includes forest cover type classification in terms of identifying cover types, as well as agroforestry mapping. This section also covers forest monitoring, such as deforestation (rain forest) and REDD+, burned forests and reforestation and clear-cut mapping and regeneration.

8.2.1 Forest Cover Type Classification

8.2.1.1 Identifying Cover Types

Remote sensing has played a crucial role in forest cover type classification. This field has steadily increased the classification accuracy with the development of robust algorithms and dramatically decreased the financial cost of acquiring and editing the relative data. One of the most dynamic and reliable procedure is connected to remote sensing data fusion. The combination of sophisticated algorithms, such as Support Vector Machine (SVM) with additional airborne platforms like LiDAR may provide effective discrimination of forest types with high accuracy (almost 90%). This may happen in two stages, namely the classification of primary fuel types to low height vegetation (grass, shrubs) and trees, as well as the discrimination of additional fuel types based on vertical information by LiDAR data (García et al., 2011).

Another approach of forest types classification involves differentiation techniques and data sources. Immitzer et al. (2012) classified ten tree species using very high-resolution data (WorldView-2). The high accuracy of classification (> 80%) achieved through the combination of object-based classification of 8-bands data and the adoption of random forest classifier. It should be highlighted that the contribution of all the bands in classifying a high number of forest species is significant, whereas their usability decreases when classifying just several types of forest (Immitzer et al., 2012). In a similar context, Puletti, et al. (2017) performed a multitemporal classification of forest types highlighting the significant issue of phenology, without which the classification of forest types by only one season was considered infeasible. The satisfying result of classification accuracy (> 83%) indicates the feasibility of forest species classification in a very cost-effective manner using data of medium spatial resolution, namely, Sentinel-2 data of 10 and 20m bands.

The classification of forest areas with just relatively fine resolution (Landsat satellites) without temporal information constitutes a challenge. This happens because of the similar phenology of different vegetation species in one season. This limitation has been tackled by Jia et al. (2014) integrating multiple sources of remote sensing data. Specifically, they combined the Normalized Difference Vegetation Index (NDVI) coming from high spatial resolution data derived from the Landsat Enhanced Thematic Mapper Plus with the NDVI produced by temporal data of Moderate-resolution Imaging Spectroradiometer (MODIS). The conjunction of these indices increased the accuracy of classification by 5%, reaching 94% accuracy (Figure 8.1). The trend of remote sensing data fusion is growing, since the outcomes are characterized by higher efficiency compared to one-source data. Ke et al. (2010) proceeded to a comparative assessment of different methods of forest types classification. They implemented an object-based classification of very high spatial resolution data (QuickBird, 2.4 m), followed by the same type of classification for the LiDAR data. When applying the classification to the combined data (QuickBird and LiDAR), the highest accuracy result was achieved.

Finally, the contribution of ancillary spatial data in terms of different types of texture data, i.e. stem from geostatistics, should be highlighted that is used in conjunction with the adoption of dynamic and effective machine learning algorithms (e.g. SVM algorithm) in order to adequately increase the overall classification accuracy of forest types (Wijaya and Gloaguen, 2007).

8.2.1.2 Agroforestry Mapping

Agroforestry is a general term, which refers to the land-use system of cultivating woody perennials, either monoculture or polyculture, i.e. plantation, on agricultural land, regardless of the current

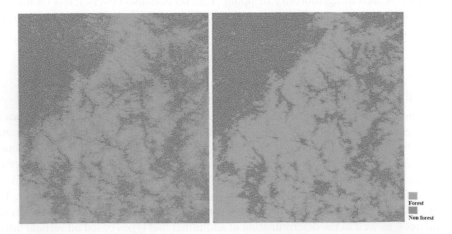

FIGURE 8.1 Forest cover classification of MLC using only Landsat ETM+ data (left) and the composited ETM+ spectral bands with time series NDVI features bands (right) (Jia et al., 2014).

existence of crops or animals. Over 10 million km² of agricultural lands have greater than 10% tree cover globally (Chen et al., 2016). Agroforestry has been adopted in many parts of the world to offer economic, social and ecological benefits: (1) mitigating the impacts of climate variability and change; (2) conserving biodiversity; (3) improving soil fertility and land productivity; (4) increasing the per capita farm income by planning high-value tree products; (5) improving air and water quality; (6) increasing household resilience.

Classification map. Remotely sensed data can be used to develop classification maps of the identifiable features or classes of the land-cover types in a scene. In fuzzy or soft classification, a pixel is usually associated with many land-cover classes, whereas in hard classification, each pixel is assigned with only one class. However, in fuzzy or soft classification, a pixel is associated with many land-cover classes. GIS and remote sensing applications in agroforestry include estimating agroforestry areas, suitability assessment for agroforestry systems, or monitoring of agroforestry parks (Rizvi et al., 2016).

An example is presented to delineate and estimate the agroforestry area in the district of Bathinda, India, using geospatial technologies (Rizvi et al., 2016). Multispectral remote sensing images of Resoursesat-2/ LISS III (resolution- 23.5 m) were procured from National Remote Sensing Centre, Hyderabad, India. LISS III scenes of path 93 and rows 49 and 50 (Date of pass: March 21, 2012) for Bathinda district was analyzed for land uses and land covers. Pre-processing of the scene includes layer stacking, mosaicking and sub-setting with district boundary, where the shape file of the district boundary was obtained from Survey of India, Dehradun. The agroforestry area was estimated using sub-pixel classifier on agriculture land of Bathinda district. Specifically, maximum likelihood classifier (MLC) was used to obtain land uses and land covers for masking agriculture land (crop land and fallow land). Indeed, with this method the pixels are classified into different classes of percent tree cover within pixel. In other words, pixels having minimum 20% to maximum 100% tree cover are classified as agroforestry. Moreover, scattered trees and boundary plantations are depicted by pixel having 20%–29% and 30%–39% tree cover. Table 8.1 shows the estimated agroforestry area in Bathinda district using sub-pixel classifier and Figure 8.2 presents the area under tree cover (%) on agricultural land in Bathinda district.

Aboveground (AGB) biomass. The biomass and C stock of different vegetation types, including agroforestry systems, can be estimated and monitored effectively through remote sensing (Chen et al., 2016). Indeed, airborne lidar (Light Detection and Ranging) has been recently considered as the most accurate technology to quantify forest aboveground biomass (AGB) at the landscape level. Specifically, Lidar is powerful over forests of high biomass, where passive optical imaging or radar

TABLE 8.1

Estimated agroforestry area in Bathinda district using sub-pixel classifier (from Rizvi et al., 2016)

Agroforestry systems	Tree cover (%)	Area in ha
Scattered trees/boundary plantations	20–29	7279.78
	30–39	8707.39
	40–49	8557.11
Agrisilviculture/Agri-horticulture system	50–59	7627.22
	60–69	6490.08
	70–79	5026.06
Block plantations	80–89	3291.90
	90–100	2497.77
	Agroforestry area	49477.31
	Geographical area	335152.15
	Agroforestry area—14.76%	

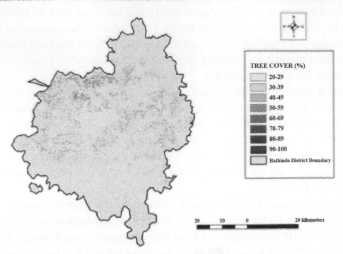

FIGURE 8.2 Area under tree cover (%) on agricultural land in Bathinda district (from Rizvi et al., 2016).

sensors have saturation problems. Indeed, Lidar can detect vertical structure and extract ground elevation by penetrating small canopy gaps. The height information derived from lidar is strongly related to biomass for most tree species. Several studies have reported the use of airborne lidar for mapping AGB in boreal, temperate and tropical forests (Chen et al., 2016).

A case study is presented, which refers to an agroforestry system in the Brazilian Amazon. At first, prediction of the plot-level AGB is considered based on fixed-effects regression models that assumed the regression coefficients to be constants (Chen et al., 2016). Then, the model prediction errors were analyzed from the perspectives of tree DBH (diameter at breast height) – height relationships and plot-level wood density. The results suggested the need for stratifying agroforestry fields to improve plot-level AGB modeling. The variation of AGB-height relationship across agroforestry types were integrated by mixed-effects models for prediction of AGB in order to separate teak plantations from other agroforestry types (Chen et al., 2016). Specifically, at the plot scale, it was found that mixed-effects models can have better model prediction performance than the fixed-effects models. Moreover, it was found that, at the landscape level, the difference between AGB densities from the two types of models was about 10% on average and up to about 30% at the pixel level. In summary, the analysis results indicate the significance of stratification based on tree AGB allometry and the use of mixed-effects models in modeling and mapping AGB of agroforestry systems. Figure 8.3 presents agroforestry and forest fields in airborne lidar.

Tree Hedge Rows (THR). In another case study, the estimation of the extent of THR in an Italian agroforestry landscape was mapped and the influence of THRs on the crop yields was assessed at the plot-farm scale (Chiocchini et al., 2018). In this study, the scattered or linear trees of the agroforestry, can be located either inside the field or along the field boundaries as tree hedge rows. Remote sensing, GIS spatial analysis and field surveying can be used to detect the complexity of the landscape patterns in order to understand the interactions between woody and crop components, and to assess, map and quantify the socioeconomic values of the agroforestry systems services. Specifically, photo interpretation of high-resolution multispectral Sentinel-2 (HRS2) images was used to identify the THRs. Indeed, based on NDVI from HRS2, it was possible to discriminate between areas with dense vegetation coverage (0.6 <NDVI <0.9, tree covered areas) and areas with low/zero vegetation cover. Narrow and long polygons corresponding to the crowns of the tree rows were identified from the 10 m spatial resolution of HRS2 scenes. THRs were identified in the two test areas (Tas) (Figure 8.4) and validated by comparison with the GPS surveys. This procedure was then applied throughout the study area to estimate the incidence of THRs per hectare.

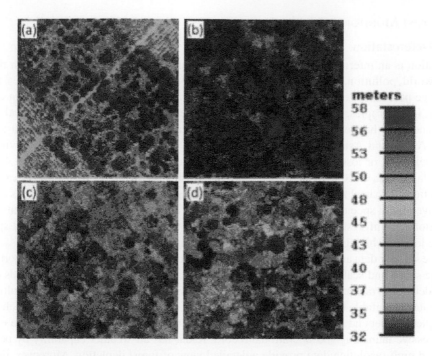

FIGURE 8.3 Examples (a–d) of agroforestry (left) and forests (right) fields shown in airborne lidar. Color (from red to blue) represents the elevation of the laser points (from Chen et al., 2016).

FIGURE 8.4 a) Detection of vegetated area and of tree hedgerows in the two test areas; b) Experimental sites with tree codes (Tx) along field margin and sampling transects (orange squares) (from Chiocchini et al., 2018).

8.2.2 Forest Monitoring

8.2.2.1 Deforestation

Deforestation is an international problem with significant implications. For instance, in the industrialized world, pollution, in terms of acid rain and chemicals from factory smoke plumes, has damaged a large percentage of forested land, such as in Central European and Scandinavian countries. Moreover, in the tropics, valuable rainforest is being destroyed in order to clear pastural and agricultural land. This has resulted in huge losses of tropical rainforest throughout Central and South America, Africa and Asia. Moreover, mangrove forest protects sensitive estuarine environments, and it is being cut or lost to urban development, aquaculture or damaged by pollution or siltation (Chamberlain et al., 2020). On the other hand, monitoring the health of mangrove forest supports the protection of coastal zones from erosion and flooding. In addition, the loss of forests affects the genetic diversity of species on Earth, which controls our adaptation ability to changing environmental conditions. Rainforests account for approximately one half of the plant and animal species on Earth, the loss of which is expected to reduce the gene and species pool.

Remote sensing data and methods consist of a multitude of technological tools to support the analysis of the scope and scale of the deforestation problem. Indeed, remote sensing methods can measure and detect the rate and extent of deforestation, as well as monitor regeneration. Multitemporal satellite data are suitable for change detection analyses. Global monitoring initiatives, such as in rain forest depletion, depend on large area coverage and data continuity, leading to the use of sensors with successive launching and operational record. Radar has all-weather capability by penetrating clouds, and high-resolution data provide a detailed view of forest depletion. Moreover, illegal cutting or damage could be identified through remote sensing.

REDD+ refers to reduction of emissions from forest degradation and deforestation, and includes conservation, sustainable management of forests, and enhancement of forest carbon stocks. In certain regions around the world, such as in Indonesia and Brazil, deforestation and forest degradation together are by far the main source of national greenhouse gas emissions. Moreover, besides climate change mitigation, reduction of deforestation and forest degradation, as well as sustainable forest management, contribute to several environmental aspects, such as effective water resources management, runoff and river siltation reduction, prevention of flooding, soil erosion control, protection of fisheries and biodiversity preservation, among others. Thus, forests are more important left standing, than cut. The Forest Carbon Partnership Facility was established from such an understanding.

Mangroves. Mangroves are coastal forests in wetland ecosystems, which provide many highly valued ecosystem services, such as coastal protection, water purification, the maintenance of fisheries, carbon sequestration, erosion control, raw materials and food, tourism, recreation, education and research. In a case study in central Queensland, Australia (Chamberlain et al., 2020), the emphasis was on the quantitative assessment of spatiotemporal changes over a period of 14 years (2004–2017) in a coastal landscape and addressing biodiversity of a national park (Figure 8.5). Four methods of change analysis are used, namely post-classification change analysis with supervised classification, visual interpretation, thematic change dynamics and trend analysis. The Sentinel-2B image is used, captured on January 31, 2018 (spatial resolution of 10 m) (Chamberlain et al., 2020) (Figure 8.5). For change detection, the Landsat satellite archive images were used, captured in April, August and September for the years 2004, 2006, 2009, 2013, 2015 and 2017, respectively. These images were acquired from the United States Geological Survey (USGS) Earth Explorer Landsat Archive, which has systematic radiometric and geometric correction applied to the data. A thematic change analysis was used to convert to pasture the cumulative area of open forest, estuarine wetland and saltmarsh grass (Chamberlain et al., 2020).

The case of the Amazon basin is also presented. The Amazon basin is home to millions of plants and animals and produces a significant amount of the world's oxygen, which contributes to the regulation of global warming as the forests absorb millions of tonnes of carbon emissions every year. From the Copernicus Sentinel-3 World Fires Atlas, 3951 fires burning in the Amazon were detected

FIGURE 8.5 Study site – Rocky Dam Creek and Cape Palmerston National Park Central Queensland – Sentinel-2B composite image visualized by using the red, green and blue wavelength bands, captured January 31, 2018 at 00:22:57 provided by United States Geological Survey (USGS) (from Chamberlain et al., 2020).

at night from August 1 to August 24, 2019, compared to 1110 fires detected in 2018 during the same period by processing 249 images for August 2018 and 275 images for August 2019 (Figure 8.6). Similarly, Figure 8.7 shows wildfires on the border between Bolivia, Paraguay and Brazil from Copernicus Sentinel-2. Specifically, the fires breaking out on the border between Bolivia, Paraguay and Brazil are shown on this false-color animation, which contains three separate images from 8, 18 and 23 August 2019. On August 23, the smoke from the fire is visible in blue, while clouds can be seen in white. The orange areas show the burned land.

8.2.2.2 Burned Forests and Reforestation

Forest fires are considered part of the natural reproductive cycle of many forests, which can revitalize growth by releasing nutrients from the soil and opening seeds. Remote sensing can be used to detect and monitor forest fires and the regrowth following a fire. There is always the risk, however, that fires can threaten settlements and wildlife, eliminate timber supplies, or damage conservation areas. Remote sensing can provide information to support fire monitoring, as well as to assess the recovery of a forest following a burn. Moreover, remote sensing data and methods can be used for surveillance, since routine sensing facilitates observing remote and inaccessible areas, as well as monitoring agencies, can be informed about the presence and extent of a fire. Specifically, thermal data is mainly used to detect and map ongoing fires, however, multispectral data are preferred for observing phenology and growth stages in a previous burn area. Indeed, NOAA/AVHRR thermal data and GOES or METEOSAT meteorological data have been extensively used to delineate active fires and remaining "hot spots" when optical sensors are hindered by haze, smoke or darkness. Haze and smoke reflect a large amount of energy at shorter wavelengths. In addition, the fire rate and direction of movement can be detected by comparing burned areas to active fire areas.

In fire monitoring and fighting, remote sensing data can also facilitate route planning for both access to, and escape from, a fire. In addition, the fire rate and direction of movement can be detected

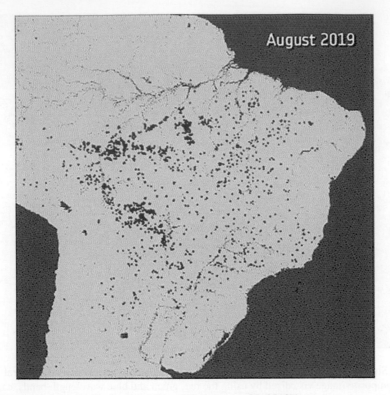

FIGURE 8.6 Number of fires in the Amazon (from ESA, August 27, 2019b).

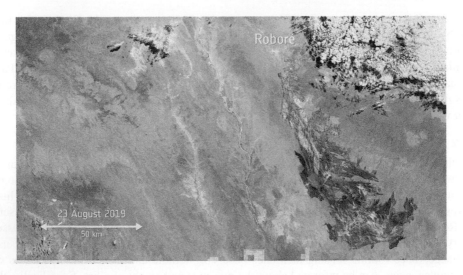

FIGURE 8.7 Wildfires on the border between Bolivia, Paraguay and Brazil from Copernicus Sentinel-2 (from ESA, August 27, 2019b).

by comparing burned areas to active fire areas. Indeed, burn mapping requires high-to-moderate resolution, moderate spatial coverage and a low turnaround time. On the other hand, fire detection and monitoring require moderate resolution, a large spatial coverage and a very quick turnaround to facilitate response. For burn mapping, when cloud cover precludes the use of optical images, radar can be used to monitor previous burn areas. Moreover, radar is effective from the second

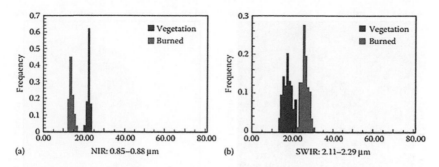

FIGURE 8.8 Histogram data plots of burned and vegetation areas in (a) NIR and (b) SWIR part of the spectrum (from Dalezios et al., 2018).

year following a burn, onwards. Single and multitemporal images can illustrate the progression of vegetation from pioneer species back to a full forest cover, as well as updates on the health and regenerative status of a burned forest.

Burned area mapping at local, regional and global scales has achieved very high classification accuracies (Pleniou and Koutsias, 2013). The post-fire spectral signal of burned areas is determined (Pereira et al., 1997) by: (1) the deposition of charcoal as the direct result of burning that is a unique consequence of fire burning, and (2) the removal of photosynthetic vegetation that may be also caused by other factors than fires (Robinson, 1991). Consequently, the spectral pattern of burned areas is characterized first by a strong decrease in reflectance in NIR region of the spectrum because of the destruction of the leaf cell structure, and second by a strong increase in reflectance in SWIR because of the reduction of water content, which absorbs radiation in this spectral region (Pereira et al., 1997; Koutsias and Pleniou, 2015). This spectral behavior of burned areas in NIR and SWIR (Figure 8.8) is the basis for the development of the Normalized Burn Ratio (NBR) index (Key and Benson, 2006). NBR is a modification of NDVI by replacing Red with SWIR (Koutsias and Karteris, 2000; Ji et al., 2011; Pleniou et al., 2012; Pleniou and Koutsias, 2013).

Despite any potential limitations, remote sensing technology has been used at a global scale to map and monitor burned areas. Global fire products, based for example on MODIS (Justice et al., 2002), have been utilized in studies referring mainly to continental scales for characterizing global fire regimes (Chuvieco et al., 2008) or for estimating global biomass burning emissions (Korontzi et al., 2004; Nioti et al., 2011). Annually resolved fire perimeters based on MODIS data are provided by the European Forest Fire Information System of the European Commission in order to provide consistent fire statistics over Europe (Nioti et al., 2011). Systematic fire products using medium-to-high resolution satellite data (e.g. Landsat) are not common on a global basis, mainly due to cost constraints on gathering and processing medium- or high-resolution satellite data series (Koutsias et al., 2013). The recent developments in informatics technology together with the freely available appropriate satellite data (e.g. Landsat, Sentinel-2) offer new possibilities to develop such global products (Figure 8.9).

Some strategies are followed in burned land mapping studies, such as: (1) the use of multitemporal versus single-date satellite data; (2) the use of multispectral transformations with emphasis on principal component analysis (PCA); and (3) the application of post-classification processing by a 3×3 majority filter (Nioti et al., 2011). For the multi- versus single-date satellite images, research findings demonstrated that methods using a multitemporal data set are more effective than those using only a single post-fire image. In the multidate approach, the confusions between burned areas and other land-cover types with similar spectral behavior that remain unchanged are minimized (Koutsias et al., 2000). However, in multitemporal data, radiometric and geometric misregistration

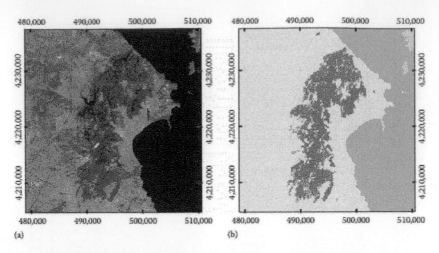

FIGURE 8.9 A Landsat image (RGB: 742) showing (a) a fire scar in red color and (b) the result of mapping the burned area (from Dalezios et al., 2018).

may result in under- or overestimation of the burned areas. Mallinis and Koutsias (2012) observed that most of the errors are distributed in the borders between burned and unburned vegetation in which mixed pixels are found. In that sense, problems can be enhanced when geometric misregistration is considered (Nioti et al., 2011).

Post-fire Vegetation Recovery Monitoring. Several factors determine the rate of post-fire vegetation recovery including climate, soil characteristics, degree of soil disturbance, initial plant mortality, topographic influences and vegetation composition (Szpakowski and Jensen, 2019). Remote sensing offers an alternative means for post-fire estimating and monitoring vegetation recovery over large areas in a more time efficient and less costly manner. These remote sensing methods can be grouped into three categories: (1) image classification; (2) vegetation indices (VIs); and (3) spectral mixture analysis (SMA). Specifically, image classification attempts to use spectral responses to determine the presence of healthy vegetation in individual pixels. Moreover, supervised classification can also be used to identify patterns of tree recovery. VIs are the most widely used vegetation recovery monitoring method. Indeed, VIs based on NIR reflectance are the most commonly used. Pixel size often exceeds the size of individual flora leading to methods, such as SMA, which addresses the mixed pixel issue and are also used in vegetation recovery monitoring. SMA determines the fraction of a pixel, which belongs to a specific endmember. These endmembers are assumed to be representative of the cover types identified in the image.

Figure 8.10 shows a post-fire vegetation recovery in Argentina. Specifically, ESA minisatellite Proba-V shows the rapid regeneration of South American grasslands from wildfire burn scars. Specifically, the fertile pampas, grasslands located in northern Argentina, Uruguay and southern Brazil, are frequently struck by wildfires (Figure 8.10). During the southern hemisphere summer of 2016–2017 fires burned across 30,000 km^2 in the La Pampa and Rio Negro provinces of Argentina. In Figure 8.10, the animated pair of Proba-V images shows the pampas recovering from these wildfires. The first 100-m resolution image, acquired on January 6, 2017 shows burned areas as brown/blackish patches, with some wildfire smoke plumes visible in blue. The second image, from July 24, 2017, reveals the recovery of these grasslands.

8.2.2.3 Forest Depletion: Clear-Cut Mapping and Regeneration

Clear-cut mapping and monitoring require regional scale images and moderate or high-resolution data depending on whether cuts are to be simply detected or delineated. Comparison of past and

FIGURE 8.10 Post-fire vegetation recovery in Argentina (from ESA, July 24, 2019c – ESA/Belspo – processed and produced by VITO Remote Sensing, Belgium).

recent satellite images can be used to assess the differences in the size and extent of the clear-cuts or loss of forest. As for many multitemporal applications, a higher resolution image can be used to define the baseline, and coarser resolution images can be used to monitor changes to that baseline. Optical sensors are preferred for clear-cut mapping and monitoring, because forest vegetation, cuts and regenerating vegetation have distinguishable spectral signatures, and optical sensors can collect cloud-free data. Moreover, radar, merged with optical data, can be used to efficiently monitor the status of existing clear-cuts or the emergence of new ones, and even assess regeneration conditions.

In cases where cutting is controlled and regulated, remote sensing serves as a monitoring tool to ensure that cut guidelines and specifications are followed. Radar has been proven to differentiate mangrove from other land covers, and some bands have long wavelengths capable of penetrating cloud and rain. The only limitation is in differentiating different mangrove species. Similarly, clear-cuts produce less backscatter than the forest canopy and as a result they can be defined on radar images, as well as forest edges being enhanced by shadow and bright backscatter. Radar is more suitable for applications in the humid tropics, since, due to its all-weather imaging capability, it can monitor all types of depletion, including clear-cuts, in areas prone to cloudy conditions. However, regenerating cuts are not easily detectable, since regeneration and mature forest canopy cannot be separated.

Figure 8.11 shows the Copernicus Sentinel-2 image features in an area in Santa Cruz, Bolivia, where part of the tropical dry forest has been cleared for agricultural use. Each patterned field is approximately 20 km^2 and each side is around 2.5 km long. This composite image was created by combing three NDVI images from the Copernicus Sentinel-2 mission. The first image, from April 8, 2019, is visible in red; the second from June 22, 2019, can be seen in green; and the third from September 5, 2019 can be seen in blue. Specifically, since the eighties, the area has been rapidly deforested owing to a large agricultural development effort, where people from the Andean high plains (the Altiplano region) have been relocated to the lowlands of Bolivia, where the local climate allows farmers to benefit from two growing seasons. The region has been transformed from dense forest into agricultural land. This deforestation method, common in this part of Bolivia, is characterized by the radial patterns that can be seen clearly in the image.

FIGURE 8.11 Clear-cut for agricultural use in Bolivia (from ESA, January 24, 2020 – contains modified Copernicus Sentinel data (2019), processed by ESA, CC BY-SA 3.0 IGO).

8.3 FORESTRY PRODUCTION

Forestry production concerns include harvest information in terms of tree volume estimates, estimation of stand volumes, information for timber supply and infrastructure mapping, as well as vegetation features, which involves vegetation types and distribution and biomass estimation.

8.3.1 HARVEST INFORMATION

8.3.1.1 Tree Volume Estimates

Throughout history forest has played a vital role in humans' economy, which results in actions towards sustainable management of forests for timber needs (Sheppard et al., 2020). Forest harvesting is a very significant activity for landowners, while forest managers dealing with this task require up-to-date and accurate information about forest attributes. Remote sensing, both active and passive, can provide the necessary data based on the current technologies, i.e. airborne and terrestrial laser scanners and digital photography and multispectral data from satellite optical sensors. The most promising results of forest measurements are provided by airborne lidar. Lidar data consist of a discrete point measurement of ranges, i.e. distance between the sensor and the target. From these measurements one can calculate elevations coupled with the strength of the return signal, and the forest structure can be described in three dimensions (Figure 8.12). Many researchers have explored the usage of airborne lidar to extract tree measurements. Most of the initial studies based on Lidar were focused on deriving tree heights. A comparison of various methods applied in different countries revealed that the laser point density has lees impact on individual tree detection, while the extraction method is the most significant feature of the workflow.

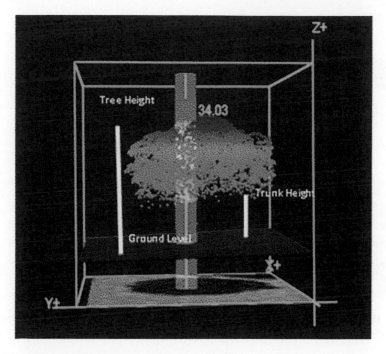

FIGURE 8.12 Tree dimensions extracted by Lidar (from Verma et al., 2016).

Lidar data have been also used in combination with other data for tree extraction and volume estimation. For example, textural and spectral data of aerial images in combination with laser scanner data have been used for tree volume estimation in different tree species, while other studies have used a combination of lidar, Inertial Measurement Unit and a High Definition video camera mounted on a UAV (Wallace et al., 2012). The results have indicated that the final product has improved the horizontal accuracy from 0.61 m to 0.34 m due to the inclusion of video. Moreover, satellite images have been used with lidar data for tree height and canopy density metrics. In a recent study, tree and canopy metrics are derived based on pixel values, texture features and vegetation indices obtained from a WorldView-3 imagery (Kanja et al., 2019). They found that the adjusted R^2 obtained by the best regression models, using LiDAR-derived metrics alone, were lower for the volume per hectare as opposed to the metrics derived by a combination of lidar and WorldView-3 imagery band values and vegetation indices. Furthermore, the texture feature was not such a significant parameter in metrics extraction.

Terrestrial laser scanning (TLS) was also applied in forest inventory. Biophysical tree parameters, such as diameter at breast height (DBH), tree height, basal area, and volume can be estimated with the use of a point cloud data retrieved by TLS. However, multiple acquisitions are essential for accurate results of tree volume. Besides TLS, mobile- or vehicle-based laser scanning (MLS or VLS), has been also explored and tested for larger coverage within a shorter acquisition time (Liang, et al., 2018b). In a recent benchmarking study, it was shown that TLS methods can estimate DBH and stem curve at 1–2 cm accuracy, but in a multi-layer and dense forest stands the bottleneck occurs at stem detection and tree height estimation (Liang, et al., 2018a).

Except from the costly lidar technology, point clouds can be produced by aerial photogrammetry based on small unmanned aerial vehicle (UAV). In this case, the point clouds can be combined with multi and hyperspectral information for accurate tree identification. However, metric information is limited especially in dense forests. Thus, this technology is suitable mostly for sparse forests. In this situation, a Digital Terrain Model (DTM) can be estimated by filtering non-ground points. The Canopy Height Model (CHM) can then be estimated by subtracting the DTM from Digital Surface

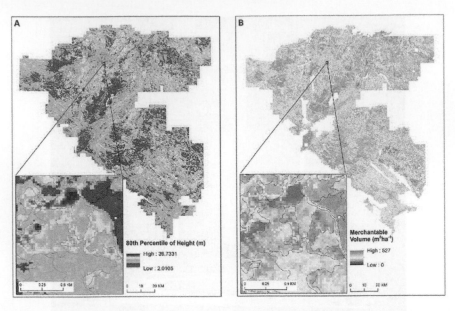

FIGURE 8.13 Height measurements (A) derived from ALS point cloud and (B) estimated merchantable volume (from White et al., 2016).

Model. Results of the aerial photogrammetry method can be enhanced by using oblique photographs (Díaz et al., 2020).

8.3.1.2 Stand Volume Estimates

Most of the studies that aim to extract forest metrics are focused on stand-level rather than tree-level extend. A comparison between these two methods has shown that both the stand-based method and the individual tree-based method produced promising and compatible results, while an increase in in point density leads to increased accuracy of the tree-based technique over that of the area-based (Yu et al., 2010). Mean stand height, the mean height of all laser pulses within a stand, and canopy cover density are the metrics that usually are computed within a stand and are correlated with ground-truth data, while different footprints and multitemporal acquisitions are usually tested (Figure 8.13).

In recent studies, ALS were combined with Landsat satellite multispectral imagery. More specifically, Bolton et al. (2018) combined ALS data and Landsat time-series metrics to produce estimates of top height, basal area and net stem volume for two timber supply areas in Canada, using an imputation approach. It was found that in the case of spatially limited ALS data acquisitions their approach with Landsat time series can effectively produce accurate estimates of key inventory attributes. This approach was further optimized by finding the optimal time series length and the most significant indices for forest attribute estimation (Bolton et al., 2020). Results revealed that the optimal time series length should be determined on a case-by-case basis, while the most important spectral indices were the indices that depend on short-wave infrared bands. Finally, a recent study explored the need of local field measurements in small-scale forest management inventories (Kotivuori et al., 2020). Authors used photogrammetrically derived drone-based image point clouds in an area-based approach and they found that height metrics for the dominant tree layer from the point cloud data showed strong correlations with similar metrics computed from the ALS data.

8.3.1.3 Information for Timber Supply

Remote sensing is widely used for extracting qualitative and quantitative information regarding harvesting monitoring and planning and is part of the modern wood supply chain integrated within Industry 4.0 (I 4.0) framework (Müller et al., 2019). A study on a large geographical extent used Landsat and SPOT images to study the forest regeneration dynamics of semi-deciduous forest

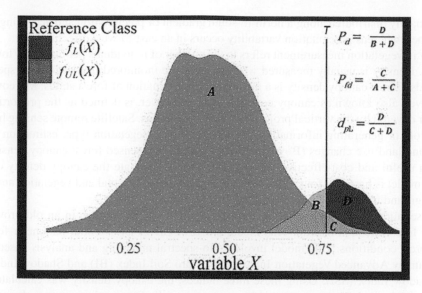

FIGURE 8.14 Machine learning approach towards selective logging (from Hethcoat et al., 2019).

located in the south-western part of the Central African Republic. Logging activities were identified because of the concentrated gaps and the linear feature of the skid roads (de Wasseige and Defourny, 2004). The extent of selective logging has also been estimated based on a time series of annual Landsat images or based on texture measures and a machine leaning approach (Figure 8.14) (Matricardi et al., 2010). Lower spatial resolution sensors have also been used for timber volume estimation. The Geosciences Laser Altimeter System (GLAS) space LiDAR data are used in conjunction with MODIS data for the estimation of timber volume for 16 different forest classes (four forest types of four different canopy density classes) in south-central Siberia (Nelson et al., 2009).

Compared to satellite images, change detection of boreal forest due to forest growth and harvested trees was performed with the use of small footprint, high sampling density ALS by applying object-oriented algorithms, where 61 out of 83 field-checked harvested trees were correctly identified (Yu et al., 2004). A comparison of ALS and TLS for the estimation of timber assortments and stem distribution was performed using as a reference single clear-cut stand (Kankare et al., 2014). From the results, it was clear that tree-level timber assortments and diameter distributions were more accurate by using TLS than a combination of TLS and ALS. The results have indicated that, by using TLS or a combination of TLS and ALS, accurate tree-level timber assortments and diameter distributions can be obtained, while the saw log volumes were estimated with higher accuracy than pulpwood volumes.

With the exception of tree and stand measurements, remote sensing technology is used to support activities related to forest harvesting, such as accessibility to harvesting areas. Lidar data have been used for detailed mapping of wheel rut depths at harvesting sites (Salmivaara et al., 2018). Specifically, the Lidar sensor was mounted on a forest and provided estimates of rut depth with an accuracy approximately 3.5 cm. UAV and TLS surveys were also used in order to create a 3D forest model that was used for RFID tree marking (Pichler et al., 2017). Moreover, orientation, navigation and performance enhancing of machinery within a forest stand can be supported by remote sensing technology within the I 4.0 framework (Müller et al., 2019).

8.3.2 Vegetation Features

8.3.2.1 Vegetation Types and Density Mapping

There are six types of forest, namely: (1) equatorial moist evergreen or rainforests; (2) tropical deciduous forests; (3) Mediterranean forests; (4) temperate broad-leaved deciduous and mixed forests; (5) warm temperate broad-leaved deciduous forests; and (6) coniferous forests. Vegetation types are mainly distributed along a north-to-south gradient of precipitation and according to different soil types, whereas precipitation variability occurs in an east-to-west direction.

Density in vegetation measurement refers to the number of individuals per unit area (for example plants/m^2). Density is usually measured when changes are monitored in a vegetation species over long periods. Forest canopy density is a major factor in evaluation of forest status. Moreover, forest canopy cover, also known as canopy coverage or crown cover, is defined as the proportion of the forest floor covered by the vertical projection of the tree crowns. Satellite remote sensing has played a pivotal role in generating information about forest cover, vegetation type, estimation of crown coverage and land-use changes (Boyd et al., 2002). Remotely sensed forest canopy density (FCD) model is a useful and cost-effective method to detect and estimate the canopy density over large area. This model is based on four indices, namely soil, shadow, thermal and vegetation and requires very little ground truths, purely for accuracy checks.

In a case study, the FCD model has been tested by using IRS image in an old growth forest of a north forest division of Iran. This model utilizes FCD as an essential parameter for characterizing forest conditions. This model involves bio-spectral modeling and analysis based on three indices, namely Advanced Vegetation Index (AVI), Bare Soil Index (BI) and Shadow Index (SI) or Scaled Shadow Index (SSI). Using these three indices, the canopy density was calculated in percentage for each pixel. The vegetation index responses to all of vegetation types, such as forest and grassland. Specifically, AVI is sensitive to the vegetation quantity as compared to NDVI. Moreover, SI increases as the forest density increases. Thermal index increases as the vegetation quantity increases. Black-colored soil area shows a high temperature. BI increases as the bare soil exposure in degrees of ground increases. These index values are calculated for every pixel. For ground-truth accuracy assessment, the distance between classes is used as follows: class 1: High Forest = HF (71%–100% density); class 2: Middle Forest = MF (41%–70% density); class 3: Low Forest = LF (5%–40% density); class 4: Grass Land = GL; class 5: Bare soil = Bs. Figure 8.15a illustrates the IRS LISS III image of the study area and Figure 8.15b indicates the classes of the forest canopy density map. In a similar study, tropical forest cover density mapping was developed using LANDSAT TM data image analysis (Rikimaru et al., 2002).

8.3.2.2 Biomass Estimation

Forest biomass is a key focus of environmental monitoring concerning global biogeochemical cycles and loss of biodiversity. Field methods are the most accurate approach to estimate forest biomass. However, they are unsuitable for calculating biomass across large areas and time periods due to their lack of spatial and temporal coverage (Soenen et al., 2010). Airborne lidar surveys (ALS) have also been used to map forest aboveground biomass (AGB) (Hudak et al., 2012). However, ALS generally tends to represent only a partial sample in space and time (Kennedy et al., 2018). As an alternative, multispectral satellite-based Earth Observations, such as Landsat imagery, can be used for estimating forest biomass across large areas due to their wide spatial and temporal coverage. Forest AGB and Landsat spectral data are highly correlated (Lu, 2006). AGB and its dynamics are often estimated by combining Landsat images with reference data extracted from field inventory plots. Traditionally, such studies often only used single-date Landsat images to estimate forest AGB at a single or limited number of points in time (Pflugmacher et al., 2014).

Remote sensing and forest research communities with a critical review of recent approaches utilize Landsat time-series (LTS) for estimating AGB and its dynamics across space and time (Nguyen et al., 2020). In contrast to using single-date images, the use of LTS can benefit forest AGB estimation in two broad areas. First, using LTS allows for the filling of spatial and temporal data gaps in AGB predictions, improving the quality of AGB products and enabling the estimation of AGB across large areas and long time periods. Second, studies have demonstrated that spectral information extracted from LTS analysis, including forest disturbance and recovery metrics, can significantly improve the accuracy of AGB models. A general trend is that methods have evolved as demonstrated through recent studies, becoming more advanced and robust. However, most of these

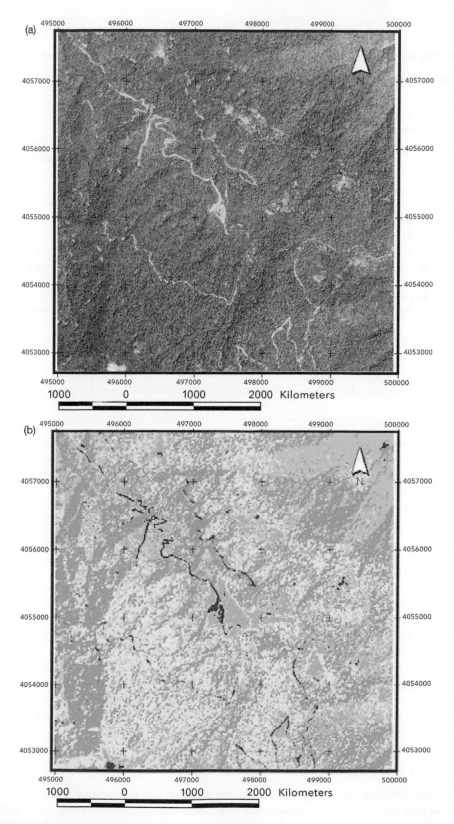

FIGURE 8.15 (a) IRS LISS III image of study area. (b) Classes of the forest canopy density map (from Azizia et al., 2003).

TABLE 8.2

Landsat spectral indices commonly used for forest AGB estimates. (Nguyen et al., 2020)

Landsat Spectral Index	Calculation
Normalized Difference Vegetation Index (NDVI)	$NDVI = (NIR - R)/(NIR + R)$
Normalized Burn Ratio (NBR)	$NBR = (NIR - SWIR)/(NIR + SWIR)$
Normalized Difference Moisture Index (NDMI)	$NDMI = (NIR - SWIR)/(NIR + SWIR)$
Enhanced Vegetation Index (EVI)	$EVI = G * ((NIR - R)/(NIR + C1 * R - C2 * B + L))$
	L = value to adjust for canopy background, C = coefficients for atmospheric resistance, B = the blue band
Soil Adjusted Vegetation Index (SAVI)	$SAVI = ((NIR - R)/(NIR + R + L)) * (1 + L)$
Chlorophyll Vegetation Index (CVI)	$CVI = (NIR \times R)/G$ G = the green band
Difference Vegetation Index (DVI)	$DVI = NIR - R$
Linear transform of multiple bands	$VIS123 = B + G + R$
	$MID57 = TM\ band\ 5 + TM\ band\ 7\ (SWIR)$
Integrated Forest Z-score (IFZ)	z-score measure of a pixel likelihood of being forested, using TM bands 3, 5 and 7
Tasseled Cap (TC) transformations: TC brightness (TCB); TC greenness (TCG); TC wetness (TCW)	TCW, TCB, and TCG are calculated by multiplying Landsat band pixel values with TC coefficients. See the coefficients in references.
TC angle (TCA)	$TCA = arctan(TCG/TCB)$

TC distance (TCD)= $TCD = TCB^2 + TCG^2$
$DI = TCB_r - (TCG_r + TCW_r)$
TC Disturbance Index (DI) r = denotes rescaled TC indices based upon the mean and standard deviation of the scene's forest values

methods have been developed and tested in areas that are either supported by established forest inventory programs and/or can rely on Lidar data across large forest areas. Further investigations should focus on tropical forest areas, where inventory data are often not systematically available and/or are out of date. Table 8.2 shows Landsat spectral indices commonly used for forest AGB estimates (Nguyen et al., 2020).

8.4 ENVIRONMENTAL MONITORING

Environmental monitoring includes the quantity, health and diversity of the Earth's forests in terms of deforestation, such as rainforest or mangrove colonies, species inventory, watershed protection, coastal protection from mangrove forests, as well as forest health and vigor. For applications requirements, high information accuracy, multispectral information, fine resolution and data continuity are the most important. There is a need to balance spatial resolution with the required accuracy and data costs. Resolution capabilities of 10 m to 30 m are deemed adequate for forest cover mapping, identifying and monitoring clear-cuts, burn and fire mapping, collecting forest harvest information, and identifying general forest damage. Spatial coverage of 1–100 km^2 is the most appropriate for site specific vegetation density and volume studies, whereas 100–10000 km^2 coverage is appropriate for district to provincial scale forest cover and clear-cut mapping. Tropical forest management must focus on a reliable data source, capable of imaging during critical time periods.

8.4.1 FOREST SPECIES INVENTORY: IDENTIFICATION AND TYPING

Forest conservation and forest supply inventory are based on forest cover typing and species identification. Indeed, reconnaissance mapping over large areas can contribute to forest cover typing, and species inventories consist of highly detailed measurements of stand contents and characteristics, such as tree type, height or density. Remote sensing multispectral, hyperspectral, or air-photo and recently UAV data are available for large-scale species identification, whereas radar or multispectral data interpretation can be used for small-scale cover type delineation. Indeed, hyperspectral imagery can provide a very high spatial resolution of forest cover and capture extremely fine radiometric resolution data to generate signatures of vegetation species and certain stresses on trees, such as infestations. For monitoring biophysical properties of forests, multispectral information and finely calibrated data are usually employed. For detailed species identification associated with forest stand analysis, very high-resolution, multispectral data is required. In general, data requirements depend on the scale of the analysis to be conducted. Specifically, for regional reconnaissance mapping, sensors are required to be sensitive to differences in forest cover, such as canopy texture, leaf density and spectral reflection. Moreover, multitemporal data can contribute to phenology information through seasonal changes of different species.Traditionally, Landsat data are the most appropriate for conducting reconnaissance mapping of forests, whereas aerial photography and digital orthophoto have been used for extracting stand and local inventory information. Air-photos and recently UAV are the most appropriate operational data source for stand level measurements including species typing, whereas SAR sensors, such as RADARSAT or Sentinel-1, are useful where persistent cloud cover limits the usefulness of optical sensors. In humid tropical areas, forest resource assessments and measurements are difficult to be conducted due to cloudy conditions and difficult terrain impeding ground surveys. In such cases, an active sensor, such as radar, can serve this purpose and an airborne sensor or UAV can cover high-resolution requirements, such as cover typing. This type of data can be used for a baseline map, whereas coarser resolution data can provide updates to any changes in the baseline.

Forest inventory constitutes a systematic process for collecting, analyzing and utilizing critical information of forests promoting the sustainable planning of this viable natural resource. These critical features may be related to specific tree or forest stands features, such as the form, weight, growth, volume and age of trees (FAO, 2020). The advent of sophisticated remote sensing technologies has broadly replaced the traditional time-consuming and costly methods of forest inventory development. Some of the most promising and reliable technologies in order to obtain the most critical parameters of forest species are airborne laser scanning; terrestrial laser scanning; digital aerial photogrammetry; and high spatial resolution/very high spatial resolution satellite optical imagery. The combination of these technologies may provide accurate information to be exploited for the sustainable use of forest in the long run (White et al., 2016).

The integration of Light Detection and Ranging (LiDAR) technology to capture accurate critical information of forests has been increasingly used despite the relatively high cost of adopting this type of technologies (Wulder and Seemann, 2003; Reutebuch et al., 2005; Wallace et al., 2012; Kelly and Di Tommaso, 2015). The variety of the captured data is high and may consist of trees and forest features, such as trees height, forest biomass and volume, ecosystems features, which in turn may provide vital information for the prevention of forest fires, e.g. recording of canopy fuels, as well as the rationalization of forest exploitation (Wulder and Seemann, 2003; Reutebuch et al., 2005; Kelly and Di Tommaso, 2015). However, the integration of LiDAR to Unmanned Aerial Vehicles (UAV) has significantly reduced the financial cost (Wallace et al., 2012). The capability of producing a denser network of point clouds and the contribution of high-resolution video camera yielded to higher accuracy of trees height/location and crown width determination (Wallace et al., 2012). Figure 8.16 depicts the spatiotemporal analysis of tree features using LiDAR technology.

However, the recording of lower canopy features may potentially not be visible or accurately estimated by the airborne laser scanning (ALS). To this end, the Terrestrial Laser Scanning (TLS) has come to complement this missing information (Liang et al., 2016; White et al., 2016). This has been confirmed by Hilker et al. (2012), where they estimated the tree-level features with higher accuracy when using TLS compared to ALS (Figure 8.17).

On the other hand, a relatively new technology used for forest inventory purposes has emerged to enrich the information required by forest managers. Specifically, the impact of digital aerial photogrammetry could act as a supplement to ALS (White et al., 2013; Rahlf et al., 2017). The main

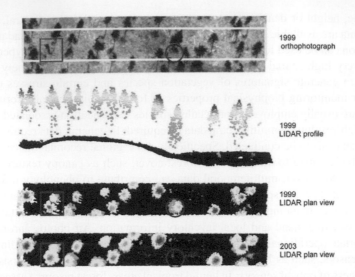

FIGURE 8.16 Comparison of 1999 and 2003 LIDAR crown measurements in a heavily thinned strip of mature forest in the Capitol State Forest study area. (From top to bottom) 1999 orthophotograph; profile view of all 1999 LIDAR points (color-coded by height aboveground) measured within the yellow box shown in the ortho-photograph; plan view of all 1999 points; and, plan view of all 2003 points. Note the crown expansion between 1999 and 2003 that is apparent in the red square and the tree that was removed (because of windthrow) apparent in the red circle (from Reutebuch et al., 2005).

two drawbacks of the digital aerial photogrammetry (DAP) compared to ALS are: (1) it requires the Digital Elevation Model which is frequently provided by the ALS; (2) it primarily delivers information regarding the outer envelope of forest canopy, whereas ALS technology retrieves data from higher depths of forest canopy. However, the DAP may provide additional information, which may not be available through ALS, such as the vegetation health and maturity (White et al., 2013). Figure 8.18 presents a comparison of ALS and DAP point clouds in a coastal temperate forest in British Columbia, Canada.

Finally, with the advent of very high spatial resolution images, new opportunities have emerged (Falkowski et al., 2009; Mora et al., 2013). Specifically, very high spatial resolution satellite images have been used (Mora et al., 2013) in order to estimate stand height, volume and biomass with high accuracy. In addition, they showed that the outcomes of these indices can be suitable for operational use by forest managers. The calibration of the results has been conducted by LiDAR images. In conclusion, the availability and interaction of multiple data sources and technologies may provide cost-effective solutions for the identification and estimation of crucial parameters, such as trees locations, height, canopy structure, biomass volume and similar aspects, useful for forest inventory purposes. The forest inventory goals may be versatile including the rationalization for forest resources exploitation as a socioeconomic good, as well as the integrated protection of forests from destructive events, such as forest fires, e.g. through fuel treatments.

8.4.2 FOREST HEALTH AND VIGOR

Forest health (FH) is normally used in forestry to describe the forest stand condition. Stress in forest ecosystems (FES) occurs as a result of land-use intensification, disturbances, resource limitations or unsustainable management (Lausch et al., 2017). Forests have short-term effects on local ecosystems and landscapes, balance global carbon stock and influence global climate (Pause et al., 2016). For example, in Europe, two aspects are of major concern for influencing forest health: (1) forest damage from air pollution, and (2) the impact of climate change. For mapping, quantifying and monitoring regional and global forest health, satellite remote sensing provides fundamental data for the observation of spatial and temporal forest patterns and processes. The emphasis is usually

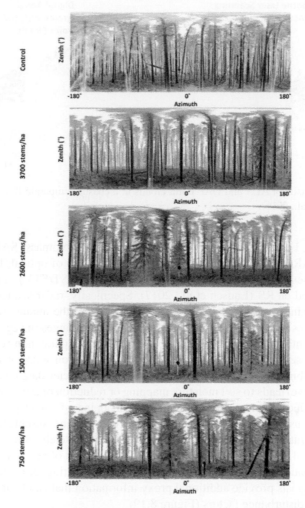

FIGURE 8.17 Full waveform image of the thinning experiments observed from full waveform LiDAR. The distortion is caused by the projection of the hemispherical data onto a two-dimensional plane (from Hilker et al., 2012).

on biotic and abiotic disturbances, such as drought stress, pest infestation and environmental pollutants. New satellite remote sensing technologies, namely hyperspectral imaging spectroscopy and full polarimetric SAR data, promise to extend the database of forest observations with new potential for forest ecosystem assessment (Wang et al., 2010).

New satellite-based analyses of soil and vegetation are increasing temporal coverage and sensor innovations, such as FLEX, TanDEM-L. Optical remote sensing imagery is the backbone of vegetation ecosystem studies, such as forests, greenland, agriculture, urban vegetation. In forest stands, information is useful for spatial and temporal species distributions and de- and afforestation. Changes to forest environmental boundary conditions including temperature, precipitation and air pollution affect foliar nutrient concentration, which in turn affect the spectral signatures in VNIR and SWIR. Moreover, the value of long-term thermal infrared observations of LST and its contribution to any biodiversity-related analyses is recognized. Optical remote sensing applications, such as for growing stock, tree species change maps, leaf and needle color, are greatly limited by atmospheric dust and clouds and require atmospheric correction for the spectral signals (Pause et al., 2016).

Airborne LiDAR data provide highly accurate geometric information on the vertical and horizontal vegetation distribution in forests, such as tree height and crown shape, vertical vegetation

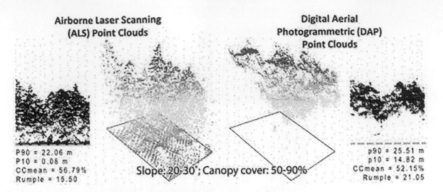

FIGURE 8.18 Comparison of ALS and DAP point clouds in a coastal temperate forest in British Columbia, Canada (from White et al., 2016).

layer, its observation. For frequent satellite-based monitoring of biomass, SAR data is an alternative to airborne LiDAR biomass estimations. The implementation of optical, RADAR and LiDAR RS-techniques to assess spectral traits/spectral trait variations (ST/STV) can be used to record indicators of FH based on RS (Lausch et al., 2017). Specifically, radar scattering components are weather and daytime independent observations that vary during the annual growing cycle mainly due to biomass changes in terms of geometrical properties and changes in vegetation water content.

Combining frequently available SAR observations with temporal non-regular (due to cloud cover) optical remote sensing data may provide a sound information source even at the inner forest level. Moreover, combining L-band SAR and the L-band radiometer data concept of SMAP (Soil Moisture Active Passive) leads to soil moisture mapping or reduction of atmospheric noise on FLEX vegetation vigor products using Sentinel-3 data. The mapping of forest parameters using imaging satellite remote sensing is performed by completely different measurement technology, namely physical signals integrated over varying time and space, and completely different spatial representativeness, such as spatial and spectral signal integration. The benefit of novel satellite remote sensing observations is the spatial and process integrated perspective covering large areas, Therefore, satellite remote sensing signals provide additional proxy information that can be interlinked with forest health indicators and disturbance factors (Figure 8.19).

Urban forest. An urban forest is defined as a forest or a collection of trees that grow within a city, town or suburb. Urban forests provide several benefits, such as saving energy and reducing air pollution. With the emergence of LiDAR methods, it has been anticipated that the urban forest inventory can be automated. LiDAR has the ability to "see" the ground through openings in canopies and to detect the 3D structure of trees. Early work during the eighties mainly focused on the quantification of a forest using profiling LiDAR systems at the stand level (Zhang et al., 2015). Stand level measurements are more important than the individual tree level for natural forest management. However, an urban forest is a mosaic of many different species and ages, and usually has a higher degree of spatial heterogeneity.

Several fundamental tree parameters, such as height, base height, crown depth and crown diameter can be estimated from the LiDAR point clouds (Figure 8.20). Tree height is the length along the main axis from the treetop to the base of a tree. Crown depth is the length along the main axis from the treetop to the base of the crown. Base height is the length from the base of the crown to the base of the tree. Crown diameter, also known as crown width, is the span of the crown of a tree. The challenges for urban forest inventory using LiDAR mainly come from three sources: (1) the complexity of urban areas; (2) the spatial heterogeneity of urban forests; and (3) the diverse structure and shape of urban trees. New methods are being developed to isolate individual trees and estimate tree metrics, such as tree height, base height, crown depth and crown diameter from LiDAR point clouds, which can bring the automated urban forest inventory down to the individual tree level.

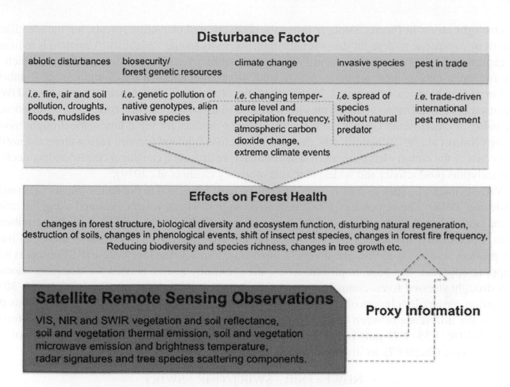

FIGURE 8.19 General impacts on forests, their origins and their effects on forest health together with remote sensing observations for identifying proxy information for factors relevant to forest health. The disturbance groups and factors are modified after FAO (Food and Agriculture Organization of the United Nations) (from Pause et al., 2016).

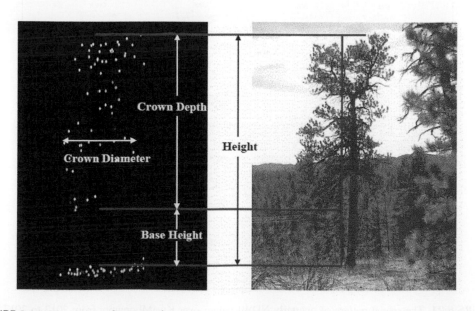

FIGURE 8.20 Illustration of tree metrics estimation using raw LiDAR data (from Zhang et al., 2015).

Drought stress in forests. Drought limits the production of plantation forests, but the identification of tree response to drought stress is uncertain, due to its spatial variability. A method is required to capture drought patterns and identify trees with similar reactions to drought stress, which is key for forest management planning. A case study of drought stress in the drought-prone Zululand region of South Africa is presented. In this study, the Normalized Difference Water Index (NDWI) was used in compartments of eucalyptus trees with similar drought characteristics and demonstrated the value of cloud-based Google Earth Engine (GEE) resources for rapid landscape drought monitoring (Xulu et al., 2019). Specifically, the most productive Zululand forestry region along the northeast coast of the country has been hit by a series of severe droughts, causing a corresponding decline in eucalyptus productivity and widespread tree mortality (Xulu et al., 2019).

Various methods, such as remote sensing indices, have been used to detect the impact of drought (Dalezios, 2018). The response of forest trees to drought has been widely explored using the ratios of the near-infrared (NIR), red and short-wave infrared (SWIR) bands and their various combinations. Specifically, the indices that are based on the NIR and SWIR bands have shown to have the greatest sensitivity to drought stress. In this study, a novel Random Forest (RF) unsupervised mapping approach has been employed, where the proximity matrix is used for detecting anomalous drought stressed forest compartments in KwaZulu-Natal, South Africa. The NDWI, with 30 m resolution, assembles clusters of compartments with similar responses to drought and represents the ratio of the difference between the near-infrared (NIR; 0.76–0.90 μm) and the short-wave infrared band (SWIR; 1.55–1.75 μm) reflectance over the combined reflectance in these two parts of the spectrum (see Chapter 7), as illustrated in the following Equation (8.1):

$$NDWI = (NIR - SWIR) / (NIR + SWIR) \tag{8.1}$$

The NDWI ranges from 0 to 1, depending on the water content in the plants, where high NDWI values correspond to high plant water content and low values correspond to low water content in the vegetation. As a result, the NDWI values are expected to decrease during the period of water stress, which is 2015 in this case study. Figure 8.21 shows the spatial pattern of quarterly NDWI values over KwaMbonambi from 2013 to 2017 (from Xulu et al., 2019).

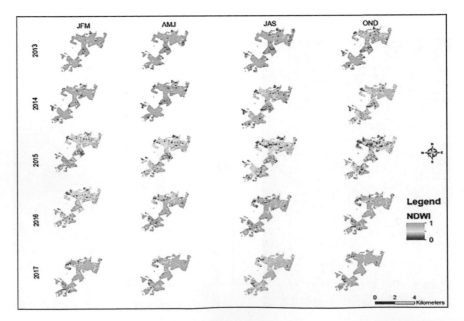

FIGURE 8.21 The spatial pattern of quarterly NDWI values over KwaMbonambi from 2013 to 2017 (from Xulu et al., 2019).

8.5 SUMMARY

In this chapter, forest management based on remote sensing has been considered, involving forest depletion due to fires and infestations or human activity, such as land conversion, burning or clear-cutting. Specifically, forests are clear-cut or burned to facilitate access to the land in order to achieve long-term sustainability. Indeed, depletion of forest species affects the local and regional hydrological regime, whereas burning trees causes atmospheric pollution and greenhouse effect, since more CO_2 is added to the atmosphere. It is well known that forests contribute to balancing the Earth's CO_2 exchange and supply, being a key component in one of the most significant current environmental issues, namely global warming and greenhouse effect.

The key aspects of forest health and growth for effective commercial exploitation and conservation have also been considered. Indeed, forestry production is an important global industry. There is a continuous effort for forest cropping and re-harvesting, and the new areas continually provide a new source of lumber. It is, thus, for the benefit of the industrial sector to ensure a healthy regeneration of trees, where forests are extracted, which ensure adequate wood supplies to meet the demands of a growing population. Remote sensing technology can be part of the information solutions internationally, which may include general cover type mapping, protection of coastal zones and watersheds through mapping and monitoring, monitoring of cutting practices and regeneration, as well as forest fire/burn mapping.

REFERENCES

Azizia, Z., Najafia, A., and Sohrabi, H., 2003. Forest canopy density estimating, using satellite images. UN Commission VIII, WG VIII/11, 4p.

Bolton, D.K., White, J.C., Wulder, M.A., Coops, N.C., Hermosilla, T., and Yuan, X., 2018. Updating stand-level forest inventories using airborne laser scanning and Landsat time series data. *International Journal of Applied Earth Observation and Geoinformation*, 66, 174–183, https://doi.org/10.1016/j.jag.2017.11.016

Bolton, D.K., Tompalski, P., Coops, N.C., White, J.C., Wulder, M.A., Hermosilla, T., Queinnec, M., et al., 2020. Optimizing Landsat time series length for regional mapping of lidar-derived forest structure. *Remote Sensing of Environment*, 239, https://doi.org/10.1016/j.rse.2020.111645

Boyd, D.S., Foody, G.M., and Ripple, W.J., 2002. Evaluation of approaches for forest cover estimation in the Pacific Northwest, USA, using remote sensing, *Applied Geography*, 22, 292–375.

Chamberlain, D., Phinn, S., and Possingham, H., 2020. Remote sensing of mangroves and estuarine communities in Central Queensland, Australia. *Remote Sensing*, 12, 197, doi:10.3390/rs12010197

Chen, Q., Lu, D., Keller, M., dos-Santos, M.N., Bolfe, E.L., Feng, Y., and Wang, C., 2016. Modeling and mapping agroforestry aboveground biomass in the Brazilian Amazon using airborne lidar data. *Remote Sensing*, 8, 21, doi:10.3390/rs8010021

Chiocchini, F., Ciolfi, M., Sarti, M., Lauteri, M., Cherubini, M., Leonardi, L., Nahm, M., Morhart, C., and Paris, P., 2018. *Inventory of tree hedgerows in an Italian agroforestry landscape by remote sensing and GIS-based methods. Proceedings, 4th European Agroforestry Conference – Agroforestry as Sustainable Land Use*, 217–221.

Chuvieco, E., Giglio, L., and Justice, C., 2008. Global characterization of fire activity: Toward defining fire regimes from Earth observation data. *Global Change Biology*, 14, 1488–1502.

Dalezios, N.R., 2018. Drought and Remote Sensing: An Overview, in *Remote Sensing of Hydrometeorological Hazards*. Editors: G.P. Petropoulos and T. Islam. Taylor & Francis, pp. 3–32

Dalezios, N.R., Kalabokidis, K., Koutsias, N., and Vasilakos, C., 2018. Wildfires and Remote Sensing: An Overview, in *Remote Sensing of Hydrometeorological Hazards*. Editors: G.P. Petropoulos and T. Islam. Taylor & Francis, pp. 211–236

de Wasseige, C., and Defourny, P., 2004. Remote sensing of selective logging impact for tropical forest management. *Forest Ecology and Management*, 188, 1–3, 161–173, https://doi.org/10.1016/j.foreco.2003.07.035

Díaz, G.M., Mohr-Bell, D., Garrett, M., Muñoz, L., and Lencinas, J.D., 2020. Customizing unmanned aircraft systems to reduce forest inventory costs: Can oblique images substantially improve the 3D reconstruction of the canopy? *International Journal of Remote Sensing*. https://doi.org/10.1080/01431161.2 019.1706200ESA, 2019a. *Recovering burn scars in Argentina*, https://www.esa.int/ESA_Multimedia/ Images/2019/07/Recovering_burn_scars_in_ArgentinaESA, 2019b. *Number of wildfires in the Amazon*, https://www.esa.int/ESA_Multimedia/Images/2019/08/Number_of_wildfires_in_the_Amazon

ESA, 2019c. *Wildfires on the border between Bolivia, Paraguay and Brazil from Copernicus Sentinel-2*, https://www.esa.int/ESA_Multimedia/Images/2019/08/Wildfires_on_the_border_between_Bolivia_Paraguay_and_Brazil_from_Copernicus_Sentinel-2

ESA, 2020. *Deforestation in Bolivia*, https://www.esa.int/ESA_Multimedia/Images/2020/01/Deforestation_in_Bolivia

Falkowski, M.J., Wulder, M.A., White, J.C., and Gillis, M.D., 2009. Supporting large-area, sample-based forest inventories with very high spatial resolution satellite imagery. *Progress in Physical Geography*, 33, 3, 403–423.

FAO, 2020. *Forest Inventory*. Available at: http://www.fao.org/sustainable-forest-management/toolbox/modules/forest-inventory/basic-knowledge/en/?type=111 [accessed: February 28, 2019].

García, M., Riaño, D., Chuvieco, E., Salas, J., and Danson, F.M., 2011. Multispectral and LiDAR data fusion for fuel type mapping using Support Vector Machine and decision rules. *Remote Sensing of Environment*, 115, 6, 1369–1379.

Hethcoat, M.G., Edwards, D.P., Carreiras, J.M.B., Bryant, R.G., França, F.M., and Quegan, S., 2019. A machine learning approach to map tropical selective logging. *Remote Sensing of Environment*, https://doi.org/10.1016/j.rse.2018.11.044

Hilker, T., Coops, N.C., Newnham, G.J., van Leeuwen, M., Wulder, M.A., Stewart, J., and Culvenor, D.S., 2012. Comparison of terrestrial and airborne LiDAR in describing stand structure of a thinned lodgepole pine forest. *Journal of Forestry*, 110, 2, 97–104.

Hudak, A.T., Strand, E.K., Vierling, L.A., Byrne, J.C., Eitel, J.U., Martinuzzi, S., and Falkowski, M.J., 2012. Quantifying aboveground forest carbon pools and fluxes from repeat LiDAR surveys. *Remote Sensing of Environment*, 123, 25–40.

Immitzer, M., Atzberger, C., and Koukal, T., 2012. Tree species classification with random forest using very high spatial resolution 8-band WorldView-2 satellite data. *Remote Sensing*, 4, 9, 2661–2693.

Ji, L., Zhang, L., Wylie, B.K., and Rover, J., 2011. On the terminology of the spectral vegetation index (NIRSWIR)/(NIR+SWIR). *International Journal of Remote Sensing*, 32, 6901–6909.

Jia, K., Liang, S., Zhang, L., Wei, X., Yao, Y., and Xie, X., 2014. Forest cover classification using Landsat ETM+ data and time series MODIS NDVI data. *International Journal of Applied Earth Observation and Geoinformation*, 33, 32–38.

Justice, C., Giglio, L., Korontzi, S., Owens, J., Morisette, J., Roy, D., et al., 2002. The MODIS fire products. *Remote Sensing of Environment*, 83, 1, 244–262.

Kanja, K., Karahalil, U., and Çil, B., 2019. Modeling stand parameters for *Pinus Brutia* (Ten.) using airborne LiDAR data: A case study in Bergama. *Journal of Applied Remote Sensing*, 14, 02, 1, https://doi.org/10.1117/1.jrs.14.022205

Kankare, V., Vauhkonen, J., Tanhuanpää, T., Holopainen, M., Vastaranta, M., Joensuu, M., Krooks, A., et al., 2014. Accuracy in estimation of timber assortments and stem distribution – A comparison of airborne and terrestrial laser scanning techniques. *ISPRS Journal of Photogrammetry and Remote Sensing*, 97, 89–97, https://doi.org/10.1016/j.isprsjprs.2014.08.008

Ke, Y., Quackenbush, L.J., and Im, J., 2010. Synergistic use of QuickBird multispectral imagery and LIDAR data for object-based forest species classification. *Remote Sensing of Environment*, 114, 6, 1141–1154.

Kelly, M., and Di Tommaso, S., 2015. Mapping forests with Lidar provides flexible, accurate data with many uses. *California Agriculture*, 69, 1, 14–20.

Kennedy, R.E., Ohmann, J., Gregory, M., Roberts, H., Yang, Z., Bell, D.M., Kane, V., Hughes, M.J., Cohen, W.B., Powell, S., et al., 2018. An empirical integrated forest biomass monitoring system. *Environmental Research Letters*, 13, 025004.

Key, C.H., and Benson, N.C., 2006. Landscape Assessment: Ground Measure of Severity, the Composite Burn Index; and Remote Sensing of Severity, the Normalized Burn Ratio, in *FIREMON: Fire Effects Monitoring and Inventory System*. Editors: D.C. Lutes, R.E. Keane, J.F. Caratti, C.H. Key, N.C. Benson, S. Sutherland and L.J. Gangi. USDA Forest Service, Rocky Mountain Research Station, Ogden, UT, 1–51

Korontzi, S., Roy, D.P., Justice, C.O., and Ward, D.E., 2004. Modeling and sensitivity analysis of fire emissions in southern Africa during SAFARI 2000. *Remote Sensing of Environment*, 92, 255–275.

Kotivuori, E., Kukkonen, M., Mehtätalo, L., Maltamo, M., Korhonen, L., and Packalen, P., 2020. Forest inventories for small areas using drone imagery without in-situ field measurements. *Remote Sensing of Environment*, https://doi.org/10.1016/j.rse.2019.111404

Koutsias, N., and Karteris, M., 2000. Burned area mapping using logistic regression modeling of a single post-fire Landsat-5 Thematic Mapper image. *International Journal of Remote Sensing*, 21, 673–687.

Koutsias, N., and Pleniou, M., 2015. Comparing the spectral signal of burned surfaces between Landsat-7 ETM+ and Landsat-8 OLI sensors. *International Journal of Remote Sensing*, 36, 3714–3732.

Koutsias, N., Karteris, M., and Chuvieco, E., 2000. The use of Intensity-Hue-Saturation transformation of Landsat-5 Thematic Mapper data for burned land mapping. *Photogrammetric Engineering & Remote Sensing*, 66, 829–839.

Koutsias, N., Pleniou, M., Mallinis, G., Nioti, F., and Sifakis, N.I., 2013. A rule-based semi-automatic method to map burned areas: Exploring the USGS historical Landsat archives to reconstruct recent fire history. *International Journal of Remote Sensing*, 34, 7049–7068.

Lausch, A., Erasmi, S., King, D.J., Magdon, P., and Heurich, M., 2017. Understanding forest health with remote sensing – Part II: A review of approaches and data models. *Remote Sensing*, 9, 129, doi:10.3390/rs9020129.

Liang, X., Hyyppä, J., Kaartinen, H., Lehtomäki, M., Pyörälä, J., Pfeifer, N., Holopainen, M., et al., 2018a. International benchmarking of terrestrial laser scanning approaches for forest inventories. *ISPRS Journal of Photogrammetry and Remote Sensing*, https://doi.org/10.1016/j.isprsjprs.2018.06.021

Liang, X., Kukko, A., Hyyppä, J., Lehtomäki, M., Pyörälä, J., Yu, X., Kaartinen, H., Jaakkola, A., and Wang, Y., 2018b. In-situ measurements from mobile platforms: An emerging approach to address the old challenges associated with forest inventories. *ISPRS Journal of Photogrammetry and Remote Sensing*, https://doi.org/10.1016/j.isprsjprs.2018.04.019

Liang, X., Kankare, V., Hyyppä, J., Wang, Y., Kukko, A., Haggrén, H., and Holopainen, M., 2016. Terrestrial laser scanning in forest inventories. *ISPRS Journal of Photogrammetry and Remote Sensing*, 115, 63–77.

Lu, D., 2006. The potential and challenge of remote sensing-based biomass estimation. *International Journal of Remote Sensing*, 27, 1297–1328.

Mallinis, G. and Koutsias, N., 2012. Comparing ten classification methods for burned area mapping in a Mediterranean environment using Landsat TM satellite data. *International Journal of Remote Sensing*, 33, 4408–4433.

Matricardi, E.A.T., Skole, D.L., Pedlowski, M.A., Chomentowski, W., and Fernandes, L.C., 2010. Assessment of tropical forest degradation by selective logging and fire using Landsat imagery. *Remote Sensing of Environment*, https://doi.org/10.1016/j.rse.2010.01.001

Mora, B., Wulder, M.A., White, J.C., and Hobart, G., 2013. Modeling stand height, volume, and biomass from very high spatial resolution satellite imagery and samples of airborne LiDAR. *Remote Sensing*, 5, 5, 2308–2326.

Müller, F., Jaeger, D., and Hanewinkel, M., 2019. Digitization in wood supply – A review on how Industry 4.0 will change the forest value chain. *Computers and Electronics in Agriculture*, 162, 206–2218, https://doi.org/10.1016/j.compag.2019.04.002

Nelson, R., Ranson, K.J., Sun, G., Kimes, D.S., Kharuk, V., and Montesano, P., 2009. Estimating Siberian timber volume using MODIS and ICESat/GLAS. *Remote Sensing of Environment*, https://doi.org/10.1016/j.rse.2008.11.010

Nguyen T.H., Jones, S., Soto-Berelov, M., Haywood, A., and Hislop, S., 2020. Landsat time-series for estimating forest aboveground biomass and its dynamics across space and time: A review. *Remote Sensing*, 12, 98, doi:10.3390/rs12010098

Nioti, F., Dimopoulos, P., and Koutsias, N., 2011. Correcting the fire scar perimeter of a 1983 wildfire using USGS-archived Landsat satellite data. *GIScience & Remote Sensing*, 48, 4, 600–613, doi: 10.2747/1548-1603.48.4.600

Pause, M., Schweitzer, C., Rosenthal, M., Keuck, V., Bumberger, J., Dietrich, P., Heurich, M., Jung, A., and Lausch, A., 2016. In situ/remote sensing integration to assess forest health – A review. *Remote Sensing*, 8, 471, doi:10.3390/rs8060471

Pereira, J.M.C., Chuvieco, E., Beaudoin, A., and Desbois, N., 1997. Remote sensing of burned areas: A review, in *A Review of Remote Sensing Methods for the Study of Large Wildland Fires*. Editor: E. Chuvieco. Universidad de Alcala, Alcala de Henares, Spain.

Pflugmacher, D., Cohen, W.B., Kennedy, R.E., and Yang, Z., 2014. Using Landsat-derived disturbance and recovery history and lidar to map forest biomass dynamics. *Remote Sensing of Environment*, 151, 124–137.

Pichler, G., Lopez, J.A.P., Picchi, G., Nolan, E., Kastner, M., Stampfer, K., and Kühmaier, M., 2017. Comparison of remote sensing based RFID and standard tree marking for timber harvesting. *Computers and Electronics in Agriculture*, https://doi.org/10.1016/j.compag.2017.05.030

Pleniou, M., and Koutsias, N., 2013. Sensitivity of spectral reflectance values to different burn and vegetation ratios: A multi-scale approach applied in a fire affected area. *ISPRS Journal of Photogrammetry and Remote Sensing*, 79, 199–210, ISSN 0924-2716, https://doi.org/10.1016/j.isprsjprs.2013.02.016

Pleniou, M., Xystrakis, F., Dimopoulos, P., and Koutsias, N., 2012. Maps of fire occurrence – Spatially explicit reconstruction of recent fire history using satellite remote sensing. *Journal of Maps*, 8, 499–506.

Puletti, N., Chianucci, F., and Castaldi, C., 2017. Use of Sentinel-2 for forest classification in Mediterranean environments. *Annals of Silvicultural Research*, 42, 1.

Rahlf, J., Breidenbach, J., Solberg, S., Næsset, E., and Astrup, R., 2017. Digital aerial photogrammetry can efficiently support large-area forest inventories in Norway. *Forestry: An International Journal of Forest Research*, 90, 5, 710–718.

Reutebuch, S.E., Andersen, H.E., and McGaughey, R.J., 2005. Light detection and ranging (LIDAR): An emerging tool for multiple resource inventory. *Journal of Forestry*, 103, 6, 286–292.

Rikimaru, A., Roy, P.S., and Miyatake, S., 2002. Tropical forest cover density mapping. *Tropical Ecology*, 43, 1, 39–47.

Rizvi, R.H., Ram, N., Karmakar, P.S., Saxena, A., and Dhyan, S.K., 2016. Remote sensing analysis of agroforestry in Bathinda and Patiala Districts of Punjab using sub-pixel method and medium resolution data. *Journal of the Indian Society of Remote Sensing*, doi: 10.1007/s12524-015-0463-3

Robinson, J.M., 1991. Fire from space: Global fire evaluation using infrared remote sensing. *International Journal of Remote Sensing*, 12, 3–24.

Salmivaara, A., Miettinen, M., Finér, L., Launiainen, S., Korpunen, H., Tuominen, S., Heikkonen, J., et al., 2018. Wheel rut measurements by forest machine-mounted LiDAR sensors – Accuracy and potential for operational applications? *International Journal of Forest Engineering*, https://doi.org/10.1080/1494211 9.2018.1419677

Sheppard, J.P., Chamberlain, J., Agúndez, D., Bhattacharya, P., Chirwa, P.W., Gontcharov, A., Sagona, W.C.J., Shen, H.L., Tadesse, W., and Mutke S., 2020. Sustainable forest management beyond the timber-oriented status quo: Transitioning to co-production of timber and non-wood forest products – A global perspective. *Current Forestry Reports*, https://doi.org/10.1007/s40725-019-00107-1

Soenen, S.A., Peddle, D.R., Hall, R.J., Coburn, C.A., and Hall, F.G., 2010. Estimating aboveground forest biomass from canopy reflectance model inversion in mountainous terrain. *Remote Sensing of Environment*, 114, 1325–1337.

Szpakowski, D.M., and Jensen, J.L.R., 2019. A review of the applications of remote sensing in fire ecology. *Remote Sensing*, 11, 2638, 10.3390/rs11222638

Verma, N.K., Lamb, D.W., Reid, N., and Wilson, B., 2016. Comparison of canopy volume measurements of scattered eucalypt farm trees derived from high spatial resolution imagery and LiDAR. *Remote Sensing*, 8, 5, art. no. 388.

Wallace, L., Lucieer, A., Watson, C., and Turner, D., 2012. Development of a UAV-LiDAR system with application to forest inventory. *Remote Sensing*, 4, 6, 1519–1543.

Wang, J., Sammis, T.W., Gutschick, V.P., Gebremichael, M., Dennis, S.O., and Harrison, R.E., 2010. Review of satellite remote sensing use in forest health studies. *Open Geography Journal*, 3, 28–42.

White, J.C., Wulder, M.A., Vastaranta, M., Coops, N.C., Pitt, D., and Woods, M., 2013. The utility of image-based point clouds for forest inventory: A comparison with airborne laser scanning. *Forests*, 4, 3, 518–536.

White, J.C., Coops, N.C., Wulder, M.A., Vastaranta, M., Hilker, T., and Tompalski, P., 2016. Remote sensing technologies for enhancing forest inventories: A review. *Canadian Journal of Remote Sensing*, 42, 5, 619–641.

Wijaya, A., and Gloaguen, R., 2007. *Comparison of multisource data support vector machine classification for mapping of forest cover*. In *2007 IEEE International Geoscience and Remote Sensing Symposium*, 1275–1278.

Wulder, M.A., and Seemann, D., 2003. Forest inventory height update through the integration of lidar data with segmented Landsat imagery. *Canadian Journal of Remote Sensing*, 29, 5, 536–543.

Xulu, S., Peerbhay, K., Gebreslasie, M., and Ismail, R., 2019. Unsupervised clustering of forest response to drought stress in Zululand Region South Africa, *Forests*, 10, 531, doi:10.3390/f10070531

Yu, X., Hyyppä, J., Holopainen, M., and Vastaranta, M., 2010. Comparison of area-based and individual tree-based methods for predicting plot-level forest attributes. *Remote Sensing*, https://doi.org/10.3390/rs2061481

Yu, X., Hyyppä, J., Kaartinen, H., and Maltamo, M., 2004. Automatic detection of harvested trees and determination of forest growth using airborne laser scanning. *Remote Sensing of Environment*, https://doi.org/10.1016/j.rse.2004.02.001

Zhang C., Zhou, Y., and Qiu, F., 2015. Individual tree segmentation from LiDAR point clouds for urban forest inventory. *Remote Sensing*, 7, 7892–7913, 10.3390/rs70607892

9 Geology[1]

9.1 INTRODUCTION: REMOTE SENSING CONCEPTS OF GEOLOGY

Remote sensing was recognized as a powerful tool for geological structural analyses almost immediately after the launch of ERTS-1, later renamed Landsat-1, in 1972. The three main advantages of remotely sensed data over conventional air-photography are : (1) a synoptic view of the Earth's surface is obtained so that structures of interest and at a variety of scales can be mapped and correlation between various structure sets can be made; (2) this view is offered at regular time intervals, depending on the orbital characteristics of the sensor, so that spaceborne imagery can be obtained over most of the globe, and global processes (for instance climate change) can be studied over longer timescales; and (3) these data are produced in digital form, so that computer processing may enhance and extract features of interest.

The data supplied to the user are digital images of the Earth's surface, comprising regularly sampled scalar intensities representing scene radiance or backscatter of emitted microwaves. These physical quantities are encoded as integers or digital numbers (DN), mostly into an 8-bit or 16-bit linear scale. These data are processed by computers to enhance features of interest and finally interpreted by visual inspection (see for example Figure 9.1 for a clear depiction of a left-lateral strike-slip fault). Image processing involves applying an algorithm or operation to each stored DN and creating a new image from the processed data.

9.2 STRUCTURAL MAPPING

9.2.1 TERRAIN ANALYSIS

The objective for the geologist is to produce a structural and/or a lithological map from surface reflectance or backscatter variations occurring in the scene, by using various image enhancement or image transformation algorithms. Most commonly faced problems in the image processing of raw data are: (1) rock surface weathering which differs with elevation and surface mineralogy; (2) signal attenuation due to atmospheric effects which depends on the illumination geometry and atmospheric composition during scene acquisition; and (3) seasonal vegetation/biomass cover (extent, height).

For each of those problems there are appropriate techniques to reduce their effects on the quality of the interpretation (e.g. Drury, 1986; Drury, 1987; Crippen, 1989; Crósta and Moore, 1989). The selection of raw data (season, wavelength) is itself an influential factor in geological interpretations. For example, structural mapping benefits from "shadowing" casted by relief and therefore optimum images are those with the lowest elevation angle of solar illumination (winter imagery; Drury, 1986; Ganas, 1997; Ganas et al., 2005a; Ganas et al., 2005b). Similarly, because of the physics of human vision monochromatic imagery is better suited to structural investigations than a color composite image which may be more useful in a lithological mapping exercise (Drury, 1986).

This sub-chapter presents a set of established approaches in the remote sensing of minerals, geological structure and geohazards, using both optical and radar sensors. For example, fault populations are an important element in mineral and petroleum exploration, environmental applications and in seismic hazard assessments. We need to understand how faults grow, what their role is in shaping the Earth's surface and how fast elastic strain energy accumulates along them. From remote sensing we can only map fault patterns on the surface of planets. We can also map fault growth after large earthquakes where the elastic energy released is so large that it results in m-size displacements on the surface (Figure 9.2) that are identifiable from space (Figure 9.2).

---

[1] Contributors: Athanassios Ganas and Nicolas R. Dalezios

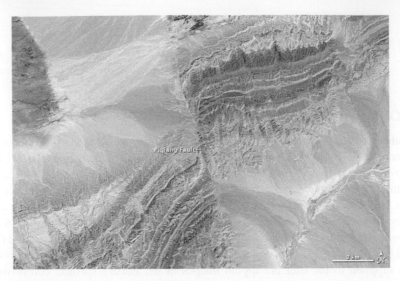

FIGURE 9.1 Strike-slip faulting in the Earth's crust forms a remarkable scenery in western China. This is Image of the NASA Day for January 16, 2014 (original in Turner, 2011). Instrument: Landsat 8 – OLI. Sense of movement is left-lateral.

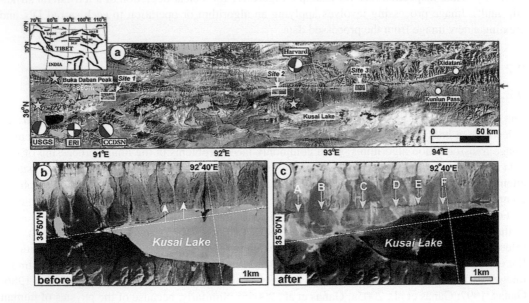

FIGURE 9.2 a) Landsat ETM/TM mosaic image shows the Kusai Lake segment of the Kunlun strike-slip fault zone (indicated by red arrows) in China. The locations of epicenter (indicated by stars) and focal mechanisms of the 2001 M = 8.1 Central Kunlun earthquake are shown as beachballs; b) Landsat ETM image of site 2 before the earthquake (April 23, 2000); (c) Landsat ETM image of site 2 after the earthquake (December 7, 2001) (from Fu and Lin, 2003).

An additional approach I endorse is the use of digital topography, reconstructed from space imagery, to quantify throw (the vertical component of slip) distribution along faults. To estimate throw we need to know the relative offset of a specific geological horizon across a fault. If fault age can be constrained by dating the oldest deformed rocks, then the throw rate (and possibly slip rate) distribution along strike may be calculated. This slip rate could be used to derive a seismic moment rate that is a measure of fault seismicity. The implementation of "structural logic" into the processing of remotely sensed data may yield critical information on the seismic potential of large, active faults.

A final important point: observations made on satellite imagery and interpretations on faulting patterns need to be validated by field work. Field work is necessary in remote sensing projects of faulting and in geohazards because of the uncertainties involved in their detection from space.

Optical data may be used to identify neotectonic faults and fault segments, based on the identification of erosion patterns, drainage and tonal anomalies of a linear nature in satellite images or from direct observation of offsets (e.g. Ganas and White, 1996; Ganas et al., 1996; Ganas et al., 2005a; Figures 9.1 and 9.2). In regions of crustal extension (e.g. Tibet, central Greece, Nevada; Figure 9.2) the linear features interpreted as fault segments are usually more than 3–5 km long and are produced by scarps and mountain ranges which develop in response to repeated relief displacements.

Neotectonic fault escarpments are readily recognizable in optical images, as abrupt and imposing linear fronts when they are aligned with solar azimuth, or as long, rectangular, bright stripes if they face the sun, and/or as long, trapezoidal or cuspate shadows, if they face away (e.g. Ganas, 1997; Ganas et al., 2005a). The shape of the shadow is, in most cases, indicative of range crest morphology, i.e. sharp terminations on either tip (trapezoidal shape) may suggest an abrupt increase of relief gradient that may have been produced by cross-faulting or some degree of mechanical interaction with neighboring faults. Multiple geometrical patterns of landforms can also be recognized on the footwall of large faults, and can be used to constrain the length of segments; these patterns include triangular facets truncating mountain spurs (e.g. Armijo et al., 1991; Ganas et al., 2005b; Tsimi and Ganas, 2015), and elongated "wine-glass" valleys (e.g. Wallace, 1978; Ganas et al., 1996) extending from the fault line into the hinterland. Other fault lines are usually depicted on satellite images by sharp tonal alignments, without any resultant relief, and may be interpreted as relict structures from older stress systems.

The most popular satellites among geologists are the Landsat and SPOT missions (e.g. Peltzer et al., 1989; Chorowicz et al., 1994; Ganas et al., 2001; Meyer et al., 2002; Ganas et al., 2005b; Turner, 2011; Figures 9.1 and 9.2). The spatial resolution of the Landsat TM, ETM and ETM+ and OLI sensors is 30 m while the SPOT size is 10–20 m. The OLI sensor is an improvement of past Landsat sensors and is based on a technical approach demonstrated by a sensor flown on NASA's experimental EO-1 satellite. OLI is a push-broom sensor with a four-mirror telescope, 12-bit quantization, which collects data for visible (VIS), near-infrared(NIR), and short-wave infrared (SWIR) spectral bands, as well as a panchromatic band (see Table 9.1 for Landsat sensor characteristics). Such products are provided with a minimum level of radiometric and geometric pre-processing, including their transformation to the UTM projection system and "along track" orientation. The optical scene can be filtered with a Fast Fourier transform (FFT) in order to remove radiometric noise (Ganas et al., 2001). The best visualization of landforms can be provided by a non-linearly, stretched monochromatic image in the short-wave infrared part of the spectra (for example the Landsat TM band 5; OLI band 6).

The processed satellite imagery can be combined with digital processing of topographic data (digital elevation model or DEM) as tectonic activity shapes topography at a variety of scales. For geological mapping the required resolution of DEMs is in the range of 5–30 m depending on the application. A resolution of 20–30 m is adequate to map active faults. Such DEMs can be produced by on-screen digitizing of elevation contours of 1:50,000 map sheets (contour interval 20 m). Then, shaded relief images in grayscale can be produced using various illumination conditions in order to study the geometry and long-term evolution of landforms (Ganas et al., 2001). Shaded relief images can also be used as a raster background to overlay vector files, like seismic aftershock sequences, or landslides.

9.2.2 MINERAL EXPLORATION

Of increasing importance to exploration and ore geology is the application of imaging spectroscopy – hyperspectral remote sensing techniques (e.g. Ferrier et al., 2016; Ferrier et al., 2019a; Ferrier et al., 2019b; Figure 9.3). Main application fields for imaging spectroscopy include precision agriculture,

TABLE 9.1

Characteristic of the Landsat sensors. Orbital height is 705 km. A full Earth's surface scan by Landsat takes 232 turns, or 16 days. Scene size is approximately 185 x 185 km. The terrain survey takes place at approximately 10 am (± 15 minutes) according to local solar time

Satellite sensor	Launch date	Band	Wavelength- μm	Resolution-m
LANDSAT (4, 5) MSS	16/07/1982	4	0.5–0.6	79 × 56
		5	0.6–0.7	
		6	0.7–0.8	
		7	0.8–1.1	
LANDSAT (4, 5) TM	01/03/1984	1	0.45–0.52	30
		2	0.52–0.60	30
		3	0.63–0.69	30
		4	0.76–0.90	30
		5	1.55–1.75	30
		6-TIR	10.4–12.5	120
		7	2.08–2.35	30
LANDSAT 7 ETM+	15/04/1999	1	0.45–0.52	30
		2	0.52–0.60	30
		3	0.63–0.69	30
		4	0.77–0.90	30
		5	1.55–1.75	30
		6-TIR	10.40–12.50	60
		7	2.09–2.35	30
		8-PAN	0.52–0.90	15
LANDSAT 8 OLI	11/02/2013	1	0.43–0.45	30
		2	0.45–0.51	30
		3	0.53–0.59	30
		4	0.63–0.67	30
		5	0.85–0.88	30
		6	1.57–1.65	60
		7	2.11–2.29	30
		8-PAN	0.50–0.68	15
		9	1.36–1.38	30
		10-TIRS1	10.60–11.19	100
		11-TIRS2	11.50–12.51	100

geological and soil mapping, forest and water management as well as the monitoring of fragile ecosystems.

Remote sensing provides information on the properties of the surface of geological exploration targets, which is potentially significant in mapping alteration zones. Specifically, broadband sensors, such as Landsat, SPOT and Sentinel-2, are capable of distinguishing between some rock types such as iron-rich and iron-poor lithologies (Ferrier and Wadge, 1996). However, narrow-band sensors such as the AVIRIS (Green et al., 1998; https://aviris.jpl.nasa.gov/; Figure 9.4) and DAIS 7915 (DLR; Mueller et al., 2002) imaging spectrometers are capable of detecting spectral absorptions in the visible to short-wave infrared specific to individual minerals (Vane and Goetz, 1988). The AVIRIS flight system is a whisk-broom imager that acquires data in 224 narrow, contiguous spectral bands covering the solar reflected portion of the electromagnetic spectrum, which is flown aboard the NASA high altitude ER-2 research aircraft. As a result, the imaging spectrometers offer

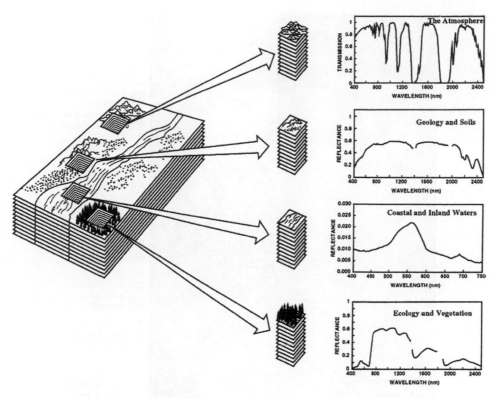

FIGURE 9.3 The concept of imaging spectroscopy is shown with a spectrum measured for each spatial element in an image. The spectra are analyzed for science research and applications in a range of disciplines (from Green et al., 1998).

the prospect of a valuable, additional source of data for the exploration geology, since these detected minerals are indicative of the type of alteration and are present in sufficient quantities at the surface from which solar radiation can be reflected to the sensor (Ferrier and Wadge, 1996).

Reflectance spectroscopy of rocks and minerals can provide valuable diagnostic information on their elemental and mineralogical composition (Hunt, 1977; Ferrier and Wadge, 1996; Ferrier et al., 2012). The remote sensing implementation of this method (e.g. Vane and Goetz, 1988) finds its most apparent application in the mapping of hydrothermally altered rocks (Rast et al., 1991). This possibility occurs, since many of the alteration minerals have distinctive absorption features caused by the presence of OH, and other hydroxyl-bonds: Mg-OH and Al-OH, particularly in the short-wave infrared (SWIR) part of the spectrum (2000–2400 nm) (Ferrier and Wadge, 1996). Specifically, the wavelength positions, depths and number of absorption features occur mainly in the 1480 to 1800 nm and the 2100 to 2500 nm regions of the spectra, respectively, which enables the determination of the minerals present. This finding constitutes the basis for mapping the zonation from an equivalent remote sensor, such as AVIRIS or DAIS-7915 (Ferrier and Wadge, 1996).

9.3 LITHOLOGICAL MAPPING

For lithological and soil mapping, optical satellite data is usually collected and processed in order to enhance the feature of interest. A systematic collection of multispectral Sentinel-2 available every 5 days (cloud-free conditions) with spatial resolution up 20–30 m and a spectral range consistent with in situ and UAV measured data (if available) can be implemented to derive thematic maps including: (1) change detection maps; (2) supervised classification maps (through support-vector

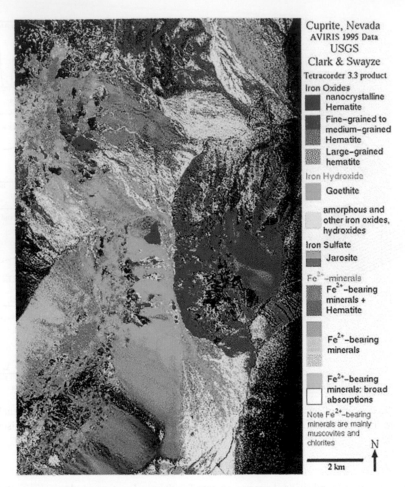

FIGURE 9.4 Hyperspectral mapping by the AVIRIS sensor. Legend on the right shows the mapped minerals (image source NASA).

machines SVM and neural networks); and (3) a systematic collection of EO1-Hyperion archive and hyperspectral PRISMA (Italian Space Agency; available from summer 2019). Mission data can be implemented to characterize soil composition and to map indirect indicators such as vegetation stress, for example induced by heavy metal. Main steps of the image processing for each sensor (EO1-Hyperion and PRISMA) include: (1) data calibration from raw data to produce intermediate radiance product; (2) atmospheric correction to produce at surface reflectance; and (3) supervised classification by using fine absorption spectral features to be linked with composition. One of the outputs of such mapping projects is to produce a Soil Spectra Library (SSL) consisting of both laboratory and in situ spectra measurements to be used as ground truth for supervised classification methods (SVM), to train neural networks, for cross comparison with UAV data and to validate other satellite data. An instrument usually used for field measurements is a portable, battery powered spectroradiometer operating in the VNIR-SWIR spectral range (0.35–2.5 μm) which permits the detection of individual absorption features due to specific chemical bonds in a solid, liquid or gas. An adequate number of samples is collected and later measured in laboratory. The field measurements are ideally done in conjunction with satellite (Sentinel-2) overpass to provide a validation of satellite data. Spectra features derived by field spectra can be used to characterize surface composition.

A useful technique to map compositional differences on the Earth's surface is image ratioing. Many ratios have been developed for multispectral scanners, such as Landsat TM (ETM), the most important of which are provided below:

- the red/blue channel and 3/2 red/green channel ratios are important for delineating ferric iron-rich rocks (light-tones) and ferric iron-poor rocks (dark tones);
- the mid/short-wave infrared ratio is useful for identifying clay-rich rocks (light-tones), since clay minerals exhibit strong absorption in the 2.2 μm region (TM band 7) and high reflectance in the 1.6 μm region (TM band 5);
- the ratio near IR/red uniquely defines the distinguishing different types of vegetation. Generally, the lighter the tone, the greater the amount of vegetation present;
- the 5/2 ratio mid IR/green is useful for distinguishing different types of vegetation. Its reciprocal is useful for identifying water bodies and wetlands;
- the 3/7 ratio red/mid IR is useful for observing differences in water turbidity. It is also useful for identifying the roads and other cultural features, showing lighter tone due to their relatively high reflectance in the red band (TM3) and low reflectance in the mid IR band (TM7).

In addition, data reduction methodologies can be applied to integrate a wide variety of complex spatial data sets for the analysis of the distribution of mineral deposits and model their spatial association (Ferrier et al., 2019a). This may provide insights into the controls on ore deposition, as well as contributing to a better understanding of local geology.

Thermal remote sensing. Thermal infrared radiation of an object is controlled mainly by the characteristics of the surface, namely temperature, emissivity and geometry of the object (Prasun et al., 2006). Thermal remote sensing can detect surface mineralogy that other wavelengths cannot. It can detect particularly silicates and Al-Si rocks, and it can also detect and differentiate different types of carbonate and clastic rocks again that cannot be detected using other wavelengths. In some mines thermal hot spots can occur with chemical changes again that can be detected using the thermal wavelengths.

Another useful approach with thermal data is the detection of mine waste water, which is usually hotter/warmer than background water temperature offering pollution monitoring approach. Moreover, the upwelling of groundwater and the inflow of groundwater into surface water, such as streams or lakes, can be detected and quantified due to the temperature differences between groundwater and surface water.

In cases of high temperature object conditions, such as forest, mine fire or volcanic activity, remote sensing can provide a good synoptic view of the area under consideration (Prasun et al., 2006). Moreover, night-time thermal data are quite useful to isolate the warm areas from the background. Suitable thermal data can be used from Landsat-8 (bands 10 & 11; 100-m resolution, resampled to 30-m; 16-days revisit time) and ASTER TIR sensors (5-bands; 90-m resolution; 16-days revisit time).

Regarding new technologies the UAV-mounted hyperspectral cameras can offer the capability of automatic, remote detection of surface mineral composition at ultra-high spatial (cm) and temporal (daily) resolution at very low cost.

9.4 ENVIRONMENTAL GEOLOGY

9.4.1 GEOHAZARD MAPPING AND MONITORING

Synthetic Aperture Radar Interferometry (InSAR) is a remote sensing method, which is applied to mapping Earth surface deformations with centimeter to millimeter accuracy. When orbital radar[2] systems are used the technique is commonly known as INSAR (INterferometric SAR). Synthetic

---

[2] Radar is the acronym for RAdio Detection And Ranging.

Aperture Radar is a technology to "synthesize" the "aperture" of the radar antenna in order to obtain a high-resolution image (e.g. 10 m or 20 m ground resolution). The InSAR technique works by exploiting the differences in round-trip phase components of SAR images relative to an investigated area. The spatial resolution of the derived phase information depends on the wavelength of the SAR: in C-band interferograms one color cycle is 28 mm while in L-band the phase cycle equals 11.8 cm. The spatial pattern of these cycles (or fringes) as well their gradient (manifested by fringe spacing) is indicative of the intensity and map extent of ground deformation. The technique is used to map (and monitor) displacements (deformations) occurring at basin scales (100 m × 100 m) and at outcrop scales (20 m × 20 m), thus being appropriate to analyze local deformations that may affect, for example, single buildings or structures. The SAR data originate from image acquisitions from ESA (such as ERS-1, ERS-2, ENVISAT and Sentinel-1) or other space agencies (for example, Japan and Canada). For long-term deformation monitoring, where the monitoring of deformations with a low velocity (cm/yr and mm/yr) is required, it is necessary to allow for long time gaps between the image acquisitions. Consequently, large areas are phase de-correlated (e.g. densely vegetated surfaces), making the interferometric phase useless. An additional limitation for all interferograms is atmospheric artifacts.

Since 1992, the availability of InSAR geodesy data has provided unprecedented detail in mapping ground deformation due to seismicity (e.g. Massonnet et al., 1993; Meyer et al., 1996; Wright et al., 1999; Fielding et al., 2003). Strong and shallow earthquakes cause permanent ground deformation (mainly co-seismic), extensive damages to the ground and multiple ground failures. Satellite missions such as ENVISAT/ASAR (not in operation since 2012), RADARSAT and SENTINEL are used to process co-seismic interferograms. Standard SAR processing packages (ROI-PAC, SNAP, DIAPASON, GAMMA™ etc.) are used for interferogram generation. In cases of strong and shallow earthquakes ground deformation, several fringes in the line of sight of the SAR satellites, can be mapped on co-seismic interferograms (Figure 9.5). By inversion of the data from the observed fringes (wrapped interferogram method), best fitting models of the activated faults can be calculated assuming rectangular dislocations in an elastic half space (e.g. Lefkada M = 6.5 2015 earthquake; 2017 Kos M = 6.6 earthquake, Ganas et al., 2019). Another method involves the unwrapping of the phase to first establish the LOS displacement and then to decompose the displacements along the horizontal components and along the vertical (Figure 9.6a and b Elazig earthquake in 2020). The preliminary unwrapped result for the Elazig shallow earthquake of January 24, 2020 (S1 ascending orbit; Figure 9.6b) showed 55 cm LOS relative displacement along a 40-km long rupture. In principle, the InSAR-derived dislocation models also need to fit observed displacements of permanent GNSS stations of public and private networks near the epicenter.

For multitemporal studies (3–5 years and beyond), the Small Baseline Subset (SBAS) technique (Berardino et al., 2002) is widely used. The SBAS method is based on an appropriate combination of differential interferograms, which are produced by using SAR image pairs depicting a small orbital separation (baseline). This reduces the spatial de-correlation phenomena. However, for the ESA satellites, the available acquisitions are generally distributed in several small baseline subsets, separated by large baselines. Specifically, the singular value decomposition (SVD) technique is used to "link" the separate small baseline subsets, increasing, thus, the temporal sampling rate. Moreover, atmospheric artifacts are filtered out based on the availability of both spatial and temporal information. The topographic phase contribution is removed by using a phase simulation, using a global digital elevation model from the SRTM/ASTER missions. Nevertheless, the SBAS method is based on the availability of unwrapped differential interferograms. In order to facilitate the phase unwrapping, only the phase in coherent pixels is used. The final unwrapping is usually conducted by using the SNAPHU software (Chen and Zebker, 2001). Indeed, the dense spatial coverage of deformation provided by the SAR data processed with the SBAS-InSAR algorithm can be used to better constrain the source of deformation, in accordance with seismological and geological research findings.

A second well-established technique is the Persistent Scatterer Interferometry (PSI) (Ferretti et al., 2001), which is used to generate and analyze the time series of spaceborne SAR data during the

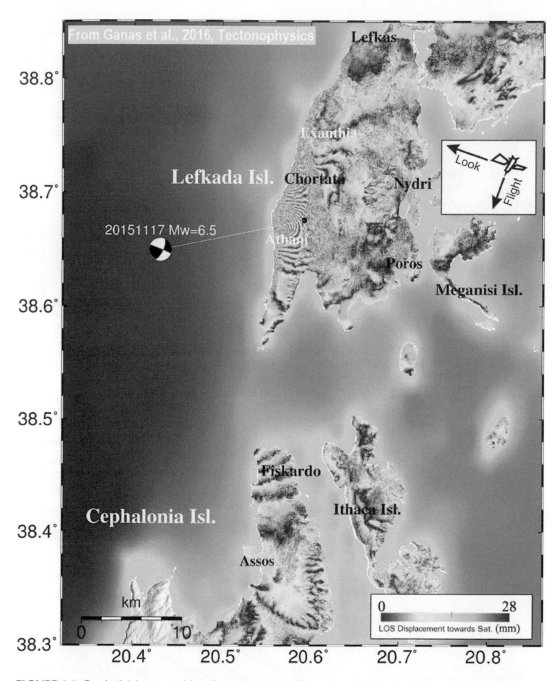

FIGURE 9.5 Sentinel 1A wrapped interferogram descending orbit of the Lefkada (Greece) November 2015 M = 6.5 earthquake (from Ganas et al., 2016; inset box shows the look-direction). Beachball indicates focal mechanism of the earthquake (GCMT solution; http://www.globalcmt.org/).

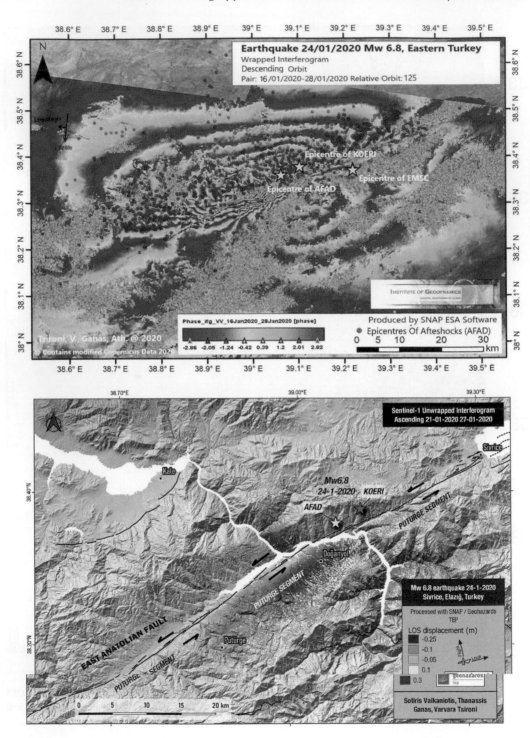

FIGURE 9.6 a) Sentinel 1A wrapped interferogram descending orbit of the Elazig (Turkey) January 2020 earthquake; b) Sentinel 1A unwrapped interferogram (ascending orbit) showing LOS displacements (in cm). Positive indicates motion towards the satellite. White color on maps indicates no data and/or water bodies.

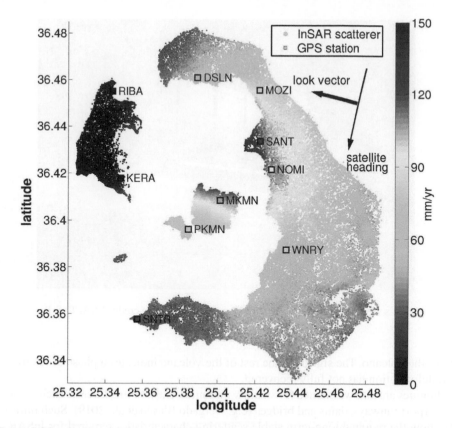

FIGURE 9.7 LOS velocities for the merged PSI and SBAS pixel cloud showing deformation (uplift) of the Santorini volcano (Aegean Sea, Greece), during the period 2011–2012. The MOZI station square corresponds to the reference area, and the colored squares represent the GPS velocities projected to the ENVISAT line of sight (from Papoutsis et al., 2013).

period of interest, thus, resolving and quantifying the deformation history (Papoutsis et al., 2013). The interferometric time series analysis (PSI and SBAS) can be conducted by using the Stanford Method for Persistent Scatterers (StaMPS), where this approach was developed to suit volcanic areas and other natural terrains (Hooper et al., 2004). An application of the mixed pixel cloud from the two techniques was done to map the deformation of the Santorini volcano in Greece by Papoutsis et al. (2013; Figure 9.7).

The Santorini volcano is a well-known touristic destination and the volcano's last major explosive eruption was about 3600 years ago, forming a large crater or caldera, which is now flooded by the sea. For the past 2000 years, Santorini has shown patterns with small eruptions of lava every few tens or hundreds of years, slowly building a new volcanic edifice from the sea floor (Lauknes et al., 2006). The last eruption of the Kameni islands was in 1950, while for the following 60 years, Santorini has been in a "quiet" phase. However, in January 2011, a series of small earthquakes began beneath the islands. Ground deformation was also detected using GPS receivers and an island-wide network of triangulation stations. Envisat data confirmed that the islands had risen as much as 14 cm from January 2011 to March 2012. The whole group of islands had been inflating – slowly rising and moving outward – almost systematically around a point just north of the Kameni islands (Papoutsis et al., 2013; Papoutsis et al., 2014). The post-2012 satellite data (data sets from COSMO-SkyMed and TerraSAR-X) indicate mainly subsidence at Nea Kameni Island and horizontal motion at the northern part of the caldera (Kaskara et al., 2016). These movements may be due to the smooth

FIGURE 9.8 Artist's view of the deployed Sentinel-1 spacecraft (image credit: ESA, TAS-I).

deflation of the volcano. The signal for the rest of the volcano indicates a phase of relative stability. However, deformation has not fully recovered.

PS techniques are also well suited for stability measurements of infrastructure such as motorways, railways, airport runways, dams and bridges (e.g. Delgado Blasco et al., 2019). Such infrastructures typically show the required long-term stable scattering characteristics required for InSAR. A comprehensive review of the subject has been published by ESA in 2012 (Satellite Earth Observation for Geohazard Risk Management, The Santorini Conference, Santorini, Greece, May 21–23, 2012).

Furthermore, Sentinel-1 (see Figure 9.8) interferometric products can be used to map and analyze change detection in the studied areas. Sentinel-1A/B acquire interferometric pairs every ~5 days (systematically). This enables the availability of radar images both pre-event and post-event. The short spatial baseline in Sentinel-1 also leads to an improved overall quality of interferometric coherence estimation. The following methods are applied: 1) Examination of <u>Multi-Temporal Coherence</u> (MTC): Multi-Temporal Coherence (MTC) combines one pair of SAR images acquired in interferometric mode and is useful for change detection as the use of coherence facilitates the detection of land cover classes (e.g. Liao et al., 2008, Nakmuenwai et al., 2016). This product is generated by composing RGB images. An example: amplitude, collected on the first date (red channel), amplitude, collected on the second date (green channel), Coherence between the two images (blue channel); 2) construction of a <u>Multi-Coherence Map</u> (MCM), the Multi-Coherence Map (MCM) combines two pairs of SAR images acquired in interferometric mode and is useful for change detection associated with significant coherence losses/gains (e.g. building destruction/construction; Liao et al., 2008, Nakmuenwai et al., 2016).) The product is generated by composing RGB images with the following base data: 1) coherence between two pre-event acquisitions (red channel); 2) coherence between one pre-event and one post-event acquisitions (green channel and blue channel); and 3) Damage Proxy Map (DPM), the Damage Proxy Map (DPM) is a derived product from the Multi-Coherence Map where coherence difference values are normalized, filtered and classified to generate a proxy map of damage assessment (Fielding et al., 2003, Matsuoka and Yamazaki, 2004, Karimzadeh and Matsuoka, 2017; Figure 9.9). The Damage Proxy Map is a point map obtained by analyzing the coherence between two pairs of radar images. Damage Proxy Map is generated with the following steps: (1) In the coherence from 1st image pair all pixels with higher coherence are extracted; (2) For all these pixels the coherence from 2nd image pair is analyzed. Only pixels with

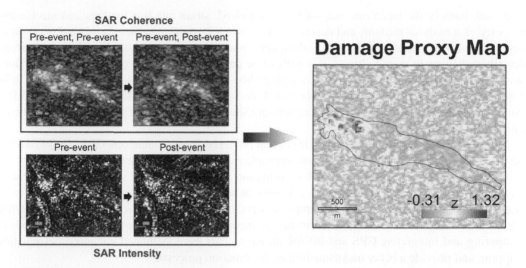

FIGURE 9.9 Coherence images (left) and damage proxy map (right) for the area of Amatrice (Italy) following the disastrous M = 6.2 earthquake of August 24, 2016 (from Karimzadeh and Matsuoka, 2017).

high coherence decrease are extracted and inserted in the Damage Proxy Map. The process also includes histogram matching to make the two coherence histograms comparable.

9.4.2 GNSS Synergies with Radar Remote Sensing

A form of "remotely sensed" geophysical variable is the continuous change of the Earth's surface. This variable can be accurately mapped by use of space geodesy, GNSS and InSAR (see previous section). This information is crucial in Solid Earth Science, as well in other fields such as climate change.

Since the nineties, the Global Positioning System (GPS) and Interferometric Synthetic Aperture Radar (InSAR) have revolutionized the study and monitoring of crustal deformation that quantifies the rate of elevation changes on the surface of the planet. Specifically, these techniques have documented patterns of deformation before, during and after earthquakes, landslides, eruptions of volcanoes, and enabled exploration of quantitative physical models to understand the processes. Moreover, the capabilities of the two techniques complement each other, where GPS can provide a 3D deformation vector at each GPS station with an accuracy of a few millimeters, while InSAR can image the line-of-sight component of ground deformation over a large area at spatial resolution of tens of meters with an accuracy of centimeters to sub-centimeters. Indeed, the two methods work in a similar fashion, both utilizing phases of electromagnetic waves to resolve the precise distance between the satellites and ground targets. Specifically, in GPS/GNSS observations, dual-frequency microwaves and temporal averaging enable the removal of ionospheric anomalies and the reduction of tropospheric artifacts, both of which affect InSAR deformation measurements. Furthermore, InSAR measurements suffer from the loss of interferometric coherence due to modification of the imaged surface characteristics caused by vegetation, snow, ice, and other environmental factors. Nevertheless, the combination of both GPS measurements and InSAR deformation images can enhance mapping, modeling and interpretation of ground deformation.

As the ancient Greek philosopher Heraclitus declared "*Panta rhei*", or "everything flows", it is expected that all bodies on the surface of the planet are expected to modify their shapes, growing or diminishing over the geological history. A number of large European initiatives, such as EPOS (https://epos-ip.org/), have invested significant resources in measuring changes in the shape of the Earth (deformation). Specifically, in the EPOS-IP project, a high-level product is foreseen to be delivered, namely the strain rate map of Eurasia. Indeed, strain rate is a fundamental mechanical property of a body or medium and describes the changes in shape, geometry and configuration of any medium of any size (and thus of a lithospheric plate) with time. Although geology beautifully describes changes in size of objects, it is with space geodesy that the rate of change can be quantified with unprecedented accuracy. To accomplish this, three open-source software packages can be implemented, namely STIB (Strain Tensor from Inversion of Baselines) (Masson et al., 2014), VISR (Velocity Interpolation for Strain Rate) algorithm (Shen et al., 2015), and STRAINTOOL (https://dsolab.github.io/StrainTool/).

Moreover, the perceptible water-vapor content retrieved from Continuous GPS (CGPS/GNSS) networks presents a challenge to estimate atmospheric water-vapor content as a means to correct atmospheric delay anomalies in InSAR deformation measurements and monitor severe weather events (http://gnss4swec.knmi.nl/ for details; also the video https://www.youtube.com/watch?v=t1i nZaRdWY4&app=desktop). The measurement accuracy of InSAR images can be improved, when water-vapor values from CGPS measurements are modeled and interpolated. Innovative methods of comparing and integrating GPS and InSAR measurements have facilitated enhanced deformation mapping and provide a better understanding of deformation processes.

9.4.3 Remote Sensing for Seismic Hazard

Specific inputs from Remote Sensing/GNSS on seismic hazard assessments include:

1. use InSAR techniques to monitor interseismic strain accumulation across major fault zones of the Earth (for example, San Andreas California, North Anatolian Fault, west China, Sumatra etc.; e.g. Hussain et al., 2016);
2. provide quickly co-seismic interferograms that are necessary for earthquake rapid response (for example, Elazig earthquake, Figure 9.6);
3. use InSAR strain maps to invert for seismic coupling along oceanic subduction zones. There are many places on Earth (South America, Pacific, Mediterranean) where GPS/GNSS station density is far from satisfactory for strain mapping;
4. Satellite Radar has provided a wealth of data about mapping co-seismic and post-seismic deformation from hundreds of strong, shallow earthquakes since 1992. Space geodesy helped to constrain the seismic source geometry and slip distribution and to identify the active fault and its surface projection (e.g. Ganas et al., 2016; Ganas et al., 2019);
5. since the early 2000s PS-type products have been also delivered for seismic hazard purposes in several seismogenic areas of the Earth. This PS supply of surface motion patterns has to be supported by independent validation and modeling tools.

9.5 SUMMARY

Remote sensing in geosciences is important in numerous fields such as mineral mapping, fault mapping, ground deformation mapping and monitoring, for assessing the risk of urban sites to natural hazards such as land subsidence and earthquakes and several other areas. Remote sensing data also contribute to an inherently spatial and temporal problem in environmental monitoring, satisfying an emerging need for integrated monitoring schemes at an urban scale that can be readily employed by end users and decision-making authorities. Given the growing urbanization of modern cities and the associated increasing exposure to natural hazards – geohazards – the implementation of multi-sensor monitoring networks is the key for modern society to prevent and mitigate, combining multi-temporal and highly accurate data sets from space- or air-borne observation systems.

The combination of GNSS and remote sensing can contribute to seismic hazard assessments and to volcanic hazards. Moreover, the combination can present new research opportunities in tectonics and geophysics/seismology that makes use of space geodesy products such as integration of high-rate, real-time GNSS measurements and real-time seismic data.

The launching of SENTINEL-1 satellites (see Figure 9.10 for a timeline) provides a large amount of data suitable for geohazard studies exploiting the methodology of SAR interferometry (InSAR). The expected impact on a continually updated ground motion product is expected to be large.

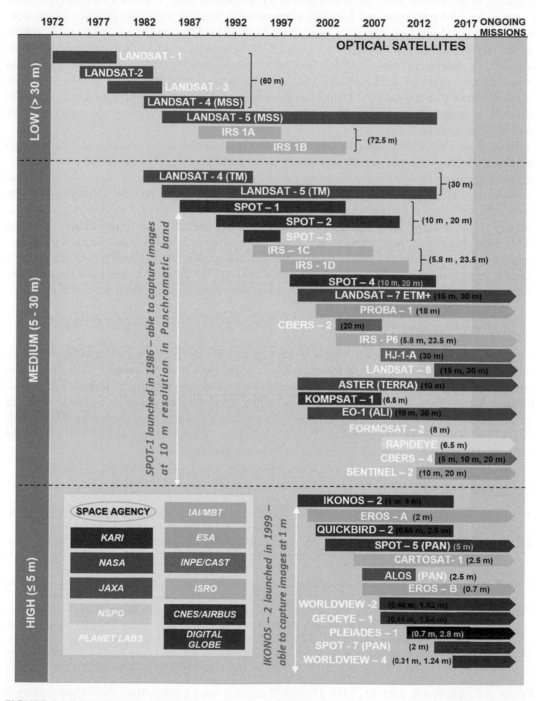

FIGURE 9.10 Timeline of operation and development of optical satellite sensors that are used in resources mapping and hazard studies (from Fan et al., 2019).

Moreover, the GNSS arrays can provide measurements in a wide range of elevations and locations of vast areas which can provide validation tools for multitemporal InSAR products. By using both space geodesy techniques (S1A, InSAR and GNSS) the ground motion product is strengthened. The GNSS products (3D velocities) can be injected routinely into an existing multitemporal inter-ferometry processing chain to extract the displacement rate of the ground with higher accuracy. It is proposed to incorporate these methodologies in an operational basis, towards a monitoring service to serve the civil protection and scientific community in areas of high risk, in terms of tectonic, volcanic and land subsidence hazards, near active faults and volcano and so on. Strong relationships should be maintained with the ongoing or past FP7/H2020 projects of EU, the Copernicus initiative https://www.copernicus.eu/en and ESA-GEP https://geohazards-tep.eu/#! , and the corresponding infrastructure, databases and knowledge should be exploited. The SENTINEL-1 archive should be fully exploited (for the areas of interest in geosciences). The impact will be to foster scientific research, to support the task of civil protection and to assist the design of new (and maintenance of existing) infrastructures.

Remote sensing will continue to have an important impact on future research and educational activity in Geology-Geophysics-Geodynamics, since it will provide: (1) state-of-the-art opportuni-ties to integrate and analyze multidisciplinary data/information for geology, mineral exploration and tectonics; (2) multinational (Figure 9.6; Figure 9.7) and overlapping abilities to improve current knowledge for disaster prevention and risk-management activities; and (3) the scientific and opera-tionally oriented cooperation with regional, national (state) and international institutions operating in the field of earth sciences, climate and coastal monitoring activities. Further impacts of remote sensing products will be on: (1) the promotion of technological/scientific transfer between research centers, private companies and end-user organizations; and (2) the improvement and the enhance-ment of training programs/joint doctorates for young scientists benefiting from new methodologies.

ACKNOWLEDGMENTS

We thank Graham Ferrier, Goeff Wadge, Stefano Salvi, Martha Stefouli, Panagiotis Elias, Pierre Briole, Sotiris Valkaniotis, Gerald Roberts, Kevin White, Salvatore Stramondo, Christina Tsimi, George Papathanassiou, Efthimios Lekkas, Varvara Tsironi, Helena Partheniou, Ioannis Karamitros, Elisavet Kolia, Vassilis Sakkas, Vassilis Kapetanidis, Dimitris Anastasiou and Ioannis Kassaras for comments and editorial assistance. Part of the author's research was funded by EU projects PREVIEW, RASOR, EPOS-PP, EPOS-IP, by ESA projects TERRAFIRMA & ALTAMIRA and the Government of Greece (General Secretariat of Research and Technology).

REFERENCES

Armijo, R., Lyon-Caen, H., and Papanastassiou, D., 1991. A possible normal-fault rupture for the 464 BC Sparta earthquake. *Nature*, 351, 137–139, doi:10.1038/351137a0

Berardino, P., Fornaro, G., Lanari, R., and Sansosti, E., 2002. A new algorithm for surface deformation moni-toring based on small baseline differential interferograms. *IEEE Transactions on Geoscience and Remote Sensing*, 40, 11 2375–2383.

Chen, C.W., and Zebker, H.A., 2001. Two-dimensional phase unwrapping with use of statistical models for cost functions in nonlinear optimization. *Journal of the Optical Society of America A*, 18, 2, 338–351.

Chorowicz, J., Luxey, P., Lyberis, N., Carvalho, J., Parrot, J.F., Yürür, T., and Gündogdu, N., 1994. The Maras Triple Junction (southern Turkey) based on digital elevation model and satellite imagery interpretation. *Journal of Geophysical Research*, 99, 20225–20242.

Crippen, R.E., 1989. *Development of Remote Sensing Techniques for the Investigation of Neotectonic Activity, Eastern Transverse Ranges and Vicinity, Southern California*. Ph.D Dissertation (Unpublished), University of California, Santa Barbara, 304p.

Crósta, A.P., and Moore, J. M. M., 1989. Geological mapping using Landsat Thematic Mapper imagery in Almeria Province, south-east Spain. *International Journal of Remote Sensing*, 10, 3, 505–514.

Delgado Blasco, J.M., Foumelis, M., Stewart, C., and Hooper, A., 2019. Measuring urban subsidence in the Rome Metropolitan Area (Italy) with Sentinel-1 SNAP-StaMPS Persistent Scatterer Interferometry. *Remote Sensing*, 11, 129.

Drury, S.A., 1986. Remote sensing of geological structure in temperate agricultural terrains. *Geological Magazine*, 123, 2, 113–121.

Drury, S.A., 1987. *Image Interpretation in Geology*. Allen and Unwin, London, 243.

ESA, 2012. *Satellite Earth Observation for Geohazard Risk Management. The Santorini Conference, Santorini*, Greece, Editor: P. Bally, http://esamultimedia.esa.int/docs/EarthObservation/Geohazards/esa-geo-hzrd-2012.pdf

Fan, X., Scaringi, G., Korup, O., West, A.J., van Westen, C.J., Tanyas, H., et al., 2019. Earthquake-induced chains of geologic hazards: Patterns, mechanisms, and impacts. *Reviews of Geophysics*, 57, 421–503, https://doi.org/10.1029/2018RG000626

Ferretti, A., Prati, C., and Rocca, F., 2001. Permanent scatterers in SAR interferometry. *Institute of Electrical and Electronics Engineers, Transactions Geoscience and Remote Sensing*, 39, 1, 8–20, doi:10.1109/IGARSS.1999.772008

Ferrier, G., and Wadge, G., 1996. The application of imaging spectrometry data to mapping alteration zones associated with gold mineralization in southern Spain, *International Journal of Remote Sensing*, 17, 2, 331–350, doi: 10.1080/01431169608949009

Ferrier, G., Naden, J., and Ganas, A., 2012. *Application of field-based thermal infrared spectroscopy in mapping volcanic terrains. 4th Workshop on Remote Sensing and Geology*, Mykonos, Greece, May 24–25.

Ferrier, G., Ganas, A., and Pope, R., 2019a. Prospectivity mapping for high sulfidation epithermal porphyry deposits using an integrated compositional and topographic remote sensing dataset. *Ore Geology Reviews*, 107, 353–363, https://doi.org/10.1016/j.oregeorev.2019.02.029

Ferrier, G., Ganas, A., Pope, R., Jo Miles, A., 2019b. Prospectivity mapping for epithermal deposits of western milos using a fuzzy multi criteria evaluation approach parameterized by airborne hyperspectral remote sensing data. *Geosciences*, 9, 3, 116, https://doi.org/10.3390/geosciences9030116

Ferrier, G., Naden, J., Ganas, A., Kemp, S., and Pope, R., 2016. Identification of multi-style hydrothermal alteration using integrated compositional and topographic remote sensing datasets. *Geosciences*, 6, 3, 36, doi:10.3390/geosciences6030036

Fielding, E.J., Talebian, M., Rosen, P.A., Nazari, H., Jackson, J.A., Ghorashi, M., and Walker, R., 2003. Surface ruptures and building damage of the 2003 Bam, Iran earthquake mapped by satellite synthetic aperture radar interferometric correlation, *Journal of Geophysical Research: Solid Earth*, 10.1029/2004JB003299

Fu, B., and Lin, A., 2003. Spatial distribution of the surface rupture zone associated with the 2001 Ms 8.1 Central Kunlun earthquake, northern Tibet, revealed by satellite remote sensing data. *International Journal of Remote Sensing*, 24, 10, 2191–2198, doi: 10.1080/0143116031000075918

Ganas, A., 1997. Fault Segmentation and Seismic Hazard Assessment in the Gulf of Evia Rift, central Greece. Unpublished PhD thesis, University of Reading, November, http://ethos.bl.uk/OrderDetails.do?uin=uk.bl.ethos.363718

Ganas, A., and White, K., 1996. Neotectonic fault segments and footwall geomorphology in Eastern Central Greece from Landsat TM data. *Geological Society of Greece Special Publication*, 6, 169–175.

Ganas, A., Wadge, G., and White, K., 1996. *Fault Segmentation and Tectonic Geomorphology in eastern central Greece from satellite data. 11th Thematic Conference and Workshops on Applied Geologic Remote Sensing*, Las Vegas, Nevada, February 27–29, 1996, 119–124.

Ganas, A., Papadopoulos, G.A., and Pavlides, S.B., 2001. The 7th September 1999 Athens 5.9Ms earthquake: Remote sensing and digital elevation model inputs towards identifying the seismic fault. *International Journal of Remote Sensing*, 22, 1, 191–196.

Ganas, A., Pavlides, S., and Karastathis, V., 2005a. DEM-based morphometry of range-front escarpments in Attica, central Greece, and its relation to fault slip rates. *Geomorphology*, 65, 301–319.

Ganas, A., Shanov, S., Drakatos, G., Dobrev, N., Sboras, S., Tsimi, C., Frangov, G., and Pavlides, S., 2005b. Active fault segmentation in southwest Bulgaria and Coulomb stress triggering of the 1904 earthquake sequence. *Journal of Geodynamics*, 40, 2–3, 316–333.

Ganas, A., Elias, P., Bozionelos, G., Papathanassiou, G., Avallone, A., Papastergios, A., Valkaniotis, S., Parcharidis, I., and Briole, P., 2016. Coseismic deformation, field observations and seismic fault of the 17 November 2015 M = 6.5, Lefkada Island, Greece earthquake. *Tectonophysics*, 687, 210–222, http://dx.doi.org/10.1016/j.tecto.2016.08.012

Ganas, A., Elias, P., Kapetanidis, V., Valkaniotis, S., Briole, P., Kassaras, I., Argyrakis, P., Barberopoulou, A., and Moshou, A., 2019. The July 20, 2017 M6.6 Kos earthquake: Seismic and geodetic evidence for an active north-dipping normal fault at the western end of the Gulf of Gökova (SE Aegean Sea). *Pure and Applied Geophysics*, 176, 10, 4177–4211, https://doi.org/10.1007/s00024-019-02154-y

Green, R.O., Eastwood, M.L., Sarture, C.M., Chrien, T.G., Aronsson, M., Chippendale, B.J., Faust, J.A., Pavri, B.E., Chovit, C.J., Solis, M., Olah, M.R., and Williams, O., 1998. Imaging spectroscopy and the Airborne Visible/Infrared Imaging Spectrometer (AVIRIS). *Remote Sensing of Environment*, 65, 3, 227–248, https://doi.org/10.1016/S0034-4257(98)00064-9

Hooper, A., Zebker, H., Segall, P., and Kampes, B., 2004. A new method for measuring deformation on volcanoes and other natural terrains using InSAR persistent scatterers. *Geophysical Research Letters*, 31, L23611, doi:10.1029/2004GL021737

Hunt, G.R., 1977. Spectral signatures of particulate minerals in the visible and near-infrared. *Geophysics*, 42, 501–503.

Hussain, E., Hooper, A., Wright, T.J., Walters, R.J., and Bekaert, D.P.S., 2016. Interseismic strain accumulation across the central North Anatolian Fault from iteratively unwrapped InSAR measurements. *Journal of Geophysical Research: Solid Earth*, 121, 9000–9019, doi:10.1002/2016JB013108

Karimzadeh, S., and Matsuoka, M., 2017. Building damage assessment using multisensor dual-polarized synthetic aperture radar data for the 2016 M 6.2 Amatrice Earthquake, Italy. *Remote Sensing*, 9, 330, doi:10.3390/rs9040330

Kaskara, M., Simone, A., Ioannis, P., Charalampos, K., Stefano, S., and Athanassios, G., 2016. *Geodetic analysis and modeling of the Santorini volcano, Greece, for the period 2012–2015, Paper 2341, European Space Agency, (Special Publication) ESA SP-740, Living Planet Symposium 2016*, Prague, Czech Republic, May 9–13, 2016.

Lauknes, T.R., Dehls, J., Yngvar, L., Kjell, H., and Weydahl, D., 2006. A comparison of SBAS and PS ERS InSAR for subsidence monitoring in Oslo, Norway. P.O. Box. 25.

Liao, M., Jiang, L., Lin, H., Huang, B., and Gong, J., 2008. Urban change detection based on coherence and intensity characteristics of SAR imagery. *Photogrammetric Engineering & Remote Sensing*, 74, 8, 999–1006

Masson, F., Lehujeur, M., Ziegler, Y., and Doubre, C., 2014. Strain rate tensor in Iran from a new GPS velocity field. *Geophysical Journal International*, 197, 1, 10–21, https://doi.org/10.1093/gji/ggt509

Massonnet, D., Rossi, M., Carmona, C., Adragna, F., Peltzer, G., Feigl, K., and Rabaute, T., 1993. The displacement field of the Landers earthquake mapped by radar interferometry. *Nature*, 364, 138–142.

Matsuoka, M., and Yamazaki, F., 2004. Use of satellite SAR intensity imagery for detecting building areas damaged due to earthquakes. *Earthquake Spectra*, 20, 975–994.

Meyer, B., Armijo, R., and Dimitrov, D., 2002. Active faulting in SW Bulgaria: Possible surface rupture of the 1904 Struma earthquakes. *Geophysical Journal International*, 148, 2 246–255.

Meyer, B., Armijo, R., Massonnet, D., de Chabalier, J.B., Delacourt, C., Ruegg, J.C., Achache, J., Briole, P., and Papanastasiou, D., 1996. The 1995 Grevena (northern Greece) earthquake: Fault model constrained with tectonic observations and SAR interferometry. *Geophysical Research Letters*, 23, 2677–2680.

Mueller, A.A., Hausold, A., and Strobl, P., 2002. HySens-DAIS/ROSIS imaging spectrometers at DLR. Proc. SPIE 4545, Remote Sensing for Environmental Monitoring, GIS Applications, and Geology, https://doi.org/10.1117/12.453677

Nakmuenwai, P., Yamazaki, F., and Liu, W., 2016. Multi-temporal correlation method for damage assessment of buildings from high-resolution SAR images of the 2013 Typhoon Haiyan. *Journal of Disaster Research*, 11, 3, 577–592, doi: 10.20965/jdr.2016, p0577

Papoutsis, I., Kontoes, C., Ganas, A., Karastathis, V., Solomos, S., and Amiridis, V., 2014. *BEYOND Center of Excellence: Geophysical activity 'seen' from space*, In *Proceedings of the 1st International Geomatics Application Conference*, Skiathos, September 8–10, 2014, pp. 15–21, ISBN: 978-960-88490-9-9.

Papoutsis, I., Papanikolaou, X., Floyd, M., Ji, K.H., Kontoes, C., Paradissis, D., and Zacharis, V., 2013. Mapping inflation at Santorini volcano, Greece, using GPS and InSAR, *Geophysical Research Letters*, 40, doi:10.1029/2012GL054137

Peltzer, G., Tapponnier, P., and Armijo, R., 1989. Magnitude of Late Quaternary left-lateral displacements along the North Edge of Tibet. *Science*, 246, 4935, 1285–1289.

Prasun, K.G., Lahiri-Dutt, K., and Saha, K., 2006. Application of remote sensing to identify coalfires in the Raniganj Coalbelt, India. *International Journal of Applied Earth Observation and Geoinformation*, 8, 3, 188–195, ISSN 0303-2434, https://doi.org/10.1016/j.jag.2005.09.001

Rast, M., Hook, S.J., Elvidge, C.D., and Alley, R.E., 1991. An evaluation of techniques for the extraction of mineral absorption features from high spectral resolution remote sensing data, *Photogrammetric Engineering & Remote Sensing*, 57, 1303–1309.

Shen, Z.-K., Wang, M., Zeng, Y., and Wang, F., 2015. Strain determination using spatially discrete geodetic data. *Bulletin of the Seismological Society of America*, 105, 4, 2117–2127, doi: 10.1785/0120140247

Tsimi, C., and Ganas, A., 2015. Using the ASTER global DEM to derive empirical relationships among triangular facet slope, facet height and slip rates along active normal faults. *Geomorphology*, 234, 171–181.

Turner, S., 2011. Structural evolution of the Piqiang Fault Zone, NW Tarim Basin, China. *Journal of Asian Earth Sciences*, 1, 4, 394–402.

Vane, G., and Goetz, A.F.H., 1988. Terrestrial imaging spectroscopy. *Remote Sensing of Environment*, 24, 1–29.

Wallace, R.E., 1978. Geometry and rates of change of fault-generated range fronts, North-Central Nevada. *Journal Research U.S. Geological Survey*, 6, 5, 637–650.

Wright, T.J., Parsons, B.E., Jackson, J.A., Haynes, M., Fielding, E.J., England, P.C., and Clarke, P.J., 1999. Source parameters of the 1 October 1995 Dinar (Turkey) earthquake from SAR Interferometry and seismic bodywave modeling. *Earth and Planetary Science Letters*, 172, 1–2), 23–37.

10 Renewable Energy Sources

10.1 INTRODUCTION: REMOTE SENSING CONCEPTS OF RENEWABLE ENERGY SOURCES

There is a steadily increasing international demand for new energy sources around the world, since fossil fuels are depleting. Renewable energy sources constitute a growing energy field, which has received significant attention during the last decades and it is expected to continue offering sustainable energy solutions. On the other hand, remote sensing is a fast-growing and promising scientific and technological field, which can significantly contribute to renewable energy research and applications. Remote sensing data and methods have already been used to explore renewable energy resource sites and there are several case studies from different parts of the world. Thus, implementing renewable energy projects based on remote sensing technology provides significant advantages to such systems. This chapter covers applications of renewable energy sources based on remote sensing data and methods. The applications include solar radiation, solar energy, aeolian (wind) energy, biomass, hydro power and other forms of renewable energy sources.

10.2 SOLAR RADIATION

The sun always remains the most important energy source for the earth system. The motion of the earth around its polar axis and the angle between the earth's equator is one of the most important parameters regarding solar radiation (Iqbal, 1983). It is, thus, necessary to analyze some of the basic astronomical-related parameters.

10.2.1 SUN-EARTH DISTANCE r

It is well known that the motion of the earth is elliptical having the sun in the middle. The amount of solar radiation that reaches the earth's surface is inversely proportional to the square of its distance from the sun. This is why this distance r_0 should be determined using the concept of the astronomical unit (AU) based on the following formula: 1 AU = 1.496×10^8 km. The astronomical unit (AU) is the mean sun-earth distance r_0. For this reason, it is considered accurate to define this distance as 149,597,890+/− 500 km. In other words, the mean sun-earth distance is between the minimum sun-earth distance (about 0.983 AU) and the maximum sun-earth distance (about 1.017 AU). Each year, on approximately July 4, the earth is at its farthest point from the sun, and on approximately January 3 it is at its closest point. The first is called "aphelion" and the second "perihelion", respectively. The mean distance of the earth from the sun is achieved on approximately April 4 and October 5 every year. Figure 10.1 shows the motion of the earth around the sun.

Despite the existence of other astronomical bodies and the leap year cycle, the sun-earth distance r for any day of the year can be estimated with considerable accuracy. "American Ephemeris and Nautical Almanac" published by the US Naval Observatory describes that distance with satisfactory accuracy. However, a mathematical expression would be more practical for estimating that distance, but unfortunately, there is not a simple and unique equation to describe that distance. Usually, the distance can be expressed in terms of a Fourier series type using several coefficients. Spencer (1971) introduced the eccentricity correction factor of the earth's orbit, E_0 using the reciprocal of the square of the radius vector of the earth:

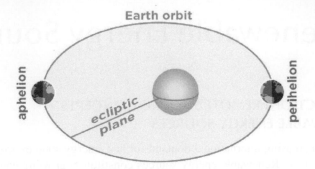

FIGURE 10.1 Motion of the earth around the sun. The point at which a planet is closest to the sun is called perihelion. The furthest point is called aphelion (taken from Spacepedia: https://www.solarsystemscope.com/spacepedia/).

$$E_o = \left(\frac{r_0}{r}\right)^2 = 1.000110 + 0.034221\cos\Gamma + 0.001280\sin\Gamma + 0.000719\cos 2\Gamma + 0.000077\sin 2\Gamma \quad (10.1)$$

where Γ stands for the day angle in radians:

$$\Gamma = 2\pi\left(d_n - 1\right)/365 \quad (10.2)$$

where d_n is the day number of the year starting from 1 on January 1 to 365 on December 31 of each year. Equation (10.3) should lose accuracy on leap years, since February is always assumed to have 28 days at this computation. For that reason, a very simple expression was introduced by Duffie and Beckman (2013) as follows:

$$E_o = \left(\frac{r_0}{r}\right)^2 = 1 + 0.033\cos\left[\left(2\pi d_n / 365\right)\right] \quad (10.3)$$

Values computed using Equation (10.3) are consistent with those at almanac tables, concluding that is safe to use them for more engineering calculations, but Equation (10.2) remains accurate for general use.

10.2.2 Solar Declination Δ

The solar altitude angle measured at noon differs from the corresponding equinoctial angle by an angle of up to ± 23° 17′, which is called "solar declination". Solar declination δ is defined as the angular distance from the zenith of the observer at the equator and the sun during solar noon. Solar declination is positive when the observer is in the north and negative when the observer is in the south.

The declination is zero at the equinoxes, positive during the northern hemisphere summer and negative during the northern hemisphere winter. The declination reaches a maximum of 23.49° on June 22 (summer solstice in the northern hemisphere) and a minimum of −23.46° on December 21–22 (winter solstice in the northern hemisphere). The opposite happens in the southern hemisphere.

The declination for any given day may be calculated in degrees using the equation (Cooper, 1969):

$$\delta = 23.45\sin\left[\left(360/365\right)d_n - 284\right] \quad (10.4)$$

where d_n represents d is the day of the year with Jan 1 as $d_n = 1$. Table 10.1 shows the solar declination for zero-hour Greenwich Mean Time (GMT) (after Spencer, 1971).

TABLE 10.1
Solar declination for zero-hour GMT

Date	Jan	Feb	Mar	Apr	May	Jun	Jul	Aug	Sept	Oct	Nov	Dec
1	−23.07	−17.28	−7.78	4.36	14.93	22.02	23.20	18.20	8.51	−2.95	−14.26	−21.74
2	−22.99	−17.00	−7.40	4.75	15.24	22.15	23.13	17.94	8.14	−3.33	−14.58	−21.90
3	−22.90	−16.71	−7.02	5.13	15.54	22.29	23.06	17.69	7.78	−3.72	−14.90	−22.05
4	−22.80	−16.41	−6.63	5.51	15.83	22.41	22.98	17.42	7.41	−4.11	−15.22	−22.19
5	−22.70	−16.11	−6.25	5.89	16.12	22.53	22.89	17.16	7.04	−4.50	−15.53	−22.32
6	−22.59	−15.81	−5.86	6.27	16.41	22.64	22.80	16.89	6.67	−4.88	−15.83	−22.45
7	−22.47	−15.50	−5.47	6.65	16.69	22.74	22.70	16.61	6.30	−5.27	−16.13	−22.57
8	−22.34	−15.18	−5.08	7.03	16.96	22.84	22.59	16.33	5.93	−5.65	−16.43	−22.68
9	−22.21	−14.87	−4.69	7.40	17.24	22.93	22.48	16.05	5.55	−6.03	−16.72	−22.79
10	−22.07	−14.54	−4.30	7.77	17.50	23.01	22.36	15.76	5.17	−6.41	−17.01	−22.28
11	−21.92	−14.22	−3.90	8.14	17.77	23.09	22.23	15.46	4.80	−6.79	−17.29	−22.98
12	−21.76	−13.89	−3.51	8.51	18.02	23.16	22.10	15.17	4.42	−7.17	−17.57	−23.06
13	−21.60	−13.55	−3.12	8.87	18.28	23.23	21.96	14.87	4.03	−7.55	−17.84	−23.13
14	−21.43	−13.22	−2.72	9.24	18.52	23.28	21.81	14.56	3.65	−7.92	−18.11	−23.20
15	−21.25	−12.87	−2.33	9.60	18.77	23.33	21.66	14.25	3.27	−8.30	−18.37	−23.26
16	−21.07	−12.53	−1.93	9.95	19.00	23.38	21.50	13.94	2.88	−8.67	−18.62	−23.31
17	−20.88	−12.18	−1.54	10.31	19.23	23.41	21.34	13.62	2.50	−9.04	−18.87	−23.36
18	−20.68	−11.83	−1.14	10.66	19.46	23.44	21.17	13.30	2.11	−9.40	−19.12	−23.39
19	−20.48	−11.47	−0.74	11.01	19.68	23.47	20.99	12.98	1.72	−9.77	−19.36	−23.42
20	−20.27	−11.12	−0.35	11.35	19.90	23.48	20.81	12.66	1.34	−10.13	−19.59	−23.44
21	−20.05	−10.76	0.05	11.70	20.10	23.49	20.63	12.33	0.95	−10.49	−19.82	−23.46
22	−19.83	−10.39	0.44	12.04	20.31	23.49	20.43	11.99	0.56	−10.85	−20.04	−23.46
23	−19.60	−10.03	0.84	12.37	20.51	23.49	20.23	11.66	0.17	−11.21	−20.25	−23.46
24	−19.37	−9.66	1.23	12.71	20.70	23.47	20.03	11.32	−0.22	−11.56	−20.46	−23.45
25	−19.13	−9.29	1.63	13.04	20.88	23.46	19.82	10.98	−0.61	−11.91	−20.67	−23.43
26	−18.88	−8.91	2.02	13.36	21.07	23.43	19.60	10.63	−1.00	−12.25	−20.86	−23.40
27	−18.63	−8.54	2.41	13.68	21.24	23.40	19.38	10.28	−1.39	−12.60	−21.05	−23.37
28	−18.37	−8.16	2.80	14.00	21.41	23.36	19.15	9.93	−1.78	−12.94	−21.23	−23.33
29	−18.11	0.0	3.19	14.32	21.57	23.31	18.92	9.58	−2.17	−13.27	−21.41	−23.28
30	−17.84	0.0	3.58	14.63	21.73	23.26	18.68	9.22	−2.56	−13.61	−21.58	−23.22
31	−17.56	0.0	3.97	0.0	21.87	0.0	18.44	8.87	0.0	−13.94	0.0	−23.16

10.2.3 SOLAR ZENITH ANGLE – SOLAR ELEVATION ANGLE

The solar zenith angle (z) is the angle between the zenith and the center of the sun's disc. The solar elevation angle (h), or altitude of the sun, is the angle between the horizon and the center of the sun's disc. These two angles are complementary, and the cosine of either one of them equals the sine of the other (Figure 10.2). They can both be calculated using spherical trigonometry.

10.2.4 SOLAR CONSTANT AND SOLAR RADIATION AT GROUND LEVEL-RADIATIVE TRANSFER

The solar irradiance or solar constant is defined as that quantity of solar energy (W/m^2) at normal incidence outside the atmosphere (extraterrestrial) at the mean earth distance. The value of solar constant is 1367.7 W/m^2 referring to all wavelengths of the electromagnetic spectrum.

The earth receives varying amounts of solar radiation due to variations in seasons, which are related to the sun-earth distance. Absorption occurs in the atmosphere due to interactions at very specific wavelengths, leading to significant depletion of solar radiation when passing through the

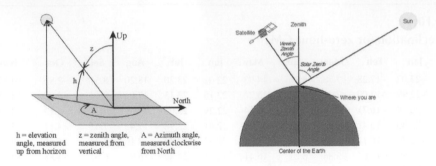

FIGURE 10.2 Sun zenith angle, sun azimuth angle (left), Solar Zenith angle – Viewing zenith angle (right).

atmosphere (Iqbal, 1983). In addition, scattering is associated with matter occurring at all wavelengths. Scattering depends on the size and shape of the particle or molecule dealing with radiation. As a result, on average, less than half of the extra-terrestrial radiation reaches ground level (Wald, 2018). All those energy processes affecting solar radiation within the atmosphere are what is called radiative transfer. The solar zenithal angle plays a major role in radiative transfer as it influences the optical path of radiation. If the receiving plane is inclined, it may receive the direct radiation only partly and the fraction of sky viewed by the plane must be considered for computing the diffuse part impinging on the plane (Wald, 2018). As a simple rule one may consider that the smaller the solar zenithal angle, the smaller the optical path, and the smaller the extinction of the radiation (Martini et al., 2003; Wald, 2018).

Direct radiation is the radiation coming from the direction of the sun. Only direct radiation is present at the top of the atmosphere. A horizontal surface at ground level receives only a portion of this direct radiation. An interesting definition is the clearness index, which is the ratio of radiation received at ground level to that received at the top of the atmosphere. Clearness index is a dimensionless index defined for any wavelength for any time duration. Typical values are 0.1 for very cloudy conditions and 0.7–0.8 for cloud-free conditions. In cloud-free skies, aerosols and water vapor are the main contributors to depletion. In such conditions, approximately 20–30% of the total extra-terrestrial radiation is lost during its downwelling path by scattering from the constituents of the atmosphere and absorption phenomena caused by aerosols and molecules. This amount, of course, differs with wavelength, thus, the spectral distribution of the solar radiation is modified as the radiation crosses the atmosphere downwards. Dealing with clouds, optically thin clouds allow a noticeable proportion of radiation to reach the ground, whereas optically thick clouds create obscurity by stopping the radiation downwards. When considering the spectral distribution of the radiation, clouds may be considered as spectrally neutral, since their presence does not have a significant effect on the spectral distribution of the radiation, even if they play a major significant role as a whole, since they are the major depleting constituents in the atmosphere.

Currently, a Solar Energy Nowcasting SystEm (SENSE) has been developed by the National Observatory of Athens in Greece in collaboration with the World Radiation Center in Switzerland (Masoom et al., 2020). SENSE helps with investments, applications and technologies related to energy (SolarHub, 2020). SENSE methodology is based on near real-time atmospheric inputs and real-time (RTM) simulations (Masoom et al., 2020). More technical details can be found in Kosmopoulos et al., 2018. Relevant studies have been conducted mainly in Europe, where the climate is more favorable in comparison with Asia or America, having no extreme atmospheric phenomena (Masoom et al., 2020). Figure 10.3 illustrates an application of SENSE in India.

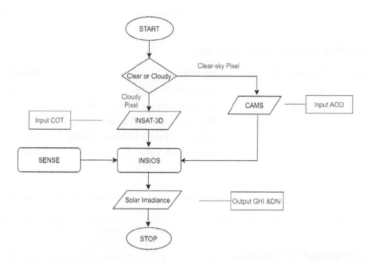

FIGURE 10.3 Application of SENSE: Flowchart of the Indian Solar Irradiance Operational System (INSIOS) technique. COT: cloud optical thickness, INSAT-3D: Indian National Satellite System, CAMS: Copernicus Atmosphere Monitoring Service, AOD: aerosol optical depth, SENSE: Solar Energy Nowcasting System (SENSE), GHI: global horizontal irradiance, DNI: direct normal irradiance (taken from Masoom et al., 2020).

10.2.5 Remote Sensing and Electromagnetic Spectrum

Remote sensing (RS) as a science is dedicated to the acquisition of information about an object or phenomenon without making physical contact with the object. RS takes advantage of the power of the sun as a source of energy or radiation. The energy of the sun can be reflected, absorbed or re-emitted. The direct sunlight as well as the reflected one is well known as the visible part of the electromagnetic spectrum, while the absorbed and re-emitted stands for the infrared wavelengths (Figures 1.1 and 1.2). The entire reflected energy takes place only during daytime, when the sun illuminates the earth. This is the situation when passive sensors are in operation. Passive sensors can only be used to detect energy when the naturally occurring energy is available, but energy is also emitted from the earth. This happens at the thermal infrared part of the spectrum, and as a result, infrared radiation can be detected day or night, as long as a sufficient amount of energy can be recorded. This is the situation where active sensors (i.e. radar systems) could be used.

The electromagnetic spectrum can be divided into several wavelength/frequency bands (Figures 1.1 and 1.2). It is worth mentioning that from the whole spectrum only a narrow band from about 400–700 nm is visible to the human eye, and of course there is no specific boundary between these regions. The boundaries shown in Figures 1.1 and 1.2 are only approximate, since there are overlaps between adjacent regions.

Parts of the electromagnetic spectrum. The most interesting part for remote sensing applications is the visible part of electromagnetic spectrum extending from about 400 nm (violet) to 700 nm (red). The various color components of the spectrum fall approximately within the regions described in Figures 1.1 and 1.2. The infrared part of the spectrum extends from 0.7 to 300 μm wavelength. This part is further divided into the bands described in Table 10.2. The near-infrared (NIR) and short-wave infrared (SWIR) are also known as the "Reflected Infrared", referring to the main infrared component of the solar radiation reflected from the earth's surface. The Mid Wavelength Infrared (MWIR) and Long Wavelength Infrared (LWIR) are the "Thermal Infrared".

The next part of the electromagnetic spectrum, which is related to microwaves and radio-waves, is presented in Table 10.3. Ultraviolet, as well as X-Rays and Gamma Rays, are not usually used in remote sensing and consequently are not analyzed further.

TABLE 10.2

Parts of visible and infrared spectrum

Visible Light	Wavelength (nm)	Infrared	Wavelength (nm)
Red	610–700	Near-Infrared (NIR)	0.7–1.5
Orange	590–610	Short Wavelength Infrared (SWIR)	1.5–3
Yellow	570–590	Mid Wavelength Infrared (MWIR)	3–8
Green	500–570	Long Wavelength Infrared (LWIR)	8–15
Blue	450–500	Far Infrared (FIR)	>15
Indigo	430–450		
Violet	400–430		

TABLE 10.3

Parts of radio-waves and microwaves

Radio-Waves	Frequency	Microwaves	Wavelength (nm)
Extremely High Frequency (EHF)	1 mm–1 cm	P band	0.3–1
Super High Frequency (SHF)	1 cm–1 cm	L band	1–2
Ultra High Frequency (UHF)	10 cm–1 m	S band	2–4
Very High Frequency (VHF)	1 m–10 m	C band	4–8
High Frequency (HF)	10 m–100 m	X band	8–12.5
Medium Frequency (MF)	100 m–1 km	Ku band	12.5–18
Low Frequency (LF)	1 km–10 km	K band	18–26.5
Very Low Frequency (VLF)	10 km–100 km	Ka band	26.5–40
Extremely Low Frequency (ELF)	>100 km		

10.2.6 SENSORS FOR MEASURING RADIANCE/REFLECTANCE

Radiance is the "flux of energy" (primarily irradiant or incident energy) per solid angle leaving a unit surface area in a given direction. When working with sensors the most important task is the measurement of radiance, which corresponds to the brightness at a given direction toward the sensor. On the other hand, reflectance is defined as the ratio of reflected versus total power energy (Congedo, 2016). Images, such as Landsat or Sentinel-2, are composed of several bands and a meta-data file, which contains information required for the conversion to reflectance. Landsat images are provided in radiance, but they are scaled prior to output. When working with Landsat images, spectral radiance at the sensor's aperture (L_λ, measured in [watts/(meter squared * ster * μm)]) is given by Equation (10.6) (Adeyeri et al., 2017):

$$L_\lambda = M_L * Q_{cal} + A_L \qquad (10.5)$$

where M_L is radiance multiplicative scaling factor from the metadata, A_L is radiance additive scaling factor from the metadata and Q_{cal} is quantized and calibrated standard product pixel values (DN).

TOA Landsat OLI radiance can be converted to TOA Reflectance (combined surface and atmospheric reflectance). Equation (10.6) is used to convert Level 1 DN values to TOA reflectance (Landsat 8 Data Users Handbook, 2019):

$$\rho_\lambda' = M_\rho * Q_{cal} + A_\rho \qquad (10.6)$$

where ρ_λ' (unitless) is TOA Planetary Spectral Reflectance, without correction for solar angle, M_ρ is a separate reflectance multiplicative scaling factor for each band taken from the metadata file, A_ρ is a separate reflectance additive scaling factor for each band from metadata and Q_{cal} is Level 1 pixel value in (DN).

The effects of the atmosphere should be considered in order to measure the reflectance on the ground. Therefore, several atmospheric corrections are needed in order to calculate reflectance (ρ). As described by Moran et al. (1992), the land surface reflectance (ρ) is:

$$\rho = [\pi * (L_{\lambda} - L_p) * d2] / [T_v * ((ESUN_{\lambda} * cos\theta s * T_z) + Edown)] \tag{10.7}$$

where L_p is the path radiance, T_v is the atmospheric transmittance in the viewing direction, T_z is the atmospheric transmittance in the illumination direction, $ESUN_{\lambda}$ is the Exo-atmospheric Spectral Irradiance, and E_{down} is the downwelling diffuse irradiance.

Finally, it is worth pointing out that while Landsat 8 images are provided with band-specific rescaling factors that allow for the direct conversion from DN to TOA reflectance, Sentinel-2 images are already provided with scaled TOA reflectance, which can be converted to TOA reflectance with a simple calculation using the Quantification Value provided in the metadata (see Sentinel-2 MSI User Guides, 2020).

10.3 SOLAR ENERGY

The sun can be considered as the vital source of every renewable energy (Datta and Karakoti, 2010). It is well known that historically heat was produced from the sun using glass or mirrors. The main advantage of solar energy is the fact that it is endless and costless. Additionally, it requires little or no maintenance, and its production is silent. One disadvantage of thermal energy is that it has a relatively high cost, but technology improvements can lead to more cost-effective systems in the near future. Solar energy is a good alternative in remote locations, where traditional energy sources are difficult to obtain (Wiginton et al., 2010).

It has already been analyzed that only a small portion of sun radiation is received on the earth as visible and infrared radiations (Figure 10.2). Despite this, the energy suffices for many purposes. According to Dr. Gerhard Knies from DESERTEC Foundation, in only six hours, all the deserts combined from our planet receive more energy from the sun than what humankind can consume within a year (Wan et al., 2015; Avtar et al., 2019). Furthermore, according to the International Energy Agency, the sun might be the biggest source of electricity by the year 2050, followed by fossil fuels, wind energy, hydropower and nuclear energy (Cano et al., 1986; Kazem et al., 2016).

Short-wave radiant energy emitted by the sun is scattered by the atmosphere and the clouds, and for this reason radiation can be divided into two categories. The first is direct radiation, responsible for creating shadows, and the second is diffuse radiation, responsible for skylight. The intensity of the sun's radiation remains constant through time, and the only barrier to receiving that energy comes from the clouds, and various particles in the atmosphere (e.g. aerosols). Remote sensing is a great technological tool for measuring all these quantities using specific methodologies for each parameter (Lukač et al., 2013; Szabó et al., 2016). Jakubiec and Reinhart (2013) have also used LiDAR and GIS data, to predict city-wide electricity gains from photovoltaic panels.

10.3.1 SOLAR RADIATION FOR THERMAL APPLICATIONS

It is common sense that solar energy should be widely used rather than other alternative energy forms, even when the related costs are slightly higher. Solar energy is assumed clean and can be supplied without any environmental pollution compared with other forms of energy. The major feature of any solar system is the solar collector. Various types of collectors exist to collect solar energy, with different thermal analysis and performance. Generally, there are two types of solar collectors:

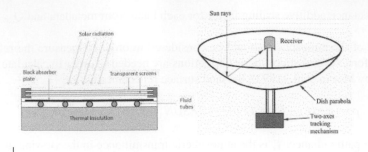

FIGURE 10.4 Overview of a flat plate solar collector with liquid transport medium (left) and a parabolic receiver (right) (Credit: Mark Fedkin (modified after Duffie and Beckman, 2013).

non-concentrating or stationary and concentrating (Singh et al., 2019). A non-concentrating collector uses the same area for intercepting and absorbing solar radiation. These sensors are permanently fixed in position and do not track the sun, whereas with the other type (concentrating), a sun-tracking concentrating solar collector usually has concave reflecting surfaces to intercept and focus on the sun's beam radiation. In that way by limiting the receiving area, the radiation flux is finally increased. Solar collectors are available on the market in large quantities.

Stationary solar energy collectors according to their motion and tracking can be:

- Flat Plate Collectors (FPC)
- Compound Parabolic Collectors (CPC)
- Evacuated Tube Collectors (ETC)

The most common category remains the flat plate collector, where solar radiation is absorbed by a black plate and transfers heat to the fluid inside the tubes. The thermal insulation then prevents heat loss during fluid transfer (Figure 10.4).

Very high temperatures can occur if a large amount of solar radiation is concentrated on a relatively small collection area. This is the idea behind the concept of the sun-tracking concentrating collectors. The working fluid here can achieve higher temperatures compared to a flat-plate system of the same solar energy collecting surface. Also, the thermal efficiency is greater, because of the small heat loss area relative to the receiver area. The collectors that fall into this category are: Parabolic Trough Collectors, Linear Fresnel Reflectors, Parabolic Dish Collector (Singh et al., 2019).

10.3.2 Solar Radiation for Electrical Power Generation

Thermal energy can be converted into electricity using an engine generator coupled directly to the receiver, or it can be transported through pipes to the central power-conversion system (Singh et al., 2019). In simpler terms, in the first method, solar energy is used as a source of heat and then the heat is further used to produce steam, which drives the steam turbine. This method is called solar thermal power generation. Applying the second method, solar energy is converted directly into electricity using Photovoltaic (PV) and/or solar cells made with silicon semiconductor material (Figure 10.5). Such a semiconductor device converts the sunlight energy into electricity without any energy conversion steps. This conversion only takes place by the photovoltaic effect, and that is why those cells are called Photovoltaic (PV). Finally, voltage and current is generated when sunlight occurs. Figure 10.6 presents the photovoltaic electricity potential (by ESMAP, 2000).

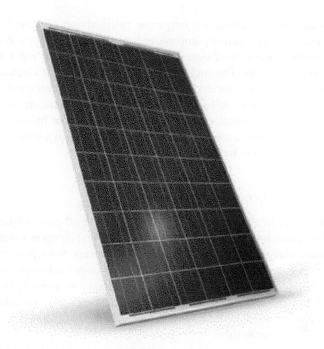

FIGURE 10.5 Figure PV cell.

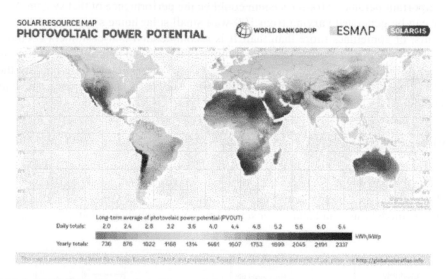

FIGURE 10.6 World Bank: Photovoltaic Electricity Potential. © 2019 The World Bank, Source: Global Solar Atlas 2.0, Solar resource data: Solargis.

10.3.3 SOLAR RADIATION FOR AGRICULTURAL APPLICATIONS

World population is continuously increasing and the demand for food products is rising. An agricultural community can mitigate climate change by using solar power and can find feasible energy solutions. Specifically, greenhouses normally use solar energy to operate. Usually, in order to regulate the temperature inside the greenhouse, a heat sink is used. A thermal mass is used as a sink, absorbing energy and storing it for later use. Water can be a perfect thermal mass, trapping heat and regulating the inside temperature.

PV solar panels can also be applied for irrigation, refrigeration of agricultural products. Solar energy can also be used to trap heat and pump water for livestock and crops. It can also be used for the heating of greenhouses and/or farmhouses and livestock yards. Solar water heaters can also provide hot water to dairy operations. The sunlight can be used to generate electricity to power homes and agriculture houses. Using the solar drying method can dry crops faster and also keep them from being outside where they are prone to birds, insects and worms. There are different types of solar dryers, such as direct drying, indirect drying, mixed mode drying or hybrid drying (hybrid solar/biomass cabinet dryer). Typically, a solar drying apparatus consists of a solar collector, a shed and a drying rack. The shed is needed to trap the solar heat.

10.3.4 Industrial Applications

Solar power systems can generally be divided into two types: the off-grid and the on-grid systems. In off-grid systems, the system has no link with the utility grid, and components used in such systems tend to be solar panels, charge controllers, inverters and batteries. Solar panels convert solar radiations into electric current and this current passes through charge controllers, which in turn charge the battery. Charge controllers maintain control on the voltage. As a result, DC power is stored in batteries and then this stored energy can be converted into AC power to operate the AC appliances. The DC appliances can be operated directly with the battery. That is how DC and AC load can be operated through a solar power system.

One of the most important parameters in solar power systems is the quality of the design systems, which should guarantee a good operation of the whole system for at least 20 years. Another important parameter for the system could be the performance of that system. Solar electric power can be generated varying from 100-watt small solar home systems to 100-MW Solar Power Plants, or sometimes more than that. It is recommended that customers should not buy watts, but watt-hours, which is the real energy unit. In the second category, on-grid systems are those systems which are connected to utility grids. User/Power producer are paid by the utility according to their specific agreement. On-grid power can be cheaper and does not require any batteries (Figure 10.7).

10.4 WIND ENERGY

A simple way to remotely determine the wind speed is by observing marked cloud drift aloft from the ground on a sunny day. Remote sensed wind speed measurements are needed to supplement and extend tall meteorological mast measurements, on and offshore, and within research to evaluate various wind flow models and wind atlases for several purposes, including: wind resource assessments;

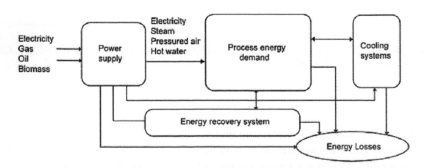

FIGURE 10.7 Block diagram of a typical industrial energy system (Mekhilef et al., 2011).

wind park development projects; power curve measurements; bankability; wind model and wind resource (wind atlas) uncertainty evaluation.

Wind energy is a renewable and clean source of energy. More quantitative and accurate remote sensing measurement techniques for wind energy applications now include sound and light wave propagation and backscatter detection-based instruments, such as sodar, lidar and satellite-based sea surface wave scatterometry. Sodar (sound detection and ranging) is based on probing the atmosphere by sound propagation; lidar (light detection and ranging) is based on probing the atmosphere by electromagnetic radiation (microwaves or laser light); and satellite RS is based on microwave scatterometry on the sea surface and synthetic aperture radar (SAR) methods (Johannessen and Korsbakken, 1998). The first two (sodar and lidar) are direct measurements of wind speed based on Doppler shift, whereas the satellite scatterometry is based on proxy-empirical calibration methods. There is motivation and demand in the wind energy market for wind lidars and wind sodars. At a continuously increasing rate today wind turbines are being installed offshore, in hilly and forested areas, and even in complex or mountainous terrain. Wind lidars on the market for vertical mean and turbulence profile measurements are available based on two different measurement principles, namely continuous wave (CW) lidars and pulsed lidars. Wind lidars on the market today offer the wind energy industry with RS instruments, for wind speed, wind direction and turbulence profiling (Mikkelsen et al., 2013); wind resource assessments, on and offshore; wind turbine performance testing (power curves); and wind resource assessment via horizontal scanning over complex terrain.

Wind energy sources are amongst the world's fastest developing energy sources (Karthikeya et al., 2016). Wind energy can be easily converted into electrical energy at "wind farms" providing electricity to isolated locations (Harris et al,. 2014). The topography of the wind farms, the elevation above mean sea level, the distance from public transportation and the distance from the local electric grids are among the most significant parameters affecting the success of a wind energy project (Chang et al., 2016; Hasager et al., 2016). A detailed economical and technical study is required before the installation in order to reduce the uncertainty of such an energy project (Kilic and Unluturk, 2015; Zhang et al., 2016. Satellite imagery can play a very significant role in decision-making in the case of wind farm projects (Harris et al., 2014; Kim et al., 2016). Additionally, the Light Detection And Ranging (LiDAR), SOnic Detection And Ranging (SODAR), and Synthetic Aperture Radar (SAR) are also used extensively in those cases (Calaudi et al., 2013). In offshore wind energy, remote sensing can be used in three different ways: Ground-based, airborne and satellite-borne (Avtar et al., 2019).

Ground-based techniques are efficient in cases whereby large wind turbines are to be installed and meteorological masts do not enable observations across the rotor plane (Mikkelsen et al., 2013). Figure 10.8 shows the frames of the SAR images used in this work in the Mediterranean Sea. S1A/B: Sentinel-1 A/B; Envisat: ASAR WS; CSK: COSMO-SkyMed; Radarsat-2: Scansar Narrow. Figure 10.9 shows backscatter cells related to the wind (left) and the aliased wind direction obtained from the orientation of the major axis of the backscatter cells showed in the left panel (right) (Zecchetto, 2018). Recently, Aeolus satellite was launched (August 2018), being the first satellite mission to acquire profiles of Earth's wind on a global scale (https://www.esa.int). Aeolus wind mission will demonstrate global wind-profiles from space, using laser technology, achieving near real-time observations (McMahon, 2019).

10.4.1 Satellite-based Wind Fields

Wind measurements can be conducted through numerous in-situ observations over land, which provide a good overview of the predominant wind conditions, whereas over the oceans and seas wind measurements are rare. On the other hand, there is no satellite instrument that can measure the

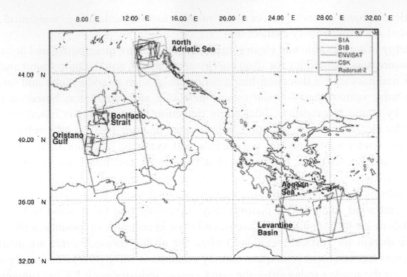

FIGURE 10.8 Frames of the Synthetic Aperture Radar images used in Mediterranean Sea (taken from Zecchetto, 2018).

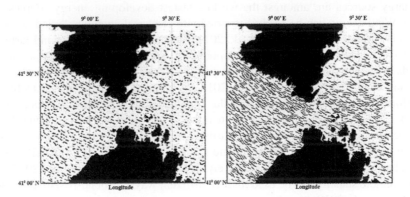

FIGURE 10.9 Backscatter cells related to the wind (left) and the aliased wind direction obtained from the orientation of the major axis of the backscatter cells showed in the left panel right) (Zecchetto, 2018).

wind field directly, however, there are several techniques that can be used to derive wind observations indirectly. As a result, wind data obtained by satellites plays a major role in data assimilation. Satellite-derived wind data and products are used for a wide range of applications in the fields of oceanography, meteorology, air/sea interactions, including wind energy.

Atmospheric motion vectors (AMV). Knowledge of atmospheric motion is essential for many applications. Winds in the upper levels can be observed using radiosondes or aircraft measurements, but those observations are limited in time and space. As satellites provide worldwide and continuous data, they are the ideal data source for regular upper atmospheric wind information. AMVs are computed by tracking clouds or moisture features from a sequence of satellite images and are thus the only observation type that provides good coverage of upper tropospheric wind data over oceans and at high latitudes. Moreover, the roughness of the sea can be used to determine the wind conditions over oceans either by active radar scatterometers (e.g. ASCAT) or by passive microwave radiometers (e.g. WindSat). Geostationary satellites are best suited to tracking clouds as they provide data from the surface and the atmosphere with a high temporal resolution

TABLE 10.4

List of agencies currently producing cloud motion winds (EUMETSAT: geostationary satellites, NOAA: polar orbiting satellites)

EUMETSAT	Europe	Meteosat-10 (GEO), Meteosat-7 (GEO), MetOp-A (P), MetOp-B (P)
NOAA/NESDIS, CIMSS	USA	GOES-13 (GEO), GOES-15 (GEO), Aqua (P), Terra (P), NOAA-15 (P), NOAA-18 (P), NOAA-19 (P), NPP (P)
JMA	Japan	MTSAT-2 (GEO) HIMAWARI 6 - 8
IMD	India	Kalpana (GEO), INSAT-3D (GEO)
CMA	China	FY-2D (GEO), FY-2E (GEO)
KMA	South Korea	COMS (GEO)

and from a non-varying satellite perspective. In the IR and VIS ranges AMVs are derived from clouds, whereas in absorption bands like the water vapor channels they are derived from moisture fields. The routine production of AMV data started as early as the seventies. In the beginning, AMVs were produced using data from geostationary satellites only, while nowadays winds are also computed from polar orbiting satellite data.

Cloud motion winds. A differentiation is necessary between motion vectors and winds. Motion vectors provide the direction and propagation speed of a selected cloud feature, while winds represent the current velocity of an air parcel at a certain height. In practice, this is achieved by a careful selection of atmospheric tracers and the use of model winds as a first guess for wind speed and direction at a given atmospheric level. The computation of cloud motion winds requires model data, not only for the first guess comparison and validity check, but also for the height assignment of the derived winds. This is achieved by assigning the brightness temperature of the tracer cloud to a given height by using a vertical temperature profile from a model. Table 10.4 presents an overview of the agencies currently producing and disseminating cloud motion winds and the satellites used for that purpose. Figure 10.10 provides the location of all AMVs used in the data assimilation for the UK Met Office model in 2013 (from UK Met Office).

Methodology. The methodology is explained with the help of AMVs derived from the SEVIRI instrument onboard MSG. Various satellite channels are used to calculate cloud motion. The IR window channels (IR10.8 μm, IR12.0 μm) and the visible channels (VIS0.6 μm, VIS0.8 μm) are used for cloud tracking. In contrast, both WV channels (WV6.2 μm, WV7.3 μm) provide information for cloudy and cloudless areas. For the latter, moisture patterns or gradients are tracked instead of clouds. The algorithm to derive cloud motion winds mainly consists of the following steps: (1) **Initialization of the data**: this is the pre-processing step, where the algorithm provides all the auxiliary data required in the following stages for computing the winds; (2) **Tracer selection**: the EUMETSAT target extraction scheme mainly depends on the contrast and the entropy as selection criteria for cloud selection. As a result, suitable targets have the highest contrast and the largest amount of standard deviation (entropy); (3) **Tracking**: the motion of the cloud/moisture feature is extracted by analyzing two or more consecutive images, where at EUMETSAT a cross-correlation is used; (4) **Height assignment:** In this step a pressure level is assigned to the vector based on a temperature/pressure profile derived either from radiative transfer calculations in the environment of the target or from model data. When computing AMVs from WV images in a region with clear sky conditions, the brightness temperature from the latter is used for the height assignment; (5) **Quality control**: this step is part of the post-processing procedure to check whether a derived motion vector fulfills the criteria of satellite-derived wind. During this process each motion vector is automatically set to a quality flag. This quality flag is the result of several independent

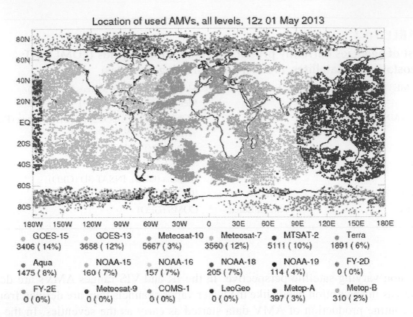

FIGURE 10.10 Location of all AMVs used in the data assimilation for the UK Met Office model in 2013 (Source: UK Met Office).

FIGURE 10.11 Satellite-derived wind product from EUMETSAT (from EUMETSAT).

quality tests; (6) **Orographic flag calculation:** this is another post-processing step in which tracers affected by land are detected and finally rejected. For example, tracers whose flow is blocked (e.g. barrage cloudiness) or affected by mountain ranges (e.g. lee waves) are not representative of atmospheric motion.

Figure 10.11 shows the result of the algorithm for a satellite-derived wind product from EUMETSAT.

10.4.2 Local and Daily Winds

Local winds, such as winds in valleys and mountains, as well as daily winds, such as breeze, can be exploited for many purposes. A few representative examples are considered.

Wind farms. As a widely used renewable energy, migrating to wind power can help reduce air pollution and CO2 emissions. However, the construction of wind farms could have certain impacts on the local ecology and meteorology. A case is considered in XilinGol, China (Figure 10.12a) based on satellite observations, along with the ERA-Interim reanalysis meteorology, which examines the difference in vegetation characteristics, local wind speed, and surface temperature between the wind power regions and the adjacent comparison areas (Wu et al., 2019). The results show that the construction of wind farms results in significant differences of normalized difference vegetation index (NDVI) values in the study areas compared to the adjacent areas in summer. The construction of the wind farms in the study area has considerable influence on the change of local vegetation growth and related meteorological variables (Wu et al., 2019). It reduces the fraction of grassland, disrupts the wind field when speed is less than 4 m/s, and changes the spatial pattern of temperature during daytime and night-time (Figure 10.12b).

Offshore wind farms have started to contribute important supplies of renewable energy. The energy production of a wind farm depends upon the local wind climate, which can be predicted in advance. Satellite Synthetic Aperture Radar (SAR) wind mapping can be a useful tool in selecting optimal sites and may therefore increase the cost-effectiveness of planning wind farms, e.g. in feasibility studies. In the WEMSAR project, wind fields from SAR, in situ measurements and model output from three test-sites have been analyzed (WEMSAR Consortium, 2003). As a result, a WEMSAR tool for effectively retrieving wind data from SAR images and utilizing them in the WAsP micro-siting model has been developed (Furevik et al., 2003).

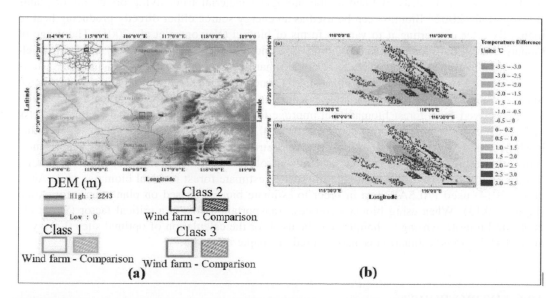

FIGURE 10.12 (a) Administrative map of XilinGol and the location of interested areas (from Wu et al., 2019), (b) Distribution of the temperature difference between (a) day and (b) night for the period before (2002–2004) and after (2011–2013) the construction of wind farm (from Wu et al., 2019).

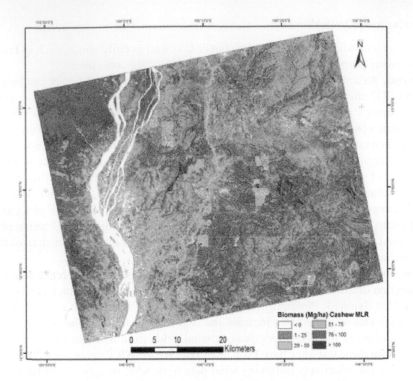

FIGURE 10.13 Natural forests biomass based on cashew biomass (Avtar et al., 2014).

10.5 BIOMASS

Biomass is one of the most commonly used energy sources in less industrialized and less developed countries and can be defined as the biological material from living organisms or plants (Lefsky et al., 2002). Biomass can be used as an energy source either directly producing heat or indirectly after converting it to various forms of biofuel (Voivontas et al., 2001). The usage of biomass varies between regions, providing 35% and 3% of the primary energy needs of developing and developed countries, respectively (Kumar and Mutanga, 2017). Agricultural waste is commonly used in Mauritius and Southeast Asia, wood residues are commonly used in the United States, while animal husbandry residue is common in the British Isles (Le Toan et al., 1992). Energy derived from forest biomass can improve forest health and dramatically reduce the risk of forest fires by removing forest debris (Avtar et al., 2019). Roy and Ravan (1996) showed that brightness and wetness parameters show a very strong relationship with biomass values, indicating that remote sensing can play an efficient role in providing biomass estimation. García-Martín et al. (2012) used Landsat data in order to estimate forest residual biomass, while Avtar et al. (2014) used PALSAR data in order to estimate biomass based on plantation information (Figure 10.13). When using biomass sources, transportation cost is a critical factor (Shi et al., 2008) and remote sensing techniques can be used for the calculation of optimal sites (Lefsky et al., 2014). Biomass estimation is also covered in Chapters 7 and 8.

10.6 HYDRO POWER

Hydropower energy sources seem to be more reliable and cleaner among all renewable energy sources, since they emit less greenhouse gases (Kusre et al., 2010; Ellabban et al., 2014). The basic concept of generating hydropower is the conversion of water's kinetic energy to electricity

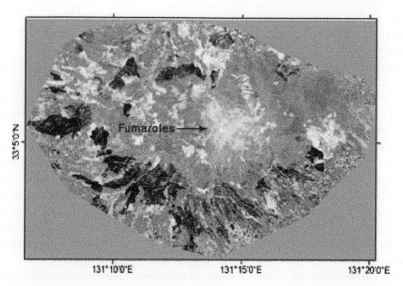

FIGURE 10.14 Hydrothermal alteration mapping using Crosta techniques. taken from Mia and Fujimitsu (2012).

using turbines. However, when selecting a location for a hydropower plant, dealing with biodiversity issues, land use, land degradation, stream network, or flow characteristics is complicated (Biggs et al., 2009). GIS and remote sensing can provide real-time imaging related to the existing terrain, the current hydrological phenomena and the local climate, which are necessary for the assessment of the potential hydropower plant (Al-Mukhtar and Al-Yaseen, 2019). Mia and Fujimitsu (2012) used Landsat 7 ETM+ imagery to map hydrothermal altered mineral deposits around the Kuju volcano, Kyushu, Japan. Spatial modeling has also been examined using thermal infrared data to quantify heat flow and hydrothermal change at Yellowstone National Park (Kratt et al., 2010; Vaughan et al., 2012), while Yi et al. (2010) used geospatial information in order to make site location analysis for a small hydropower. Figure 10.14 shows hydrothermal alteration mapping using Crosta techniques (from Mia and Fujimitsu, 2012).

10.7 OTHER FORMS OF RENEWABLE ENERGY SOURCES

Thermal/geothermal energy is a very important energy source physically generated from the heat coming from the decay of naturally occurring radioactive isotopes in the earth. Geothermal energy could play a significant role in substantially reducing greenhouse gas emissions (van der Meer et al., 2014). Geothermal energy could save an amount of energy equivalent to the burning of 100 million barrels of oil per day (Forsberg, 2009). Lund et al. (2007) support the notion that geothermal energy could theoretically supply energy requirements for six million years. According to Avtar et al. (2019), geothermal energy resources can be broadly classified into three categories: hydrothermal or convective systems, conductive systems, and deep aquifers. Pruess (2006) analyses that hydrothermal (convective) systems, such as hot springs, fumaroles, and chemically altered rocks, may either be vapor-dominated or water-dominated. On the other hand, conductive systems (hot rock magmatic resources and deep aquifers) contain moving fluids in porous media at depths more than 3 km, but without a local magmatic heat source (Avtar et al., 2019). Electric power can be generated using steam and turbine-generators set to produce electrons. Geothermal energy accounts for about 0.4% of the world's global power generation, with a growth rate of 5%, while solar energy provides less than 0.2% of the global power generation (Martini et al., 2003). Presently, the total installed capacity for geothermal resources worldwide corresponds to approximately 67,246 GWh of electricity (Avtar et al., 2019). Van der Meer et al. (2014) stated that some parameters can be directly related to geothermal activity, while others could be estimated indirectly.

Remote sensing can make a valuable contribution to mapping the spatial extent of surface geology and mineralogy either directly or indirectly (Pickles et al., 2004). Direct evidence for geothermal activity can be retrieved from remote sensing through surface features or anomalous mineral assemblages. Geothermal areas often have a distinct signature related to surface mineralogy, thus, multispectral broad band remote sensing can be used to map broad classes of surface mineral surfaces. Features of geothermal activity that can be retrieved and mapped from remote sensing are caldera structures, hot springs, steaming ground and fumaroles (Avtar et al., 2019). On the other hand, hyperspectral optical remote sensing can also be used to produce surface mineralogy maps (van der Meer et al., 2014). Variations in surface temperature in relation to the presence of heat sources, as well as heat fluxes, can be readily measured and indirectly indicate geothermal activities. In addition, surface deformation related to geothermal activity can be retrieved using SAR interferometry. Carnec and Fabriol, 1999 and Sarychikhina et al. (2011, 2018) used SAR to monitor land subsidence at Cerro Prieto, Mexico.

Some of the early studies on remote sensing in relation to geothermal resources date back to the eighties. The first projects focused on helicopter-based mapping of geothermal area (Sekioka, 1985), while several early studies focused on the geothermal areas in New Zealand (Mongillo,

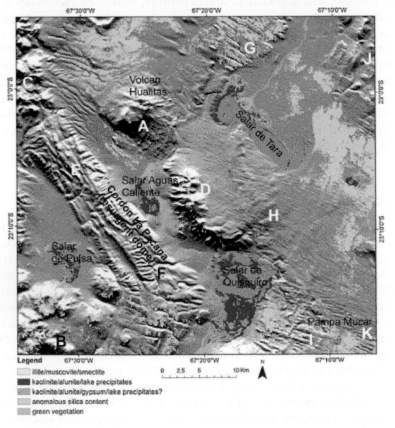

FIGURE 10.15 Classification of the ASTER daytime imagery showing various groups of minerals and vegetation. Hill shaded SRTM image used as background (symbols A–K are explained in the text). Taken from van der Meer et al., 2014.

1994; Deroin et al., 1995, Romaguera et al., 2018). Specifically, Mongillo (1994) used night-time airborne thermal infrared data focusing on temperature mapping of hot springs, while Deroin et al. (1995) used Landsat and data to map structures related to geothermal processes. Several studies of hydrothermal systems have taken advantage of the launch of ASTER, the Advanced Spaceborne Thermal Emission and Reflectance Radiometer, and provided enhanced mineral mapping capabilities for the geologic remote sensing community, due to its increased spectral resolution of hyperspectral remote sensing instruments (Rowan et al., 2003; Carranza et al., 2008). Coolbaugh et al. (2007) detected geothermal anomalies at Bradys Hot Springs, Nevada, USA, using ASTER thermal infrared images. Gutiérrez et al. (2012) worked with thermal anomaly detection using satellite images at volcanic/geothermal complexes in the Andes of Central Chile. Landsat ETM+ thermal infrared data has also been used for geothermal area detection in China (Watson et al., 2008; Qin et al., 2011; Mia et al., 2012) to develop radiometric techniques for estimating geothermal heat flux.

Savage et al. (2010) used Landsat imagery to make a review of alternative methods for estimating terrestrial emittance and geothermal heat flux for Yellowstone National Park, while Kruse (2012) used imaging spectrometry in order to map surface mineralogy. Reath and Ramsey (2013) used Visible near-infrared (VNIR), short-wave infrared (SWIR) and thermal infrared (TIR) to explore geothermal systems. During the same year, Haselwimmer et al. (2013) developed a methodology to quantify the heat flux and outflow rate of hot Sarychikhina springs using airborne thermal imagery applied in Alaska, USA. Figure 10.15 shows a Chilean case study produced through an ASTER-based temperature map.

10.8 SUMMARY

Renewable energy has received significant attention during the last decades. It has been recognized that, since fossil fuels are depleting, there is a constant need for new energy sources around the world. Several case studies from different parts of the world use remote sensing techniques in exploring renewable energy resource sites. Several remote sensing and GIS techniques dealing with renewable energy and future perspectives and potential solutions are suggested in this chapter. Nevertheless, it is important to note that only the combination of a different set of techniques is the key to the success of each task.

REFERENCES

Adeyeri, O.E., Akinsanola, A.A., and Ishola, K.A., 2017. Investigating surface urban heat island characteristics over Abuja, Nigeria: Relationship between land surface temperature and multiple vegetation indices. *Remote Sensing Applications: Society and Environment*, 7, 57–68.

Al-Mukhtar, M., and Al-Yaseen, F., 2019. Modeling water quality parameters using data-driven models, a case study Abu-Ziriq Marsh in South of Iraq. *Hydrology*, 6, 24.

Avtar, R., Suzuki, R., and Sawada, H., 2014. Natural forest biomass estimation based on plantation information using PALSAR data. *PLoS ONE*, 9, e86121.

Avtar, R., Sahu, N., Aggarwal, A.K., Chakraborty, S., Kharrazi, A., Yunus, A.P., Dou, J., and Kurniawan, T.A., 2019. Exploring renewable energy resources using remote sensing and GIS – A review. *Resources*, 8, 149, doi:10.3390/resources8030149

Biggs, D., Miller, F., Hoanh, C.T., and Molle, F., 2009. The delta machine: Water management in the Vietnamese Mekong Delta in historical and contemporary perspectives. In *Contested Waterscapes in the Mekong Region; Hydropower, Livelihoods and Governance*. Routledge, Abingdon-on-Thames, UK, 203–225.

Calaudi, R., Arena, F., Badger, M., and Sempreviva, A.M., 2013. Offshore wind mapping Mediterranean area using SAR. *Energy Procedia*, 40, 38–47.

Cano, D., Monget, J.M., Albuisson, M., Guillard, H., Regas, N., and Wald, L., 1986. A method for the determination of the global solar radiation from meteorological satellite data. *Solar Energy*, 37, 31–39.

Carnec, C., and Fabriol, H., 1999. Monitoring and modeling land subsidence at the Cerro Prieto Geothermal Field, Baja California, Mexico, using SAR interferometry. *Geophysical Research Letters*, 26, 1211–1214.

Carranza, E.J.M., Wibowo, H., Barritt, S.D., and Sumintadireja, P., 2008. Spatial data analysis and integration for regional-scale geothermal potential mapping, West Java, Indonesia. *Geothermics*, 37, 3, 267–299.

Chang, R., Zhu, R., and Guo, P., 2016. A case study of land-surface-temperature impact from large-scale deployment of wind farms in China from Guazhou. *Remote Sensing*, 8, 790

Climate.weatheroffice.gc.ca, 2013. Calculation of the 1971 to 2000 Climate Normals for Canada. July 10, 2013. Archived from the original on June 27, 2013, [accessed: August 9, 2013].

Congedo, L., 2016. Semi-automatic classification plugin documentation. Release 6.1.1.1, 10.13140/RG.2.2.29474.02242/1

Coolbaugh, M.F., Kratt, C., Fallacaro, A., Calvin, W.M., and Taranik, J.V., 2007. Detection of geothermal anomalies using Advanced Spaceborne Thermal Emission and Reflection Radiometer (ASTER) thermal infrared images at Bradys Hot Springs, Nevada, USA. *Remote Sensing of Environment*, 106, 350–359.

Cooper, P.I., 1969. The absorption of radiation in solar stills. *Solar Energy*, 12, 333–346.

Datta, A., and Karakoti, I., 2010. Solar resource assessment using GIS & remote sensing techniques. ESRI India, 9560272741, 1–20.

Deroin, J.P., Cochrane, G.R., Mongillo, M.A., and Browne, P.R.L., 1995. Methods of remote sensing in geothermal regions – the geodynamic setting of the Taupo volcanic zone (North Island, New Zealand). *International Journal of Remote Sensing*, 16 9, 1663–1677.

Duffie, J.A., and Beckman, W.A., 2013. *Solar Engineering of Thermal Processes*, 4th ed. John Wiley & Sons, Hoboken, NJ.

Ellabban, O., Abu-Rub, H., and Blaabjerg, F., 2014. Renewable energy resources: Current status, future prospects and their enabling technology. *Renewable and Sustainable Energy Reviews*, 39, 748–764.

ESMAP, 2000. *Global Photovoltaic Power Potential by Country*. Washington, DC: World Bank.

Forsberg, C.W., 2009. Sustainability by combining nuclear, fossil, and renewable energy sources. *Progress in Nuclear Energy*, 51, 192–200.

Furevik, B.R., Espedal, H.A., Hamre, T., Hasager, C.B., Johannessen, O.M., Jørgensen, B.H. and Rathmann, O., 2003. Satellite-based wind maps as guidance for siting offshore wind farms. *Wind Engineering*, doi: 10.1260/030952403322770931

García-Martín, A., de la Riva, J., Pérez-Cabello, F., and Montorio, R., 2012. Using remote sensing to estimate a renewable resource: Forest residual biomass. In *Remote Sensing of Biomass – Principles and Applications*. IntechOpen, London, UK.

Gutiérrez, F.J., Lemus, M., Parada, M.A., Benavente, O.M., and Aguilera, F.A., 2012. Contribution of ground surface altitude difference to thermal anomaly detection using satellite images: Application to volcanic/geothermal complexes in the Andes of Central Chile. *Journal of Volcanology and Geothermal Research*, 237, 69–80.

Harris, R., Zhou, L., and Xia, G., 2014. Satellite observations of wind farm impacts on nocturnal land surface temperature in Iowa. *Remote Sensing*, 6, 12234–12246.

Hasager, C.B., Astrup, P., Zhu, R., Chang, R., Badger, M., and Hahmann, A.N., 2016. Quarter-Century offshore winds from SSM/I and WRF in the North Sea and South China Sea. *Remote Sensing*, 8, 769, doi:10.3390/rs8090769

Haselwimmer, C., Prakash, A., and Holdmann, G., 2013. Quantifying the heat flux and outflow rate of hot springs using airborne thermal imagery: case study from Pilgrim Hot Springs, Alaska. *Remote Sensing of Environment*, 136, 37–46.

Iqbal, M., 1983. *An Introduction to Solar Radiation*. Academic, Toronto.

Jakubiec, J.A., and Reinhart, C.F., 2013. A method for predicting city-wide electricity gains from photovoltaic panels based on LiDAR and GIS data combined with hourly Daysim simulations. *Solar Energy*, 93, 127–143.

Johannessen, O., and Korsbakken, E., 1998. Determination of wind energy from SAR images for siting windmill locations. Earth Observation *Quarterly*, 59, 2–4.

Karthikeya, B.R., Negi, P.S., and Srikanth, N., 2016. Wind resource assessment for urban renewable energy application in Singapore. *Renewable Energy*, 87, 403–414.

Kazem, H.A., Yousif, J.H., and Chaichan, M.T., 2016. Modeling of daily solar energy system prediction using support vector machine for Oman. *International Journal of Applied Engineering Research*, 11, 10166–10172.

Kilic, G., and Unluturk, M.S., 2015. Testing of wind turbine towers using wireless sensor network and accelerometer. *Renewable Energy*, 75, 318–325.

Kim, D., Kim, T., Oh, G., Huh, J., and Ko, K., 2016. A comparison of ground-based LiDAR and met mast wind measurements for wind resource assessment over various terrain conditions. *Journal of Wind Engineering & Industrial Aerodynamics*, 158, 109–121.

Kosmopoulos, P.G., Kazadzis, S., El-Askary, H., Taylor, M., Gkikas, A., Proestakis, E., Kontoes, C., and El-Khayat, M.M., 2018. Earth-observation-based estimation and forecasting of particulate matter impact on solar energy in Egypt. *Remote Sensing*, 10, 1870.

Kratt, C., Calvin, W.M., and Coolbaugh, M.F., 2010. Mineral mapping in the Pyramid Lake basin: Hydrothermal alteration, chemical precipitates and geothermal energy potential. *Remote Sensing of Environment*, 114, 2297–2304.

Kruse, F.A., 2012. Mapping surface mineralogy using imaging spectrometry. *Geomorphology*, 137, 41–56.

Kumar, L., and Mutanga, O., 2017. Remote sensing of above-ground biomass. *Remote Sensing*, 9, 935.

Kusre, B., Baruah, D., Bordoloi, P., and Patra, S., 2010. Assessment of hydropower potential using GIS and hydrological modeling technique in Kopili River basin in Assam (India). *Applied Energy*, 87, 298–309.

Landsat 8 Data Users Handbook, 2019. Owned by Vaughn Ihlen, approved by Karen Zanter, Version 5. November, USGS, USA.

Le Toan, T., Beaudoin, A., Riom, J., and Guyon, D., 1992. Relating forest biomass to SAR data. *IEEE Transactions on Geoscience and Remote Sensing*, 30, 403–411.

Lefsky, M.A., Cohen, W.B., Parker, G.G., and Harding, D.J., 2014. Lidar remote sensing for ecosystem studies. *BioScience*, 52, 19–30.

Lefsky, M.A., Cohen, W.B., Harding, D.J., Parker, G.G., Acker, S.A., and Gower, S.T., 2002. Lidar remote sensing of above-ground biomass in three biomes. *Global Ecology and Biogeography*, 11, 393–399.

Lukač, N., Žlaus, D., Seme, S., Žalik, B., and Štumberger, G., 2013. Rating of roofs' surfaces regarding their solar potential and suitability for PV systems, based on LiDAR data. *Applied Energy*, 102, 803–812.

Lund, J.W., Bjelm, L., Bloomquist, G., and Mortensen, A.K., 2007. Characteristics, development, and utilization of geothermal resources. In *Geo-Heat Center Bulletin*. Oregon Institute of Technology, Klamath Falls, OR, USA.

Martini, B., Silver, E., Pickles, W., and Cocks, P., 2003. *Hyperspectral Mineral Mapping in Support of Geothermal Exploration: Examples from Long Valley Caldera, CA and Dixie Valley, NV, USA*. Lawrence Livermore National Lab (LLNL), Livermore, CA, USA, Volume 27.

Masoom, A., Kosmopoulos, P., Bansal, A., and Kazadzis, S., 2020. Solar energy estimations in India using remote sensing technologies and validation with sun photometers in urban areas. *Remote Sensing*, 12, 254, https://doi.org/10.3390/rs12020254

McMahon, B.B., 2019. Measuring winds from space: The European Space Agency's Aeolus mission. *Weather*, 74 9, 312–315.

Mekhilef, S., Saidur, R., and Safari, A., 2011. A review on solar energy use in industries. *Renewable and Sustainable Energy Reviews*, 15 4, 1777–1790.

Mia, M.B., and Fujimitsu, Y., 2012. Mapping hydrothermal altered mineral deposits using Landsat 7 ETM+ image in and around Kuju volcano, Kyushu, Japan. *Journal of Earth System Science*, 121, 1049–1057.

Mia, M.B., Bromley, C.J., and Fujimitsu, Y., 2012. Monitoring heat flux using Landsat TM/ETM+ thermal infrared data – A case study at Karapiti ("Craters of the Moon") thermal area, New Zealand. *Journal of Volcanology and Geothermal Research*, 235–236, 1–10.

Mikkelsen, T., Angelou, N., Hansen, K., Sjöholm, M., Harris, M., Slinger, C., Hadley, P., Scullion, R., Ellis, G., and Vives, G., 2013. A spinner-integrated wind lidar for enhanced wind turbine control: Spinner-integrated wind lidar for enhanced steering and control. *Wind Energy*, 16, 625–643.

Mongillo, M.A., 1994. Aerial thermal infrared mapping of the Waimangu-Waiotapu geothermal region, New Zealand. *Geothermics*, 23, 5–6, 511–526.

Moran, S., Jackson, R.D., Slater, P.N., and Teillet, P.M., 1992. Evaluation of simplified procedures for retrieval of land surface reflectance factors from satellite sensor output. *Remote Sensing of Environment*, 41, 2–3, 169–184.

Pickles, W.L., Martini, B.A., Silver, E.A., and Cocks, P.A., 2004. *Hyperspectral Mineral Mapping in Support of Geothermal Exploration: Examples from Long Valley Caldera, CA and Dixie Valley, NV, USA*. N. p. Web, United States.

Pruess, K., 2006. Enhanced geothermal systems (EGS) using CO2 as working fluid – A novel approach for generating renewable energy with simultaneous sequestration of carbon. *Geothermics*, 35, 351–367.

Qin, Q., Zhang, N., Nan, P., and Chai, L., 2011. Geothermal area detection using Landsat ETM+ thermal infrared data and its mechanistic analysis – A case study in Tengchong, China. *International Journal of Applied Earth Observation and Geoinformation*, 13, 4, 552–559.

Reath, K.A., and Ramsey, M.S., 2013. Exploration of geothermal systems using hyperspectral thermal infrared remote sensing. *Journal of Volcanology and Geothermal Research*, 265, 27–38.

Romaguera, M., Vaughan, R.G., Ettema, J., Izquierdo-Verdiguier, E., Hecker, C.A., and van der Meer, F.D., 2018. Detecting geothermal anomalies and evaluating LST geothermal component by combining thermal remote sensing time series and land surface model data. *Remote Sensing of Environment*, 204, 534–552.

Rowan, L.C., Hook, S.J., Abrams, M.J., and Mars, J.C., 2003. Mapping hydrothermally altered rocks at Cuprite, Nevada, using the advanced Spaceborne Thermal Emission and Reflection Radiometer (ASTER), a new satellite-imaging system. *Economic Geology*, 98, 5, 1019–1027.

Roy, P.S., and Ravan, S.A., 1996. Biomass estimation using satellite remote sensing data – An investigation on possible approaches for natural forest. *Journal of Biosciences*, 21, 535–561.

Sarychikhina, O., Glowacka, E., and Robles, B., 2018. Multi-sensor DInSAR applied to the spatiotemporal evolution analysis of ground surface deformation in Cerro Prieto basin, Baja California, Mexico, for the 1993–2014 period. Natural Hazards: Journal of the International Society for the Prevention and Mitigation of Natural Hazards, Springer, *International Society for the Prevention and Mitigation of Natural Hazards*, 92, 1, 225–255

Sarychikhina, O., Glowacka, E., Mellors, R., and Vidal, F.S., 2011. Land subsidence in the Cerro Prieto Geothermal Field, Baja California, Mexico, from 1994 to 2005: An integrated analysis of DInSAR, leveling and geological data: *Journal of Volcanology and Geothermal Research*, 204, 76–90.

Savage, S.L., Lawrence, R.L., Custer, S.G., Jewett, J.T., Powell, S.L., and Shaw, J.A., 2010. Review of alternative methods for estimating terrestrial emittance and geothermal heat flux for Yellowstone National Park using Landsat imagery. *GIScience & Remote Sensing*, 47, 4, 460–479.

Scapepedia, 2018. https://www.solarsystemscope.com/spacepedia/earth/orbital-and-rotational-characteristics-of-earth

Sekioka, M., 1985. Geothermal observations by use of a helicopter-borne remote sensing system. *Remote Sensing of Environment*, 18, 2, 193–203.

Shi, X., Elmore, A., Li, X., Gorence, N.J., Jin, H., Zhang, X., and Wang, F., 2008. Using spatial information technologies to select sites for biomass power plants: A case study in Guangdong Province, China. *Biomass Bioenergy*, 32, 35–43.

Singh, I., Vardhan, S., and Singh, S., 2019. *A Review on Solar Energy Collection for Thermal Applications. Conference: International Conference on Current Trends in Engineering, Sciences and Management*, Kuala Lumpur, Malaysia.

SolarHub, 2020. Available: http://beyond-eocenter.eu/index.php/web-services/solarhub [accessed: March 5, 2020].

Spencer, J.W., 1971. Fourier series representation of the position of the sun. *Search*, 2, 5, 172.

Szabó, S., Enyedi, P., Horváth, M., Kovács, Z., Burai, P., Csoknyai, T., and Szabó, G., 2016. Automated registration of potential locations for solar energy production with Light Detection And Ranging (LiDAR) and small format photogrammetry. *Journal of Cleaner Production*, 112, 3820–3829.

Van der Meer, F., Hecker, C., van Ruitenbeek, F., van der Werff, H., de Wijkerslooth, C., and Wechsler, C., 2014. Geologic remote sensing for geothermal exploration: A review. *International Journal of Applied Earth Observation and Geoinformation*, 33, 1, 255–269.

Vaughan, R.G., Keszthelyi, L.P., Lowenstern, J.B., Jaworowski, C., and Heasler, H., 2012. Use of ASTER and MODIS thermal infrared data to quantify heat flow and hydrothermal change at Yellowstone National Park. *Journal of Volcanology and Geothermal Research*, 233, 72–89.

Voivontas, D., Assimacopoulos, D., and Koukios, E., 2001. Assessment of biomass potential for power production: A GIS based method. *Biomass Bioenergy*, 20, 101–112.

Wald, L., 2018. Basics in solar radiation at earth surface. Lecture Notes. Edition 1. January 3, 2018. MINES ParisTech, PSL Research.

Wan, C., Zhao, J., Song, Y., Xu, Z., Lin, J., and Hu, Z., 2015. Photovoltaic and solar power forecasting for smart grid energy management. *Chinese Society for Electrical Engineering (CSEE), Journal of Power and Energy Systems*, 1, 4, 38–46

Watson, F.G.R., Lockwood, R.E., Newman, W.B., Anderson, T.N., and Garrott, R.A., 2008. Development and comparison of Landsat radiometric and snowpack model inversion techniques for estimating geothermal heat flux. *Remote Sensing of Environment*, 112, 471–481.

WEMSAR Consortium, 2003. WEMSAR Final Report, NERSC Technical Report no. 237, Nansen Environmental and Remote Sensing Center, Edv. Greigsvei 3a, N-5059 Bergen, Norway, March 2003.

Wiginton, L.K., Nguyen, H.T., and Pearce, J.M., 2010. Quantifying rooftop solar photovoltaic potential for regional renewable energy policy. *Computers, Environment and Urban Systems*, 34, 345–357.

Wu, X.-L., Zhang, L.-X., Zhao, C., Gegen, T., Zheng, C.W., Shi, X.-Q., Geng, J., and Letu, H., 2019. Satellite-based assessment of local environment change by wind farms in China. Earth and Space Science, doi: 10.1029/2019EA000628, 1–12.

Yi, C.S., Lee, J.H., and Shim, M.P., 2010. Site location analysis for small hydropower using geo-spatial information system. *Renewable Energy*, 35, 852–861.

Zecchetto, S., 2018. Wind direction extraction from SAR in coastal areas. *Remote Sensing*, 10, 2, 261

Zhang, H., Wang, J., Xie, Y., Yao, G., Yan, Z., Huang, L., Chen, S., Pan, T., Wang, L., Su, Y., et al., 2016. Self-powered, wireless, remote meteorologic monitoring based on triboelectric nanogenerator operated by scavenging wind energy. *ACS Applied Materials & Interfaces – ACS Publications*, 8, 32649–32654.

11 Land Use and Land Cover (LULC)

11.1 INTRODUCTION: REMOTE SENSING CONCEPTS OF LAND USE – LAND COVER

Landscape is formed and shaped by natural and human processes. Specifically, land evolution is produced through complex dynamics based on the interactions between abiotic, biotic and anthropogenic factors and their impact on different scales (Castellanos et al., 2018). Moreover, the dynamics of landscape morphology are subject to gradual change caused by factors, such as rainfall amount, type of rock and soil, land shape, vegetation type, river shape, size and flow, slope effect and drainage pattern, which may act independently or jointly. As a result, landscape topography integrates the complex interactions between all these factors. Several scientific disciplines have contributed to the understanding of these complex dynamics (Castellanos et al., 2018).

Remote sensing data and methods can implement detailed analysis of Land Use/Land Cover (LULC) in complex rural watersheds, which results in a complete profile for better rural management and planning, as well as for its natural resources (Usman et al., 2015). This procedure is called change detection, which identifies these temporal modifications in LULC. Specifically, there is a need for improved and updated LULC data sets, which address these changes leading to effective planning and production management. Indeed, it is mainly developing countries that are faced globally with rapid changes in LULC, since they rely heavily on agricultural production and increasing population (Anderson, 1977; Usman et al., 2015). Contemporary remote sensing data for agricultural use can be applied to overcome the deficiency in LULC data. Specifically, satellite remote sensing data have been used since the seventies to monitor LULC changes at coarse spatial scales (Usman et al., 2015). Nevertheless, its use in rural environments has gained much popularity in recent years (Dalezios, 2015). LULC analysis in rural environments may include land use changes and land use classification for natural resources management.

On the other hand, urban environments are heterogeneous and complex ecosystems characterized by mixed LULC. In evaluating urban environments, LULC mapping using multispectral classification methods has been widely utilized. Indeed, urban remote sensing can help improve our understanding of cities. However, remote sensing technology has the capability of acquiring images that cover large areas, which constitutes the largest benefit and can also be applied to urban studies. Specifically, remote sensing can provide synoptic views, which allow the identification of objects, patterns and human-land interactions (Mitraka and Chrysoulakis, 2018). LULC analysis in urban environment may involve urban planning and monitoring, land use change for urbanization, urban vegetation and water, as well as the study of the Urban Heat Island (UHI). In this chapter, archaeological resource mapping is also considered using UAV.

11.2 RURAL ENVIRONMENT

In this section of rural environment, land use changes are considered, which involve forest change detection, LULC change in rural coastal zones, subtropical forest change detection, as well as agricultural land-use changes. Moreover, land use classification is considered in terms of natural resource management, which involves LULC classification and land cover mapping.

11.2.1 LAND USE CHANGES

Forest change detection. Forest cover is assessed to be around 31% of the total land surface area globally, which includes habitat provision to about 75% of terrestrial biodiversity, as well as protection of waterbodies to control climate change (FAO, 2014; Mensah et al., 2019). Moreover, forest ecosystem is characterized by high carbon density, being considered potential carbon sink, with estimates of about 80% terrestrial aboveground carbon and 70% of soil organic carbon (Mensah et al., 2019). In addition, forests products and services contribute and promote the social and economic growth of developing and developed countries. On the other hand, there is a decline in the global rate of forest area loss since 2010 to 3.3 million hectares or 0.08% per annum, which accounts for half the rate in the nineties (Mensah et al., 2019; FAO, 2014). The importance of forest area reservation is also recognized as multifunctional forest management through reduction of emissions from deforestation and forest degradation, as well as conservation, sustainable forest management and enhancement of forest carbon stocks (REDD+).

The case of tropical deforestation in Ghana is examined, which has recently attracted the interest of policy makers and development practitioners (Mensah et al., 2019). Analysis of LULC changes is considered critical for understanding the patterns and interactions of diachronic anthropogenic activities and has become essential for sustainable forest management in Ghana. Indeed, improved forest management could be supported by such a powerful decision-making tool. Specifically, land cover and change detection analysis are considered extremely significant for sustainable forest management due to changing conditions from increasing forest fragmentation. In this study, supervised classification has been applied using maximum likelihood algorithm in Quantum GIS to detect LULC changes in the Bosomtwe Range Forest Reserve (BRFR), Ghana, from 1991, 2002 and 2017 using Landsat 4 – TM, Landsat 7 – ETM from the USGS-GloVis platform and Sentinel-2 satellite imageries, respectively, as well as very high-resolution imagery provided by the Worldview-3 satellite platform for the accuracy assessment (Mensah et al., 2019).

For the assessment of forest vegetation, the NDVI is used, which is one of the most widely applied indices for understanding and characterizing vegetation dynamics (Dalezios, 2015; Mensah et al., 2019). At first, all features were extracted to predetermine land cover classes. Four land cover classes were obtained and characterized, namely: (1) closed forest canopy (forest area with a canopy exceeding 60%); (2) opened/sparse forest canopy (forest area with a canopy cover between 10 and 60%); (3) built-up/bare landscapes and; (4) dense grass/herbaceous cover (includes grass and shrub lands and also represents landscape that is potentially suitable for agricultural purposes) (Mensah et al., 2019). Then, the NDVI was computed in different years using the red (R) and near-infrared (NIR) bands in the Landsat and Sentinel images. The classified map of BRFR is shown in Figure 11.1.

LULC change in rural coastal zones. LULC changes in rural coastal zones refer to changes from agricultural land to rural settlement, fallow land, or vacant land. Nevertheless, land use change is a common phenomenon in coastal zones, which affects the coastal livelihood, and has been recognized as an important driver of environmental change on all spatial and temporal scales (Das et al., 2018). Indeed, forest ecosystem change is fully caused by land cover change. The case of Sharankhola Upazila, one of the Upazilas of the coastal region of Bangladesh, is examined (Das et al., 2018). The study involves detection of LULC changes based on GIS and remote sensing methods and its impact on the coastal livelihood. For this study, Landsat satellite images of 1989 (Landsat 5 TM), 1999 (Landsat 7 ETM) and 2010 (Landsat 7 ETM+), respectively, have been collected for analysis and interpretation from the official website of US Geological Survey (USGS). Pre-processing and processing of these images has been conducted. Specifically, interactive supervised classification has been applied for classifying land-cover types into six classes, namely rural settlements, water, agricultural land, vegetation, sundarbans and vacant land (Das et al., 2018). The images are projected to WGS 1984 and UTM Zone-45N Coordinate system, and land-cover maps of 1989, 1999 and 2010, respectively, are prepared. Land cover changes are analyzed by modeling

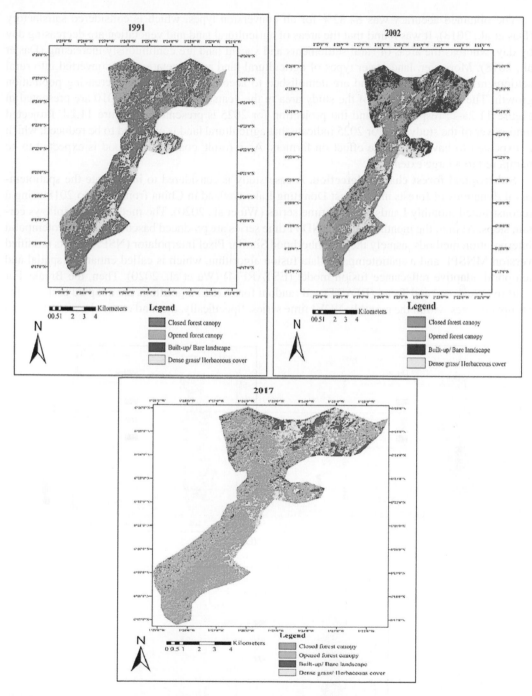

FIGURE 11.1 Classified maps of BRFR (1991-Landsat 4TM, 2002-Landsat 7ETM and 2017-Sentinel 2A) (from Mensah et al., 2019).

the ecological sustainability through IDRISI Selva software. Digital elevation model (DEM) for basic road layer of the study area is the driving force for this analysis. Land change modeling is conducted for change detection from 1989 to 2010, which could be projected up to 2025. The projection has been performed through multi-layer perception (MLP) Markov chain Neural Network built in IDRISI Selva (Das et al., 2018).

The obtained accuracy was 84.82% for all conversion types, which is considered satisfactory (Das et al., 2018). It was found that the areas of agricultural land and vegetation are decreasing day by day, whereas the areas of rural settlements and vacant land are continuously increasing (Das et al., 2018). Moreover, land-cover types of agricultural land and vegetation are converted into rural settlements and vacant land, and are demolished to accommodate the ever-increasing population growth. The land-cover maps of the study area in the years 1989, 1999 and 2010 are presented in Figure 11.2a–c, respectively, and the projection for 2025 is presented in Figure 11.2d. Projected land-cover of the study area for 2025 indicate that agricultural land is expected to be reduced, which is expected to have a negative effect on farmers. As a result, coastal livelihood is expected to be hampered to a large extent.

Subtropical forest change detection. A case study is considered to investigate the spatiotemporal dynamics of forests in the West Dongting Lake wetland in China from 2000 to 2018 using a reconstructed monthly Landsat NDVI time series (Wu et al., 2020). The methodology follows certain steps. At first, the monthly Landsat NDVI time series are produced based on two spatiotemporal interpolation methods, namely the Neighborhood Similar Pixel Interpolator (NSPI) and its modified version MNSPI, and a spatiotemporal data fusion algorithm, which is called enhanced spatial and temporal adaptive reflectance fusion model (ESTARFM) (Wu et al., 2020). Then, the Breaks For Additive Seasonal and Trend (BFAST) and random forest (RF) algorithms are adopted to detect the abrupt changes within the monthly NDVI time series. Specifically, the land cover classes before and

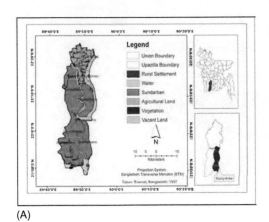

(A)

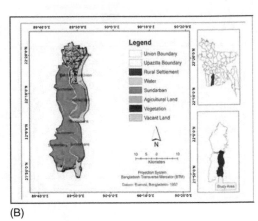

(B)

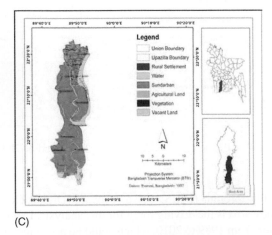

(C)

FIGURE 11.2 (a) Land- cover Map for 1989; (b) Land- cover Map for 1999; (c) Land- cover Map for 2010;

(Continued)

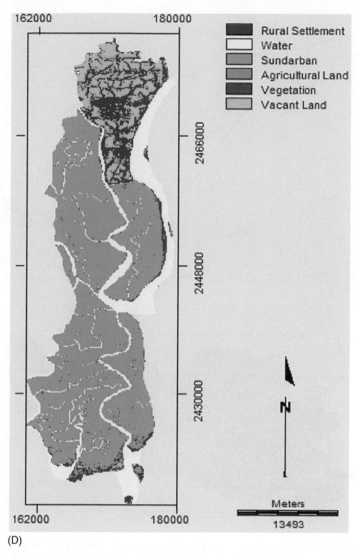

(D)

FIGURE 11.2 (*Continued*) (d) Projected Land- cover map of Study Area for 2025 (from Das et al., 2018).

after abrupt changes are identified through setting the coefficients of the time series model as the input parameters of the RF algorithm (Wu et al., 2020). Moreover, a forest mask is developed for each image in the time series and these masks are merged to produce an "anytime" forest mask for extracting all the forest pixels, which leads to computation of forest extent.

The spatiotemporal distributions of the number and timing of abrupt changes, which are related to forest are analyzed after accuracy assessment of change detection and classification. Finally, the variation of forest area in West Dongting Lake is presented and the potential drivers are discussed (Wu et al., 2020). The multi-type forest changes, including conversion from forest to another land cover category, conversion from another land cover category to forest, and conversion from forest to forest (such as flooding and replantation post-deforestation), and land cover categories before and after change are effectively detected, with an overall accuracy of 87.8% (Wu et al., 2020). Figure 11.3 shows the geographical location, Landsat image of the study area: (a) general location within China; (b) within Hunan Province; (c) within the Dongting Lake region; (d) Landsat 8 OLI image (RGB = near-infrared, red, and green band) (Wu et al., 2020).

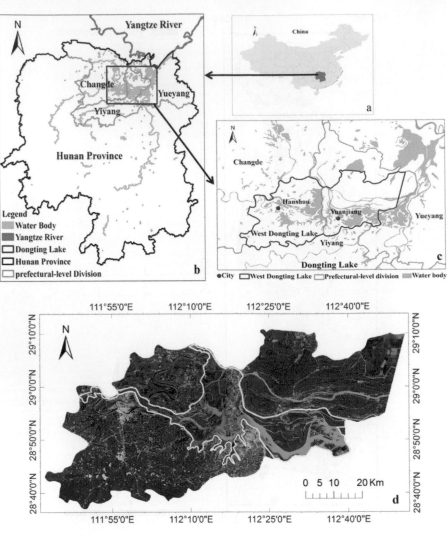

FIGURE 11.3 Geographic location, Landsat and DEM image of the study area: (a) general location within China, (b) within Hunan Province and (c) within the Dongting Lake region; (d) Landsat 8 OLI image (RGB = near-infrared, red, and green band) (from Wu et al., 2020).

Agricultural land-use changes. Agricultural land-use changes in trading countries rely heavily on international food trade, which has affected food production and the global environment (Jing et al., 2017). A case study is considered to investigate the spatial attributes of soybean land changes at the same time within and among trading countries, namely Brazil, China and the USA (Jing et al., 2017).

This study is based on an integrated framework of tele-coupling, i.e. socioeconomic and environmental interactions over distances. Figure 11.4 shows the map of the study regions. Specifically: (a) Western Corn Belt (WCB) in the United States (Alaska and Hawaii are not shown); (b) State of Mato Grosso (MT) in Brazil; (c) Heilongjiang Province in China. The world map is inserted, which is a simplified illustration showing the tele-coupling relationship between the three countries, where China is the receiving system, Brazil is the sending system, and the USA is the spillover system. Arrows indicate directions of flows of soybeans and money, which are bidirectional here. Solid arrow is direct flow between Brazil and China, and dashed arrows are indirect flow between Brazil and China by passing through the USA (from Jing, et al., 2017).

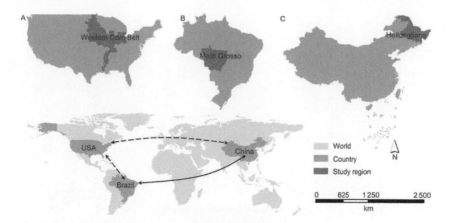

FIGURE 11.4 Map of study regions. (A) Western Corn Belt (WCB) in the United States (Alaska and Hawaii are not shown); (B), State of Mato Grosso (MT) in Brazil; (C) Heilongjiang Province in China. The insert world map is a simplified illustration showing the tele-coupling relationship between the three countries, where China is the receiving system, Brazil is the sending system and the USA is the spillover system. Arrows indicate directions of flows of soybeans and money, which are bidirectional here. Solid arrow is direct flow between Brazil and China, and dashed arrows are indirect flow between Brazil and China by passing through the USA (from Jing, et al., 2017).

The spatial distribution of the occurred changes is affected by the spatial attributes of agricultural land-use changes. Specifically, it is examined whether these changes occur clustered or evenly, if they differ at multiple scales, and whether they vary among trading countries. All these are expected to help understand the dynamics of food production and environmental issues within and among trading countries (Jing, et al., 2017). The tele-coupling framework is composed of five interrelated components: systems, agents, flows, causes, and effects. Specifically, systems are coupled human and natural systems, which can be further classified as sending, receiving and spillover systems. Moreover, agents are decision-making entities involved in the tele-coupling, and they affect flows of energy, materials and information between the systems. Furthermore, causes are drivers or factors that generate the tele-coupling and alter its dynamics, resulting in socioeconomic and environmental effects (Jing, et al., 2017). The tele-coupling framework has been applied to analyze several important issues, such as ecosystem services, food and forest sustainability, conservation, energy sustainability, water and species invasion.

Soybean maps of 2005 and 2010 were used as input. Soybean maps of WCB, MT and Heilongjiang were generated by moderate resolution imaging spectroradiometer (MODIS) data and provided at 250-m spatial resolution (Jing, et al., 2017). The data for South Dakota, Kansas, Minnesota and Missouri were not available in 2005, so 2006 WCB data was used as an approximation of 2005. Because WCB data were originally produced at 56-m in 2006 and 30-m in 2010, both were resampled to 250-m to be consistent with the data of Heilongjiang and MT (Jing, et al., 2017). A moving window analysis was then applied to the results of change detection, including soybean gain and soybean loss, to detect the change information across multi-spatial scales by varying window size. The results indicated that the spatial distribution of soybean land changes was spatially clustered, different across multi-spatial scales and varied among the trading countries (Jing, et al., 2017). This study has indicated the needs of coordination between trading countries for global sustainability. For illustrative purposes, Figure 11.5 shows the map of change detection in MT, with (a) soybean gain and (b) soybean loss (from Jing, et al., 2017).

11.2.2 LAND USE CLASSIFICATION: NATURAL RESOURCE MANAGEMENT

Land cover is the most important property of earth's surface defining its physical condition and biotic component, and land use is the modification of land cover as per human needs. A detailed analysis of LULC based on remote sensing data in complex irrigated basins is presented in order to

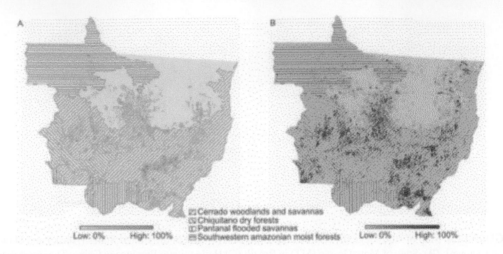

FIGURE 11.5 Map showing the results of change detection in MT. (A) soybean gain and (B) soybean loss (from Jing et al., 2017).

obtain a complete profile for better water resource management and planning (Usman et al., 2015). There are specific objectives of this study, such as: classification of major LULC and their accuracy assessment in complex irrigated lands at a higher spatial resolution; the relationship between orography and climatic factors with NDVI and estimation of the areal extents of different LULC classes for individual cropping seasons; and detection of the spatial and temporal LULC changes for exploring maximum flexibility of change for major classes or major crops (Usman et al., 2015).

The Lower Chenab Canal (LCC) irrigated region of Pakistan is analyzed from 2005 to 2012 for LULC change detection using remote sensing data. Wheat and rice are found to be the major crop types in rabi and kharif seasons, respectively (Usman et al., 2015). Temporal profiles of NDVI are developed using MODIS at 250 m × 250 m spatial resolution and temporal coverage of NDVI data after every 8 to 16 days. This spatial resolution is considered satisfactory to capture all major crop classes in the current study region for precise measurement of crop water requirements and subsequent irrigation water allocation planning (Usman et al., 2015). Accuracy assessment of the maps is performed using three different methods, namely error matrix approach, comparison with ancillary data and with previous studies. The results indicate that LULC change detection for wheat and rice has less volatility of change in comparison with both rabi and kharif fodders (Usman et al., 2015). Figure 11.6 presents a map of LCC (East), Rechna Doab, Punjab, Pakistan and ground truthing points (from Usman et al., 2015).

Land cover mapping. A case study is presented that investigates the possibility of the automation of land cover mapping by using fusion Sentinel-2 imageries and light detection and ranging (LiDAR) point clouds (Szostak et al., 2020). Two former sulfur mines are considered located in Southeast Poland, namely Jeziórko, where 216.5 ha of afforested area was reclaimed after borehole exploitation, and Machów, where 871.7 ha of dump area was reclaimed after open cast strip mining. The LULC classes at the Machów and Jeziórko former sulfur mines are generated based on Sentinel-2 image processing and confirmed the applied type of reclamation for both areas. LiDAR point clouds are also used for collecting 2D and 3D information about vegetation. Airborne laser scanning (ALS) point clouds can also be directly retrieved for other vegetation parameters, such as the light penetration index, the number of trees, the number and area of gaps, the crown size, or the vertical structure of the forest stand, which are strongly correlated with the vegetation biomass (Szostak et al., 2020).

The resulted LULC classes cover a significant spatial range, from broad-leaved forest, to coniferous forest and transitional woodland shrub. The results of this case study indicate differences in

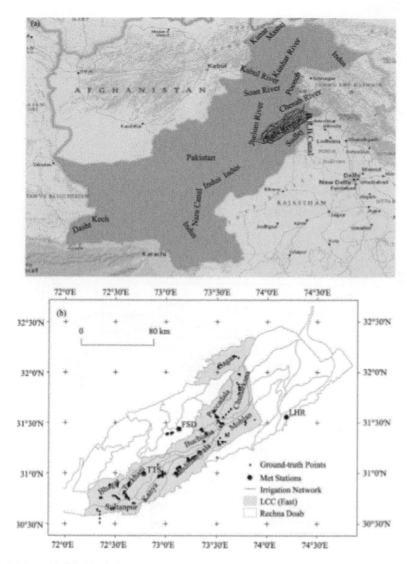

FIGURE 11.6 Map of LCC (East), Rechna Doab, Punjab, Pakistan and ground truthing points (from Usman et al., 2015).

vegetation parameters, such as height and canopy cover. Different stages of vegetation growth were also observed. Figure 11.7a presents the Machów study area, using Sentinel-2 red, green blue (RGB) color composition and Figure 11.7b shows the Machów LULC classes based on Sentinel-2 image classification.

11.3 URBAN ENVIRONMENT

Satellite and airborne EO data present an opportunity to extract significant information for urban and peri-urban environments and study the dynamics and consequences of urbanization at various spatial, temporal and spectral scales (Mitraka and Chrysoulakis, 2018). Indeed, EO data are increasingly used to produce urban land cover information as a primary boundary condition used in many spatially distributed models. Remote sensing is also used to extract information for urban surface and geometry (Mitraka and Chrysoulakis, 2018). This section covers urban planning and

(A)

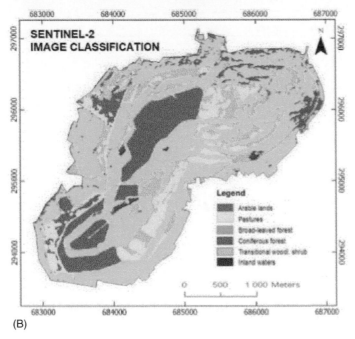

(B)

FIGURE 11.7 (a) The study area: (b) Machów, using Sentinel-2 (European Space Agency; ESA) red, green blue (RGB) color composition (red line boundaries of study areas. (b) Machów land use and land cover (LULC) classes – results of Sentinel-2 image classification.

monitoring, land use change for urbanization, urban vegetation and water, urban heat island (UHI) (urban atmospheric pollution is covered in Chapter 4).

11.3.1 Urban Planning and Monitoring

Remote sensing can connect different urban scales. Most of the urban phenomena are scale-dependent, which means that urban patterns change with the scale of observation. The selection of a specific satellite data source is a compromise between data availability, costs and the required spatial, spectral and temporal resolution.

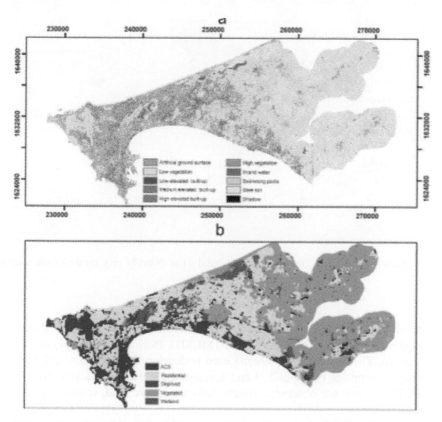

FIGURE 11.8 (a) Very high-resolution land cover (LC) of Dakar derived from a 2015 Pleiades imagery and (b) land use at the street block level for the same date (from Georganos et al., 2019).

Wealth index. A case study is presented, where satellite-derived very high-resolution (VHR) LULC data sets are developed and coupled with the Demographic and Health Surveys (DHS) Wealth Index (WI), a robust household wealth indicator, in order to provide city-scale wealth maps. Several modeling approaches are undertaken using a random forest regressor as the underlying algorithm for prediction in several geographical administrative scales (Georganos et al., 2019). The model is validated against a census database available for the city of Dakar, Senegal, and the results indicate a satisfactory performance for WI even at very fine resolutions.

The significance of remotely sensed features, such as LULC for mapping demographic and socio-economic conditions at various geographical scales has been demonstrated (Georganos et al., 2019). Satellite image features, such as the number and density of buildings, type of roads and number of cars explained roughly a high percentage of the variation in poverty models derived from census household consumption per capita estimates in Dakar. Two satellite-derived VHR LULC maps were used as predictors of the DHS WI. The LULC products are publicly available with a LC map at 0.5 m resolution and a LU map at the street block level (Figure 11.8). The overall accuracy of the LC and LU maps was 89.5% and 79%, respectively. Both were derived from Pleiades imagery collected in 2015 (Georganos et al., 2019).

Index-based LULC mapping. A case study is presented, which seeks to apply an Improved Thematic-based Remote Sensing Index (ITRSI) to map LULC classes within the city of Nairobi in Kenya (Kinoti, 2018). Nairobi is the capital and largest city in Kenya. The city covers an area of 689 km² and lies on an elevation of approximately 1700 m above sea level. It has an estimated population of 3.36 million people. Nairobi has undergone dramatic socioeconomic and infrastructural expansion since 2000, which makes it the fastest urbanizing counties in Kenya today and one of the largest

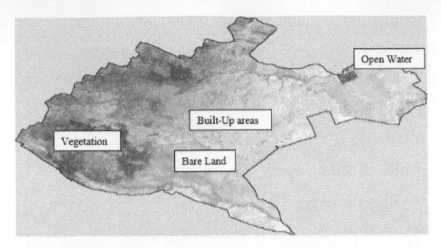

FIGURE 11.9 False color image (band 2: band 3: band 4) of Nairobi city created from Landsat (from Kinoti, 2018).

and fastest developing cities in Africa (Kinoti, 2018). The satellite image used is that of Landsat 5 Thematic Mapper (Scene ID: LT51680612010231MLK00, Path: 168, Row: 61) of 2010 covering the city of Nairobi. The classification scheme used is developed based on visual inspection of a false color image comprising of bands 2, 3 and 4, respectively. From the image, four land use and land cover (LULC) classes are observed, namely built areas, bare land, vegetation and open water (Figure 11.9).

This improved index (ITRSI) is generated from the combination of four derived thematic oriented indices of SAVI (Soil Adjusted Vegetation Index), which is sui for mapping vegetation cover in areas with low plant cover of less than 15 percent, Normalized Difference Bareness Index (NDBaI) to map bare land areas, Modified Normalized Difference Water Index (MNDWI) to delineate open water bodies, and Normalized Difference Built-Up Index (NDBI) for mapping built-up lands from urban environment, based on the equation:

$$ITRSI = SAVI + NDBaI + MNDWI + NDBI \qquad (11.1)$$

ITRSI seeks to highlight each of the four generalized LULCs of bare land, open water, vegetation and built areas typical of most urban areas as Boolean image, which are extracted independently from index-based threshold values. The equations of the four components of Equation (11.1) are as follows:

$$SAVI = \left[\left(NIR\,band - Red\,Band \right) \left(1 + L \right) \right] / \left(NIR + Red + L \right) \qquad (11.2)$$

$$NDBaI = \left(SWIR\,band - TIR\,band \right) / \left(SWIR\,band + TIR\,band \right) \qquad (11.3)$$

$$MNDWI = \left(GREEN\,band - MIR\,band \right) / \left(GREEN\,band + MIR\,band \right) \qquad (11.4)$$

$$NDBI = \left(MIR\,band - NIR\,band \right) / \left(MIR\,band + NIR\,band \right) \qquad (11.5)$$

where in Equation (11.2) NIR is the near-infrared band and L is a correction factor ranging from 0 for very high plant densities to 1 for very low plant densities, in Equation (11.3) SWIR is band 5 and TIR is band 6, respectively, in Equation (11.4) GREEN is the green band in visible and MIR is the middle infrared band, respectively. The index offers an alternative method to map LULC and classify urban areas in Kenya based on LULC data.

11.3.2 Land Use Change for Urbanization

Several case studies are considered and briefly presented as follows:

Case study 1. A case study is considered for the effect of land cover change on LST in the rapidly urbanizing Lagos metropolis. Using spatiotemporal Landsat imageries with thermal bands and ancillary data, land cover and LST changes were assessed from 1984 to 2015. The spatial patterns of LST and LC were generated to examine the response of LST to urban growth (Obiefuna et al., 2018).

Case study 2. Anthropogenic disturbances cause changes in land cover, which are estimated, within the central part of Ulaanbaatar, the capital of Mongolia (Tsutsumida et al., 2013). The breaks for additive seasonal and trend (BFAST) method is used, which is a powerful tool, since it can robustly and automatically generate the timing and locations of land cover changes from spatiotemporal data sets. The BFAST method was applied for the first time to urban expansion analysis, with NDVI time series calculated from MODIS (MOD09A1 product) during the period 2000–2010 (Tsutsumida et al., 2013).

Case study 3. Rapid urbanization and the risk of climatic variations, including temperature rise and rainfall increase, have urged the need to develop methods for monitoring the modification of LULC. A recent study used the NDVI and semi-supervised image classification (SSIC) integrated with high-resolution Google Earth images of the Kuantan River Basin (KRB) in Malaysia (Zaidi et al., 2017). The Landsat-5 (TM) images for the years 1993, 1999 and 2010 were selected. The results from both classifications indicated a consistent assessment accuracy with a reasonable level of agreement. However, SSIC was found to be more precise than NDVI (Zaidi et al., 2017).

Case study 4. The district Gujrat, a large district of Punjab, Pakistan, is considered, which faces a high growth rate of population resulting in expansion in infrastructure (Anwar and Siddiqui, 2018). Ninety-one percent of Gujrat's total land area is arable land, which constitutes the most valuable resource. However, a gradual increase in built-up land is directly decreasing arable land and permeable surfaces. The current research focuses on the identification of the expansion of built-up areas mainly on arable land and the estimation of the loss of arable land using GIS and RS methods. Remotely sensed data (Landsat TM-1993, ETM+ 2003 and TIRS 2017) is used to evaluate the LULC classification. In addition, secondary data are collected from soil surveys of Pakistan. Quantitative and spatial attribute data have been analyzed by GIS and statistical methods. Results indicate that the Gujrat district is losing 0.1% arable land annually, and if such loss continues, then Gujrat will lose all its arable land within the next 500 years (Anwar and Siddiqui, 2018).

Case study 5. This case study analyzes the association between built-up, green cover and LST. Specifically, district-level analysis of the normalized differential built-up index (NDBI), NDVI and LST has been conducted over the urban area of Delhi (Kumari et al., 2018). Landsat 7 (ETM+ SLC) for 2003, Landsat 5 (TM) for 2010 and Landsat 8 (OLI/TIRS) for 2017 have been used along with Survey of India (SOI) toposheet of Delhi at 1:25,000. Specifically, NDBI, NDVI and LST are computed for 2003, 2010 and 2017, respectively, using the spectral radiance model (SRM), the mono-window algorithm (MWA) and the split window algorithm (SWA). The analyses indicate a change in the distribution of vegetation cover and gradual increase in the built-up land, which results in an increase in LST of about 3.31°C in the last 14 years. Moreover, temporal analysis of LST using all three algorithms shows the increase in LST in Delhi between 2003 and 2017 (Figure 11.10).

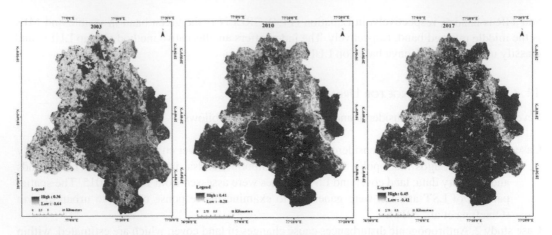

FIGURE 11.10 LST image of NCT of Delhi, 2003–2017 (from Kumari et al., 2018).

11.3.3 URBAN VEGETATION AND WATER

Urban vegetation. Urban vegetation presents several benefits for urban inhabitants, including providing shade, improving air quality and enhancing the look and feel of communities (Ramdani, 2013). However, vegetation cover in urban areas is changing rapidly due to urbanization. Google Earth Engine has been used to map vegetation cover in all urban areas larger than 15 km^2 in 2000 and 2015, which covered 390,000 km^2 and 490,000 km^2, respectively (Richards and Belcher, 2020). In 2015, urban vegetation covered a substantial area, equivalent to the size of Belarus. Most urban areas that increased in vegetation cover also increased in size.

The Landsat 7 Surface Reflectance Tier 1 image archive, which provides atmospherically corrected surface reflectance at a resolution of 30 m x 30 m, was used for classification. Three years of image data were processed for each period (1999–2001 and 2014–2016) using Google Earth Engine to generate cloud-free composite images (Richards and Belcher, 2020). Specifically, compositing was conducted on a per-pixel basis, based on a temporal stack of observed values for each pixel, the CFMASK algorithm.

The results indicated mapping of 176,000 km^2 of urban vegetation cover in 2000 and 210,000 km^2 in 2015. Thus, there is a rapid increase in the total area of urban vegetation, at a rate of more than 1% of the 2000 extent every year. Figure 11.11 presents the net difference in vegetation cover percentage between 2000 and 2015, in all urban areas larger than 15 km^2 that existed in both these years. Inset subfigures show regions with large numbers of urban areas; (a) eastern North America, (b) Europe, and (c) East Asia. Specifically, significant vegetation cover change was shown for 4,093 urban areas and no significant change was shown for 163 urban areas (from Richards and Belcher, 2020).

Urban vegetation provides several ecosystem services on a local scale and, thus, constitutes a potential adaptation strategy that can be used in global warming to offset the increasing impacts of human activity on urban environments. Specifically, holistic approaches are required to assess the health impacts of climate change mitigation and adaptation policies with regards to increasing public green spaces. A case study is considered, where a set of urban green ecological metrics has been used to evaluate urban green ecosystem services. The metrics were produced from two complementary surveys: a remote sensing survey of multispectral images and LiDAR data, and a survey using proximate sensing through images made available by the Google Street View database (Barbierato et al., 2020). Moreover, the extraction of urban vegetation was conducted with satisfactory results, especially for mapping and monitoring the urban tree heights, using fused hyperspectral image and LiDAR data (Ramdani, 2013). The spectral angle mapper method was used for endmember distribution mapping.

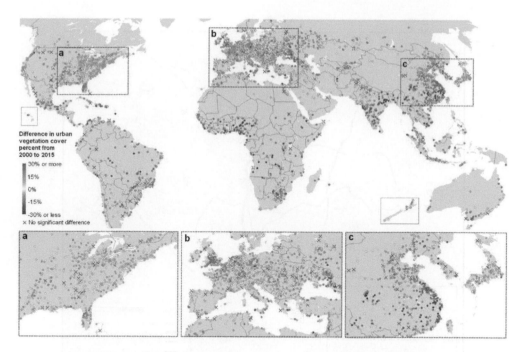

FIGURE 11.11 Net difference in vegetation cover percentage between 2000 and 2015, in all urban areas larger than 15 km² that existed in both these years. Inset subfigures show more detail for regions with large numbers of urban areas; (a) eastern North America, (b) Europe, and (c) East Asia (from Richards and Belcher, 2020).

Urban water. Urbanization in coastal areas is an increasing worldwide problem, where 10% of the total population and 13% of the urban population are located at 10 m or less above sea level. Specifically, in the United States, 52% of the population lives along the coast, and North Carolina is in the top 10 fastest growing states (Halls and Magolan, 2019). A case study is considered in the southeastern coast of North Carolina. Specifically, a methodology has been developed that investigates the complex relationship between urbanization, land cover change and potential flood risk and tested the approach in a rapidly urbanizing region (Halls and Magolan, 2019). A variety of data, including satellite (PlanetScope) and airborne imagery (NAIP and LiDAR) and vector data (C-CAP, FEMA floodplains, and building permits), were used to assess spatiotemporal changes.

The methods included: (1) matrix change analysis, (2) a new approach to analyze shorelines by computing adjacency statistics for changes in wetland and urban development, and (3) risk assessment using a fishnet, where hexagons of equal size (15 ha) were ranked into high, medium and low risk and these results were compared with the amount of urbanization (Halls and Magolan, 2019). The combination of the developed methods has resulted in data that are used by the local authorities and by the US Federal Emergency Response System. Figure 11.12 shows the study area is Pender County in Southeastern North Carolina (NC), in the southeastern United States. Specifically, the study area included two scales: (1) county-wide analysis, and (2) higher spatial resolution along the coast (outlined in yellow) east of Highway 17 and west of the Intracoastal Waterway (from Halls and Magolan, 2019).

11.3.4 Urban Heat Island (UHI)

The urban heat island (UHI) phenomenon refers to higher air temperatures in built-up urban areas than those of the surrounding countryside. In the framework of intensifying urbanization, the UHI effect is expected to influence nearly 70% of the global population by 2050 (Leal Filho et al., 2017).

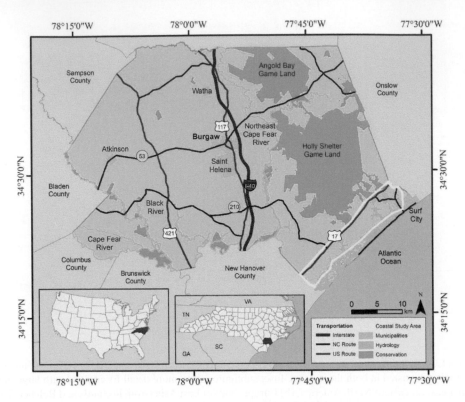

FIGURE 11.12 The study area is Pender County in southeastern North Carolina (NC), in the southeastern United States. The study area included two scales: (1) county-wide analysis and (2) higher spatial resolution along the coast (outlined in yellow) east of Highway 17 and west of the Intracoastal Waterway. (Data sources: NC Department of Transportation, Pender County zoning, and US Geological Survey) (from Halls and Magolan, 2019).

This phenomenon occurs mostly in the summer, during calm nights in the mid-latitude settlements. One risk factor of UHI is the occurrence of heatwaves (see Chapter 7). Due to climate change, heatwaves are likely to increase in frequency and intensity (Leal Filho et al., 2017). The UHI has negative effect on the urban population causing different impacts, such as: (1) increasing mortality rates; (2) elevating the concentration of ground-level tropospheric ozone; (3) increasing the formation of urban smog; and (4) increasing the energy consumption (Konopacki and Akbari, 2002).

The spatial and temporal pattern of the UHI depends on the characteristics of the urban environment, such as topography, building density, urban pollution by anthropogenic activities, air-conditioning systems, microclimatic characteristics and similar aspects. Moreover, meteorological conditions facilitate the occurrence of the UHI effect, such as during hot sunny summer days under cloudless skies and light wind (Morris et al., 2001). The measurement of the UHI can be conducted through air temperature observations in the urban and the surrounding rural environment. A network of meteorological stations can measure the differences between the built-up areas and the countryside. It been observed that during the UHI the difference in air temperature between urban and the surrounding rural areas may reach 11°C (Valsson and Bharat, 2009). Furthermore, Earth Observation (EO) satellites with thermal sensors provide information about thermal patterns. Specifically, satellites, such as Sentinel-3 (ESA's Copernicus program), can provide evening and night images of LST. The spatial resolution is low (1 km), in contrast to high temporal resolution (twice daily). Over the last few years very promising synergistic methods have been proposed that combine high-resolution visible and near-infrared observations with low-resolution, but more frequent thermal imagery. These methods pave the way for monitoring UHI at a global scale (Ravanelli et al., 2018).

Satellite-derived LST reveals the variations and impacts on the terrestrial thermal environment on a broad spatial scale. Several research efforts have been conducted to investigate LST intensity response to urban LULC by examining the thermal impact on urban settings, such as in Nepal (Baniya et al., 2018), as well as in ten Chinese megacities (i.e. Beijing, Dongguan, Guangzhou, Hangzhou, Harbin, Nanjing, Shenyang, Suzhou, Tianjin and Wuhan) (Liu et al., 2020). Surface urban heat island (SUHI) footprints were analyzed and compared by magnitude and extent. The causal mechanism among land cover composition (LCC), population and SUHI was also considered. Spatial patterns of the thermal environments were identical to those of LULC. In addition, most impervious surface materials (greater than 81%) were labeled as heat sources, whereas water and vegetation functioned as heat sinks. SUHI for all megacities showed spatially gradient decays between urban and surrounding rural areas (Liu et al., 2020).

Similarly, several studies have addressed UHI in several Mediterranean urban areas (Polydoros et al., 2018), and specifically the thermopolis campaign in the urban area of Athens Greece (Daglis et al., 2010; Kourtidis et al., 2015; Rapsomanikis et al., 2015; Loupa et al., 2016). An episode of the UHI phenomenon in Athens is illustrated in Figure 11.13. The UHI effect was analyzed by a Sentinel-3 image of the city of Athens Greece on August 10, 2018. The temperature of the center of the city is 8° Celsius higher than the rural surroundings. Moreover, with the boundaries of the metropolitan area, the population density effected by higher temperature can be estimated to be almost one million inhabitants. Mitigation of the UHI effect could be achieved by changing the LULC character of the urban territory. Specifically, the increase in the percentage of the green areas, the existence of water bodies and the reduction of the anthropogenic heat are some of the measures that urban planners should consider as possible solutions, especially in the framework of climate change and urban sprawl (Ali et al., 2016). Moreover, further analysis seems to be required to better understand the negative effects of UHI on each city. In summary, for effective urban design and planning, knowledge of UHI should be obtained and public awareness should increase in order to implement successful adaptation strategies against UHI phenomena.

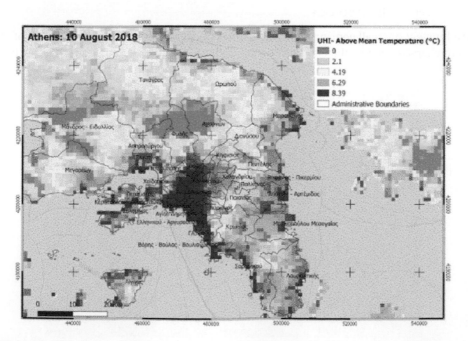

FIGURE 11.13 UHI effect in Athens on a hot summer day of August 10, 2018 in Athens, Greece.

11.4 ARCHAEOLOGICAL RESOURCE MAPPING USING UAV

There are diachronic efforts for image processing and analysis of archaeological sites. Protecting and preserving cultural and natural landscapes is vital to the preservation of cultural and natural heritage that hides centuries or even millennia of knowledge. This knowledge must be preserved for future generations. Helping in this direction is its technology and achievements through digital imaging of monuments using the appropriate tools, one of which is UAV (Unmanned Aerial Vehicles). In this section, specific methods are considered using Unmanned Aerial Vehicles (UAVs) for archaeological resource mapping.

The term UAV refers to an aircraft designed to operate without human presence on board. In addition to the official definition mentioned earlier, in the international literature, UAVs are also referred to as "Unmanned Platforms", or "Unmanned Aerial Vehicles". Unmanned Aerial System (UAS), as a general concept, includes several subsystems, such as: unmanned aerial vehicle, base station or communication systems (Austin 2010). UAVs technology enables the collection of airborne data and the detection of changes occurring at the surface of the ground. Their widespread use, especially in geoinformatics, is mainly due to their ability to capture images of the area of interest with many advantages, such as speed during the overhead process, ability to map inaccessible areas and to obtain high-resolution terrain image (Brutto et al., 2014). Their initial use was focused on military purposes, such as environmental identification and observation, maritime surveillance and various mining activities. But in recent years, in the field of UAVs, significant changes have taken place: (1) reducing the cost of acquiring UAVs and sophisticated technology used while creating an international community of UAV-based open-source software and applications; (2) the ability to adjust different sensors (e.g. Visible, Infrared); (3) Great remote control or stand-alone pre-programmed capabilities. Figure 11.14 shows a fully computer-controlled, ultra-lightweight UAV with autonomous flight capability. According to the international literature, there is a rapid increase of recorded UAV applications in several research and operational fields, such as crop mapping, crop water stress mapping, monitoring of LULC changes, monitoring disasters, floods, storms and managing their impacts, identifying sources of pollution, archaeological mapping and similar topics.

11.4.1 MAPPING METHODOLOGY

The development of small unmanned aerial platforms is significant. These fly in a range of 20–200 m and are characterized by the ability to obtain both quantitative and qualitative data with a high

FIGURE 11.14 Ultra-light Quadcopter with custom camera

level of detail and low cost (Eisenbeiss and Sauerbier, 2011). An example of recording and mapping of cultural and natural heritage (using UAV) is considered, namely the archaeological site of "Magoula Zerelia", located in central Greece. It is a well-known prehistoric site and landmark of the wider area. The site has already been inhabited from the Middle Neolithic Age to the Late Bronze Age, and at the same time eight archaeological layers have been identified, with the latter being located at a depth of about 6–8 meters. The aim of this application is to create a high-resolution digital orthomosaic map and digital surface model (DSM) covering an area of approximately five hectares. For this purpose, several overlapping aerial photographs were taken from ultra-light UAV at a height of approximately 100 meters, following a specific methodological approach. There are two basic steps to successfully create the digital imaging of the study area:

1. **Aerial photos acquired by UAV.** Initially, the Flight Plan is implemented in the office. It refers to the automated route that the UAV will perform by recording parameters, such as altitude, speed, orientation from the time of take-off to that of take-off during the aerial photography process. An open-source software is used to record the flight plan. The software introduced parameters, such as focal length (5 mm), flight altitude (100 m) and flight speed (5 m/s) to calculate the required flight time and the total number of flights of aerial photographs that can be taken. Followed by aerial photography field in the area of the five hectares, ground targets are established with known coordinates, which serve as ground control points for the transformation of the final orthomosaic map to the local projection system. Some targets are also used as control points for quality control of the products. Finally, wireless communication of the base station (laptop) with the UAV is installed to monitor the performance of the stand-alone flight. Figure 11.15 shows the base station (left), from which the operator monitors the parameters during the UAV stand-alone flight.
2. **Photogrammetric process of aerial imagery.** Photogrammetric software is then used to process the aerial photos. Image processing is implemented, such as: (1) the orientation of the photos, (2) the creation of the 3D model (polygon mesh), and (3) the texture of the images (Verhoeven, 2011). Figure 11.16 shows the camera position of the UAV in the 3D space at the time of aerial photography and the 3D model of the archaeological site. Finally, the quality of the outputs is evaluated based on the ground control points.

11.4.2 Analysis of Results

The creation of a 2 cm pixel orthomosaic map is the main product. The area of about five hectares is mapped very precisely, capturing elements of both the natural and cultural environments. Figure 11.17 illustrates the archaeological site hill with the excavated sections. On a scale of 1:200, the orthomosaic map captures in detail all the elements in the space (trees, roads, anthropogenic

FIGURE 11.15 Base station (left) for autonomous UAV flight (right)

FIGURE 11.16 Photogrammetric processing of Aerial photos

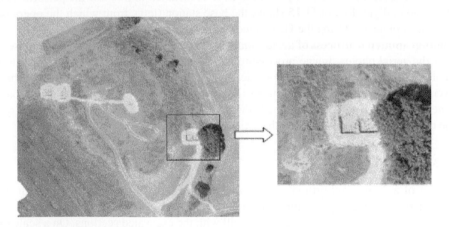

FIGURE 11.17 Detailed orthomosaic map of the archaeological site

interventions, etc.). Furthermore, the digital surface model (DSM) of the case study area, with a pixel size of 1.6 meters, is created.

The combination of the aforementioned two products enables the creation of 3D representations using GIS software. Three-dimensional interactive representations can be a key tool in understanding the elements recorded in the space, while also being able to assist archaeologists for interpretation. Figure 11.18 illustrates a 3D representation of the archaeological area. Overlaying additional spatial information for the study area onto the 3D interactive model may constitute a basic tool for archaeologists to understand the spatial distribution of the elements. In summary, the Unmanned Aircraft Systems (UAS) have revolutionized the field of mapping cultural resources. The low cost, the capture speed and the ultra-high spatial resolution provided are expected to serve as an operation tool for digitized and photographic documentation.

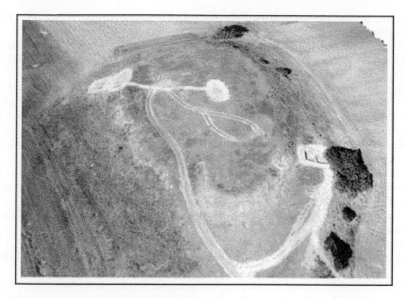

FIGURE 11.18 3D representation of the archaeological site

11.5 SUMMARY

In this chapter, remote sensing data and methods have indicated the significance of LULC change as a global phenomenon, where its accurate and updated delineation plays a major role for detailed ecosystem analysis. Specifically, the importance of remote sensing becomes obvious in rural regions dominated by complex agricultural lands, where rapid changes occur from season to season. Moreover, it has been demonstrated that LULC change analysis using remote sensing data and methods in complex rural and urban basins provides a complete profile for better management and planning. Indeed, remote sensing can support our understanding of urban environments and can recognize many benefits of using EO data for urban studies. In evaluating urban environments, LULC mapping using multispectral classification methods has been widely utilized. In this chapter, LULC analysis in rural environments has involved land use changes and land use classification for natural resources management. Similarly, LULC analysis in urban environment has included urban planning and monitoring, land use change for urbanization, urban vegetation and water, as well as the Urban Heat Island (UHI). In addition, in this chapter, archaeological resource mapping has also been analyzed using UAV.

REFERENCES

Ali, J.M., Marsh, S.H., and Smith, M.J., 2016. Modelling the spatiotemporal change of canopy urban heat islands. *Building and Environment*, 107, 64–78.

Anderson, J.R., 1977. Land use and land cover changes: A framework for monitoring. *Journal of Research by the Geological Survey*, 5 143–153.

Anwar, N., and Siddiqui, S., 2018. Loss of arable land due to rapid urbanization: A remote sensing-based study on Gujrat District, Pakistan. *Journal of Biodiversity and Environmental Sciences (JBES)*, 13, 3, 38–47.

Austin, R., 2010. *Unmanned Aircraft Systems: UAVs Design, Development and Deployment*. United Kingdom, Wiley.

Baniya, B., Techato, K., Ghimire, S.K., Chhipi-Shrestha, G., 2018. A review of green roofs to mitigate urban heat island and Kathmandu Valley in Nepal. *Applied Ecology and Environmental Sciences*, 6, 4, 137–152.

Barbierato, E., Bernetti, I., Capecchi, I., and Saragosa, C., 2020. Integrating remote sensing and street view images to quantify urban forest ecosystem services. *Remote Sensing*, 12, 329.

Brutto, L., Garraffa, M., and Meli, A.P., 2014. UAV platforms for cultural heritage survey: First results. *Annals of the Photogrammetry, Remote Sensing and Spatial Information Sciences*, 2 5, 227–234.

Castellanos, M.T., Morató, M.C., Aguado, P.L., del Monte, J.P., and Tarquis, A.M., 2018. Detrended fluctuation analysis for spatial characterisation of landscapes. *Biosystems Engineering*, 168, 4–25, ISSN1537-5110, https://doi.org/10.1016/j.biosystemseng.2017.09.016

Daglis, I.A., Rapsomanikis, S., Kourtidis, K., Melas, D., Papayannis, A., Keramitsoglou, I., Giannaros, T., Amiridis, V., Petropoulos, G., Georgoulias, A., Sobrino, J.A., Manunta, P., Gröbner, J., Paganini, M., and Bianchi, R., 2010. *Results of the due thermopolis campaign with regards to the urban heat island (UHI) effect in Athens*. Proc. "ESA Living Planet Symposium", Bergen, Norway.

Dalezios, N.R., 2015. Agrometeorology: Analysis and Simulation (in Greek). KALLIPOS: Libraries of Hellenic Universities (also e-book), ISBN: 978-960-603-134-2, 481p , November 2015.

Das, S., Ashikuzzaman, M., and Esraz-Ul-Zannat, M., 2018. Investigating the coastal livelihood in relation to land use-land cover change modeling: A case study of Sharankhola, Bagerhat, *Journal of Bangladesh Institute of Planners*, 9, 59–68.

Eisenbeiss, H., and Sauerbier, M., 2011. Investigation of UAV systems and flight modes for photogrammetric applications. *The Photogrammetric Record*, 26, 136, 400–421.

FAO, 2014. Global Forest Land-Use Change 1990–2010. An update to FAO Forestry Paper No. 169. The Food and Agricultural Organization of the United Nations (FAO) with the E.U. Joint Research Centre (JRC). Rome.

Georganos, S., Gadiaga, A.N., Linard, C., Grippa, T., Vanhuysse, S., Mboga, N., Wolff, E., Dujardin, S., and Lennert, M., 2019. Modelling the wealth index of demographic and health surveys within cities using very high-resolution remotely sensed information. *Remote Sensing*, 11, 2543.

Halls, J.N., and Magolan, J.L., 2019. A methodology to assess land use development, flooding, and wetland change as indicators of coastal vulnerability. *Remote Sensing*, 11, 2260.

Jing, S., Tong, Y.X., and Liu, J., 2017. Telecoupled land-use changes in distant countries. *Journal of Integrative Agriculture*, 16, 2, 368–376.

Kinoti, K.D., 2018. Monitoring nature of Nairobi City land features from Landsat 5 images using index-based mapping. *Journal of Urban Design*, 1, 1–11.

Konopacki, S., and Akbari, H., 2002. Energy savings for heat-island reduction strategies in Chicago and Houston, Available at: http://repositories.cdlib.org/lbnl/LBNL-49638

Kourtidis K., Georgoulias, A.K., Rapsomanikis, S., Amiridis, V., Keramitsoglou, I., Hooyberghs, H., Maiheu, B., and Melas, D., 2015. A study of the hourly variability of the urban heat island effect in the Greater Athens Area during summer. *Science of the Total Environment*, 517, 162–177.

Kumari, B., Tayyab, M., Shahfahad, S., Mallick, J., Khan, M.F., and Rahman, A., 2018. Satellite-driven Land Surface Temperature (LST) using Landsat 5, 7 (TM/ETM+ SLC) and Landsat 8 (OLI/TIRS) data and its association with built-up and green cover over urban Delhi, India. *Remote Sensing in Earth Systems Sciences*, 1, 63–78.

Leal Filho, W., Icaza, L.E., Neht, A., Klavins, M., and Morgan, E.A., 2017. Coping with the impacts of urban heat islands. A literature-based study on understanding urban heat vulnerability and the need for resilience in cities in a global climate change context. *Journal of Cleaner Production*, 171, 1140–1149, doi: 10.1016/j.jclepro.2017.10.086

Liu, F., Zhang, X., Murayama, Y., and Morimoto, T., 2020. Impacts of land cover/use on the urban thermal environment: A comparative study of 10 megacities in China. *Remote Sensing*, 12, 307.

Loupa, G., Rapsomanikis, S., Trepekli, A., and Kourtidis, K., 2016. Energy flux parametrization as an opportunity to get urban heat island insights: The case of Athens, Greece (Thermopolis 2009 Campaign). *Science of the Total Environment*, 542, 136–143.

Mensah, A.A., Sarfo, D.A., and Partey, S.T., 2019. Assessment of vegetation dynamics using remote sensing and GIS: A case of Bosomtwe Range Forest Reserve, Ghana. *The Egyptian Journal of Remote Sensing and Space Sciences*, 22, 145–154.

Mitraka, Z., and Chrysoulakis, N., 2018. *Chapter 7: Earth Observation for Urban Climate Monitoring: Surface Cover and Land Surface Temperature*, IntechOpen, 125–145, http://dx.doi.org/10.5772/Intechopen.71986.

Morris, C.J.G., Simmonds, I., and Plummer, N., 2001. Quantification of influences of wind and cloud on the nocturnal urban heat island of a large city. *Journal of Applied Meteorology*, 40, 169–182.

Obiefuna, J.N., Nwilo, P.C., Okolie, C.J., Emmanuel, E.I., and Daramola, O.E., 2018. Dynamics of land surface temperature in response to land cover changes in Lagos Metropolis. *Nigerian Journal of Environmental Sciences and Technology (NIJEST)*, 2, 148–159.

Polydoros, A., Mavrakou, T., and Cartalis, C., 2018. Quantifying the trends in land surface temperature and surface urban heat island intensity in Mediterranean cities in view of smart urbanization. *Urban Science*, 2, 16, doi: 10.3390/urbansci2010016

Ramdani, F., 2013. Urban vegetation mapping from fused hyperspectral image and LiDAR data with application to monitor urban tree heights. *Journal of Geographic Information System*, 5, 404–408, 10.4236/jgis.2013.54038

Rapsomanikis, S., Trepekli, A., Loupa, G., and Polyzou, C., 2015. Vertical energy and momentum fluxes in the centre of Athens, Greece during a heatwave period (Thermopolis 2009 Campaign). *Boundary-Layer Meteorology*, 154, 497–512, doi: 10.1007/s10546-014-9979-2

Ravanelli, R., Nascetti, A., Cirigliano, R.V., Di Rico, C., Leuzzi, G., Monti, P., and Crespi, M., 2018. Monitoring the impact of land cover change on surface urban heat island through Google Earth Engine: Proposal of a global methodology, first applications and problems. *Remote Sensing*, 10, 1488.

Richards, D.R., and Belcher, R.N., 2020. Global changes in urban vegetation cover. *Remote Sensing*, 12, 23.

Szostak, M., Pietrzykowski, M., and Likus-Cieślik, J., 2020. Reclaimed area land cover mapping using Sentinel-2 imagery and LiDAR point clouds. *Remote Sensing*, 12, 261.

Tsutsumida N., Saizen, I., Matsuoka, M., and Ishii, R., 2013. Land cover change detection in Ulaanbaatar using the breaks for additive seasonal and trend method. *Land*, 2, 534–549, doi:10.3390/land2040534

Usman, M., Liedl, R., Shahid, M.A., and Abbas, A., 2015. Land use/land cover classification and its change detection using multi-temporal MODIS NDVI data. *Journal of Geographical Sciences*, 25, 12, 1479–1506, doi: 10.1007/s11442-015-1247-y

Valsson, S., and Bharat, A., 2009. Urban heat island: cause for microclimate variations. *Architecture – Time Space and People*, 20–25.

Verhoeven, G., 2011. Taking computer vision aloft – archaeological three-dimensional reconstructions from aerial photographs with photoscan. *Archaeological Prospection*, 18, 1, 67–73.

Wu, L., Li, Z., Liu, X., Zhu, L., Tang, Y., Zhang, B., Xu, B., Liu, M., Meng, Y., and Liu, B., 2020. Multi-type forest change detection using BFAST and monthly Landsat time series for monitoring spatiotemporal dynamics of forests in subtropical wetland. *Remote Sensing*, 12, 341, doi: 10.3390/rs12020341

Zaidi, S.M., Akbari, A., Samah, A.A., Kong, N.S., and Gisen, J.I.A., 2017. Landsat-5 time series analysis for land use/land cover change detection using NDVI and semi-supervised classification techniques. *Polish Journal of Environmental Studies*, 26, 10.15244/pjoes/68878

Index

Page numbers in *italics* refer to figures; **bold** refer to tables.

Printed and bound by CPI Group (UK) Ltd, Croydon, CR0 4YY

Printed and bound by CPI Group (UK) Ltd, Croydon, CR0 4YY

24/10/2024

01778292-0009